AF739767

Free Convection Film Flows and Heat Transfer

Deyi Shang

Free Convection Film Flows and Heat Transfer

With 109 Figures and 69 Tables

Springer

Prof. Dr. Deyi Shang
286 Woodfield Dr.
Ottawa ON
Canada K2G 3W9
e-mail:deyishang@yahoo.ca

ISBN 978-3-642-06710-5 e-ISBN 978-3-540-29128-2

Springer is a part of Springer Science+Business Media
springer.com

Softcover reprint of the hardcover 1st edition 2006

Cover design: *design & production* GmbH, Heidelberg

Preface

Welcome to *Free Convection Film Flows and Heat Transfer*! Free convection film flows occur in many industrial processes. However, engineers still have to deal with many unresolved problems. This book systematically summarizes my recent research results that have been referred to and cited by many other researchers in this field. The purpose of this book is to provide a practical guide to university students, graduate students, design engineers, researchers, and scientists who wish to further understand the characteristics of free convection film flows and heat transfer. I hope this book will serve as a useful tool for them, as well as a guide to future research.

This book includes three related parts (1) accelerating convective boundary layers of Newtonian fluids, (2) accelerating film boiling and condensation of Newtonian fluids, and (3) accelerating film flows of non-Newtonian power-law fluids. These phenomena are all caused by buoyancy or gravity, and can be summed up in terms of the free convection film flows. In addition, the free convection film flows of Newtonian fluids can be taken as a special case of non-Newtonian power-law fluids.

In this book, I present my recent studies of free convection film flows and heat transfer on both vertical and inclined plates. Because of a lack of related books presenting the effects of variable thermophysical properties on heat and mass transfer, these effects are especially emphasized in this book with respect to free convection, film boiling, and film condensation of Newtonian fluids. A system of models for the treatment of variable thermophysical properties is introduced in this book, with an innovative temperature parameter method for gases and temperature-dependent models for liquids. A novel system of analysis and transformation models with an innovative velocity component method is applied throughout the book. This is a better alternative to the traditional Falkner-Skan type transformation. The new analytical system and models led to simplification for treatment of variable thermophysical properties of fluids, as well as hydrodynamics and heat transfer analysis. A system of reliable and rigorous computations solving the problems for two-point or three-point boundary values is provided in this book. In the analyses and calculations of the first two parts of this book, I focus on clarifying the effects of variable thermophysical properties on heat and mass transfer. A system of numerical solutions is formulated to predict heat and mass transfer simply and reliably. In the last part of this book, heat and mass transfer of the accelerating film flows of Newtonian fluids are extended to that of non-Newtonian power-law fluids. So far, there has been a lack of such information and analysis for advanced heat and mass transfer of accelerating film flows of non-Newtonian power-law fluids.

In addition, a collection of novel terminologies has arisen in this book, e.g., *velocity component method, temperature parameter method, thermal conductivity parameter, viscosity parameter, specific heat parameter, overall temperature parameters, thermal physical property factors, boundary temperature ratio, buoyancy factor, wall superheated grade, wall subcooled grade, vapor bulk superheated grade, liquid bulk subcooled grade, computation for three-point boundary value problem, temperature gradient on the wall, velocity components at the interface, vapor film thickness, liquid film thickness, mass flow rate through the interface, mass flow rate parameter, Non-Newtonian power-law fluids, length of boundary layer region, boundary layer thickness, local Prandtl number, critical local Prandtl number, critical boundary layer thickness*, and so on. These terminologies reflect the recent developments on my study of *free convection film flows and heat transfer*. Therefore, I strongly urge readers to pay particular attention to the special physical significance of these terminologies. Readers will find them beneficial to understanding the essence of this book.

I am greatly indebted to Professor B.X. Wang, Academician of Chinese Academy of Science, and member of the Executive Committee of the International Center for Heat and Mass Transfer. He was my supervisor in the period of my Ph.D studies of Tsinghua University, China. The recent developments devoted to Part 1 and Part 2 of this book relied on our long-term research cooperation. Besides suggesting the title of this book, he carefully proofread the second chapter of this book and provided many valuable suggestions to the whole book.

I am very grateful to Professor H.I. Andersson, Department for Energy and Process Engineering, Norwegian University of Science and Technology, Norway, for his highly effective cooperation related to the research developments shown in Part 3 of this book. As my host professor and a distinguished researcher in the field of accelerating film flows of non-Newtonian fluids, his erudite and honorable character deeply impressed me. At the same time, I gratefully acknowledge the Norwegian Research Council for awarding me the very prestigious title of intenational scientist and providing financial support for my extensive research there in cooperation with Professor Andersson.

In addition, many friends and colleagues have contributed to this book. Here, I would particularly like to thank Professor Liangcai Zhong, Northeastern University, China, as well as some of my previous students, notably Yu Quan, Yang Wang, Yue Yuan, Hongyi Wang, and Li Ren. Their contributions are presented in the book. Without their collaborative research efforts this book would not have been possible.

I would like to offer my sincere gratitude to Professor Hongtan Liu, Department of Mechanical and Aerospace Engineering, University of Miami, USA, and Professor Ben Q. Li, School of Mechanical and Materials Engineering, Washington State University, USA. As good friends in my academic circles in North America, their warm encouragement gave me the full confidence to complete this book.

I would like to thank my respectable friend, Professor Pran Manga, School of Management, Department of Economy, University of Ottawa, who spent time going through parts of the manuscript. Owing to his generous help, this book could be completed in time. Meanwhile, my sincere thanks should be given to Professor H. S. Takhar, University of Manchester also, who took his valuable time for going through several chapters at the beginning of my book writing.

Last and most of all, I offer a special word of thanks to my wife, Shihua Sun. During most of the past one and half years when I devoted to writing this book, she provided the loving family environment that offered me the tranquility and peace of mind that made writing it possible. The book is dedicated to her.

Ottawa, ON
Canada
January, 2006

Deyi Shang

Contents

1

Introduction

1.1 Scope

This book systematically presents recent developments in hydrodynamics and heat and mass transfer in accelerative boundary layers and film flow. The range of research in this book involves three related parts. The first part is devoted to the presentation of the studies related to accelerating boundary layers. It involves free convection of Newtonian gases and liquids. Also, all temperature-dependent physical properties of fluids are considered for phenomena with large temperature differences. The second part is devoted to the presentation of studies related to accelerating film boiling and condensation of Newtonian fluids. The temperature-dependent physical properties of fluids are considered for phenomena with large temperature differences. In the third part, the development of studies for hydrodynamics and heat transfer for falling film flow of non-Newtonian power-law fluids (FFNF) is presented. The boundary layers and film flows we deal with are all caused by buoyancy or gravity, both of which lead to acceleration of the fluid in boundary layers and film flows. Because of the similar flow situation, the studies in these three parts can be summed up in terms of the laminar free convection film flows caused by acceleration. In addition, even the studies related to the free convection film flows for Newtonian fluids can be taken as a special case of those related to non-Newtonian power-law fluids.

1.2 Application Backgrounds

Heat transfer in boundary layers and film flows caused by acceleration often involves large temperature differences. Its practical applications exist widely in various branches of industry, such as the metallurgical, chemical, mechanical, and food industries. The heat transfer on surfaces of various industrial furnaces (such as boilers, heating, and smelting furnaces) is caused by various forms of free convection under large temperature differences, except for the radiation heat transfer. The heat transfer rate affects the heating process and heat efficiency of the furnaces. On the surface of an ingot mold in metal casting there exists free convection heat transfer, and this transfer affects the solidification and crystalline process and therefore the quality of the product. In the process of the surface hardening of metal, in the initial stage, the film boiling free convection is produced on the surface and in the final stage on the surface there exists liquid free convection. These processes will improve the mechanical function of the metal surface. In the electronic industry, cooling process occurs with free convection on the surface integrated circuits. This cooling process tends to restrict the surface temperature to below the allowable temperature. In addition, it is widely known that film condensation free convection has significant applications in various condensators. The suitable design of the corresponding heating equipment and the optimal control of the corresponding heat transfer depends on correct prediction of these processes related to heat transfer mentioned earlier.

Non-Newtonian power-law fluid behavior is encountered in a great variety of everyday life as well as in industrial operations. By far the largest effort has been devoted to Newtonian fluid mechanics. Recently, modest attention has been devoted to gravity-driven thin film flow of the non-Newtonian power-law fluids, as compared with its Newtonian counterpart. Yet, the free surface flow of the non-Newtonian power-law fluids is a widely occurring phenomenon in various industrial applications, for instance in polymer and plastics fabrication, food processing, and in coating equipment. The heat transfer from the solid surface to a liquid film is of practical importance in various types of heat and mass transfer equipment such as coolers, evaporators, and trickling filters. The obvious advantage with the falling film principle is that the short residence time for heat transfer can be realized, which is most desirable for heat-sensitive materials.

1.3 Previous Developments

1.3.1 For Accelerating Boundary Layers and Film Flow of Newtonian Fluids

The basic ideas underlying the approximation that yield the boundary layer equations were developed by Prandtl [1]. The essential idea is to divide a flow into two major parts. The larger part concerns a free stream of fluid far from any solid surface. The smaller part constitutes a thin layer next to a solid surface in which the effects of molecular transport properties (viscosity and thermal conductivity) are considered using some approximation. Prandtl initiated the study of free convection by means of boundary layer theory. For a long time, the study was based on the Boussinesq approximation [2, 3]. In this approximation, the temperature-dependent properties of fluids are neglected in the governing partial equations of the boundary layer, except for density in the buoyancy term of the momentum equation. Pohlhausen [4] solved partly the governing equations of boundary layer. Ostrach [5] supplied a more detailed numerical solution for free convection. Ede [6] also provided a numerical solution for the dimensionless temperature gradient for various values of Prandtl number. LeFevre [7] proposed an approximation for the prediction of the Nusselt number. However, since these research results are based on the Boussinesq approximation, they are only suitable for the case of small temperature difference between the body surface and the ambient fluid. However, for the case of large temperature differences, these results are not appropriate.

Therefore, it is important to study free convection with larger temperature differences, and should include free convection with and without phase change, such as free convection of fluids, film boiling free convection, and film condensation free convection. Free convection with a small temperature difference dealt with by the Boussinesq approximation is only a special case of free convection with larger temperature differences.

Due to the universality of free convection with large temperature, the consideration of variable temperature-dependent properties is very important in the corresponding studies. The earliest theoretical consideration of variable thermophysical properties for free convection is the perturbation analysis of Hara [8] for air. The solution is applicable for small values of the perturbation parameter, $\varepsilon_{\mathrm{H}} = (T_{\mathrm{w}} - T_{\infty})/T_{\infty}$. Tataev [9] also investigated the free convection of a gas with variable viscosity. A well-known analysis of the variable fluid property problem for laminar free convection on an isothermal vertical flat plate has been presented by Sparrow and Gregg [10]. They considered five different gases and provided the corresponding solutions of the boundary layer equations. They proposed a reference temperature and suggested that the problem of variable thermophysical properties can be treated as a constant property problem, i.e., Boussinesq approximation. Gray and Giogini [11] discussed the validity of the Boussinesq approximation and proposed a method for analyzing natural convection flow with fluid properties assumed to be a linear function of the temperature and pressure. Clausing and Kempka [12] reported their experimental study of the influence of property variations on natural convection and calculated it for the laminar region. The Nusselt number Nu_{f} will be a function of Rayleigh number $Ra_{\mathrm{f}}(= Gr_{\mathrm{f}}Pr_{\mathrm{f}})$ only with the reference temperature, T_{f}, taken as the average temperature in the boundary layer.

In [13–22], studies of the effects of variable thermophysical properties of liquid on the laminar free convection with larger temperature difference were carried out. Fujii et al. [13] examined two methods of correlating the effects of variable thermophysical properties on heat transfer for free convection from vertical surfaces in liquids. The first method of correlating the data consisted of using the constant property correlations for the Nusselt number and evaluating all physical properties at a reference temperature, $T_{\mathrm{r}} = T_{\mathrm{w}} - (T_{\mathrm{w}} - T_{\infty})/4$. They noted that the choice of the reference temperature corresponds with the solution provided by two previous studies [14, 15]. The second method that they used to correlate their data for oils was first proposed by Akagi [15] and applies only to liquids for which the viscosity variation is dominant. The similarity analysis of Piau [16] also treated variable property effects in free convection from vertical surfaces with high Prandtl number liquids. It was indicated that the main property variations in water at moderate temperature levels are in the viscosity, μ, and the volumetric coefficient of thermal expansion, β, and that for higher Prandtl number liquids, the variation of β is often negligible. Piau [17] also included the effect of thermal stratification of the ambient fluid in an analysis which also includes variable μ and β for water. Brown [18] used an integral method and studied the effect of the coefficient of volumetric expansion on laminar free convection heat transfer. Carey and Mollendorf [19] have shown the mathematical forms of viscosity variation with temperature and gave similarity solutions for laminar free convection from a vertical isothermal surface in liquids with temperature-dependent viscosity. Sabhapathy and Cheng [20] studied the effects of temperature-dependent viscosity and coefficient of thermal expansion on the stability of laminar free convection boundary-layer flow of a liquid along an isothermal, vertical surface, employing linear stability theory for Prandtl numbers between 7 and 10. Qureshi and Gebhart [21] studied the stability of vertical thermal buoyancy-induced flow in cold and saline water. They showed that the anomalous density behavior of cold water (for example, a density extremum at about 4°C in pure water at atmospheric pressure) has very large effects on flow and transport. Meanwhile, Herwig and Wickern [22] studied the effect of variable thermophysical properties on laminar boundary layer flows.

Different gases and liquids have different thermophysical properties. The effects of the different thermophysical properties on the laminar free convection and heat transfer are complicated. The results reported so far are not convenient for the prediction of free convection heat transfer due to the difficulty of treating the variable thermophysical properties in the governing equations. Consequently, it is necessary to do more research related to rigorous and reliable prediction of heat transfer coefficient of free convection with large temperature differences.

Bromley [23] first treated laminar film boiling heat transfer of saturated liquid around a horizontal cylinder in a pool. Some other researchers [24–30] have analyzed pool film boiling on a vertical plate. However, only a few analyses took into account the temperature dependence of the fluid's thermophysical properties. McFadden and Grash [27] developed the analysis of saturated film boiling in a pool where the temperature dependence of density and specific heat were considered. Nishikawa et al. [28, 29] made an analysis of pool film boiling as a variable property problem on the basis of the two-phase boundary layer theory, considering only the effect of variation of the vapor's thermophysical properties with temperature in the lower range of subcooling, i.e., ($T_s - T_\infty$ = 0, 20, 40°C). Herwig [30] provided an asymptotic analysis of laminar film boiling on vertical plate including variable property effect. In fact, the temperature difference between heating surface and bulk liquid may be very large, and the thermophysical property variations of the medium in the condensate and vapor films with temperature can have great influences on the pool film boiling free convection.

For film condensation free convection, Nusselt [31] first treated the condensation of saturated steam on a vertical isothermal flat plate. In his theory the inertia and thermal convection of condensate film, the dependence of the thermophysical properties of the condensate medium on temperature, and the effect of surface tension were all neglected. Bromley [32] and Rohsenow [33] first investigated the effect of thermal convection. Later on, the study of Sparrow and Gregg [34] included the effects of thermal convection and inertia forces in the liquid film by using the boundary layer analysis. Koh et al. [35] further solved numerically a boundary-layer model for both the condensate and vapor films. Chen [36] has considered analytically the effect of thermal convection, inertia, and the interface shear force. On the basis of foregoing studies of the independent-temperature physical properties Drew (see [37]), Voskresenskiy [38], and Labuntsov [39] made relatively simple modifications for variable thermophysical properties. Subsequently, Poots and Miles [40] studied the effects of variable thermophysical properties on laminar film condensation of a saturated steam along a vertical flat plate. They simplified the governing equations of the liquid and vapor phases by neglecting the effects of surface tension at the liquid–vapor interface, and obtained solutions of the resulting ordinary differential equations. Late Stinnesbeck and Herwig [41] provided an asymptotic analysis of laminar film condensation on a vertical flat plate including variable property effect. Based on the research results thus far, it is necessary to provide corresponding correlations for the prediction of heat and mass transfer of the film condensation with the large temperature differences.

Generally, there are two problems that hindered the development of studies of the laminar free convection with single and two-phase boundary layers under large temperature differences. The first difficulty is the traditional Falkner–Skan transformation [42]. With this transformation one encounters a large difficulty in the treatment of variable thermophysical properties. So it is necessary to carry out the study of an improved transformation method in order to suit the development of the free convection with a large temperature differences. The second difficulty is the traditional treatment of the variable thermophysical properties. Since Sparrow and Gregg [10] proposed the treatment model of the variable thermophysical properties with the five different gases in 1958, the treatment method of the variable thermophysical properties has not been improved much. Thus, for a long time, there has been an absence of studies of the free convection with large temperature difference by means of model involving the variable thermophysical properties.

1.3.2 For Gravity-Driven Film Flow of Non-Newtonian Power-Law Fluids

Non-Newtonian power-law fluid behavior has been the subject of many recent books [43–47] and useful numerical calculation techniques for non-Newtonian fluid flow have been reviewed by Crochet and Walters [48] and Crochet et al. [49].

The study of the hydrodynamics of falling film flow of power-law fluids was reviewed by Andersson and Irgens [50]. However, the initial studies were carried out experimentally. The experiments of Astarita et al. [51], Therien et al. [52] and Sylvester et al. [53] all include measurements of film thickness as a function of the volumetric rate. The hydrodynamics of gravity-driven power-law films has been studied theoretically by means of the integral method approach [54–57] and similarity analysis [58,59]. Yang and Yarbrough [54,55] and Narayana Murthy and Sarma [56] extended the conventional integral analysis for Newtonian films to cover power-law fluids. Later, Narayana Murthy and Sarma [57] included the effect of interfacial drag at the liquid–vapor interface in a similar analysis, while Tekic et al. [58] presented results which accounted for the streamwise pressure gradient and surface tension. More recently, Andersson and Irgens [59] explored the influence of the rheology of the film on the hydrodynamic entrance length.

A different approach was adopted by Andersson and Irgens [59,60], namely to divide the accelerating film flow into three regions, the boundary layer region, the fully viscous region and the developed flow region. While the boundary layer region is divided into a developing viscous boundary layer and an external inviscid freestream. They further demonstrated that a similarity transformation exists, such that the boundary layer momentum equation for power-law fluids is exactly transformed into a Falkner–Skan type ordinary differential equation. The resulting two-point boundary-value problem was solved numerically with a standard shooting technique based on classical fourth-order Runge–Kutta integration in combination with a Newton iteration procedure. Numerical results were obtained for values of the power-law index n in the range $0.5 \leq n \leq 2.0$. It was conjectured that converged results could have been obtained also for highly pseudo-plastic fluids, i.e., for $n < 0.5$, by using a different integration technique, for instance a finite-difference scheme.

So far, there has been a lack of research work on heat and mass transfer in falling film flow of power-law fluids in comparison with that on the hydrodynamics. The dissolution of a soluble wall and the subsequent penetration of the solute into the non-Newtonian liquid film were considered by Astarita [61], who provided the mass transfer rate between the wall and the hydrodynamically fully developed film, with an assumption of velocity near the wall to vary linearly with the distance from the wall. Mashelkar and Chavan [62] provided a more general solution of this problem. Van der Mast et al. [63] indicated that for accelerating film flow the heat transfer coefficient for the inlet section considerably higher than further downstream. In this connection, Yih and Lee [64] used an integral method and provided a corresponding solution of the heat transfer in the thermal entrance region of a non-Newtonian, laminar, falling liquid film, without consideration of properties of the non-Newtonian fluids. Narayana Murthy and Sarma [56] provided an integral approach for investigation of the problem of heat transfer for the transition and developed regions of the thin, non-Newtonian falling liquid films. Unfortunately, it is readily verified that their solutions based on the integral methods do not induce to the exact analytic solution. As for the effect of injection/suction on the heat transfer, so far there has been only one study of Pop, Watanabe and Komishi [65] on the steady of laminar gravity-driven film flow along a vertical wall for Newtonian fluids, which is based on Falkner–Skan type transformation.

However, it is seen that even the Falkner–Skan type transformation has its limitations. As we know, it is necessary to introduce a stream function for using the Falkner–Skan type transformation. As a consequence, the variables in the resulting dimensionless governing equations are so abstract that their relationships with the flow variables are complicated. Therefore, it is difficult for the Falkner–Skan type transformation to find solutions to some key problems related to hydrodynamics and heat and mass which are rigorous and convenient for predicting the mass flow rate entrained into the boundary layer at any position of the hydrodynamic entrance region, the critical thickness of the film flow, and the resultant heat and mass transfer. On the other hand, it is very difficult to treat variable thermophysical properties in the models based on the Falkner–Skan type transformation.

1.4 Recent Development

1.4.1 A Novel System of Analysis Models

There is a long history of using Falkner–Skan type transformation for treatment of governing differential equations of the boundary layers and film flows of Newtonian and power-law fluids caused by acceleration. In view of some difficulties produced in using the Falkner–Skan type transformation, a new transformation method, velocity component method, is presented in this book, in which the velocity components are directly transformed instead of inducing the flow function ψ. With this method our new system of theoretical and mathematical models are provided for the laminar free boundary layer of Newtonian fluids, gases by Shang and Wang [66–68] and liquids by Shang, Wang, Wang, and Quan [69], for film boiling by Shang, Wang, and Zhong [70] and condensation by Shang and Adamek [71] and Sang and Wang [72] of Newtonian fluids, and for gravity-driven FFNF by Andersson and Shang [73], Shang and Andersson [74], and Shang and Gu [75], and the earlier difficult situations are avoided. In these models, it is noted that the new variables in the new transformations have obvious physical meanings. Then, by means of the velocity component method, it is convenient to treat the problems of the hydrodynamics and heat and mass transfer, even those with variable thermophysical properties and complicated physical factors. In this book, it can be found that all the theoretical models both for Newtonian and non-Newtonian fluid are based on the same similarity transformation, using the dimensionless velocity component method.

1.4.2 A New Approach for the Treatment of Variable Thermophysical Properties

The effect of large temperature differences on heat transfer of the free convection and accelerating film boiling and condensation reflects the influence of the variable thermophysical properties. The thermophysical properties of most fluids vary with temperature. For gases, although the specific heat varies only slightly with temperature, the variation of other thermophysical properties cannot be neglected. The density varies inversely with the first power of the absolute temperature, and absolute viscosity μ and thermal conductivity λ increase with different powers of the absolute temperature. Generally, for a gas with an increase of the atomic number, the exponent of the power increases. According to the recent study of Shang and Wang [66,67], a temperature parameter method for the treatment of variable thermophysical properties of gas was presented. For example, if we express the variations of μ and λ with $\mu \propto T^{n_\mu}$ and $\mu \propto T^{n_\lambda}$, respectively, n_μ and n_λ are 0.649 and 0.71 for a monatomic gas Ne, 0.694 and 0.86 for diatomic gas O_2 and, and 0.88 and 1.3 for polyatomic gas CO^2, respectively. In addition, this temperature-dependent thermophysical property is especially pronounced for liquids, even for viscous oils and pseudo plastic-liquids. The μ and λ values of these liquids are highly temperature-dependent, and the Prandtl number thus varies with temperature in the same manner as μ and λ.

With the temperature parameter method the variations of gas thermophysical properties can be described in the form of powers of absolute temperature. Consequently each temperature parameter, i.e., the temperature exponent, represents the variation of the corresponding thermophysical property of gas with temperature. Also, the temperature parameters of thermal conductivity and viscosity for a series of monatomic and diatomic gases, air and water vapor are proposed based on the typical experimental data. In addition, it has been found that the variation of specific heat with temperature of a polyatomic gas is so important that it cannot be neglected in the study of the effect of variable thermophysical properties on the gas free convection. In this context, the temperature parameter of the specific heat was proposed and the effect of variable thermophysical properties on the free convection of polyatomic gas was further studied [67]. All the temperature parameters were obtained rigorously on the basis of the typical experimental data. In addition, Shang and Wang [69] recommended a polynomial method to obtain simple and exact polynomial equations of density and thermal conductivity for treatment of variable thermophysical properties of liquids. In this book it is shown that with the advanced treatment method of variable thermophysical properties combined with the velocity component transformation, the fluid properties of the governing equations can be always transformed into the corresponding physical property factors. Such advanced method for the treatment of variable thermophysical properties has become an important part of the related theoretical models.

1.4.3 Hydrodynamics and Heat and Mass Transfer

Heat and Mass Transfer of Free Convection and Film Flows of Newtonian Fluids

Based on the new theoretical and mathematical models in this book, the studies are devoted to hydrodynamics and heat and mass transfer of fluid free convection, accelerating film boiling and condensation, as well as driven film flow of non-Newtonian power-law fluids. First, a series of developments are shown in the heat and mass transfer of gas free convection, liquid free convection, film boiling, and condensation, which belong to boundary layer and film flows of Newtonian fluids. The related developments on heat and mass transfer shown in this book can be briefly introduced as follows.

The first study is for the heat transfer of free convection of gases [66–68] with consideration of variable thermophysical properties. A serious effort is devoted to the study of effect of variable thermophysical properties on the heat transfer. According to different variations of gaseous specific heat with temperature, heat transfer problems for two kinds of gases are studies separately. The first kind of gases is monatomic and diatomic gases, air and water vapor whose specific heat variation with temperature may be taken as constant, while the second kind of gases is polyatomic gases with variable specific heat. Obviously, the first kind of gases is a special case of the second one. The temperature parameter method is used for simulation of the variations of gaseous thermophysical properties, such as thermal conductivity, viscosity, density, and specific heat with temperature. The temperature parameter methods are so simple that each gas corresponds to its special temperature parameters, such as the thermal conductivity parameter, viscosity parameter, and specific heat parameter. The simulation expressions of the variable thermophysical properties with the temperature parameter method have been conveniently coupled with the dimensionless governing equations of the boundary layers. The effects of the main physical factors including variable thermophysical properties on heat transfer of laminar free convection of gases are clarified by considering large temperature differences. On this basis, the corresponding shortcut formulae are developed for simple and practical prediction of the heat transfer coefficients of laminar free convection of gases.

The second study relates to the free convection of liquids with variable thermophysical properties [69]. A theoretically rigorous approach of the study on heat transfer of free convection of liquids is proposed with the combination of the dimensionless governing equations with the simulation expressions of the variable thermophysical properties. An essential effort is devoted to study free convection of water with large temperature difference. It is concluded that the Prandtl number Pr_∞ at the temperature of the bulk fluid dominates the heat transfer coefficient of the laminar free convection of water. This conclusion is not only simple, but also in close agreement with the rigorous numerical solutions. On this basis, the corresponding shortcut formula is developed for simple and practical prediction of the heat transfer coefficient of water free convection with large temperature differences.

The third study is for film boiling [70] and film condensation [71,72]. These studies are extensive and contain the situations of film boiling of subcooled liquid and film condensation of superheated vapor. A theoretically rigorous approach of studies on heat and mass transfer for the two-phase boundary layers problem is proposed by considering variable thermophysical properties and complicated physical factors on the interface between the liquid and vapor films. An extensive effort is devoted to the study of heat and mass transfer for film boiling of subcooled water and film condensation of superheated water vapor both with large temperature differences. For this purpose, the corresponding mathematical models are systematically developed with the combination of the dimensionless governing equations of the two-phase boundary layers and the simulation expressions of the variable thermophysical properties of gases and liquids. The numerical procedures of the three-point boundary value problem are provided for the film boiling and condensation, respectively, in which the complicated boundary conditions at the interface of the films are rigorously considered. Rigorous numerical results are obtained for large temperature differences. The dimensionless physical property factors and their effects on heat transfer coefficient and mass flow rate are demonstrated. For application purposes, shortcut formulae are developed for the simple and reliable prediction of heat and mass transfer of the film boiling and condensation.

All the earlier-mentioned studies are not only devoted to the heat transfer for vertical plate case, but also for the inclined case [76]. The dimensionless governing equations of the new mathematical models can be used directly to express the inclined plate/surface case, although these do not involve any angle explicitly. In addition, all the transformation relationships for the heat, mass, and momentum transfer from the vertical plate/surface case to the corresponding inclined plate/surface case are derived.

Hydrodynamics and Heat Transfer of Boundary Layer and Film Flows of Non-Newtonian Power-Law Fluids

More recently, Rao [77] measured experimentally the heat transfer in a developed non-Newtonian fluid films falling down a vertical tube. Andersson and Shang [73], Shang and Andersson [74], and Shang and Gu [75] continuously provided extended analysis and numerical calculation for hydrodynamics, heat transfer, and the thermal boundary layer of the boundary layer region of FFNF system on isothermal flat plate. Massoudi and Phuoc [78] supposed a fully developed flow for the FFNF system and on this basis to calculate velocity and temperature fields. Ouldhadda et al. [79, 80] investigated numerically the laminar flow of heat transfer of FFNF on horizontal cylinder with supposition of a simple developed flow region for the FFNF system. However, except a few works, such as of Andersson and Irgens [59, 60], Andersson and Shang [73], Shang and Andersson [74], and Shang and Gu [75], in the most of current studies, the hydraulic entrance region (i.e. the boundary layer region) was ignored in their analysis of modeling and simulation for the FFNF system. Without considering the existing boundary layer region of the FFNF system, it is never possible to capture the adaptive remodeling process of hydrodynamics and heat transfer, and obtain correct calculation for velocity and temperature fields, film thickness, and heat transfer coefficient of the FFNF system.

On the other hand, although a large number of industrial processes involve heat transfer of FFNF system, the related heat transfer information that can be found in the open literature is relatively scarce. The reason is that the study on a system of heat transfer is a difficult point for FFNF due to its complexity, especially its different characteristics in different regions. Additionally, overcoming the difficult point for hydrodynamics and heat transfer study in hydraulic entrance region is the essential prerequisite of the study for the following hydraulic region.

However, studies [74, 75] dealt with the heat transfer of the boundary layer region, the first part of the hydraulic entrance region. With the local Prandtl number Pr_x proposed by Shang and Andersson [74], the following dependence of the thermal boundary layer thickness was found: (1) except for the case when the local Prandtl number Pr_x equals the related critical local Prandtl number Pr_x^*, the thicknesses of velocity and temperature boundary layers are different; (2) if $Pr_x < Pr_x^*$ the velocity boundary layer thickness is less than the temperature boundary layer thickness; and (3) on the contrary, the velocity boundary layer thickness is larger than the temperature boundary layer thickness. Furthermore, they made a series of contributions to the boundary layer region: (1) a novel approach for prediction of length of boundary layer region; (2) rigorous and practical approach for prediction of mass flow rate entrained into the boundary layer; (3) novel prediction approach of friction coefficient C_{f} on surface; (4) correctly calculated the thicknesses of thermal and momentum boundary layers, and on this basis correctly calculated the velocity and temperature fields and heat transfer coefficients; (5) found the dependent factors on velocity and temperature boundary layer thicknesses and heat transfer coefficients; and (6) innovation of a curve-fitted correlation for rigorous and practical calculation of heat transfer coefficient with quite different thicknesses of temperature and velocity boundary layers.

However, the earlier achievements on heat transfer research for the boundary layer region should be extended to the entire hydrodynamic entrance region, and even further to the entire FFNF system. The study should also be extended to include the effects of various boundary conditions, e.g., porous medium, permeable, and soluble wall conditions, and inclined isothermal and constant heat flux surfaces on heat transfer. The studies should also consider the transition regulation from the laminar flow to turbulent flow of the FFNF system, and the effects of temperature-dependent properties on the system of heat transfer coefficients.

1.4.4 Recent Experimental Measurements of Velocity Field in Boundary Layer

Besides the advanced theoretical studies, in this book, we also show recently obtained experimental measurements of velocity field on the boundary layer of the laminar free convection, both of air and water. Very important is that the measurement of velocity field of the boundary layer of the laminar free convection requires a high degree of precision. The difficulty in accurate quantification is very great. In 1930, Schmidt and Beckman measured the velocity field of the laminar free convection of air [81], and hitherto this measurement is taken as classical. Their experimental results were well identical to the corresponding theoretical solutions based on the Boussinesq approximation obtained by Pohlhausen. However, for a long time, there has been a shortage of the experimental results of the velocity field of the boundary layer for the gas laminar free convection with the large temperature difference. Meanwhile, for the velocity field in the boundary layer for the liquid laminar free convection, even in the case of the small temperature differences, there has been a shortage of experimental results. Therefore, our experimental results for the velocity field on the boundary layer of laminar free convection with the large temperature differences for air [82] and water [69,83] are reported in the book. These experimental measurements have been very difficult to obtain. The velocity fields of the laminar free convection of air and water in the case of different temperature differences obtained by the experiments have not only verified the corresponding theoretical results of the free convection for gas and liquid with the large temperature difference mentioned earlier, but also filled in the gaps in the study of the measurement of velocity field of laminar free convection for gases and liquids with the large temperature differences.

These new theoretical and experimental studies introduce a new development for the study of laminar free convection in single and two-phase boundary layers and films under large temperature differences. These are all described in the following chapters of this book. The purpose of this book is to systematically express these results, to promote further development of the study of free convection film flows with large temperature differences, and to satisfy the increasing demands of industry.

References

1. L. Prandtl, ÜbDie Flussigkeitsbewegung bei Sehr Kleiner Reibung, Proc. Third Int. Math. Koug., Heidelberg, 1904
2. J. Boussinesq, Theorie analytique de la chaleur, mise en harmonie avec la Thermodynamigue et avec la Theorie mechanique de la lumiere, Vol. 11, Ganthier –Villars, Paris, 1903
3. J. Boussinesq, Calcul du poirior refroidissant des courants fluids, J. Math. Pures Appl. 60, pp. 285, 1905
4. E. Pohlhausen, Der Warmesustausch zwischen festen Karpern und Flüssigkeiten Mit kleiner Reibung und kleiner Warmeleitung, Zeitschrift fur angewandte Mathematik und Mechanik 1, pp. 115–121, 1921
5. S. Ostrach, An analysis of laminar free-convection flow and heat transfer about a plate parallel to the direction of the generating body force, NACA Report 1111, 1953
6. A. J. Ede, Advances in free convection, Adv. Heat Transfer 4, pp. 1–64, 1967
7. E. J. LeFevre, Laminar free convection from a vertical plane surface, Mech. Eng. Res. Lab., Heat 113 (Great Britain), pp. 168, 1956
8. T. T. Hara, The free-convection flow about a heated vertical plate in air, Trans. Jpn. Soc. Mech. Eng. 20, pp. 517–520, 1954
9. A. A. Tataev, Heat exchange in condition of free laminar movement of gas with variable viscosity at a vertical wall, Zh. Tekh. Fiz. 26, pp. 2714–2719, 1956
10. E. M. Sparrow and J. L. Gregg, The variable fluid-property problem in free convection, Trans. ASME, 80, pp. 879–886, 1958
11. D. D. Gray and A. Giogini, The validity of the Boussinesq approximation for liquids and gases, Int. J. Heat Mass Transfer 19, pp. 545–551, 1977
12. A. M. Clausing and S. N. Kempka, The influences of property variations on natural convection from vertical surfaces, J. Heat Transfer 103, pp. 609–612, 1981
13. T. Fujii et al., Experiments on natural convection heat transfer from the outer surface of a vertical cylinder to liquids, Int. J. Heat Mass Transfer 13, pp. 753–787, 1970
14. T. Fujii, Heat transfer from a vertical flat surface by laminar free convection – the case where the physical constants of fluids depend on the temperature and the surface has an arbitrary temperature distribution in the vertical direction, Trans. Jpn. Soc. Mech. Ens. 24, pp. 964–972, 1958
15. S. Akagi, Free convection heat transfer in viscous oil, Trans. Jpn. Soc. Mech. Ens. 30, pp. 624–635, 1964
16. J. M. Piau, Convection Natural Laminar en Regime Permanent dans les Liquids, Influence des Variations des Properties Physique avec la Temperature, C.R. Hebd, Seanc, Acad. Sci., Paris, Vol. 271, pp. 935–956, 1970
17. J. M. Piau, Influence des variations des properties physiques et la stratification en convection naturelle, Int. J. Heat Mass Transfer 17, pp. 465–476, 1974
18. A. Brown, The effect on laminar free convection heat transfer of temperature dependence of the coefficient of volumetric expansion, Trans. ASME, Ser. C, J. Heat Transfer 97, pp. 133–135, 1975
19. V. P. Carey and J. C. Mollendorf, Natural convection in liquids with temperature dependence viscosity. In Proc. Sixth Int. Heat Transfer Conf., Toronto, NC-5, Vol. 2, pp. 211–217, Hemisphere, Washington, DC 1978

20. P. Sabhapathy and K. C. Cheng, The effect of temperature-dependent viscosity and coefficient of thermal expansion on the stability of laminar, natural convective flow along an isothermal, vertical surface, Int. J. Heat Mass Transfer 29, pp. 1521–1529, 1986
21. Z. H. Qureshi and B. Gebhart, The stability of vertical thermal buoyancy induced flow in cold pure and saline water, Int. J. Heat Mass Transfer 29, pp. 1383–1392, 1986
22. H. Herwig and G. Wickern, The effect of variable properties on laminar boundary layer flows, Wärme- und Stoffübertragung, Bd. 20, 47–57, 1986
23. Bromley, Heat transfer in table film boiling, Chem. Eng. Prog. 46(5), pp. 221–227, 1950
24. J. C. Y. Koh, Analysis of film boiling on vertical surface, J. Heat Transfer 84(1), pp. 55–62, 1962
25. R. D. Cess, Analysis of laminar film boiling from a vertical flat plate, Research Report 405 FF 340- R2-X, Westinghouse Research Lab., Pittsburgh, PA, March 1959
26. M. Sparrow and R. D. Cess, The effect of subcooled liquid on laminar film boiling, J. Heat Transfer 84c(2), pp. 149–156, 1962
27. P. W. McFadden and R. J. Grash, An analysis of laminar film boiling with variable properties, Int. J. Heat Mass Transfer 1, pp. 325–335, 1961
28. K. Nishikawa and T. Ito, Two-phase boundary-layer treatment of free convection film boiling, Int. J. Heat Mass Transfer 9, pp. 103–115, 1966
29. K. Nishikawa, T. Ito, and K. Matsumoto, Investigation of variable thermophysical property problem concerning pool film boiling from vertical plate with prescribed uniform temperature, Int. J. Heat Mass Transfer 19, pp. 1173–1182, 1976
30. H. Herwig, An asymptotic analysis of laminar film boiling on vertical plates including variable property effects, Int. J. Heat Mass Transfer 31, 2013–2021, 1988
31. W. Nusselt, Die Oberflächenkondensation des Wasserdampfes, Z. Ver. D. Ing. 60, pp. 541–569, 1916
32. L. A. Bromley, Effect of heat capacity of condensate in condensing, Ind. Eng. Chem. 44, pp. 2966–2969, 1952
33. W. M. Rohsenow, Heat transfer and temperature distribution in laminar film condensation, Trans. Am. Soc. Mech. Eng. 78, pp. 1645–1648, 1956
34. E. M. Sparrow and J. L. Gregg, A boundary layer treatment of laminar film condensation, J. Heat Transfer 81, pp. 13–18, 1959
35. J. C. Y. Koh, E. M. Sparrow, and J. P. Hartnett, The two phase boundary layer in laminar film condensation, Int. J. Heat Mass Transfer 2, pp. 69–82, 1961
36. M. M. Chen, An analytical study of laminar film condensation, part 1 – flat plate, J. Heat Transfer, pp. 48–54, 1961
37. W. H. McAdams, Heat Transmission, 3rd edn. McGraw-Hill, New York, pp. 331–332, 1954
38. Voskresenskiy, Calculation of heat transfer in film condensation allowing for the temperature dependence of the physical properties of the condensate, USSR Acad. Sci., 1948
39. D. A. Labuntsov, Effect of temperature dependence of physical parameters of condensate on heat transfer in film condensation of steam, Teploenergetika 4(2), pp. 49–51, 1957

40. G. Poots and R. G. Miles, Effect of variable physical properties on laminar film condensation of saturated steam on a vertical flat plate, Int. J. Heat Mass Transfer 10, pp. 1677–1692, 1967
41. J. Stinnesbeck and H. Herwig, An asymptotic analysis of laminar film condensation on a vertical flat plate including variable property effects, Proc. nineth Int. Heat Transfer Conf., Jerusalem, 1990
42. V. M. Falkner and S. W. Skan, Some approximate solutions of the boundary layer equations, Phil. Mag. 12, pp. 865, 1931
43. G. Astarita and G. Marrucci, Principles of Non-Newtonian Fluid Mechanics, McGraw-Hill, London, 1974
44. R. Darby, Viscoelastic fluids: an introduction to their properties and behavior, Marcel Dekker, New York, 1976
45. W. R. Schowalter, Mechanics of Non-Newtonain Fluids, Pergamon Press, Oxford, 1978
46. R. I. Tanner, Engineering Rheology, Clarendon Press, Oxford, 1985
47. R. B. Bird, R. C. Armstrong, and O. Hassager, Dynamics of Polymetric Liquids, Vol. 1 Fluid Mechanics, Wiley, New York, 2nd ed., 1987
48. M. J. Crochet and K. Walters, Numerical Methods in Non-Newtonian Fluid Mechanics, Ann. Rev. Fluid Mech. 15, 241, 1983
49. M. J. Crochet, A. R. Davies, and K. Walters, Numerical Simulation of Non-Newtonian Flow, Elsevier, Amsterdam, 1984
50. H. I. Andersson and F. Irgens, Film flow of power law fluids, Encyclopaedia of Fluid Mechanics, Gulf Publishing Company, Houston, TX, 9, pp. 617–648, 1990
51. G. Astarita, G. Marrucci, and G. Palumbo, Non-Newtonian gravity flow along inclined plane surface, Ind. Eng. Chem. Fundam. 3, pp. 333–339, 1964
52. N. Therien, N. B. Coupal, and J. L. Corneille, Verification experimente de l'epaisseur du film pour des liquides non-Newtoniens s'ecoulant par gravite sur un plan incline, Can. J. Chem. Eng. 48, pp. 17–20, 1970
53. N. D. Sylvester, J. S. Tyler, and A. H. P. Skelland, Non-Newtonian film fluids: theory and experiment, Can. J. Chem. Eng. 51, pp. 418–429, 1973
54. T. M. T. Yang and D. W. Yarbrough, A numerical study of the laminar flow of non-Newtonian fluids along a vertical wall, ASME J. Appl. Mech. 40, pp. 290–292, 1973
55. T. M. T. Yang and D. W. Yarbrough, Laminar flow of non-Newtonian liquid films inside a vertical pipe, Rheol. Acta 19, pp. 432–436, 1980
56. V. Narayana Murthy and P. K. Sarma, A note on hydrodynamics entrance length of non-Newtonian laminar falling films, Chem. Eng. Ser. 32, pp. 566–567, 1977
57. V. Narayana Murthy and P. K. Sarma, Dynamics of developing laminar non-Newtonian falling liquid films with free surface. ASME J. Appl. Mech. 45, pp. 19–24, 1978
58. M. N. Tekic, D. Posarac, and D. A. Petrovic, A note on the entrance region lengths of non-Newtonian laminar falling films. Chem. Eng. Ser. 41, pp. 3230–3232, 1986
59. H. I. Andersson and F. Irgens, Hydrodynamic entrance length of non-Newtonian liquid films, Chem. Eng. Sci. 45, pp. 537–541, 1990
60. H. I. Andersson and F. Irgens, Gravity-driven laminar film flow of power-law fluids along vertical walls, J. Non-Newtonian Fluid Mech. 27, pp. 153–172, 1988
61. G. Astarita, Mass transfer from a flat solid surface to a falling non-Newtonian liquid film, Ind. Eng. Chem. Fundam. 5, pp. 14–18, 1966

62. R. A. Mashelker and V. V. Chavan, Solid dissolution in falling films of non-Newtonian liquids, J. Chem. Jpn 6, pp. 160–167, 1973
63. V. C. Van der Mast, S.M. Read, and L.A. Bromley, Boiling of natural sea water in falling film evaporators, Disalinations 18, pp. 71–94, 1976
64. S. M. Yih and M. W. Lee, Heating or evaporating in the thermal entrance region of a non-Newtonian laminar-falling region film. Int. J. Heat Mass Transfer 29, pp. 1999–2002, 1986
65. I. Pop, T. Watanabe, and H. Komishi, Gravity-deriven laminar film flow along a vertical wall with surface mass transfer, Int. Commun. Heat Mass Transfer 23(5), pp. 687–695, 1996
66. D. Y. Shang and B. X. Wang, Effect of variable thermophysical properties on laminar free convection of gas, Int. J. Heat Mass Transfer 33(7), pp. 1387–1395, 1990
67. D. Y. Shang and B. X. Wang, Effect of variable thermophysical properties on laminar free convection of polyatomic gas, Int. J. Heat Mass Transfer 34(3), pp. 749–755, 1991
68. D. Y. Shang and B. X. Wang, The deviation of heat transfer calculation for laminar free convection of gas due to ignoring the variable thermophysical properties, Warme-und Stoffubertragung 28, pp. 33–36, 1993
69. D. Y. Shang, B. X. Wang, Y. Wang, and Y. Quan, Study on liquid laminar free convection with consideration of variable thermophysical properties, Int. J. Heat Mass Transfer 36(14), pp. 3411–3419, 1993
70. D. Y. Shang, B. X. Wang, and L. C. Zhong, A study on laminar film boiling of liquid along an isothermal vertical plate in a pool with consideration of variable thermophysical properties, Int. J. Heat Mass Transfer 37(5), pp. 819–828, 1994
71. D. Y. Shang and T. Adamek, Study on laminar film condensation of saturated steam on a vertical flat plate for consideration of various physical factors including variable themophysical properties, Wärme- und Stoff Übertragung 30, Springer Berlin Heidelberg New York, pp. 89–100, 1994
72. D. Y. Shang and B. X. Wang, An extended study on steady-strate laminar film condensation of a superheated vapor on isothermal vertical plate, Int. J. Heat Mass Transfer for publication 40(4), pp. 931–941, 1997
73. H. I. Andersson and D.Y. Shang, An extended study of hydrodynamics of gravity-driven film flow of power-law fluids, Fluid Dyn. Res. 22, pp. 345–357, 1998
74. D. Y. Shang and H. I. Andersson, Heat transfer in gravity-driven film flow of power-law fluids, Int. J. Heat Mass Transfer 42(11), pp. 2085–2099, 1999
75. D. Y. Shang and J. Gu, Analyses of pseudo-similarity and boundary layer thickness for non-Newtonian falling film flow, Heat Mass Trans. 41(1), pp. 44–50, 2004
76. D. Y. Shang and H. S. Takhar, An extended study on relationships of heat, momentum and mass transfer for laminar free convection between inclined and vertical plates, J. Theor. and Appl. Fluid Mech., Inaugural Issue, 16–32, 1995
77. B. K. Rao, Heat transfer to falling power-law fluid film, Int. J. Heat Fluid Flow 20, pp. 429–436, 1999
78. M. Massoudi, T. X. Phuoc, Fully developed flow of a modified second grade fluid with temperature dependent viscosity, Acta Mech. 150(1–2) pp. 23–37, 2001
79. D. Ouldhadda, A. Idrissi, Laminar flow and heat transfer of non-Newtonian falling liquid film on a horizontal tube with variable surface heat flux, Int. Commun. Heat Mass Trans. 28(8), pp. 1125–1136, 2002

80. D. Ouldhadda, A. Idrissi, and M. Asbik, Heat transfer in non-Newtonian falling liquid film on a horizontal circular cylinder, Heat Mass Trans. 38, pp. 713–721, 2002
81. E. Schmidt and W. Beckman, Ads Temperature- und Geschwindigkeitsfeld von einer Wärme Abgebenden Senkrechten Plätte bei Naturlicher Konvection, Forsch. Ing. - Wes. 1, p. 391, 1930
82. D. Y. Shang and B. X. Wang, Measurement on velocity of laminar boundary layer for gas free convection along an isothermal vertical flat plat, Anon (Ed.), Inst. Chem. Eng. Symp. Ser. Vol. 1, No. 12, third UK National Conf. Incorporating first Eur. Conf. Therm. Sci., Birmingham Engl., Sep. 16–18, Hemisphere Publishing Corporation, pp. 484–489, 1992
83. D. Y. Shang, B. X. Wang, and H. S. Takhar, Measurements of the velocity field of the laminar boundary layer for water free convection along an isothermal vertical flat plate, Appl. Mech. Eng. 3(4), pp. 553–570, 1998

Part I

Laminar Free Convection

2

Basic Conservation Equations for Laminar Free Convection

Nomenclature

A	area, m^2
E	internal energy, J
e	internal energy per unit mass, J kg^{-1}
$\dot{E}$	internal energy per unit time in system, $\Delta\dot{E} = \dot{Q} + \dot{W}_{out}$, W
ΔE	increment of internal energy in a system, J
$\Delta\dot{E}$	increment of internal energy per unit time in system, W
F	force, N
$\vec{F}$	is mass force per unit mass, kg/s^2
F_m	mass force acting on the control, N
F_s	surface force acting on the control, N
$\dot{G}$	momentum increment per unit time in system, N
g	gravity acceleration, m s^{-2}
H	enthalpy, $E + pV$, J
h	specific enthalpy (enthalpy per unit mass), $e + pv, c_p t$, J kg^{-1}
$\dot{m}$	mass increment per unit time, kg s^{-1}
p	pressure, N m^{-2}
Q	heat, J
$\dot{Q}$	heat entering the system per unit time, W
Q_{in}	heat transferred into the system from its surroundings, J
t	temperature, °C
T	absolute temperature, K
V	volume, m^3
v	specific volume, m^3 kg^{-1}

w_x, w_y, w_z velocity components in x, y, z direction, respectively, m s^{-1}
$\vec{W}$ velocity, $w_x i + w_y j + w_z k$
$\dot{W}$ work done per unit time, W

Greek symbols
ρ mass density, kg m^{-3}
λ thermal conductivity, W $(\text{m K})^{-1}$
μ absolute viscosity, kg $(\text{m s})^{-1}$
τ time, s , or shear force, N m^{-2}
$\vec{\tau}$ surface force acting on unit area
Φ viscous dissipation function
ε deformation rate
[ ] symbol of tensor
{ } symbol of quantity grade

Subscripts
m mass force
s surface force
∞ infinity

In this chapter the basic conservation equations related to fluid convection heat transfer are introduced, and on this basis, the corresponding conservation equations of mass, momentum, and energy for steady laminar free convection in a boundary layer are obtained.

2.1 Continuity Equation

The conceptual basis for the derivation of the continuity equation of fluid flow is the mass conservation law. The control volume for the derivation of continuity equation is shown in Fig. 2.1 in which the mass conservation principle is stated as

$$\dot{m}_{\text{increment}} = \dot{m}_{\text{in}} - \dot{m}_{\text{out}}, \tag{2.1}$$

where $\dot{m}_{\text{increment}}$ expresses the mass increment per unit time in the control volume, $\dot{m}_{\text{in}}$ represents the mass flowing into the control volume per unit time, and $\dot{m}_{\text{out}}$ is the mass flowing out of the control volume per unit time. The dot notation signifies a unit time.

In the control volume, the mass of fluid flow is given by $\rho\,\mathrm{d}x\,\mathrm{d}y\,\mathrm{d}z$, and the mass increment per unit time in the control volume can be expressed as

$$\dot{m}_{\text{increment}} = \frac{\partial \rho}{\partial \tau}\mathrm{d}x\,\mathrm{d}y\,\mathrm{d}z. \tag{2.2}$$

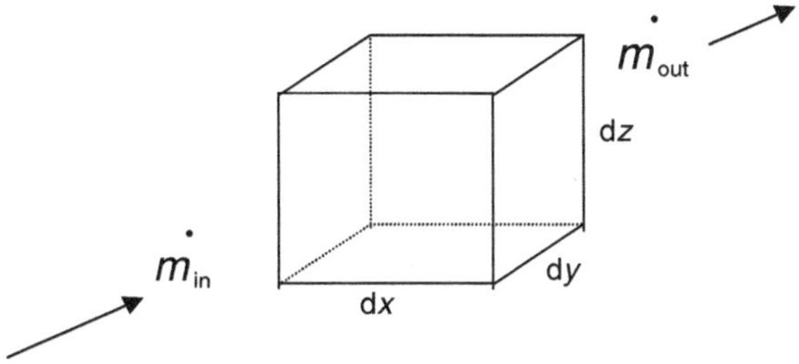

Fig. 2.1. Control volume for the derivation of the continuity equations

The mass flowing into the control volume per unit time in the x direction is given by ρw_x dx dz. The mass flowing out of the control volume in a unit time in the x direction is given by $[\rho w_x + \partial(\rho w_x)/\partial x \cdot \mathrm{d}x]\mathrm{d}y\ \mathrm{d}z$. Thus, the mass increment per unit time in the x direction in the control volume is given by

$$\frac{\partial(\rho w_x)}{\partial x}\mathrm{d}x\ \mathrm{d}y\ \mathrm{d}z.$$

Similarly, the mass increments in the control volume in the y and z directions per unit time are given by

$$\frac{\partial(\rho w_y)}{\partial y}\mathrm{d}y\ \mathrm{d}x\ \mathrm{d}z$$

and

$$\frac{\partial(\rho w_z)}{\partial z}\mathrm{d}z\ \mathrm{d}x\ \mathrm{d}y,$$

respectively. We thus obtain

$$\dot{m}_{\text{out}} - \dot{m}_{\text{in}} = \left(\frac{\partial(\rho w_x)}{\partial_x} + \frac{\partial(\rho w_y)}{\partial y} + \frac{\partial(\rho w_z)}{\partial z}\right)\mathrm{d}x\ \mathrm{d}y\ \mathrm{d}z. \tag{2.3}$$

Combining (2.1) with (2.2) and (2.3) we obtain the following continuity equation in Cartesian coordinates:

$$\frac{\partial \rho}{\partial \tau} + \frac{\partial(\rho w_x)}{\partial x} + \frac{\partial(\rho w_y)}{\partial y} + \frac{\partial(\rho w_z)}{\partial z} = 0. \tag{2.4}$$

or in the vector notation

$$\frac{\partial \rho}{\partial \tau} + \nabla \cdot (\rho \vec{W}) = 0, \tag{2.5}$$

or

$$\frac{\mathrm{D}\rho}{\mathrm{D}\tau} + \rho \nabla \cdot (\vec{W}) = 0, \tag{2.6}$$

where $\vec{W} = iw_x + jw_y + kw_z$ is the fluid velocity.

For steady state, the vector and Cartesian forms of the continuity equation are given by

$$\frac{\partial}{\partial x}(\rho w_x) + \frac{\partial}{\partial y}(\rho w_y) + \frac{\partial}{\partial z}(\rho w_z) = 0, \tag{2.7}$$

or

$$\nabla \cdot (\rho \vec{W}) = 0. \tag{2.8}$$

2.2 Momentum Equation (Navier–Stokes Equations)

The control volume for the derivation of the momentum equation of fluid flow is shown in Fig. 2.2. Meanwhile, take an enclosed surface A that includes the control volume. According to momentum law, the momentum increment of the fluid flow per unit time equals the sum of the mass force and surface force acting on the fluid. The relationship is shown as below:

$$\dot{G}_{\text{increment}} = \vec{F}_{\text{m}} + \vec{F}_{\text{s}}, \tag{2.9}$$

where $\vec{F}_{\text{m}}$ and $\vec{F}_{\text{s}}$ denote mass force and surface force, respectively.

In the system the momentum increment $\dot{G}_{\text{increment}}$ of the fluid flow per unit time can be described as

$$\dot{G}_{\text{increment}} = \frac{\mathrm{D}}{\mathrm{D}\tau}\int_V \rho \vec{W}\,\mathrm{d}V. \tag{2.10}$$

In the system the sum of mass force F_{m} and surface force F_{s} acting on the fluid is expressed as

$$F_{\text{m}} + F_{\text{s}} = \int_V \rho \vec{F}\ \mathrm{d}V + \int_A \vec{\tau}_n\ \mathrm{d}A, \tag{2.11}$$

where V and A are volume and surface area of the system, respectively, $\vec{F}$ is mass force per unit mass, $\vec{\tau}_n$ is surface force acting on unit area.

Combining (2.9) with (2.10) and (2.11), we have the following equation:

$$\frac{\mathrm{D}}{\mathrm{D}\tau}\int_V \rho \vec{W}\ \mathrm{d}V = \int_V \rho \vec{F}\ \mathrm{d}V + \int_A \vec{\tau}_n\ \mathrm{d}A. \tag{2.12}$$

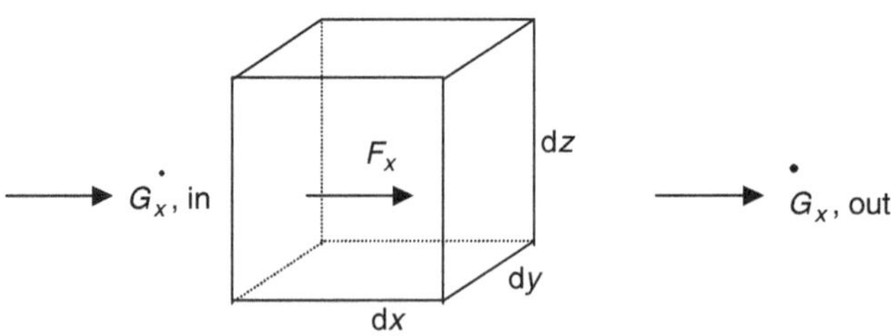

Fig. 2.2. Control volume for the derivation of momentum equations

According to tensor calculation, the right side of (2.12) is changed into the following form:

$$\int_V \rho\vec{F}\ \mathrm{d}V + \int_A \vec{\tau}_n\ \mathrm{d}A = \int_V \rho\vec{F}\ \mathrm{d}V + \int_V \nabla\cdot[\tau]\mathrm{d}V, \tag{2.13}$$

where $\nabla\cdot[\tau]$ is divergence of the shear force tensor.

The left side of (2.12) can be rewritten as

$$\frac{\mathrm{D}}{\mathrm{D}\tau}\int_V \rho\vec{W}\ \mathrm{d}V = \int_V \frac{\mathrm{D}(\rho\vec{W})}{\mathrm{D}\tau}\mathrm{d}V. \tag{2.14}$$

With (2.13) and (2.14), (2.12) can be simplified as

$$\int_V \left\{\frac{\mathrm{D}(\rho\vec{W})}{\mathrm{D}\tau} - \rho\vec{F} - \nabla\cdot[\tau]\right\}\mathrm{d}V = 0. \tag{2.15}$$

Therefore,

$$\frac{\mathrm{D}(\rho\vec{W})}{\mathrm{D}\tau} = \rho\vec{F} + \nabla\cdot[\tau]. \tag{2.16}$$

This is the Navier–Stokes equations of fluid flow. For Cartesian Coordinates, (2.16) can be expressed as

$$\rho\left(\frac{\partial w_x}{\partial\tau} + w_x\frac{\partial w_x}{\partial x} + w_y\frac{\partial w_x}{\partial y} + w_z\frac{\partial w_x}{\partial z}\right) = \frac{\partial\tau_{xx}}{\partial x} + \frac{\partial\tau_{yx}}{\partial y} + \frac{\partial\tau_{zx}}{\partial y} + \rho g_x, \tag{2.17}$$

$$\rho\left(\frac{\partial w_y}{\partial\tau} + w_x\frac{\partial w_y}{\partial x} + w_y\frac{\partial w_y}{\partial y} + w_z\frac{\partial w_y}{\partial z}\right) = \frac{\partial\tau_{xy}}{\partial x} + \frac{\partial\tau_{yy}}{\partial y} + \frac{\partial\tau_{zy}}{\partial z} + \rho g_y, \tag{2.18}$$

$$\rho\left(\frac{\partial w_z}{\partial\tau} + w_x\frac{\partial w_z}{\partial x} + w_y\frac{\partial w_z}{\partial y} + w_z\frac{\partial w_z}{\partial z}\right) = \frac{\partial\tau_{xz}}{\partial x} + \frac{\partial\tau_{yz}}{\partial y} + \frac{\partial\tau_{zz}}{\partial z} + \rho g_z, \tag{2.19}$$

where g_x, g_y, and g_z are gravity accelerations in x, y, and z directions, respectively, while, the related shear forces are given later:

$$\tau_{xx} = \left[p + \frac{2}{3}\left(\frac{\partial w_x}{\partial x} + \frac{\partial w_y}{\partial y} + \frac{\partial w_z}{\partial z}\right)\right] + 2\mu\frac{\partial w_x}{\partial x},$$

$$\tau_{yy} = \left[p + \frac{2}{3}\left(\frac{\partial w_x}{\partial x} + \frac{\partial w_y}{\partial y} + \frac{\partial w_z}{\partial z}\right)\right] + 2\mu\frac{\partial w_y}{\partial y},$$

$$\tau_{zz} = \left[p + \frac{2}{3}\left(\frac{\partial w_x}{\partial x} + \frac{\partial w_y}{\partial y} + \frac{\partial w_z}{\partial z}\right)\right] + 2\mu\frac{\partial w_z}{\partial z},$$

$$\tau_{xy} = \tau_{yx} = \mu\left(\frac{\partial w_y}{\partial x} + \frac{\partial w_x}{\partial y}\right),$$

$$\tau_{yz} = \tau_{zy} = \mu\left(\frac{\partial w_z}{\partial y} + \frac{\partial w_y}{\partial z}\right),$$

$$\tau_{zx} = \tau_{xz} = \mu\left(\frac{\partial w_x}{\partial z} + \frac{\partial w_z}{\partial x}\right).$$

Then, (2.17)–(2.19) are rewritten as follows, respectively:

$$\rho\frac{\mathrm{D}w_x}{\mathrm{D}\tau} = -\frac{\partial p}{\partial x} + 2\frac{\partial}{\partial x}\left(\mu\frac{\partial w_x}{\partial x}\right) + \frac{\partial}{\partial y}\left[\mu\left(\frac{\partial w_x}{\partial y} + \frac{\partial w_y}{\partial x}\right)\right]$$
$$+\frac{\partial}{\partial z}\left[\mu\left(\frac{\partial w_x}{\partial z} + \frac{\partial w_z}{\partial x}\right)\right] - \frac{\partial}{\partial x}\left[\frac{2}{3}\mu\left(\frac{\partial w_x}{\partial x} + \frac{\partial w_y}{\partial y} + \frac{\partial w_z}{\partial z}\right)\right] + \rho g_x. \tag{2.20}$$

Similarly, the momentum equations in the y and z directions are given by

$$\rho\frac{\mathrm{D}w_y}{\mathrm{D}\tau} = -\frac{\partial p}{\partial y} + \frac{\partial}{\partial x}\left[\mu\left(\frac{\partial w_x}{\partial y} + \frac{\partial w_y}{\partial x}\right)\right] + 2\frac{\partial}{\partial y}\left(\mu\frac{\partial w_y}{\partial y}\right)$$
$$+\frac{\partial}{\partial z}\left[\mu\left(\frac{\partial w_y}{\partial z} + \frac{\partial w_z}{\partial y}\right)\right] - \frac{\partial}{\partial y}\left[\frac{2}{3}\mu\left(\frac{\partial w_x}{\partial x} + \frac{\partial w_y}{\partial y} + \frac{\partial w_z}{\partial z}\right)\right] + \rho g_y, \tag{2.21}$$

$$\rho\frac{\mathrm{D}w_y}{\mathrm{D}\tau} = -\frac{\partial p}{\partial z} + \frac{\partial}{\partial x}\left[\mu\left(\frac{\partial w_x}{\partial z} + \frac{\partial w_z}{\partial x}\right)\right] + \frac{\partial}{\partial y}\left[\mu\left(\frac{\partial w_y}{\partial z} + \frac{\partial w_z}{\partial y}\right)\right]$$
$$+2\frac{\partial}{\partial z}\left(\mu\frac{\partial w_z}{\partial z}\right) - \frac{\partial}{\partial z}\left[\frac{2}{3}\mu\left(\frac{\partial w_x}{\partial x} + \frac{\partial w_y}{\partial y} + \frac{\partial w_z}{\partial z}\right)\right] + \rho g_z. \tag{2.22}$$

For steady state, the momentum equations (2.20)–(2.22) are given as follows, respectively:

$$\rho\left(w_x\frac{\partial w_x}{\partial x} + w_y\frac{\partial w_x}{\partial y} + w_z\frac{\partial w_x}{\partial z}\right)$$
$$= -\frac{\partial p}{\partial x} + 2\frac{\partial}{\partial x}\left(\mu\frac{\partial w_x}{\partial x}\right) + \frac{\partial}{\partial y}\left[\mu\left(\frac{\partial w_x}{\partial y} + \frac{\partial w_y}{\partial x}\right)\right] + \frac{\partial}{\partial z}\left[\mu\left(\frac{\partial w_x}{\partial z} + \frac{\partial w_z}{\partial x}\right)\right]$$
$$-\frac{\partial}{\partial x}\left[\frac{2}{3}\mu\left(\frac{\partial w_x}{\partial x} + \frac{\partial w_y}{\partial y} + \frac{\partial w_z}{\partial z}\right)\right] + \rho g_x, \tag{2.23}$$

$$\rho\left(w_x\frac{\partial w_y}{\partial x} + w_y\frac{\partial w_y}{\partial y} + w_z\frac{\partial w_y}{\partial z}\right)$$
$$= -\frac{\partial p}{\partial y} + \frac{\partial}{\partial x}\left[\mu\left(\frac{\partial w_x}{\partial y} + \frac{\partial w_y}{\partial x}\right)\right] + 2\frac{\partial}{\partial y}\left(\mu\frac{\partial w_y}{\partial y}\right) + \frac{\partial}{\partial z}\left[\mu\left(\frac{\partial w_y}{\partial z} + \frac{\partial w_z}{\partial y}\right)\right]$$
$$-\frac{\partial}{\partial y}\left[\frac{2}{3}\mu\left(\frac{\partial w_x}{\partial x} + \frac{\partial w_y}{\partial y} + \frac{\partial w_z}{\partial z}\right)\right] + \rho g_y, \tag{2.24}$$

$$\rho\left(w_x\frac{\partial w_z}{\partial x}+w_y\frac{\partial w_z}{\partial y}+w_z\frac{\partial w_z}{\partial z}\right)$$
$$=-\frac{\partial p}{\partial z}+\frac{\partial}{\partial x}\left[\mu\left(\frac{\partial w_x}{\partial z}+\frac{\partial w_z}{\partial x}\right)\right]+\frac{\partial}{\partial y}\left[\mu\left(\frac{\partial w_y}{\partial z}+\frac{\partial w_z}{\partial y}\right)\right]+2\frac{\partial}{\partial z}\left(\mu\frac{\partial w_z}{\partial z}\right)$$
$$-\frac{\partial}{\partial z}\left[\frac{2}{3}\mu\left(\frac{\partial w_x}{\partial x}+\frac{\partial w_y}{\partial y}+\frac{\partial w_z}{\partial z}\right)\right]+\rho g_z. \tag{2.25}$$

2.3 Energy Equation

The control volume for derivation of the energy equation of fluid flow is shown in Fig. 2.3. Meanwhile, take an enclosed surface A that includes the control volume. According to the first law of thermodynamics, we have the following equation:

$$\Delta \overset{\bullet}{E}=\overset{\bullet}{Q}+\overset{\bullet}{W}_{\text{out}}, \tag{2.26}$$

where $\Delta \overset{\bullet}{E}$ is energy increment in the system per unit time, $\overset{\bullet}{Q}$ is heat increment in the system per unit time, and $\overset{\bullet}{W}_{\text{out}}$ denotes work done by the mass force and surface force on the system per unit time.

The energy increment per unit time in the system is described as

$$\Delta \overset{\bullet}{E}=\frac{\mathrm{D}}{\mathrm{D}\tau}\int_V \rho(e+\frac{W^2}{2})\mathrm{d}V, \tag{2.27}$$

where τ denotes time, $W^2/2$ is the fluid kinetic energy, W is fluid velocity, and the symbol e represents the internal energy per unit mass.

The work done by the mass force and surface force on the system per unit time is expressed as

$$\overset{\bullet}{W}_{\text{out}}=\int_V \rho\vec{F}\cdot\vec{W}\,\mathrm{d}V+\int_A \vec{\tau_n}\cdot\vec{W}\,\mathrm{d}A, \tag{2.28}$$

where $\vec{F}$ is the mass force per unit mass, and $\vec{\tau_n}$ is surface force acting on unit area.

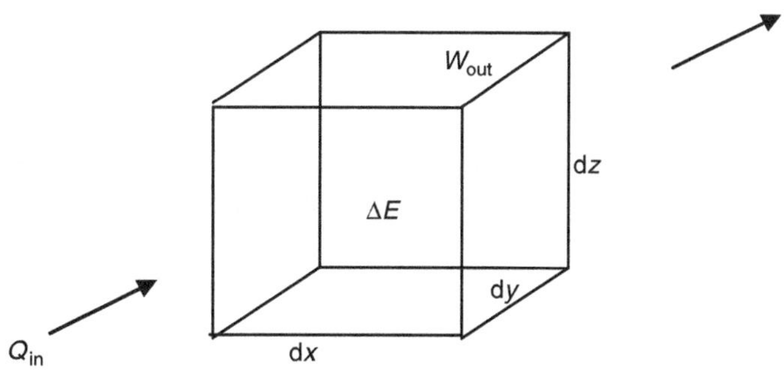

Fig. 2.3. Control volume for derivation of the energy equations of fluid flow

The heat increment entering into the system per unit time through thermal conduction is described by using Fourier's law as follows:

$$\overset{\bullet}{Q} = \int_A \lambda \frac{\partial t}{\partial n} \mathrm{d}A, \tag{2.29}$$

where n is normal line of the surface, and here only the heat conduction is considered. With (2.27)–(2.29), (2.26) is rewritten as

$$\frac{\mathrm{D}}{\mathrm{D}\tau} \int_V \rho \left(e + \frac{W^2}{2} \right) \mathrm{d}V = \int_V \rho \vec{F} \cdot \vec{W} \ \mathrm{d}V + \int_A \vec{\tau_n} \cdot \vec{W} \ \mathrm{d}A + \int_A \lambda \frac{\partial t}{\partial n} \mathrm{d}A, \tag{2.30}$$

where

$$\frac{\mathrm{D}}{\mathrm{D}\tau} \int_V \rho \left(e + \frac{W^2}{2} \right) \mathrm{d}V = \int_V \rho \frac{\mathrm{D}}{\mathrm{D}\tau} \left(e + \frac{W^2}{2} \right) \mathrm{d}V, \tag{2.31}$$

$$\int_A \vec{\tau_n} \cdot \vec{W} \ \mathrm{d}A = \int_A \vec{n}[\tau] \cdot \vec{W} \ \mathrm{d}A = \int_A \vec{n}([\tau] \cdot \vec{W}) \ \mathrm{d}A = \int_A \nabla \cdot ([\tau] \cdot \vec{W}) \mathrm{d}V, \tag{2.32}$$

$$\int_A \lambda \frac{\partial t}{\partial n} \mathrm{d}A = \int_V \nabla \cdot (\lambda \nabla t) \mathrm{d}V. \tag{2.33}$$

With (2.31)–(2.33), (2.30) is rewritten as

$$\int_V \rho \frac{\mathrm{D}}{\mathrm{D}\tau} \left(e + \frac{W^2}{2} \right) \mathrm{d}V = \int_V \rho \vec{F} \cdot \vec{W} \mathrm{d}V + \int_A \nabla \cdot ([\tau] \cdot \vec{W}) \mathrm{d}V + \int_V \nabla \cdot (\lambda \nabla t) \mathrm{d}V. \tag{2.34}$$

Then,

$$\rho \frac{\mathrm{D}}{\mathrm{D}\tau} \left(e + \frac{W^2}{2} \right) = \rho \vec{F} \cdot \vec{W} + \nabla \cdot ([\tau] \cdot \vec{W}) + \nabla \cdot (\lambda \nabla t), \tag{2.35}$$

where $[\tau]$ denotes tensor of shear force.

Equation (2.35) is the energy equation.

Through tensor and vector analysis, (2.35) can be further derived into the following form:

$$\rho \frac{\mathrm{D}e}{\mathrm{D}\tau} = [\tau] \cdot [\varepsilon] + \nabla \cdot (\lambda \nabla t). \tag{2.36}$$

Equation(2.32) is an another form of the energy equation. Here, $[\tau] \cdot [\varepsilon]$ is the scalar quantity product of force tensor $[\tau]$ and deformation rate tensor $[\varepsilon]$, and represents the work done by fluid deformation surface force. The physical significance of (2.36) is that the internal energy increment of fluid with unit volume during the unit time equals the sum of the work done by deformation surface force of fluid with unit volume, $[\tau] \cdot [\varepsilon]$, and the heat entering the system.

The general Newtonian law is expressed as

$$[\tau] = 2\mu[\varepsilon] - \left(p + \frac{2}{3}\mu\nabla\cdot\vec{W}\right)[I], \tag{2.37}$$

where $[I]$ is unit tensor.

According to (2.37) the following equation can be obtained:

$$[\tau]\cdot[\varepsilon] = -p\nabla\cdot\vec{W} - \frac{2}{3}\mu(\nabla\cdot\vec{W})^2 + 2\mu[\varepsilon]^2. \tag{2.38}$$

Then, (2.36) can be rewritten as

$$\rho\frac{\mathrm{D}e}{\mathrm{D}\tau} = -p\nabla\cdot\vec{W} + \Phi + \nabla\cdot(\lambda\nabla t), \tag{2.39}$$

where $\Phi = -2/3(\mu(\nabla\cdot\vec{W})^2) + 2\mu[\varepsilon]^2$ is viscous dissipation function, which is further described as

$$\Phi = \mu\left[2\left(\frac{\partial w_x}{\partial x}\right)^2 + 2\left(\frac{\partial w_y}{\partial y}\right)^2 + 2\left(\frac{\partial w_z}{\partial z}\right)^2 + \left(\frac{\partial w_x}{\partial y} + \frac{\partial w_y}{\partial x}\right)^2 + \left(\frac{\partial w_y}{\partial z} + \frac{\partial w_z}{\partial y}\right)^2 + \left(\frac{\partial w_z}{\partial x} + \frac{\partial w_x}{\partial z}\right)^2\right] - \frac{2}{3}\left[\mathrm{div}(\vec{W})\right]^2. \tag{2.40}$$

Equation (2.6) can be rewritten as

$$\nabla\cdot\vec{W} = -\frac{1}{\rho}\frac{\mathrm{D}\rho}{\mathrm{D}\tau} = \rho\frac{\mathrm{D}}{\mathrm{D}\tau}\left(\frac{1}{\rho}\right).$$

With the earlier equation, (2.39) is changed into the following form:

$$\rho\left[\frac{\mathrm{D}e}{\mathrm{D}\tau} + p\frac{\mathrm{D}}{\mathrm{D}\tau}\left(\frac{1}{\rho}\right)\right] = \Phi + \nabla\cdot(\lambda\nabla t). \tag{2.41}$$

According to thermodynamics equation of fluid

$$\frac{\mathrm{D}h}{\mathrm{D}\tau} = \frac{\mathrm{D}e}{\mathrm{D}\tau} + p\frac{\mathrm{D}}{\mathrm{D}\tau}\left(\frac{1}{\rho}\right) + \frac{1}{\rho}\frac{\mathrm{D}p}{\mathrm{D}\tau}. \tag{2.42}$$

Equation(2.41) can be expressed as the following enthalpy form:

$$\rho\frac{\mathrm{D}h}{\mathrm{D}\tau} = \frac{\mathrm{D}p}{\mathrm{D}\tau} + \Phi + \nabla\cdot(\lambda\nabla t), \tag{2.43}$$

or

$$\rho\frac{\mathrm{D}(c_p t)}{\mathrm{D}\tau} = \frac{\mathrm{D}p}{\mathrm{D}\tau} + \Phi + \nabla\cdot(\lambda\nabla t), \tag{2.44}$$

where $h = c_p t$, while c_p is specific heat.

In Cartesian form, the energy (2.44) can be rewritten as

$$\rho\left[\frac{\partial(c_p t)}{\partial\tau}+w_x\frac{\partial(c_p t)}{\partial x}+w_y\frac{\partial(c_p t)}{\partial y}+w_z\frac{\partial(c_p t)}{\partial z}\right]$$
$$=\frac{\mathrm{D}p}{\mathrm{D}\tau}+\frac{\partial}{\partial x}\left(\lambda\frac{\partial t}{\partial x}\right)+\frac{\partial}{\partial y}\left(\lambda\frac{\partial t}{\partial y}\right)+\frac{\partial}{\partial z}\left(\lambda\frac{\partial t}{\partial z}\right)+\Phi. \quad (2.45)$$

For steady state and nearly constant pressure processes, the viscous dissipation can be ignored, and then the Cartesian form of the energy equation (2.45) is changed into

$$\rho\left[w_x\frac{\partial(c_p t)}{\partial x}+w_y\frac{\partial(c_p t)}{\partial y}+w_z\frac{\partial(c_p t)}{\partial z}\right]=\frac{\partial}{\partial x}\left(\lambda\frac{\partial t}{\partial x}\right)+\frac{\partial}{\partial y}\left(\lambda\frac{\partial t}{\partial y}\right)+\frac{\partial}{\partial z}\left(\lambda\frac{\partial t}{\partial z}\right). \quad (2.46)$$

2.4 Basic Equations of Free Convection Boundary Layer

In Fig. 2.4 the physical model and coordinate system of boundary layer with two-dimensional laminar free convection are shown schematically. An inclined flat plate is suspended in liquid. The surface temperature is T_w and the fluid bulk temperature is T_∞. If T_w is not equal to T_∞, the laminar free convection can be produced on the inclined surface in both the cases as shown in Fig. 2.4(a) and (b), respectively.

In the following sections we will make quantitative grade analysis successively to investigate the governing equations of mass, momentum, and energy conservation for steady laminar free convection in the two-dimensional boundary layer.

2.4.1 Continuity Equation

Based on (2.7), the steady state three-dimensional continuity equation is given by

$$\frac{\partial}{\partial x}(\rho w_x)+\frac{\partial}{\partial y}(\rho w_y)+\frac{\partial}{\partial z}(\rho w_z)=0. \quad (2.47)$$

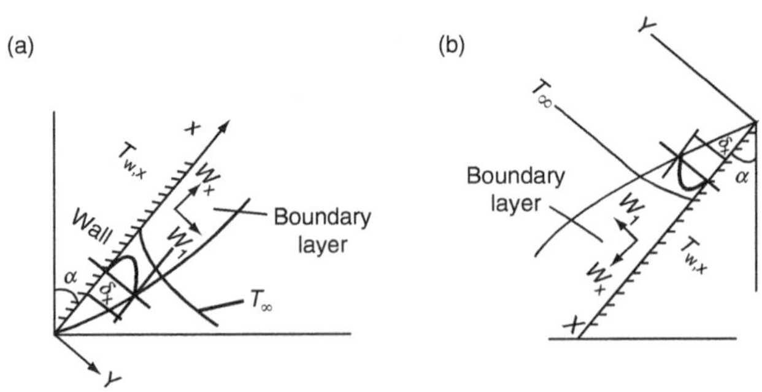

Fig. 2.4. Physical model and coordinate system. (**a**) Ascending flow on the inclined surface ($t_w > t_\infty$). (**b**) Falling flow on the inclined surface ($t_w > t_\infty$)

While, the steady state two-dimensional continuity equation is given by

$$\frac{\partial}{\partial x}(\rho w_x) + \frac{\partial}{\partial y}(\rho w_y) = 0. \tag{2.48}$$

In (2.47) and (2.48) variable fluid density with temperature is considered.

Before the quantitative grade analysis, it is necessary to define its analytical standard. A normal quantitative grade is regarded as $\{1\}$, i.e., unit quantity grade, a very small quantitative grade is regarded as $\{\delta\}$, even very small quantitative grade is regarded as $\{\delta^2\}$, and so on. The ration of the quantities is easily defined, and some examples of ratios are introduced as follows:

$$\frac{\{1\}}{\{1\}} = \{1\}, \quad \frac{\{\delta\}}{\{\delta\}} = \{1\}, \quad \frac{\{1\}}{\{\delta\}} = \{\delta^{-1}\}, \quad \frac{\{1\}}{\{\delta^2\}} = \{\delta^{-2}\}.$$

According to the theory of laminar free boundary layer, the quantities of the velocity component w_x and the coordinate x can be regarded as unity, i.e., $\{w_x\} = \{1\}$ and $\{x\} = \{1\}$. However, the quantities of the velocity component w_y and the coordinate y should be regarded as δ, i.e., $\{w_y\} = \{\delta\}$ and $\{y\} = \{\delta\}$.

For the terms of (2.48) the following ratios of quantity grade are obtained:

$$\frac{\{\rho w_x\}}{\{x\}} = \frac{\{1\}}{\{1\}} = \{1\} \text{ and } \frac{\{\rho w_y\}}{\{y\}} = \frac{\{\delta\}}{\{\delta\}} = \{1\}.$$

Therefore both the terms of (2.48) should be kept, and (2.48) can be regarded as the continuity equation of the steady state laminar two-dimensional boundary layers. Of course, (2.48) is also suitable for the steady state two-dimensional boundary layers with laminar free convection.

2.4.2 Momentum Equations (Navier–Stokes Equations)

According to (2.23) and (2.24), the momentum equations for steady two-dimensional convection are

$$\rho\left(w_x\frac{\partial w_x}{\partial x} + w_y\frac{\partial w_x}{\partial y}\right) = -\frac{\partial p}{\partial x} + 2\frac{\partial}{\partial x}\left(\mu\frac{\partial w_x}{\partial x}\right) + \frac{\partial}{\partial y}\left[\mu\left(\frac{\partial w_x}{\partial y} + \frac{\partial w_y}{\partial x}\right)\right] - \frac{\partial}{\partial x}\left[\frac{2}{3}\mu\left(\frac{\partial w_x}{\partial x} + \frac{\partial w_y}{\partial y}\right)\right] + \rho g_x. \tag{2.49}$$

$$\rho\left(w_x\frac{\partial w_y}{\partial x} + w_y\frac{\partial w_y}{\partial y}\right) = -\frac{\partial p}{\partial y} + \frac{\partial}{\partial x}\left[\mu\left(\frac{\partial w_x}{\partial y} + \frac{\partial w_y}{\partial x}\right)\right] + 2\frac{\partial}{\partial y}\left(\mu\frac{\partial w_y}{\partial y}\right) - \frac{\partial}{\partial y}\left[\frac{2}{3}\mu\left(\frac{\partial w_x}{\partial x} + \frac{\partial w_y}{\partial y}\right)\right] + \rho g_y. \tag{2.50}$$

According to the theory of boundary layer, the quantity grade of the pressure gradient $\partial p/\partial x$ can be regarded as unity, i.e., $\{\partial p/\partial x\} = \{1\}$, but the quantity grade of the pressure gradient $\partial p/\partial y$ is only regarded as very small quantity grade, i.e., $\{\partial p/\partial y\} = \{\delta\}$.

The quantity grades of the terms of (2.49) and (2.50) are expressed as follows, respectively:

$$\rho\left(w_x\frac{\partial w_x}{\partial x}+w_y\frac{\partial w_x}{\partial y}\right)=-\frac{\partial p}{\partial x}+2\frac{\partial}{\partial x}\left(\mu\frac{\partial w_x}{\partial x}\right)+\frac{\partial}{\partial y}\left[\mu\left(\frac{\partial w_x}{\partial y}+\frac{\partial w_y}{\partial x}\right)\right]$$
$$-\frac{\partial}{\partial x}\left[\frac{2}{3}\mu\left(\frac{\partial w_x}{\partial x}+\frac{\partial w_y}{\partial y}\right)\right]+\rho g_x$$

$$\{1\}\left(\{1\}\frac{\{1\}}{\{1\}}+\{\delta\}\frac{\{1\}}{\{\delta\}}\right)=\{1\}+\frac{\{1\}}{\{1\}}\{\delta^2\}\frac{\{1\}}{\{1\}}+\frac{\{1\}}{\{\delta\}}\{\delta^2\}\left(\frac{\{1\}}{\{\delta\}}+\frac{\{\delta\}}{\{1\}}\right)$$
$$-\frac{\{1\}}{\{1\}}\delta^2\left(\frac{\{1\}}{\{1\}}+\frac{\{\delta\}}{\{\delta\}}\right)+\{1\}\{1\}, \tag{2.49a}$$

$$\rho\left(w_x\frac{\partial w_y}{\partial x}+w_y\frac{\partial w_y}{\partial y}\right)=-\frac{\partial \rho}{\partial y}+\frac{\partial}{\partial x}\left[\mu\left(\frac{\partial w_x}{\partial y}+\frac{\partial w_y}{\partial x}\right)\right]+2\frac{\partial}{\partial y}\left(\mu\frac{\partial w_y}{\partial y}\right)$$
$$-\frac{\partial}{\partial y}\left[\frac{2}{3}\mu\left(\frac{\partial w_x}{\partial x}+\frac{\partial w_y}{\partial y}\right)\right]+\rho g_y$$

$$\{1\}\left(\{1\}\frac{\{\delta\}}{\{1\}}+\{\delta\}\frac{\{\delta\}}{\{\delta\}}\right)=\{\delta\}+\frac{\{1\}}{\{1\}}\{\delta^2\}\left(\frac{\{1\}}{\{\delta\}}+\frac{\{\delta\}}{\{1\}}\right)+\frac{\{1\}}{\{\delta\}}\{\delta^2\}\frac{\{\delta\}}{\{\delta\}}$$
$$-\frac{\{1\}}{\{\delta\}}\{\delta^2\}\left(\frac{\{1\}}{\{1\}}+\frac{\{\delta\}}{\{\delta\}}\right)+\{1\}\{\delta\}. \tag{2.50a}$$

The quantity grades of (2.49a) and (2.50a) are simplified as follows, respectively:

$$\rho\left(w_x\frac{\partial w_x}{\partial x}+w_y\frac{\partial w_x}{\partial y}\right)=-\frac{\partial p}{\partial x}+2\frac{\partial}{\partial x}\left(\mu\frac{\partial w_x}{\partial x}\right)+\frac{\partial}{\partial y}\left[\mu\left(\frac{\partial w_x}{\partial y}+\frac{\partial w_y}{\partial x}\right)\right]$$
$$-\frac{\partial}{\partial x}\left[\frac{2}{3}\mu\left(\frac{\partial w_x}{\partial x}+\frac{\partial w_y}{\partial y}\right)\right]+\rho g_x$$
$$\{1\}(\{1\}+\{1\})=\{1\}+\{\delta^2\}+\{1\}+\{\delta^2\}-(\{\delta^2\}+\{\delta^2\})+\{1\}, \tag{2.49b}$$

$$\rho\left(w_x\frac{\partial w_y}{\partial x}+w_y\frac{\partial w_y}{\partial y}\right)=-\frac{\partial p}{\partial y}+\frac{\partial}{\partial x}\left[\mu\left(\frac{\partial w_x}{\partial y}+\frac{\partial w_y}{\partial x}\right)\right]+2\frac{\partial}{\partial y}\left(\mu\frac{\partial w_y}{\partial y}\right)$$
$$-\frac{\partial}{\partial y}\left[\frac{2}{3}\mu\left(\frac{\partial w_x}{\partial x}+\frac{\partial w_y}{\partial y}\right)\right]+\rho g_y$$
$$\{1\}(\{\delta\}+\{\delta\})=\{\delta\}+(\{\delta\}+\{\delta^3\})+\{\delta\}-(\{\delta\}(\{1\}+\{1\}))+\{\delta\}. \tag{2.50b}$$

Observing the quantity grades in (2.49b) it is found that the terms

$$2\frac{\partial}{\partial x}(\mu\frac{\partial w_x}{\partial x}),$$
$$\frac{\partial w_y}{\partial x}$$

in term

$$\frac{\partial}{\partial y}\left[\mu\left(\frac{\partial w_x}{\partial y}+\frac{\partial w_y}{\partial x}\right)\right],$$

and

$$\frac{\partial}{\partial x}\left[\frac{2}{3}\mu\left(\frac{\partial w_x}{\partial x}+\frac{\partial w_y}{\partial y}\right)\right]$$

are very small and can be ignored from (2.49). Then, (2.49) is simplified as follows:

$$\rho\left(w_x\frac{\partial w_x}{\partial x}+w_y\frac{\partial w_x}{\partial y}\right)=-\frac{\partial p}{\partial x}+\frac{\partial}{\partial y}\left(\mu\left(\frac{\partial w_x}{\partial y}\right)\right)+\rho g_x. \tag{2.51}$$

Comparing the quantity grades of (2.49b) with that of Eq.(2.50b), it is found that the quantity grades of (2.50b) are very small. Then, (2.50) can be ignored, and only (2.51) is taken as the momentum equation of steady state laminar two-dimensional boundary layer.

From Fig. 2.4 it is found that for free convection on inclined plate the gravity acceleration component g_x is expressed as

$$g_x = g\cdot\cos\ \alpha, \tag{2.52}$$

where g is gravity acceleration and α is the inclined angle of the plate.

With (2.52), (2.51) is rewritten as

$$\rho\left(w_x\frac{\partial w_x}{\partial x}+w_y\frac{\partial w_x}{\partial y}\right)=-\frac{\partial p}{\partial x}+\frac{\partial}{\partial y}\left(\mu\left(\frac{\partial w_x}{\partial y}\right)\right)+\rho g\cdot\cos\ \alpha. \tag{2.53}$$

Suppose the direction of $g\cdot\cos\ \alpha$ is reverse to that opposite to that of the velocity component w_x, (2.53) can be rewritten as

$$\rho\left(w_x\frac{\partial w_x}{\partial x}+w_y\frac{\partial w_x}{\partial y}\right)=-\frac{\partial p}{\partial x}+\frac{\partial}{\partial y}\left(\mu\left(\frac{\partial w_x}{\partial y}\right)\right)-\rho g\cdot\cos\ \alpha. \tag{2.54}$$

Ignoring the effects of viscosity beyond the boundary layer, the momentum equation (2.54) is simplified into the following equation:

$$-\frac{\mathrm{d}p}{\mathrm{d}x}=\rho_\infty g\cdot\cos\ \alpha+\rho_\infty w_{x,\infty}\frac{\mathrm{d}w_{x,\infty}}{\mathrm{d}x}, \tag{2.55}$$

where ρ_∞ and $w_{x,\infty}$ are fluid density and velocity component beyond the boundary layer.

With (2.55), (2.54) becomes

$$\rho\left(w_x\frac{\partial w_x}{\partial x}+w_y\frac{\partial w_x}{\partial y}\right)=\frac{\partial}{\partial y}\left(\mu\frac{\partial w_x}{\partial y}\right)+g(\rho_\infty-\rho)\cos\,\alpha+\rho_\infty w_\infty\frac{\mathrm{d}w_\infty}{\mathrm{d}x}. \tag{2.56}$$

For constant w_∞ the equation (2.56) transforms to

$$\rho\left(w_x\frac{\partial w_x}{\partial x}+w_y\frac{\partial w_x}{\partial y}\right)=\frac{\partial}{\partial y}\left(\mu\frac{\partial w_x}{\partial y}\right)+g(\rho_\infty-\rho)\cos\,\alpha. \tag{2.57}$$

This is the momentum equation of two-dimensional boundary layer on an inclined plate with laminar free convection.

For the free convection of a perfect gas (ideal gas) the following simple power law can be used:

$$\frac{\rho_\infty}{\rho}=\frac{T}{T_\infty},$$

where T denotes absolute temperature. Therefore,

$$\rho_\infty-\rho=\rho\left(\frac{T}{T_\infty}-1\right). \tag{2.58}$$

Thus, for steady two-dimensional laminar free convection of a perfect gas, (2.57) can be changed into

$$\rho\left(w_x\frac{\partial w_x}{\partial x}+w_y\frac{\partial w_x}{\partial y}\right)=\frac{\partial}{\partial y}\left(\mu\frac{\partial w_x}{\partial y}\right)+g\rho\left(\frac{T}{T_\infty}-1\right)\cos\,\alpha. \tag{2.59}$$

2.4.3 Energy Equations

According to (2.46), the energy equation for steady state laminar two-dimensional convection is shown as follows:

$$\rho\left[w_x\frac{\partial(c_pt)}{\partial x}+w_y\frac{\partial(c_pt)}{\partial y}\right]=\frac{\partial}{\partial x}\left(\lambda\frac{\partial t}{\partial x}\right)+\frac{\partial}{\partial y}\left(\lambda\frac{\partial t}{\partial y}\right). \tag{2.60}$$

With the quantity grade analysis similar to that mentioned earlier, (2.60) can be changed into the following form for energy equation of steady state laminar two-dimensional boundary layer. Of course, it is suitable for that with laminar free convection.

$$\rho\left[w_x\frac{\partial(c_pt)}{\partial x}+w_y\frac{\partial(c_pt)}{\partial y}\right]=\frac{\partial}{\partial x}\left(\lambda\frac{\partial t}{\partial x}\right). \tag{2.61}$$

Up to now it is the time to summarize the basic governing equations for description of mass, momentum, and energy conservation of two-dimensional boundary layers with laminar steady state free convection as follows:

$$\frac{\partial}{\partial x}(\rho w_x) + \frac{\partial}{\partial y}(\rho w_y) = 0, \tag{2.48}$$

$$\rho\left(w_x\frac{\partial w_x}{\partial x} + w_y\frac{\partial w_x}{\partial y}\right) = \frac{\partial}{\partial y}\left(\mu\frac{\partial w_x}{\partial y}\right) + g(\rho_\infty - \rho)\cos\ \alpha, \tag{2.57}$$

$$\rho\left[w_x\frac{\partial(c_p t)}{\partial x} + w_y\frac{\partial(c_p t)}{\partial y}\right] = \frac{\partial}{\partial x}\left(\lambda\frac{\partial t}{\partial x}\right). \tag{2.61}$$

For rigorous solutions of the governing equations, the fluid temperature-dependent properties, such as density ρ, absolute viscosity μ, specific heat c_p, and thermal conductivityλ will be considered in the successive chapters of this book.

The laminar free convection with two-dimensional boundary layer belongs to two-point boundary value problem, which is the basis of three-point boundary value problem, such as for film boiling and film condensation. For isothermal plate for example, the boundary conditions for the two-point boundary problem can be expressed as follows:

$$y = 0: \quad w_x = 0, \quad w_y = 0, \quad t = t_{\mathrm{w}}, \tag{2.62}$$

$$y \to \infty: \quad w_x = w_{x,\infty}, \quad t = t_\infty, \tag{2.63}$$

where t_{w} is plate temperature, t_∞ is the fluid temperature beyond the boundary layer, and $w_{x,\infty}$ denotes the fluid velocity component in x-direction beyond the boundary layer.

The tern $g(\rho_\infty - \rho)\cos\ \alpha$ in (2.57) is regarded as buoyancy factor. For perfect gas, the buoyancy factor can be expressed as

$$g(\rho_\infty - \rho) = \rho g\left(\frac{T}{T_\infty} - 1\right). \tag{2.58}$$

In addition, although (2.58) is originally for perfect gas, it is well known that it can be very accurately applied to free convection and film flows of general gases.

It is indicated that for the governing equations and boundary conditions of general gas free convection, it is better to express the temperature by using absolute temperature T. This way is very beneficial for treatment of gas temperature-dependent properties.

3

Brief Review of Previous Method for Analysis of Laminar Free Convection

Nomenclature

c_p	specific heat at constant pressure, J (kg k)$^{-1}$
g	gravitation acceleration, m s^{-2}
Gr	Grashof number
P	pressure, N m^{-2}
Pr	Prandtl number
T	absolute temperature, K
T_{r}	reference temperature, K
T_{w}	absolute temperature on the wall, K
T_∞	absolute temperature beyond the boundary layer, K
T_{w}/T_∞	Boundary temperature ratio
w_x, w_y	velocity component in the x- and y- directions, respectively, m s^{-1}
W_x, W_y	dimensional velocity component in the x- and y- directions, respectively
x, y	dimensional coordinate variables

Greek symbols

η	dimensionless coordinate variable for boundary layer
$\theta(\eta)$	dimensionless temperature
λ	thermal conductivity, W (m K)$^{-1}$
μ	absolute viscosity, kg (m s)$^{-1}$
ν	kinetic viscosity, m^2 s^{-1}
ρ	density, kg m^{-3}
ψ	flow function

Subscripts

w	at wall
∞	far from the wall surface

3.1 Falkner–Skan Transformation for Fluid Laminar Forced Convection

In solving the laminar boundary problem, a similarity transformation method is usually used. So far the widely applied similarity transformation for the laminar film free convection problem is the Falkner–Skan transformation [1–3]. We shall first apply the Falkner–Skan transformations to the forced convection on a horizontal plate, for constant physical properties.

Let us consider the governing equations of laminar boundary layer with forced convection on a horizontal plate. By ignoring variable thermophysical properties the governing partial differential equations for laminar forced convection are given by:

$$\frac{\partial}{\partial x}(\mathrm{w_x}) + \frac{\partial}{\partial y}(\mathrm{w_y}) = 0, \tag{3.1}$$

$$\left(w_x\frac{\partial w_x}{\partial x} + w_y\frac{\partial w_x}{\partial y}\right) = \nu\frac{\partial^2 w_x}{\partial y^2} - \frac{1}{\rho}\frac{\mathrm{d}p}{\mathrm{d}x}, \tag{3.2}$$

$$\left(w_x\frac{\partial t}{\partial x} + w_y\frac{\partial t}{\partial y}\right) = \frac{\nu}{Pr}\frac{\partial^2 t}{\partial y^2}. \tag{3.3}$$

With the boundary conditions

$$y = 0: \quad w_x = 0, \quad w_y = 0, \quad t = t_{\mathrm{w}}, \tag{3.4}$$

$$y \to \infty: \quad w_x \to w_{x,\infty}, \quad t = t_\infty, \tag{3.5}$$

where w_x and w_y are velocity components of the fluids in x and y directions, respectively, ρ is the density, ν is kinetic viscosity, Pr is Prandtl number, and t is temperature. For the external flow we can replace $(-1/\rho)\mathrm{d}p/\mathrm{d}x$ by $(w_{x,\infty})\mathrm{d}w_{x,\infty}/\mathrm{d}x$ in (3.2). For given $\nu, Pr, w_{x,\infty}, t_\infty$ and $t_{\mathrm{w}}(x)$, the solution of the earlier equations can be written in the following form:

$$\frac{w_x}{w_{x,\infty}} = \phi_1(x, y), \tag{3.6}$$

$$\frac{t - t_\infty}{t_{\mathrm{w}} - t_\infty} = \phi_2(x, y). \tag{3.7}$$

In some special cases (3.6) and (3.7) can be rewritten as:

$$\frac{w_x}{w_{x,\infty}} = \phi_1(\eta), \tag{3.8}$$

$$\frac{t - t_\infty}{t_{\mathrm{w}} - t_\infty} = \phi_2(\eta). \tag{3.9}$$

where η is dimensionless similarity variable. It is a function of x and y. Later it will be shown that η is proportional to $y/(\delta(x))$ where $\delta(x)$ is function of x. Then, the number of independent variables can be reduced from two to one.

A prerequisite of the Falkner–Skan transformation is the derivation of the similarity variables. The typical method for this procedure is group theory, which was discussed at length by Hansen [4] and Na [5]. For the application of this method, a stream function $\psi(x, y)$ is assumed to satisfy (3.1).

$$w_x = \frac{\partial \psi}{\partial y}, \quad w_y = -\frac{\partial \psi}{\partial x}. \tag{3.10}$$

Using (3.10), (3.2) is transformed into

$$\frac{\partial \psi}{\partial y}\frac{\partial^2 \psi}{\partial x \partial y} - \frac{\partial \psi}{\partial x}\frac{\partial^2 \psi}{\partial y^2} = w_{x,\infty}\frac{\mathrm{d}w_{x,\infty}}{\mathrm{d}x} + \nu\frac{\partial^3 \psi}{\partial y^3}. \tag{3.11}$$

Subsequently, (3.11) is solved by using the group theory. It is important to introduce the following linear transformation:

$$x = A^{a_1}\overline{x}, \quad y = A^{a_2}\overline{y}, \quad \psi = A^{a_3}\overline{\psi}, \quad w_{x,\infty} = A^{a_4}\overline{w_{x,\infty}}, \tag{3.12}$$

where a_1 , a_2 , a_3, and a_4 are constant, and A is the transformation parameter.

With (3.12), (3.11) can be rewritten as

$$A^{2a_3-a_1-2a_2}\left(\frac{\partial \overline{\psi}}{\partial \overline{y}}\frac{\partial^2 \overline{\psi}}{\partial \overline{x}\partial \overline{y}} - \frac{\partial \overline{\psi}}{\partial \overline{x}}\frac{\partial^2 \overline{\psi}}{\partial \overline{x}^2}\right) = A^{2a_4-a_1}\overline{w_{x,\infty}}\frac{\mathrm{d}\overline{w_{x,\infty}}}{\mathrm{d}\overline{x}} + \nu A^{a_3-3a_2}\frac{\partial^3 \overline{\psi}}{\partial \overline{y}^3}. \tag{3.13}$$

From (3.13) and (3.11), due to the constancy of the variables, we compare the exponents of A in each of the terms and obtained;

$$2a_3 - a_1 - 2a_2 = 2a_4 - a_1 = a_3 - 3a_2. \tag{3.14}$$

The incomplete solution of (3.14) is

$$a_3 = a_1 - a_2, \quad a_4 = a_3 - a_2. \tag{3.15}$$

Defining a new variable $a = a_2/a_1$, then (3.15) can be rewritten as

$$\frac{a_3}{a_1} = 1 - a, \quad \frac{a_4}{a_1} = 2\frac{a_3}{a_1} - 1 = 1 - 2a. \tag{3.16}$$

From (3.12) we get

$$A = \left(\frac{x}{\overline{x}}\right)^{1/a_1} = \left(\frac{y}{\overline{y}}\right)^{1/a_2} = \left(\frac{\psi}{\overline{\psi}}\right)^{1/a_3} = \left(\frac{w_{x,\infty}}{\overline{w_{x,\infty}}}\right)^{1/a_4}. \tag{3.17}$$

Using (3.16), (3.17) can be transformed as

$$\frac{y}{x^a} = \frac{\overline{y}}{\overline{x}^a}, \quad \frac{\psi}{x^{1-a}} = \frac{\overline{\psi}}{(\overline{x})^{1-a}}, \quad \frac{w_{x,\infty}}{x^{1-2a}} = \frac{\overline{w_{x,\infty}}}{(\overline{x})^{1-2a}}. \tag{3.18}$$

The combined variables in (3.18) are called absolute variables in the case of (3.12). According to Hansen [4], if the boundary conditions of the velocity field of (3.4) and (3.5) can also be transformed and do not depend on x, these absolute variables are similarity variables. We set

$$\eta = \frac{y}{x^a}, \qquad f(\eta) = \frac{\psi}{x^{1-a}}, \tag{3.19}$$

$$h(\eta) = \frac{w_{x,\infty}}{x^{1-2a}} = \text{constant} = C. \tag{3.20}$$

The second part of (3.19) can be rewritten as

$$\psi = x^{1-a} f(\eta) \frac{x^a}{x^a} = x^{1-2a} x^a f(\eta).$$

With (3.20) this equation can be transformed into

$$\psi = \frac{w_{x,\infty}}{C} x^a f(\eta). \tag{3.21}$$

By using the first equation of (3.10), (3.21), and the first equation of (3.19), $w_{x,\infty}$ is rewritten as

$$w_x = \frac{\partial \psi}{\partial y} = \frac{\partial \psi}{\partial \eta}\frac{\partial \eta}{\partial y} = \frac{w_{x,\infty}}{C}\frac{\partial f(\eta)}{\partial \eta},$$

or

$$\frac{w}{w_{x,\infty}} = \frac{f'(\eta)}{C}. \tag{3.22}$$

Now, we set $m = 1 - 2a$. From (3.20) we obtain

$$w_{x,\infty} = C x^m. \tag{3.23}$$

Using (3.23) the first equation of (3.19) will be transformed to

$$\eta = \frac{y}{x^{(1-m)/2}} = \left(\frac{w_{x,\infty}}{Cx}\right)^{1/2} y. \tag{3.24}$$

With (3.23), (3.21) will be

$$\psi = \frac{w_{x,\infty} x^{(1-m)/2}}{C} f(\eta) = \left(\frac{w_{x,\infty} x}{C}\right)^{1/2} f(\eta), \tag{3.25}$$

where Ψ has the unit $m^2\, s^{-1}, \eta$ and $f(\eta)$ should be dimensionless. Replacing C with ν, the correct dimension of Ψ and η can be obtained:

$$\psi = (w_{x,\infty} \nu x)^{1/2} f(\eta), \tag{3.26}$$

$$\eta = \left(\frac{w_{\mathrm{x},\infty}}{\nu x}\right)^{1/2} y. \tag{3.27}$$

Equations (3.26) and (3.27) are defined as the Falkner–Skan transformations for laminar forced convection.

On this basis, similarity transformation of (3.1)–(3.3) is undertaken as follows:

Using (3.26) the velocity component w_x can be derived from (3.10) as follows:

$$w_x = (w_{x,\infty}\nu x)^{1/2} f'(\eta)\frac{\partial \eta}{\partial y} = (w_{x,\infty}\nu x)^{1/2} f'(\eta)\left(\frac{w_{x,\infty}}{\nu x}\right)^{1/2}.$$

or

$$w_x = w_{x,\infty} f'(\eta). \tag{3.28}$$

Employing (3.26) the velocity component w_y can be derived from (3.10) as follows:

$$\begin{aligned} w_y &= -(w_{x,\infty}\nu x)^{1/2} f'(\eta)\frac{\partial \eta}{\partial x} - \frac{1}{2}\left(\frac{w_{x,\infty}\nu}{x}\right)^{1/2} f(\eta) \\ &= (w_{x,\infty}\nu x)^{1/2} f'(\eta)\frac{1}{2}\left(\frac{w_{x,\infty}}{\nu x^3}\right)^{1/2} y - \frac{1}{2}\left(\frac{w_{x,\infty}\nu}{x}\right)^{1/2} f(\eta) \\ &= \frac{1}{2}\left(\frac{w_{x,\infty}\nu}{x}\right)^{1/2} f'(\eta)\eta - \frac{1}{2}\left(\frac{w_{x,\infty}\nu}{x}\right)^{1/2} f(\eta) \\ &= \frac{1}{2}\left(\frac{w_{x,\infty}\nu}{x}\right)^{1/2} [\eta f'(\eta) - f(\eta)]. \end{aligned} \tag{3.29}$$

Equation (3.9) can be rewritten as

$$\frac{t - t_\infty}{t_\mathrm{w} - t_\infty} = \theta(\eta). \tag{3.30}$$

The similarity transformation of the governing (3.2) and (3.3), can now be made. From (3.27), we have

$$\frac{\partial \eta}{\partial x} = -\frac{1}{2}\left(\frac{w_{x,\infty}}{\nu x^3}\right)^{1/2} y = -\frac{1}{2}x^{-1}\eta.$$

From (3.28), we have

$$\frac{\partial w_x}{\partial x} = w_{x,\infty} f''(\eta)\frac{\partial \eta}{\partial x} = -\frac{1}{2}x^{-1}\eta w_{x,\infty} f''(\eta).$$

$$\frac{\partial w_x}{\partial y} = w_{x,\infty} f''(\eta)\frac{\partial \eta}{\partial y} = w_{x,\infty} f''(\eta)\left(\frac{w_{x,\infty}}{\nu x}\right)^{1/2},$$

$$\frac{\partial^2 w_x}{\partial y^2} = w_{x,\infty} f'''(\eta)\left(\frac{w_{x,\infty}}{\nu x}\right).$$

On these bases, if we set $dw_{x,\infty}/dx = 0$, (3.2) is transformed into

$$-w_{x,\infty}f'(\eta)\frac{1}{2}x^{-1}\eta w_{x,\infty}f''(\eta) + \frac{1}{2}\left(\frac{w_{x,\infty}\nu}{x}\right)^{1/2}$$
$$\times[\eta f'(\eta) - f(\eta)]w_{x,\infty}f''(\eta)\left(\frac{w_{x,\infty}}{\nu x}\right)^{1/2}$$
$$= \nu w_{x,\infty}f'''(\eta)\left(\frac{w_{x,\infty}}{\nu x}\right).$$

Simplifying this equation, we obtain

$$-f'(\eta)\eta f''(\eta) + [\eta f'(\eta) - f(\eta)]f''(\eta) = 2f''(\eta),$$

i.e.,

$$2f'''(\eta) + f(\eta)f''(\eta) = 0. \quad (3.31)$$

This is well known Blasius's equation.

Using (3.27)–(3.29), (3.3) is transformed into

$$-w_{x,\infty}f'(\eta)\theta'(\eta)\frac{1}{2}\eta x^{-1} + \frac{1}{2}\left(\frac{w_{x,\infty}\nu}{x}\right)^{1/2}(\eta f'(\eta) - f(\eta))\theta'(\eta)\left(\frac{w_{x,\infty}}{\nu x}\right)^{1/2}$$
$$= \frac{\nu}{Pr}\theta''(\eta)\left(\frac{w_{x,\infty}}{\nu x}\right).$$

Simplifying this equation, we obtain

$$-f'(\eta)\theta'(\eta)\eta + (\eta f'(\eta) - f(\eta))\theta'(\eta) = \frac{2}{Pr}\theta''(\eta),$$

or

$$\frac{2}{Pr}\theta''(\eta) + f(\eta)\theta'(\eta) = 0. \quad (3.32)$$

The boundary conditions for (3.4) and (3.5) are now transformed into

$$\eta = 0: \qquad f(\eta) = f'(\eta) = 0 \qquad \theta(\eta) = 1, \quad (3.33)$$

$$\eta = \infty: \qquad f'(\eta) = 1, \qquad \theta(\eta) = 0, \quad (3.34)$$

3.2 Falkner–Skan Transformation for Fluid Laminar Free Convection

3.2.1 For Boussinesq Approximation

Let us consider the governing equations of the boundary layer of steady state fluid laminar free convection. Based on (2.48), (2.57), and (2.61) of Chap. 2, the governing equations of the boundary layer of steady state fluid laminar free convection are as follows for Boussinesq approximation:

$$\frac{\partial}{\partial x}(w_x)+\frac{\partial}{\partial y}(w_y) = 0, \tag{3.35}$$

$$\left(w_x\frac{\partial w_x}{\partial x} + w_y\frac{\partial w_x}{\partial y}\right) = \nu\frac{\partial^2 w_x}{\partial y^2} + g(\rho_\infty - \rho), \tag{3.36}$$

$$\left(w_x\frac{\partial t}{\partial x} + w_y\frac{\partial t}{\partial y}\right) = \frac{\nu}{Pr}\frac{\partial^2 t}{\partial y^2}, \tag{3.37}$$

with the boundary conditions

$$y = 0: \quad w_x = 0, \quad w_y = 0, \quad t = t_w, \tag{3.38}$$

$$y \to \infty: \quad w_x \to 0, \quad t = t_\infty. \tag{3.39}$$

Similar to the analysis in Sect. 3.1, for Falkner–Skan transformation for fluid laminar free convection, we get the following variables for describing expressions for the stream function ψ and dimensionless coordinate variable η:

$$\eta = \left(\frac{1}{4}Gr\right)^{1/4}\frac{y}{x}; \qquad \psi = 4\nu\left(\frac{1}{4}Gr\right)^{1/4} f(\eta), \tag{3.40}$$

where

$$Gr = \frac{g(\rho_\infty/\rho_\mathrm{w} - 1)x^3}{\nu^2}. \tag{3.41}$$

Combined with (3.10), the dimensionless expressions of w_x and w_y are derived from (3.40), respectively, as follows:

$$w_x = \frac{4\nu}{x}\left(\frac{1}{4}Gr\right)^{1/2} f'(\eta), \tag{3.42}$$

$$w_y = \frac{\nu}{x}\left(\frac{1}{4}Gr\right)^{1/4} [\eta f'(\eta) - 3f(\eta)]. \tag{3.43}$$

Dimensionless temperature $\theta(\eta)$ is also given by

$$\theta(\eta) = \frac{t - t_\infty}{t_\mathrm{w} - t_\infty}. \tag{3.44}$$

With (3.42)–(3.43), (3.35)–(3.37) lead to the following differential equations

$$f'''(\eta) + 3f(\eta)f''(\eta) - 2(f'(\eta))^2 + \theta(\eta) = 0, \tag{3.45}$$
$$\theta''(\eta) + 3Prf(\eta)\theta'(\eta) = 0, \tag{3.46}$$

with the boundary conditions

$$\eta = 0: \quad f(\eta) = f'(\eta) = 0, \quad \theta(\eta) = 1, \tag{3.47}$$
$$\eta \to \infty: \quad f'(\eta) = 0, \theta(\eta) = 0. \tag{3.48}$$

3.2.2 Consideration of Variable Thermophysical Properties

Consider the boundary layer of the fluid laminar free convection from a vertical plate. According to (2.48), (2.57), and (2.61) of Chap. 2, the basic equations of laminar gas free convection with consideration of variable thermophysical properties are given by

$$\frac{\partial}{\partial x}(\rho w_x) + \frac{\partial}{\partial y}(\rho w_y) = 0, \tag{3.49}$$

$$\rho\left(w_x\frac{\partial w_x}{\partial x} + w_y\frac{\partial w_x}{\partial y}\right) = \frac{\partial}{\partial y}\left(\mu\frac{\partial w_x}{\partial y}\right) + g(\rho_\infty - \rho), \tag{3.50}$$

$$\rho c_p\left(w_x\frac{\partial t}{\partial x} + w_y\frac{\partial t}{\partial y}\right) = \frac{\partial}{\partial y}\left(\lambda\frac{\partial t}{\partial y}\right), \tag{3.51}$$

with the boundary conditions

$$y = 0: \quad w_x = 0, \quad w_y = 0, \quad t = t_w, \tag{3.52}$$

$$y \to \infty: \quad w_x \to 0, \quad t = t_\infty, \tag{3.53}$$

where μ is the absolute viscosity, ρ is the density, λ is the thermal conductivity, g is the gravity acceleration, and t is temperature.

For variable thermophysical properties we set up following definition of the stream function ψ:

$$\frac{\rho}{\rho_w}w_x = \frac{\partial\psi}{\partial y}, \frac{\rho}{\rho_\mathrm{w}}w_y = -\frac{\partial\psi}{\partial x}. \tag{3.54}$$

For variable thermophysical properties we can give expressions for stream function ψ and dimensionless coordinate variable η as defined by:

$$\eta = cx^{-1/4}\int_0^y \frac{\rho}{\rho_\mathrm{w}}\mathrm{d}y, \quad \psi = 4\nu_\mathrm{w}\left(\frac{1}{4}Gr_{x,\mathrm{w}}\right)^{1/4} f(\eta),$$

where

$$G_{x,\mathrm{w}} = \frac{g(\rho_\infty/\rho_\mathrm{w} - 1)x^3}{\nu_\mathrm{w}^2}. \tag{3.55}$$

Dimensionless temperature $\theta(\eta)$ is defined by

$$\theta(\eta) = \frac{t - t_\infty}{t_\mathrm{w} - t_\infty}. \tag{3.56}$$

The function θ is a dimensionless temperature and f is related to the velocities in the following way

$$w_x = 4\nu_\mathrm{w}\mathrm{c}^2x^{1/2}f'_{(\eta)}, \qquad w_y = \left(\frac{\rho_\mathrm{w}}{\rho}\right)\left(\frac{\nu_\mathrm{w}c}{x^{1/4}}\right)(\eta f'(\eta) - 3f(\eta)), \tag{3.57}$$

where

$$C = Gr_{x,\mathrm{w}} x^{-3/4}. \tag{3.58}$$

With the variables used in (3.54)–(3.58), (3.49)–(3.51) are now transformed into the following equations:

$$\frac{\mathrm{d}}{\mathrm{d}\eta}\left[\frac{\rho\mu}{\rho_{\mathrm{w}}\mu_{\mathrm{w}}} f''(\eta)\right] + 3f(\eta)f''(\eta) - 2[f'(\eta)]^2 + \theta(\eta) = 0, \tag{3.59}$$

$$\frac{\mathrm{d}}{\mathrm{d}\eta}\left[\frac{\rho\lambda}{\rho_{\mathrm{w}}\lambda_{\mathrm{w}}} \theta'(\eta)\right] + 3Pr_{\mathrm{w}}\left(\frac{c_p}{c_{p_{\mathrm{w}}}}\right) f(\eta)\theta'(\eta) = 0, \tag{3.60}$$

where λ, ρ, c_p and Pr are thermal conductivity, density, specific heat, and Prandtl number, respectively, while subscript w denotes the temperature on the wall.

The boundary conditions, (3.52) and (3.53) are transformed into

$$\eta = 0: \quad f(\eta) = f'(\eta) = 0, \quad \theta(\eta) = 1, \tag{3.61}$$

$$\eta \to \infty: \quad f'(\eta) = 0, \theta(\eta) = 0. \tag{3.62}$$

It should be mentioned that there are some difficulties produced by the Falkner–Skan transformations. It is difficult to transform (3.49)–(3.51) to (3.59) and (3.60), since the velocity components cannot easily be replaced by the stream function ψ . In addition, great difficulty is encountered in the treatment of the variable thermophysical properties in (3.59) and (3.60), because the physical property factors $\rho\mu/\rho_{\mathrm{w}}\mu_{\mathrm{w}}$, $\rho\lambda/\rho_{\mathrm{w}}\lambda_{\mathrm{w}}$ and $c_p/c_{p_{\mathrm{w}}}$ are functions of temperature, and therefore are functions of η. Finally, it is also difficult to transform the variables $f(\eta)$ and $f'(\eta)$ to the corresponding velocity components w_x and w_y. In this case, in the following chapters, a novel system of transformation models for free convection and film flows will be presented instead of the traditional Falkner–Skan type transformation.

3.3 Some Previous Methods for Treatment of Variable Thermophysical Properties

A well-known treatment of the variable fluid property problem for laminar free convection on an isothermal vertical flat plate has been presented by Sparrow and Gregg [6] with solutions of the boundary layer equations for some special cases. Gray and Giogini [7] discussed the validity of the Boussinesq's approximation and proposed a method for analyzing natural convection flows with fluid properties assumed to be linear functions of temperature and pressure. Clausing and Kempka [8] reported their experimental study of the influence of property variations on the natural convection and concluded that, for the laminar region, Nu_{f} will be a function of $Ra_{\mathrm{f}}(= Gr_{\mathrm{f}}Pr_{\mathrm{f}})$ only with reference temperature, T_{f}, taken as the average temperature in the boundary layer.

The instability of laminar free convection flow and transition to a turbulent state has been studied by Gebhart [9] and summarized in a textbook by Eckert and Drake [10]. For variable thermophysical properties the treatment method developed by Sparrow and Gregg is representative, hence we introduce mainly their method.

Sparrow and Gregg [6] have studied the description of gas physical properties and proposed a model used to describe five kinds of assumptions as given in Table 3.1.

Numerical solutions of (3.59) and (3.60) for gas A were carried out by Sparrow and Gregg [6] for a wide range of values of T_w/T_∞ for $Pr = 0.7$ and for selected values of T_w/T_∞ for $Pr = 1.0$. The heat transfer results corresponding to these solutions are listed in Table 3.2.

For gas C the heat transfer (for $Pr = 0.7$) is given by

$$Nu_w/(Gr_w)^{1/4} = 0.353. \tag{3.63}$$

Sparrow and Gregg [6] have further suggested a reference temperature approach, in which the free convection heat transfer under variable property conditions can be calculated by using the constant properties results in conjunction with the following reference temperature relation

$$T_r = T_w - 0.38(T_w - T_\infty). \tag{3.64}$$

Table 3.1. Description of gases A, B, C, D, and E proposed by Sparrow and Gregg [6]

gas A	gas B	gas C	gas D	gas E
$P = \rho RT$	$P = \rho RT$	$P = \rho RT$	$P = \rho RT$	$P = \rho RT$
$\lambda \propto T^{3/4}$	$\lambda \propto T^{2/3}$	$\rho\lambda = \text{const.}$	$\lambda \propto \frac{T^{3/2}}{T+A_1}$	$\lambda \propto \frac{T^{3/2}}{T+A_1}$
$\mu \propto T^{3/4}$	$\mu \propto T^{2/3}$	$\rho\mu = \text{const.}$	$\mu \propto \frac{T^{3/2}}{T+A_2}$	$\mu \propto \frac{T^{3/2}}{T+A_2}$
$c_p = \text{const.}$	$c_p = \text{const.}$	$c_p = \text{const.}$	$c_p = \text{const.}$	$c_p b_0 + b_1 T$
$Pr = \text{const.}$	$Pr = \text{const.}$	$Pr = \text{const.}$	$Pr = \text{const.}$	$Pr = \text{const.}$

A_1 and A_2 are the dimensionless constants in the Sutherland's formulae

Table 3.2. Heat transfer results, $Nu_w/(Gr_w)^{1/4}$ for gas A proposed by Sparrow and Gregg [6]

T_w/T_∞	$Pr = 0.7$	$Pr = 1.0$
5/2	0.366	
3	0.368	0.418
2	0.363	
3/4	0.348	
1/2	0.339	
1/3	0.330	0.375

$Nu_w = \alpha_x x/\lambda$ and $Gr_x = \frac{gx^3(|\rho_\infty - \rho_w|/\rho_w}{\nu_m^2}$

where T_r is reference temperature. Sparrow and Gregg [6] supplied a good approach for prediction of the heat transfer for laminar free convection of gas with property variations. However, this approach has not expressed accurately the thermophysical property variations with temperature for different gases, and therefore the free convection heat transfer could not be predicted rigorously.

Actually, different gases have different thermophysical properties. The effects of the different thermophysical properties on free convection film flows and heat transfer are complicated. In order to predict free convection film flows and heat transfer more accurately and conveniently, the following very important investigations must be done (a) to provide reasonable method for reliable and simple simulation of the variable thermophysical properties, and (b) to provide a system of theoretical and mathematical models for rigorous and convenient treatment of the variable thermophysical properties. In the successive chapters of this book, these investigations will be dealt with and completed.

References

1. V.M. Falkner and S.W. Skan, Some approximate solutions of the boundary layer equations, Phil. Mag. 12, 865, 1931
2. H. Schlichting, Boundary Layer Theory, translated by J. Kestin, pp. 316–317, 1979
3. T. Cebeci and P. Bradshaw, Physical and Computational aspects of convective heat transfer, Springer, Berlin Heidelberg New York, 1984
4. A.G. Hansen, Similarity Analysis of Boundary Value Problems in Engineering, Prentice-Hall, Englewood Cliffs, NJ, 1964
5. T.Y. Na, Computational Methods in Engineering Boundary Value Problems, Academic, New York, 1979
6. E.M. Sparrow and J.L. Gregg, The variable fluid property problem in free convection, Trans. ASME 80, 879–886, 1958
7. D.D. Gray and A. Giogini, The validity of the Boussinesq approximation for liquids and gases, Int. J. Heat Mass Transfer 19, 545–551, 1977
8. A.M. Clausing and S.N. Kempka, The influences of property variations on natural convection from vertical surfaces, J. Heat Transfer 103, 609–612, 1981
9. B. Gebhart, Natural convection flow, instability, and transition, J. Heat Transfer 91, 293–309, 1969
10. E.R.G. Eckert and R.M. Drake, Analysis of Heat and Mass Transfer, McGraw-Hill, New York, 1972

4

Laminar Free Convection of Monatomic and Diatomic Gases, Air, and Water Vapor

Nomenclature

b	width of plate, m
c_p	specific heat at constant pressure, J $(\text{kg K})^{-1}$
$Gr_{x,\infty}$	local Grashof number for gas free convection, $g \left\|T_\text{w}/T_\infty - 1\right\| x^3/\nu_\infty^2$
$Nu_{x,\text{w}}$	local Nusselt number, $\alpha_x x/\lambda_\text{w}$
$\overline{Nu}_{x,\text{w}}$	average Nusselt number defined, $\overline{\alpha}_x x/\lambda_\text{w}$
n_λ	thermal conductivity parameter of gas
n_μ	viscosity parameter of gas
$n_{\mu\lambda}$	overall temperature parameter of gas
Pr	Prandtl number
q_x	local heat transfer rate at position x per unit area on the plate, W m^{-2}
Q_x	total heat transfer rate for position $x = 0$ to x with width of b on the plate, W
t	temperature, °C
T	absolute temperature, K
T_w/T_∞	boundary temperature ratio
w_x, w_y	velocity component in the x- and y- directions, respectively, m s^{-1}
W_x, W_y	dimensionless velocity component in the x- and y- directions, respectively

Greek symbols

α_x	local heat transfer coefficient, W $\left(\text{m}^2\text{K}\right)^{-1}$
$\overline{\alpha}_x$	vaverage heat transfer coefficient, W $\left(\text{m}^2\text{K}\right)^{-1}$
δ	boundary layer thickness, m
η	dimensionless coordinate variable
θ	dimensionless temperature, $\theta = (T - T_\infty)/(T_\text{w} - T_\infty)$
λ	thermal conductivity, W $(\text{m K})^{-1}$
μ	absolute viscosity, kg $(\text{m s})^{-1}$

ν	kinetic viscosity, $m^2\ s^{-1}$
ρ	density, $kg\ m^{-3}$
$\left(\frac{d\theta}{d\eta}\right)_{\eta=0}$	dimensionless temperature gradient on the plate
$\frac{1}{\rho}\frac{d\rho}{dx}$	density factor
$\frac{1}{\mu}\frac{d\mu}{d\eta}$	viscosity factor
$\frac{1}{\lambda}\frac{d\lambda}{d\eta}$	thermal conductivity factor
$\psi\left(Pr\right)$	Boussinesq approximation solution

Subscripts

w	at wall
δ	thickness of boundary layer
∞	far from the wall surface
sub	subcooling state

4.1 Introduction

The study of laminar free convection of gases with variable thermophysical properties can be traced back to the perturbation analysis of Hara [1] for air free convection. The solution is applicable for small values of the perturbation parameter, $\varepsilon_{H} = (T_{w} - T_{\infty})/T_{\infty}$. Later, Tataev [2] investigated the free convection of a gas with variable viscosity. A well known analysis of the variable fluid property problem for laminar free convection on an isothermal vertical flat plate has been presented by Sparrow and Gregg [3], giving solutions of the boundary layer equations for five assumed gases. They proposed a reference temperature and suggested that with it the problem of variable thermophysical properties can be treated as a constant property problem, i.e., using the Boussinesq approximation. Gray and Giogini [4] discussed the validity of the Boussinesq approximation and proposed a method for analyzing natural convection flow with fluid properties assumed to be a linear function of temperature and pressure. Clausing and Kempka [5] reported their experimental study of the influence of property variations on natural convection and showed that, for the laminar region, Nu_f is a function of $Ra_f\ (= Gr_f\ Pr_f)$ only, with the reference temperature T_f taken as the average temperature in the boundary layer.

In Chaps. 4 and 5 I will present, respectively, our recent studies described in [6, 7], for effect of variable thermophysical properties on laminar free convection of different kind of gases. In this chapter, the gases involved are monatomic and diatomic gases as well as air and water vapor. The variation of specific heat of these gases is very small, and so can be neglected when considering variable thermophysical properties. In Chap. 5, the gases involved are polyatomic gases in which the variation of specific heat cannot be neglected. In these studies a recently developed velocity component method is provided for the similarity transformation of the governing partial differential equations of the laminar boundary layer, instead of the traditional Falkner–Skan transformations. Additionally, a temperature parameter method for the treatment of a gas with variable thermophysical properties is proposed. With this method, the thermal conductivity, dynamic viscosity, and specific heat of gases are assumed to vary with absolute temperature according to a simple power law. The parameters of thermal conductivity, absolute viscosity, and specific heat are proposed and the corresponding values are provided from the typical experimental data of the thermophysical properties. The density is taken to be inversely proportional to the absolute temperature at constant pressure, while the Prandtl number is assumed to be constant. The governing equations for the laminar free convection of gas are transformed into the dimensionless ordinary equations by using the velocity component method, and meanwhile the variable thermophysical properties are treated by employing the temperature parameter method. The governing ordinary differential equations with the boundary conditions are solved for various boundary temperature ratios for the various gases mentioned earlier, and the rigorous numerical results are provided. These numerical results have shown that there are different velocity and temperature distributions for different boundary temperature ratios, as well as for different gases. Curve-fit formulas for the temperature gradient at the wall with very good agreement to the numerical solutions are provided, which facilitate rapid and yet accurate estimates of the heat transfer coefficient and the Nusselt number together with various boundary temperature ratios T_w/T_∞ and different gases.

4.2 Governing Partial Differential Equations

The physical analytical model and coordinate system used for laminar free convection of gas on an isothermal vertical flat plate is shown in Fig. 4.1. The boundary layer is laminar when Raleigh number, $Ra\,(= Gr\ Pr)$ is less than 10^9 [8].

According to the presentation in Chap. 2 the conservation equations for mass, momentum, and energy for a steady laminar flow in the boundary layer for vertical free convection of a gas are given by

$$\frac{\partial}{\partial x}(\rho w_x) + \frac{\partial}{\partial y}(\rho w_y) = 0, \tag{4.1}$$

$$\rho\left(w_x\frac{\partial w_x}{\partial x} + w_y\frac{\partial w_x}{\partial y}\right) = \frac{\partial}{\partial y}\left(\mu\frac{\partial w_x}{\partial y}\right) + g\rho|T/T_\infty - 1|, \tag{4.2}$$

$$\rho c_p\left(w_x\frac{\partial T}{\partial x} + w_y\frac{\partial T}{\partial y}\right) = \frac{\partial}{\partial y}\left(\lambda\frac{\partial T}{\partial y}\right), \tag{4.3}$$

The absolute value of buoyancy term $\rho g|T/T_\infty - 1|$ shows that it has always positive sign no matter which one is larger between T and T_∞. In this case, the buoyancy term $\rho g|T/T_\infty - 1|$ and the velocity component w_x have the same sign. The boundary conditions are

$$y = 0: \quad w_x = 0, \quad w_y = 0, \quad T = T_w, \tag{4.4}$$

$$y \to \infty: \quad w_x \to 0, \quad T = T_\infty. \tag{4.5}$$

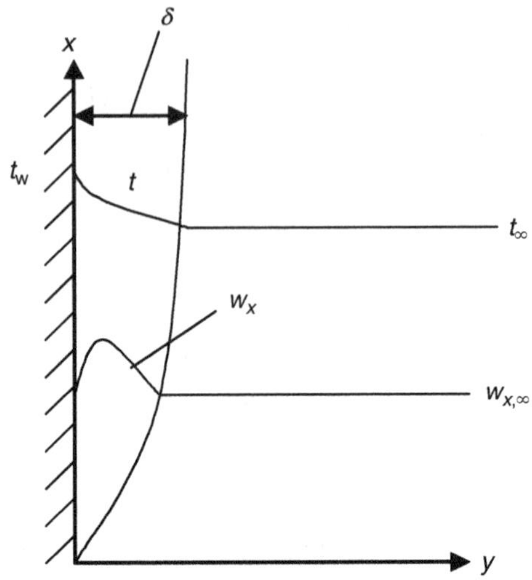

Fig. 4.1. Physical model and coordinate system of boundary layer for laminar free convection

4.3 Similarity Transformation of the Governing Equations

4.3.1 Assumed Dimensionless Variables with Velocity Component Method

From Chap. 3, it is found that with the traditional Falkner–Skan transformation there are some difficulties encountered with the similarity transformation of the governing partial differential equations for the boundary layer problem and with treatment of variable thermophysical properties. It is the reason why for a long time there has been lack of rigorous theoretical studies and numerical solutions for the laminar free convection with rigorous consideration of variable thermophysical properties. In [6] Shang and Wang proposed a velocity component method for the similarity transformation of the laminar boundary layer problem. With this method, a key work is to assume suitable forms of dimensionless velocity components. The typical method for this procedure is group theory, which was discussed at length by Hansen [9] and Na [10]. For the similarity transformation of (4.1)–(4.3) the dimensionless variables are assumed as follows:

A dimensionless similarity coordinate variable is assumed as given by

$$\eta = \frac{y}{x}\left(\frac{1}{4}\,Gr_{x,\infty}\right)^{1/4}, \tag{4.6}$$

where the local Grashof number $Gr_{x,\infty}$ is defined as

$$Gr_{x,\infty} = \frac{g\,|T_{\mathrm{w}}/T_\infty - 1|\,x^3}{\nu_\infty^2}. \tag{4.7}$$

A dimensionless temperature variable is defined as

$$\theta = \frac{T - T_\infty}{T_{\mathrm{w}} - T_\infty}. \tag{4.8}$$

Dimensionless velocity components are assumed to be:

$$W_x = \left[2\sqrt{gx}\,|T_{\mathrm{w}}/T_\infty - 1|^{1/2}\right]^{-1} w_x, \tag{4.9}$$

$$W_y = \left[2\sqrt{gx}\,|T_{\mathrm{w}}/T_\infty - 1|^{1/2}\left(\frac{1}{4}Gr_{x,\infty}\right)^{-1/4}\right]^{-1} w_y. \tag{4.10}$$

4.3.2 The Similarity Transformation

With the earlier assumed dimensionless variables the governing partial differential equations can be transformed similarly as follows:

For (4.1)

Equation (4.1) is initially changed into

$$\rho\left(\frac{\partial w_x}{\partial x} + \frac{\partial w_y}{\partial y}\right) + w_x\frac{\partial \rho}{\partial x} + w_y\frac{\partial \rho}{\partial y} = 0. \tag{4.11}$$

With the dimensionless variables assumed in (4.6), (4.7), (4.9), and (4.10) we obtain the following relations:

$$\frac{\partial w_x}{\partial x} = \left[2\sqrt{gx}\,|T_{\mathrm{w}}/T_\infty - 1|^{1/2}\right]\frac{\mathrm{d}W_x}{\mathrm{d}\eta}\frac{\partial \eta}{\partial x} + \frac{1}{2}x^{-1/2}\left[2\sqrt{g}\,|T_{\mathrm{w}}/T_\infty - 1|^{1/2}\right]W_x,$$

where

$$\begin{aligned}
\frac{\partial \eta}{\partial x} &= \frac{\partial}{\partial x}\left[\frac{y}{x}\left(\frac{1}{4}Gr_{x,\infty}\right)^{1/4}\right] \\
&= \frac{\partial}{\partial x}\left[y\left(\frac{1}{4}\frac{g\,|T_{\mathrm{w}}/T_\infty - 1|\,x^{-1}}{\nu_\infty^2}\right)^{1/4}\right] \\
&= -\frac{1}{4}\left[y\left(\frac{1}{4}\frac{g\,|T_{\mathrm{w}}/T_\infty - 1|}{\nu_\infty^2}\right)^{1/4}\right]x^{-5/4} \\
&= -\frac{1}{4}\left[y\left(\frac{1}{4}\frac{g\,|T_{\mathrm{w}}/T_\infty - 1|\,x^3}{\nu_\infty^2}\right)^{1/4}\right]x^{-2} \\
&= -\frac{1}{4}x^{-1}\eta.
\end{aligned}$$

Then,

$$\begin{aligned}\frac{\partial w_x}{\partial x} &= \left[2\sqrt{gx}\,|T_w/T_\infty - 1|^{1/2}\right]\frac{\mathrm{d}W_x}{\mathrm{d}\eta}\left(-\frac{1}{4}x^{-1}\eta\right)\\ &\quad+\frac{1}{2}x^{-1/2}\left[2\sqrt{g}\,|T_w/T_\infty - 1|^{1/2}\right]W_x\\ &= -\frac{1}{2}\left[\sqrt{\frac{g}{x}}\,|T_w/T_\infty - 1|^{1/2}\right]\eta\frac{\mathrm{d}W_x}{\mathrm{d}\eta} + \left[\sqrt{\frac{g}{x}}\,|T_w/T_\infty - 1|^{1/2}\right]W_x\\ &= \sqrt{\frac{g}{x}}\,|T_w/T_\infty - 1|^{1/2}\left(W_x - \frac{1}{2}\eta\frac{\mathrm{d}W_x}{\mathrm{d}\eta}\right),\end{aligned} \tag{4.12}$$

$$\begin{aligned}\frac{\partial w_y}{\partial y} &= \left[2\sqrt{gx}\,|T_w/T_\infty - 1|^{1/2}\left(\frac{1}{4}Gr_{x,\infty}\right)^{-1/4}\right]\frac{\mathrm{d}W_y}{\mathrm{d}\eta}\frac{\partial\eta}{\partial y}\\ &= \left[2\sqrt{gx}\,|T_w/T_\infty - 1|^{1/2}\left(\frac{1}{4}Gr_{x,\infty}\right)^{-1/4}\right]\frac{\mathrm{d}W_y}{\mathrm{d}\eta}\frac{1}{x}\left(\frac{1}{4}Gr_{x,\infty}\right)^{1/4}\\ &= 2\sqrt{\frac{g}{x}}\,|T_w/T_\infty - 1|^{1/2}\frac{\mathrm{d}W_y}{\mathrm{d}\eta},\end{aligned} \tag{4.13}$$

$$\frac{\partial\rho}{\partial x} = \frac{\mathrm{d}\rho}{\mathrm{d}\eta}\frac{\partial\eta}{\partial x} = -\frac{1}{4}x^{-1}\eta\frac{\mathrm{d}\rho}{\mathrm{d}\eta}, \tag{4.14}$$

$$\frac{\partial\rho}{\partial y} = \frac{\mathrm{d}\rho}{\mathrm{d}\eta}\frac{\partial\eta}{\partial y} = \frac{1}{x}\left(\frac{1}{4}\,Gr_{x,\infty}\right)^{1/4}\frac{\mathrm{d}\rho}{\mathrm{d}\eta}. \tag{4.15}$$

By using (4.12)–(4.15),(4.11) can be changed to

$$\begin{aligned}&\rho\left[\sqrt{\frac{g}{x}}\,|T_w/T_\infty - 1|^{1/2}\left(W_x - \frac{1}{2}\eta\frac{\mathrm{d}W_y}{\mathrm{d}\eta}\right) + 2\sqrt{\frac{g}{x}}\,|T_w/T_\infty - 1|^{1/2}\frac{\mathrm{d}W_y}{\mathrm{d}\eta}\right]\\ &+2\sqrt{gx}\,|T_w/T_\infty - 1|^{1/2}W_x\left(-\frac{1}{4}\eta x^{-1}\frac{\mathrm{d}\rho}{\mathrm{d}\eta}\right) + 2\sqrt{gx}\,|T_w/T_\infty - 1|^{1/2}\\ &\times\left(\frac{1}{4}Gr_{x,\infty}\right)^{-1/4}W_y\frac{\mathrm{d}\rho}{\mathrm{d}\eta}\left(\frac{1}{4}Gr_{x,\infty}\right)^{-1/4}x^{-1} = 0.\end{aligned} \tag{4.16}$$

Equation (4.16) is divided by $|T_w/T_\infty - 1|^{1/2}\sqrt{\dfrac{g}{x}}$ and is transformed into

$$\rho\left[\left(W_x - \frac{1}{2}\eta\frac{\mathrm{d}W_y}{\mathrm{d}\eta}\right) + 2\frac{\mathrm{d}W_y}{\mathrm{d}\eta}\right] + 2W_x\left(-\frac{1}{4}\eta\frac{\mathrm{d}\rho}{\mathrm{d}\eta}\right) + 2W_y\frac{\mathrm{d}\rho}{\mathrm{d}\eta} = 0, \tag{4.17}$$

or

$$2W_x - \eta\frac{\mathrm{d}W_x}{\mathrm{d}\eta} + 4\frac{\mathrm{d}W_y}{\mathrm{d}\eta} - \frac{1}{\rho}\frac{\mathrm{d}\rho}{\mathrm{d}\eta}(\eta W_x - 4W_y) = 0. \tag{4.18}$$

For (4.2):

Equation (4.2) can be rewritten as

$$\rho\left(w_x\frac{\partial w_x}{\partial x}+w_y\frac{\partial w_x}{\partial y}\right)=\mu\frac{\partial^2 w_x}{\partial y^2}+\frac{\partial w_x}{\partial y}\frac{\partial \mu}{\partial y}+\rho g\left|T/T_\infty-1\right|. \tag{4.19}$$

With the dimensionless variables assumed in (4.6), (4.7), (4.9) and (4.10) we obtain the following equations:

$$\frac{\partial w_x}{\partial y}=2\sqrt{gx}\left|T_\mathrm{w}/T_\infty-1\right|^{1/2}\frac{\mathrm{d}W_x}{\mathrm{d}\eta}\frac{\partial \eta}{\partial y},$$

where

$$\frac{\partial \eta}{\partial y}=x^{-1}\left(\frac{1}{4}Gr_{x,\infty}\right)^{1/4}.$$

Then,

$$\frac{\partial w_x}{\partial y}=2\sqrt{gx}\left|T_\mathrm{w}/T_\infty-1\right|^{1/2}\frac{\mathrm{d}W_x}{\mathrm{d}\eta}x^{-1}\left(\frac{1}{4}Gr_{x,\infty}\right)^{1/4}, \tag{4.20}$$

$$\begin{aligned}\frac{\partial^2 w_x}{\partial y^2}&=2\sqrt{gx}\left|T_\mathrm{w}/T_\infty-1\right|^{1/2}\frac{\mathrm{d}^2W_x}{\mathrm{d}\eta^2}x^{-1}\left(\frac{1}{4}Gr_{x,\infty}\right)^{1/4}\frac{\partial \eta}{\partial y}\\&=2\sqrt{gx}\left|T_\mathrm{w}/T_\infty-1\right|^{1/2}\frac{\mathrm{d}^2W_x}{\mathrm{d}\eta^2}x^{-1}\left(\frac{1}{4}Gr_{x,\infty}\right)^{1/4}x^{-1}\left(\frac{1}{4}Gr_{x,\infty}\right)^{1/4}\\&=2\sqrt{gx}\left|T_\mathrm{w}/T_\infty-1\right|^{1/2}\frac{\mathrm{d}^2W_x}{\mathrm{d}\eta^2}\left(\frac{1}{4}Gr_{x,\infty}\right)^{1/2}x^{-2},\end{aligned} \tag{4.21}$$

$$\frac{\partial \mu}{\partial y}=\frac{\mathrm{d}\mu}{\mathrm{d}\eta}\frac{\partial \eta}{\partial y}=\frac{\mathrm{d}\mu}{\mathrm{d}\eta}\left(\frac{1}{4}Gr_{x,\infty}\right)^{1/4}x^{-1}. \tag{4.22}$$

Using (4.9), (4.10), (4.12), (4.20)–(4.22), (4.19) becomes

$$\begin{aligned}&\rho\left[2\sqrt{gx}\left|T_\mathrm{w}/T_\infty-1\right|^{1/2}W_x\sqrt{\frac{g}{x}}\left|T_\mathrm{w}/T_\infty-1\right|^{1/2}\left(W_x-\frac{1}{2}\eta\frac{\mathrm{d}W_x}{\mathrm{d}\eta}\right)\right.\\&+2\sqrt{gx}\left|T_\mathrm{w}/T_\infty-1\right|^{1/2}\left(\frac{1}{4}Gr_{x,\infty}\right)^{-1/4}W_y\left(2\sqrt{gx}\left|T_\mathrm{w}/T_\infty-1\right|^{1/2}\right.\\&\left.\left.\times\frac{\mathrm{d}^2W_x}{\mathrm{d}\eta^2}x^{-1}\left(\frac{1}{4}Gr_{x,\infty}\right)^{1/4}\right)\right]=2\mu\sqrt{gx}\left|T_\mathrm{w}/T_\infty-1\right|^{1/2}\frac{\mathrm{d}^2W_x}{\mathrm{d}\eta^2}\left(\frac{1}{4}Gr_{x,\infty}\right)^{1/2}x^{-2}\\&+2\sqrt{gx}\left|T_\mathrm{w}/T_\infty-1\right|^{1/2}\frac{\mathrm{d}W_x}{\mathrm{d}\eta}x^{-1}\left(\frac{1}{4}Gr_{x,\infty}\right)^{1/4}\frac{\mathrm{d}\mu}{\mathrm{d}\eta}\left(\frac{1}{4}Gr_{x,\infty}\right)^{1/4}x^{-1}+\rho g\left|T/T_\infty-1\right|.\end{aligned} \tag{4.23}$$

Equation (4.23) is divided by$\rho g(T_{\mathrm{w}}/T_{\infty}-1)$, meanwhile, the definition of local Grashof number $Gr_{x,\infty}$ is considered, and then, (4.23) is simplified to

$$2W_x\left(W_x-\frac{1}{2}\eta\frac{\mathrm{d}W_x}{\mathrm{d}\eta}\right)+2W_y\left(2\frac{\mathrm{d}^2W_x}{\mathrm{d}\eta^2}\right)=2\nu\frac{\mathrm{d}^2W_x}{\mathrm{d}\eta^2}\left(\frac{1}{4}\frac{1}{\nu_{\infty}^2}\right)^{1/2}$$

$$+2\frac{1}{\rho}\frac{\mathrm{d}W_x}{\mathrm{d}\eta}\frac{\mathrm{d}\mu}{\mathrm{d}\eta}\left(\frac{1}{4}\frac{1}{\nu_{\infty}^2}\right)^{1/2}+\theta,$$

or

$$2W_x\left(W_x-\frac{1}{2}\eta\frac{\mathrm{d}W_x}{\mathrm{d}\eta}\right)+2W_y\left(2\frac{\mathrm{d}W_x}{\mathrm{d}\eta}\right)=\frac{\nu}{\nu_{\infty}}\frac{\mathrm{d}^2W_x}{\mathrm{d}\eta^2}+\frac{1}{\rho}\frac{\mathrm{d}W_x}{\mathrm{d}\eta}\frac{\mathrm{d}\mu}{\mathrm{d}\eta}\frac{1}{\nu_{\infty}}+\theta. \tag{4.24}$$

Equation (4.24) is multiplied by ν_{∞}/ν and is simplified to

$$\frac{\nu_{\infty}}{\nu}\left[W_x\left(2W_x-\eta\frac{\mathrm{d}W_x}{\mathrm{d}\eta}\right)+4W_y\frac{\mathrm{d}W_x}{\mathrm{d}\eta}\right]=\frac{\mathrm{d}^2W_x}{\mathrm{d}\eta^2}+\frac{1}{\mu}\frac{\mathrm{d}\mu}{\mathrm{d}\eta}\frac{\mathrm{d}W_x}{\mathrm{d}\eta}+\frac{\nu_{\infty}}{\nu}\theta. \tag{4.25}$$

For (4.3):

Finally, (4.3) can be rewritten as:

$$\rho c_p\left(w_x\frac{\partial T}{\partial x}+w_y\frac{\partial T}{\partial y}\right)=\lambda\frac{\partial^2T}{\partial y^2}+\frac{\partial\lambda}{\partial y}\frac{\partial T}{\partial y}, \tag{4.26}$$

where

$$T=(T_{\mathrm{w}}-T_{\infty})\theta+T_{\infty},$$

$$\frac{\partial T}{\partial x}=-(T_{\mathrm{w}}-T_{\infty})\frac{\mathrm{d}\theta}{\mathrm{d}\eta}\frac{\partial\eta}{\partial x}=-(T_{\mathrm{w}}/T_{\infty})\frac{\mathrm{d}\theta}{\mathrm{d}\eta}\left(\frac{1}{4}\right)\eta x^{-1}, \tag{4.27}$$

$$\frac{\partial T}{\partial y}=(T_{\mathrm{w}}-T_{\infty})\frac{\mathrm{d}\theta}{\mathrm{d}\eta}\frac{\partial\eta}{\partial y}=(T_{\mathrm{w}}-T_{\infty})\frac{\mathrm{d}\theta}{\mathrm{d}\eta}\left(\frac{1}{4}Gr_{x,\infty}\right)^{1/4}x^{-1}, \tag{4.28}$$

$$\begin{aligned}\frac{\partial^2T}{\partial y^2}&=(T_{\mathrm{w}}-T_{\infty})\left(\frac{1}{4}Gr_{x,\infty}\right)^{1/4}x^{-1}\frac{\mathrm{d}^2\theta}{\mathrm{d}\eta^2}\frac{\partial\eta}{\partial y}\\&=(T_{\mathrm{w}}-T_{\infty})\frac{\mathrm{d}^2\theta}{\mathrm{d}\eta^2}\left(\frac{1}{4}Gr_{x,\infty}\right)^{1/2}x^{-2},\end{aligned} \tag{4.29}$$

$$\frac{\partial\lambda}{\partial y}=\frac{\mathrm{d}\lambda}{\mathrm{d}\eta}\frac{\partial\eta}{\partial y}=\frac{\mathrm{d}\lambda}{\mathrm{d}\eta}\left(\frac{1}{4}Gr_{x,\infty}\right)^{1/4}x^{-1}. \tag{4.30}$$

Then, (4.26) will now transform into

$$\rho c_p \left(2\sqrt{gx}\left|T_{\rm w}/T_\infty - 1\right|^{1/2} W_x(-1)(T_{\rm w} - T_\infty)\frac{{\rm d}\theta}{{\rm d}\eta}\left(\frac{1}{4}\right)\eta x^{-1} \right.$$
$$\left. + 2\sqrt{gx}\left|T_{\rm w}/T_\infty - 1\right|^{1/2}\left(\frac{1}{4}Gr_{x,\infty}\right)^{-1/4} W_y(T_{\rm w} - T_\infty)\frac{{\rm d}\theta}{{\rm d}\eta}\left(\frac{1}{4}Gr_{x,\infty}\right)^{1/4} x^{-1}\right)$$
$$= \lambda(T_{\rm w} - T_\infty)\frac{{\rm d}^2\theta}{{\rm d}\eta^2}\left(\frac{1}{4}Gr_{x,\infty}\right)^{1/2} x^{-2} + \frac{{\rm d}\lambda}{{\rm d}\eta}\left(\frac{1}{4}Gr_{x,\infty}\right)^{1/4}$$
$$\times x^{-1}(T_{\rm w} - T_\infty)\frac{{\rm d}\theta}{{\rm d}\eta}\left(\frac{1}{4}Gr_{x,\infty}\right)^{1/4} x^{-1}. \tag{4.31}$$

Equation (4.31) is divided by $(T_{\rm w} - T_\infty)\left|T_{\rm w}/T_\infty - 1\right|^{1/2}\sqrt{g/x}$, meanwhile, the definition of local Grashof number $Gr_{x,\infty}$ is considered, and then, (4.31) is simplified to

$$\rho c_p \left(2W_x(-1)\frac{{\rm d}\theta}{{\rm d}\eta}\left(\frac{1}{4}\right)\eta + 2W_y\frac{{\rm d}\theta}{{\rm d}\eta}\right) = \lambda\frac{{\rm d}^2\theta}{{\rm d}\eta^2}\left(\frac{1}{4}\frac{1}{\nu_\infty^2}\right)^{1/2} + \frac{{\rm d}\lambda}{{\rm d}\eta}\frac{{\rm d}\theta}{{\rm d}\eta}\left(\frac{1}{4}\frac{1}{\nu_\infty^2}\right)^{1/2},$$

or

$$\rho c_p \left(-2W_x\frac{{\rm d}\theta}{{\rm d}\eta}\left(\frac{1}{4}\right)\eta + 2W_y\frac{{\rm d}\theta}{{\rm d}\eta}\right) = \frac{1}{2\nu_\infty}\lambda\frac{{\rm d}^2\theta}{{\rm d}\eta^2} + \frac{1}{2\nu_\infty}\frac{{\rm d}\lambda}{{\rm d}\eta}\frac{{\rm d}\theta}{{\rm d}\eta}. \tag{4.32}$$

Equation is multiplied by $2\nu_\infty/\lambda$ and simplified into

$$Pr\frac{\nu_\infty}{\nu}(-\eta W_x + 4W_y)\frac{{\rm d}\theta}{{\rm d}\eta} = \frac{{\rm d}^2\theta}{{\rm d}\eta^2} + \frac{1}{\lambda}\frac{{\rm d}\lambda}{{\rm d}\eta}\frac{{\rm d}\theta}{{\rm d}\eta}. \tag{4.33}$$

The governing ordinary differential equations can be summarized as follows:

$$\left(2W_x - \eta\frac{{\rm d}W_x}{{\rm d}\eta} + 4\frac{{\rm d}W_y}{{\rm d}\eta}\right) - \frac{1}{\rho}\frac{{\rm d}\rho}{{\rm d}\eta}(\eta W_x - 4W_y) = 0, \tag{4.18}$$

$$\frac{\nu_\infty}{\nu}\left[W_x\left(2W_x - \eta\frac{{\rm d}W_x}{{\rm d}\eta}\right) + 4W_y\frac{{\rm d}W_x}{{\rm d}\eta}\right] = \frac{{\rm d}^2W_x}{{\rm d}\eta^2} + \frac{1}{\mu}\frac{{\rm d}\mu}{{\rm d}\eta}\frac{{\rm d}W_x}{{\rm d}\eta} + \frac{\nu_\infty}{\nu}\theta, \tag{4.25}$$

$$Pr\frac{\nu_\infty}{\nu}(-\eta W_x + 4W_y)\frac{{\rm d}\theta}{{\rm d}\eta} = \frac{{\rm d}^2\theta}{{\rm d}\eta^2} + \frac{1}{\lambda}\frac{{\rm d}\lambda}{{\rm d}\eta}\frac{{\rm d}\theta}{{\rm d}\eta}. \tag{4.33}$$

With the assumed dimensionless variables the following dimensionless boundary conditions are easily obtained from (4.4) and (4.5):

$$\eta = 0: \quad W_x = 0, \quad W_y = 0, \quad \theta = 1 \tag{4.34}$$
$$\eta \to \infty: \quad W_x \to 0, \quad \theta \to 1. \tag{4.35}$$

4.4 Treatment of Variable Thermophysical Properties

4.4.1 Temperature Parameters

For treatment of variable thermophysical properties of a gas, a treatment method on variable thermophysical properties developed in [6] is presented here. With this method the viscosity and thermal conductivity are expressed as $\mu \approx T^{n_\mu}$ and $\lambda \approx T^{n_\lambda}$, respectively. The thermodynamic temperature of the gas far away from the wall, T_∞, can be taken as the reference temperature for free convection analysis. So we assume

$$\frac{\mu}{\mu_\infty} = \left(\frac{T}{T_\infty}\right)^{n_\mu}, \tag{4.36}$$

$$\frac{\lambda}{\lambda_\infty} = \left(\frac{T}{T_\infty}\right)^{n_\lambda}. \tag{4.37}$$

While the change of density with thermodynamic temperature at constant pressure can be expressed as

$$\frac{\rho}{\rho_\infty} = \left(\frac{T}{T_\infty}\right)^{-1}. \tag{4.38}$$

Combining (4.36) with (4.38) we have

$$\frac{\nu}{\nu_\infty} = \left(\frac{T}{T_\infty}\right)^{n_\mu+1}, \tag{4.39}$$

where the exponents n_μ and n_λ are named viscosity parameter, and thermal conductivity parameter, respectively.

According to the experimental values of μ and λ for several monatomic and diatomic gases, and also for air and water vapor, reported by Hisenrath et al. [11], n_μ and n_λ values are given in Table 4.1 [6]. If we take reference temperature T_0 to replace T_∞, the percentage deviations for predicted values of μ/μ_∞ and λ/λ_∞ predicted from (4.36) and (4.37) are shown in Figs. 4.2–4.6.

The Prandtl number is defined as $Pr = \mu c_p/\lambda$. Strictly speaking, Pr should also depend on temperature. However, it is well known that $Pr \approx 0.72$ for a diatomic gas, $Pr \approx 0.7$ for air, and $Pr \approx 1$ for water vapor. Hence, Pr can be taken as a constant for monatomic and diatomic gases, and for air and water vapor in the temperature range from T to T_∞. Therefore, if we assume

$$c_p/c_{p\infty} = (T/T_\infty)^{n_{c_p}}, \tag{4.40}$$

then

$$n_{c_p} = n_\lambda - n_\mu, \tag{4.41}$$

where n_{c_p} is named specific heat parameter.

Table 4.1. The value of Parameters n_μ, n_λ, and $n_{\mu\lambda}$ from experimental data for μ and λ with deviation, cited from Shang and Wang [6]

gas	T_0 (K)	n_μ	temperature range (K)	deviation (%)	n_λ	temperature range (K)	deviation (%)	$n_{\mu\lambda}$	temperature range (K)	deviation (%)
air	273	0.68	220–1,400	±3	0.81	220–1,000	±2.5	0.75	220–850	±6.5
water vapor	380	1.04	380–1,500	±4	1.185	380–800	±0.3	1.12	380–800	±5
CO	273	0.71	230–1,500	±2	0.83	220–600	±2	0.78	210–600	±4
N_2	273	0.67	220–1,500	±3	0.76	220–1,200	±3	0.72	200–1,000	±7
O_2	273	0.694	230–2,000	±2	0.86	220–600	±2	0.79	200–600	±6
H_2	273	0.68	80–1,000	±2	0.8	220–700	±1	0.75	180–700	±6
Ar	273	0.72	220–1,500	±3	0.73	210–1,500	±4	0.726	210–1,500	±4
He	273	0.66	273–873	±1	0.725	273–873	±0.3	0.7	273–873	±4
Ne	273	0.649	273–873	±0.4	0.71	273–873	±0.5	0.68	273–873	±4

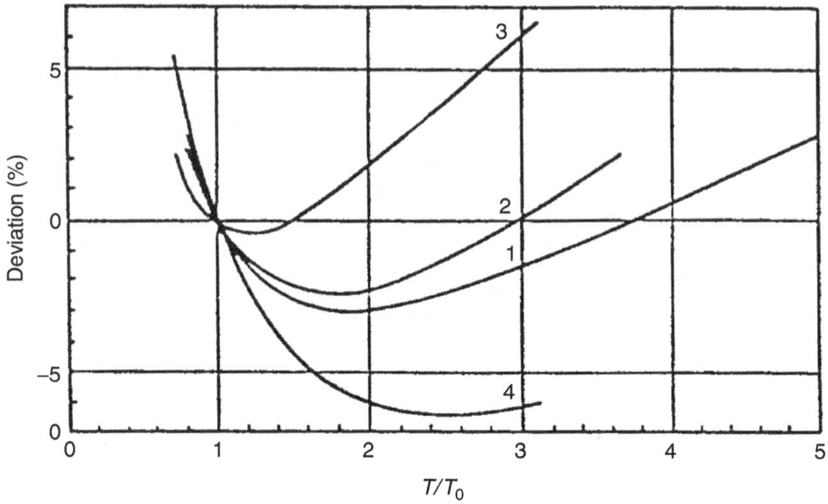

Fig. 4.2. Deviation of predicted values of μ and λ for air (1) for evaluation of μ with n_μ, (2) for evaluation of λ with n_λ, (3) for evaluation of μ with $n_{\mu\lambda}$, and (4) for evaluation of λ with $n_{\mu\lambda}$, cited from Shang and Wang [6]

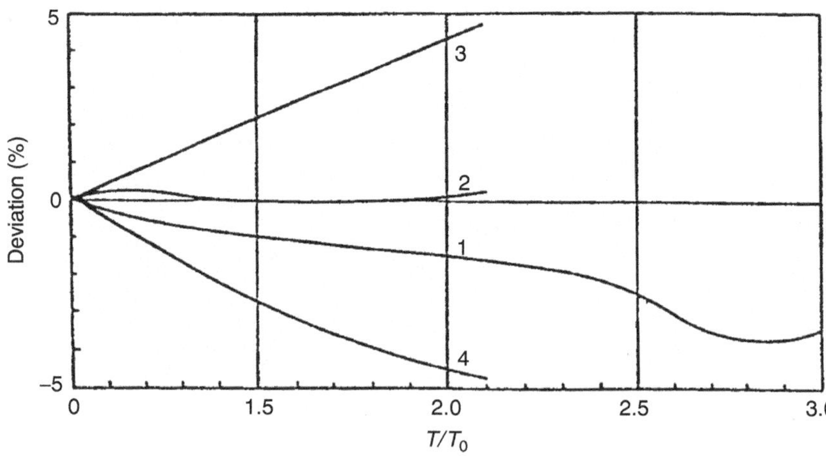

Fig. 4.3. Deviation of predicted values of μ and λ for water vapor (1) for evaluation of μ with n_μ, (2) for evaluation of λ with n_λ, (3) for evaluation of μ with $n_{\mu\lambda}$, and (4) for evaluation of λ with $n_{\mu\lambda}$, cited from Shang and Wang [6]

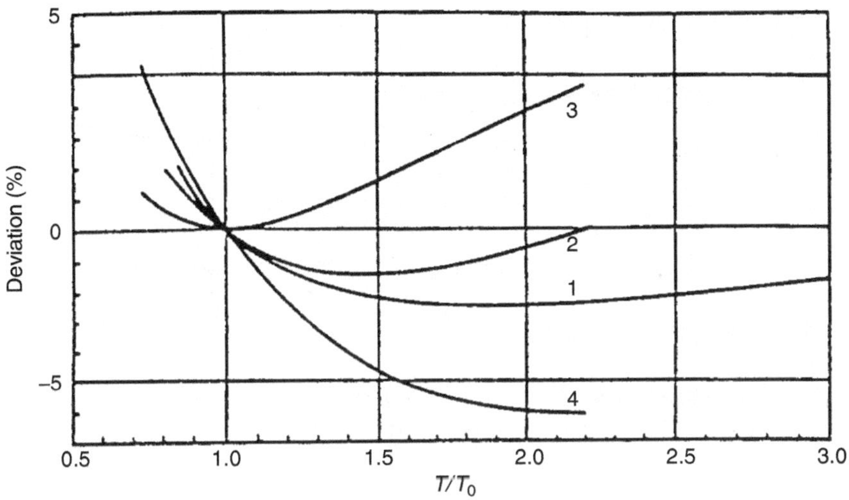

Fig. 4.4. Deviation of predicted values of μ and λ for oxygen O_2 (1) for evaluation of μ with n_μ, (2) for evaluation of λ with n_λ, (3) for evaluation of μ with $n_{\mu\lambda}$, and (4) for evaluation of λ with $n_{\mu\lambda}$, cited from Shang and Wang [6]

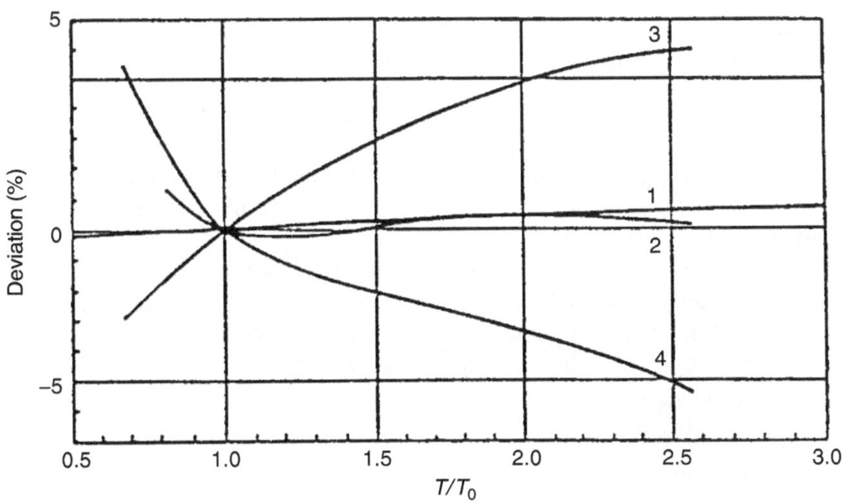

Fig. 4.5. Deviation of predicted values of μ and λ for hydrogen H_2 (1) for evaluation of μ with n_μ, (2) for evaluation of λ with n_λ, (3) for evaluation of μ with $n_{\mu\lambda}$, and (4) for evaluation of λ with $n_{\mu\lambda}$, cited from Shang and Wang [6]

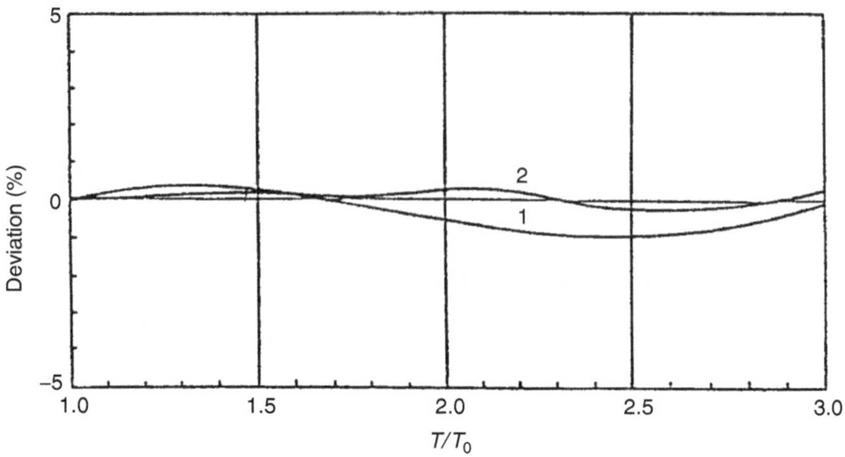

Fig. 4.6. Deviation of predicted values of μ and λ for helium He (1) for evaluation of μ with n_μ, and (2) for evaluation of λ with n_λ , cited from Shang and Wang [6]

It can be found from Table 4.1 that the values of n_{c_p} are much lower than 0.1 for monatomic gases, and around 0.1–0.16 for diatomic gases, air, and water vapor. For the case $1/2 \leq (T/T_\infty) \leq 2$, thus it is possible to treat c_p as a constant value for these gases, so as to simplify the analysis but still suit the needs of engineering application.

With Pr and c_p both assumed constant, $\mu/\lambda = \text{const.}$, and therefore it is logical, for monatomic and diatomic gases, air, and water vapor, to take some mean value of n_μ and n_λ as $n_{\mu\lambda}$, such that $n_\mu \approx n_{\mu\lambda} \approx n_\lambda$. We try to express the overall temperature parameter $n_{\mu\lambda}$ by a weighted sum of n_μ and n_λ, as given by

$$n_{\mu\lambda} \quad = \quad 0.45n_\mu \ + 0.55n_\lambda. \qquad (4.42)$$

The deviations for evaluation of μ and λ with $n_{\mu\lambda}$ for monatomic gases, diatomic gases, air, and water vapor are listed in Table 4.1 and also plotted in Figs. 4.2–4.6.

So far, four parameters, such as viscosity parameter n_μ, thermal conductivity parameter n_λ, and specific heat parameter n_λ, and overall temperature parameter $n_{\mu\lambda}$ have been presented for expression of variation of thermophysical properties of gases with temperature. In short, they can generally be called temperature parameters n_μ, n_λ, n_λ, or $n_{\mu\lambda}$.

4.4.2 Temperature Parameter Method

In the transformed governing (4.18), (4.25), and (4.33) there are four thermophysical property factors $(1/\rho)(\mathrm{d}\rho/\mathrm{d}\eta)$, $(1/\mu)(\mathrm{d}\mu/\mathrm{d}\eta)$, $(1/\lambda)(\mathrm{d}\lambda/\mathrm{d}\eta)$, and (ν_∞/ν) dominating the effects of variable thermophysical properties on free convection of gases. These factors tend to greatly increase the difficulty of getting a solution of the governing equations. However, with the provided gas temperature parameters, the thermophysical property factors can be transformed into the functions of temperature θ . Then, the governing dimensionless equations can be solved. The transformation of these thermophysical factors is expressed as follows:

Transformation of the density factor $(1/\rho)(\mathrm{d}\rho/\mathrm{d}\eta)$*:*

With (4.38) we obtain

$$\frac{1}{\rho}\frac{\mathrm{d}\rho}{\mathrm{d}\eta}=\frac{1}{\rho}\frac{\mathrm{d}}{\mathrm{d}\eta}\left(\rho_\infty\frac{T_\infty}{T}\right).$$

By using (4.8) we have

$$T=(T_\mathrm{w}-T_\infty)\theta+T_\infty.$$

Then,

$$\begin{aligned}\frac{1}{\rho}\frac{\mathrm{d}\rho}{\mathrm{d}\eta}&=\frac{\rho_\infty}{\rho}\frac{\mathrm{d}}{\mathrm{d}\eta}\left[\frac{T_\infty}{(T_\mathrm{w}-T_\infty)\theta+T_\infty}\right]\\&=\frac{T}{T_\infty}\frac{\mathrm{d}}{\mathrm{d}\eta}\left[\frac{1}{(T_\mathrm{w}/T_\infty-1)\theta+1}\right]\\&=-[(T_\mathrm{w}/T_\infty-1)\theta+1]\frac{(T_\mathrm{w}/T_\infty-1)\frac{\mathrm{d}\theta}{\mathrm{d}\eta}}{[(T_\mathrm{w}/T_\infty-1)\theta+1]^2}\\&=-\frac{(T_\mathrm{w}/T_\infty-1)\frac{\mathrm{d}\theta}{\mathrm{d}\eta}}{(T_\mathrm{w}/T_\infty-1)\theta+1}.\end{aligned}\tag{4.43}$$

Transformation of the viscosity factor $(1/\mu)(\mathrm{d}\mu/\mathrm{d}\eta)$*:*

For (4.36) we get

$$\begin{aligned}\frac{1}{\mu}\frac{\mathrm{d}\mu}{\mathrm{d}\eta}&=\frac{\mu_\infty}{\mu}\frac{\mathrm{d}}{\mathrm{d}\eta}\left(\frac{T}{T_\infty}\right)^{n_\mu}\\&=\left(\frac{T}{T_\infty}\right)^{-n_\mu}\frac{\mathrm{d}}{\mathrm{d}\eta}\left[\frac{(T_\mathrm{w}-T_\infty)\theta+T_\infty}{T_\infty}\right]^{n_\mu}\\&=((T_\mathrm{w}/T_\infty-1)\theta+1)^{-n_\mu}\frac{\mathrm{d}}{\mathrm{d}\eta}((T_\mathrm{w}/T_\infty-1)\theta+1)^{n_\mu}\\&=((T_\mathrm{w}/T_\infty-1)\theta+1)^{-n_\mu}n_\mu((T_\mathrm{w}/T_\infty-1)\theta+1)^{n_\mu-1}(T_\mathrm{w}/T_\infty-1)\frac{\mathrm{d}\theta}{\mathrm{d}\eta}.\end{aligned}$$

This leads to

$$\frac{1}{\mu}\frac{\mathrm{d}\mu}{\mathrm{d}\eta}=\frac{n_\mu(T_\mathrm{w}/T_\infty-1)\mathrm{d}\theta/\mathrm{d}\eta}{(T_\mathrm{w}/T_\infty-1)\theta+1}.\tag{4.44}$$

Transformation of the thermal conductivity factor $(1/\lambda)(\mathrm{d}\lambda/\mathrm{d}\eta)$

With a derivation similar to that for the factor $(1/\mu)(\mathrm{d}\mu/\mathrm{d}\eta)$ we can obtain

$$\frac{1}{\lambda}\frac{\mathrm{d}\lambda}{\mathrm{d}\eta} = \frac{n_\lambda(T_\mathrm{w}/T_\infty - 1)\mathrm{d}\theta/\mathrm{d}\eta}{(T_\mathrm{w}/T_\infty - 1)\theta + 1}. \tag{4.45}$$

Transformation of factor (ν_∞/ν)

By using (4.8), (4.39) can be transformed to

$$\frac{\nu_\infty}{\nu}\left[(T_\mathrm{w}/T_\infty - 1)\theta + 1\right]^{-(n_\mu+1)}. \tag{4.46}$$

It is found that the method for treatment of variable thermophysical properties of gases closely depends on the temperature parameters. Therefore, such method for treatment of variable thermophysical properties of gases can be called “temperature parameter method.”

4.5 Heat Transfer Analysis

The local heat transfer rate q_x at position x per unit area from the surface of the plate to the gas can be calculated by Fourier's law as

$$q_x = -\lambda_\mathrm{w}\left(\frac{\partial T}{\partial y}\right)_{y=0}.$$

With (4.28) we have

$$\left(\frac{\partial T}{\partial y}\right)_{y=0} = (T_\mathrm{w} - T_\infty)\left(\frac{\mathrm{d}\theta}{\mathrm{d}\eta}\right)_{\eta=0}\left(\frac{1}{4}Gr_{x,\infty}\right)^{1/4} x^{-1}.$$

Then,

$$q_x = -\lambda_\mathrm{w}(T_\mathrm{w} - T_\infty)\left(\frac{1}{4}Gr_{x,\infty}\right)^{1/4} x^{-1}\left(\frac{\mathrm{d}\theta}{\mathrm{d}\eta}\right)_{\eta=0}. \tag{4.47}$$

Total heat transfer rate for position $x = 0$ to x with width of b on the plate is a integration $Q_x = \iint\limits_A q_x\,\mathrm{d}A = \int_0^x q_x b\,\mathrm{d}x$, and hence

$$Q_x = -\lambda_\mathrm{w} b(T_\mathrm{w} - T_\infty)\left(\frac{\mathrm{d}\theta}{\mathrm{d}\eta}\right)_{\eta=0}\int_0^x\left(\frac{1}{4}Gr_{x,\infty}\right)^{1/4} x^{-1}\,\mathrm{d}x.$$

With (4.7) for definition of local Grashof number $Gr_{x,\infty}$ we obtain

$$Q_x = -\frac{4}{3}b\lambda_\mathrm{w}(T_\mathrm{w} - T_\infty)\left(\frac{1}{4}Gr_{x,\infty}\right)^{1/4}\left(\frac{\mathrm{d}\theta}{\mathrm{d}\eta}\right)_{\eta=0}. \tag{4.48}$$

The local heat transfer coefficient α_x, defined as $q_x = \alpha_x(T_\mathrm{w} - T_\infty)$, will be given by

$$\alpha_x = -\lambda_\mathrm{w}\left(\frac{1}{4}Gr_{x,\infty}\right)^{1/4} x^{-1}\left(\frac{\mathrm{d}\theta}{\mathrm{d}\eta}\right)_{\eta=0}. \tag{4.49}$$

The average heat transfer coefficient $\overline{\alpha_x}$ defined as $Q_x = \overline{\alpha_x}(T_\mathrm{w}-T_\infty)\times b\times x$ is expressed as

$$\overline{\alpha_x} = -\frac{4}{3}\lambda_\mathrm{w}\left(\frac{1}{4}Gr_{x,\infty}\right)^{1/4} x^{-1}\left(\frac{\mathrm{d}\theta}{\mathrm{d}\eta}\right)_{\eta=0}. \tag{4.50}$$

The local Nusselt number defined by $Nu_{x,\mathrm{w}} = \alpha_x x/\lambda_\mathrm{w}$ will be

$$\begin{aligned} Nu_{x,\mathrm{w}} &= -\lambda_\mathrm{w}\left(\frac{1}{4}Gr_{x,\infty}\right)^{1/4} x^{-1}\left(\frac{\mathrm{d}\theta}{\mathrm{d}\eta}\right)_{\eta=0}\frac{x}{\lambda_\mathrm{w}} \\ &= -\left(\frac{1}{4}Gr_{x,\infty}\right)^{1/4}\left(\frac{\mathrm{d}\theta}{\mathrm{d}\eta}\right)_{\eta=0}, \end{aligned} \tag{4.51}$$

or expresses as

$$\frac{Nu_{x,\mathrm{w}}}{\left(\frac{1}{4}Gr_{x,\infty}\right)^{1/4}} = -\left(\frac{\mathrm{d}\theta}{\mathrm{d}\eta}\right)_{\eta=0}. \tag{4.52}$$

The average Nusselt number is defined as $\overline{Nu_{x,\mathrm{w}}} = \overline{\alpha_x}x/\lambda_\mathrm{w}$, and hence

$$\begin{aligned} \overline{Nu_{x,\mathrm{w}}} &= -\frac{4}{3}\lambda_\mathrm{w}\left(\frac{1}{4}Gr_{x,\infty}\right)^{1/4} x^{-1}\left(\frac{\mathrm{d}\theta}{\mathrm{d}\eta}\right)_{\eta=0}\frac{x}{\lambda_\mathrm{w}} \\ &= -\frac{4}{3}\left(\frac{1}{4}Gr_{x,\infty}\right)^{1/4}\left(\frac{\mathrm{d}\theta}{\mathrm{d}\eta}\right)_{\eta=0}, \end{aligned}$$

or expressed as

$$\frac{\overline{Nu_{x,\mathrm{w}}}}{\left(\frac{1}{4}Gr_{x,\infty}\right)^{1/4}} = -\frac{4}{3}\left(\frac{\mathrm{d}\theta}{\mathrm{d}\eta}\right)_{\eta=0}. \tag{4.53}$$

It is seen that, for practical calculation of heat transfer, only $(\mathrm{d}\theta/\mathrm{d}\eta)_{\eta=0}$ dependent on numerical solution is no-given variable.

4.6 Numerical Results

It is obvious that the velocity and temperature fields can be obtained from the solution of the governing ordinary differential equations (4.18), (4.25), and (4.33) with boundary conditions, (4.34) and (4.35), combined with the property factor (4.43)–(4.46). It is expected that, for the case of constant properties, the dimensionless velocity field w_x and dimensionless temperature field θ will be functions of Pr only. But for the case of variable properties, both the dimensionless velocity field and the dimensionless temperature field will depend not only on Pr but also on the temperature parameters n_μ and n_λ, and the boundary temperature ratio T_w/T_∞.

The nonlinear two-point boundary value problem defined by equations (4.18), (4.25), and (4.33) were solved, and calculations were carried out numerically by using a shooting method. First, (4.18), (4.25), and (4.33) were written as a system of five first-order differential equations, which were solved by means of fifth-order Runge–Kutta iteration.

The Runge–Kutta integration scheme, along with Newton–Raphson shooting method is one of the most commonly used techniques for the solution of such two-point boundary value problem. Even though this method provides satisfactory result for such type of problems, it may fail when applied to problems in which the differential equations are very sensitive to the choice of the missing initial conditions.

Moreover, another serious difficulty which may be encountered in the boundary-value problems is in-linear instability. Difficulty also arises in the case in which one end of the range of integration is at infinity. The end-point of integration is usually approximated by assigning a finite value to this point, and by estimating a value at this point the solution will reach its asymptotic state. The computing time for integrating differential equations sometimes depends critically on the quality of the initial guesses of the unknown boundary conditions and the initial end-point.

Then, a Newton iteration procedure was employed to satisfy the outer boundary equations, (4.35). The present fifth-order scheme utilizes variable grid spacing. The typical results for the velocity and temperature field together with different boundary temperature ratios T_w/T_∞ are plotted in Figs. 4.7–4.10 for comparison of velocity and temperature profiles with different gases and different boundary temperature ratios T_w/T_∞, respectively.

It is found that both the velocity and temperature fields of argon laminar free convection are higher than those of oxygen laminar free convection, while, both the velocity fields of oxygen free convection are higher than those of water vapor laminar free convection. It follows that with increasing the temperature parameters n_μ and n_λ, the level both of the velocity and temperature fields of free convection will decrease.

Additionally, it is seen that with increasing the boundary temperature ratio T_w/T_∞, the temperature field will raise and the maximum of velocity field will increase and shift far from the plate.

Furthermore, from the numerical calculations it is found that, even for the diatomic gases, air, and water vapor, the modifications using n_μ and n_λ replaced by $n_{\mu\lambda}$ are unnecessary, because the numerical results obtained either with the actual n_μ and n_λ values or with the modified $n_{\mu\lambda}$ values are almost the same.

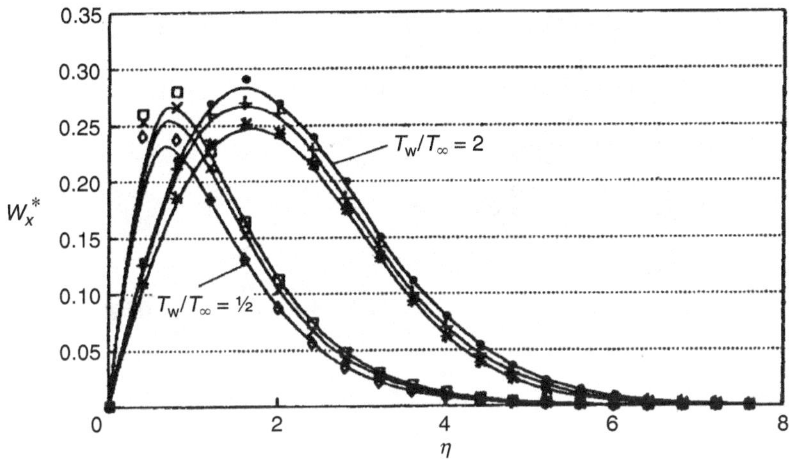

Fig. 4.7. Comparison of velocity profiles for free convection of different gases, cited from Shang and Wang [6] • − •, □ − □, *Ar* ($Pr = 0.622$, $n_\lambda \approx n_\mu \approx n_{\mu\lambda}$), + − + − ×, O_2 ($Pr = 0.733$, $n_{\mu\lambda} = 0.79$); ∗ − ∗, ◇ − ◇, water vapor ($Pr = 1$, $n_{\mu\lambda} = 1.12$)

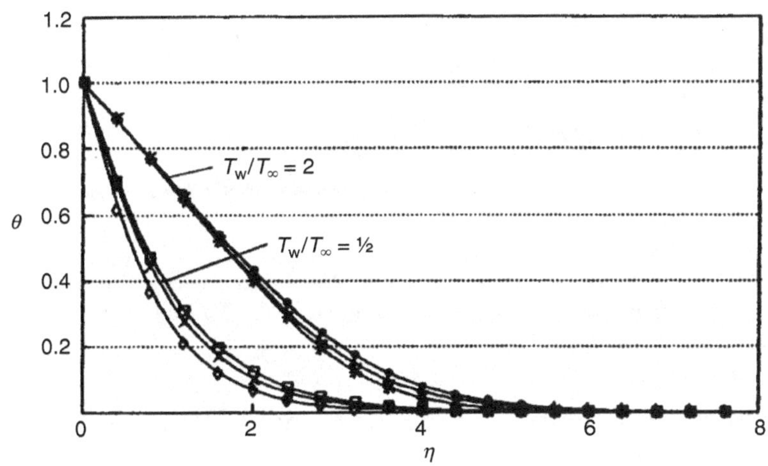

Fig. 4.8. Comparison of temperature profiles for free convection of different gases, cited from Shang and Wang [6] • − •, □ − □, *Ar* ($Pr = 0.622$, $n_\lambda \approx n_\mu \approx n_{\mu\lambda}$), + − +, × − ×, O_2 ($Pr = 0.733$, $n_{\mu\lambda} = 0.79$); ∗ − ∗, ◇ − ◇, water vapor ($Pr = 1$, $n_{\mu\lambda} = 1.12$)

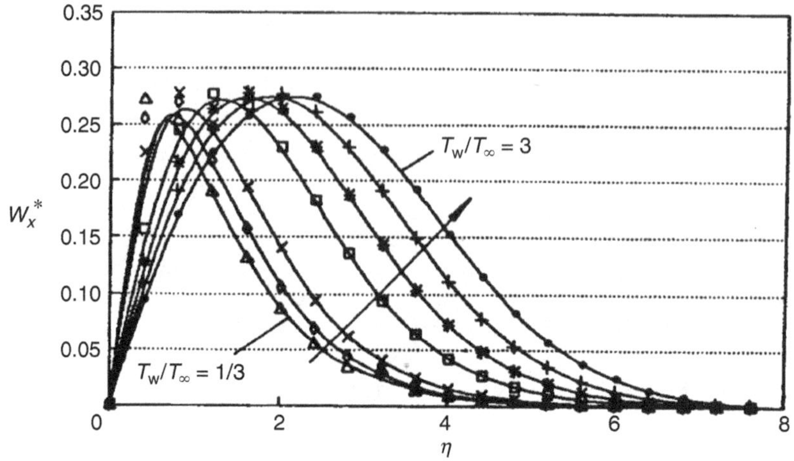

Fig. 4.9. Comparison of velocity profiles for free convection of air ($Pr = 0.7$, $n_{\mu\lambda} = 0.79$) with different T_w/T_∞, cited from Shang and Wang [6]

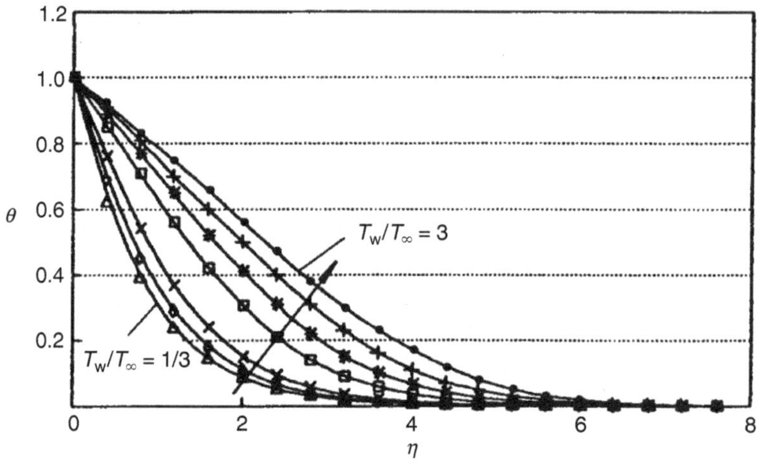

Fig. 4.10. Comparison of temperature profiles for free convection of air ($Pr = 0.7$, $n_{\mu\lambda} = 0.79$) with different temperature ratio T_w/T_∞, cited from Shang and Wang [6]

4.7 Effect of Variable Thermophysical Properties on Heat Transfer

From the heat transfer analysis we find that the dimensionless temperature gradient $(\mathrm{d}\theta/\mathrm{d}\eta)_{\eta=0}$ is only one ungiven variable which depends on numerical solution for prediction of heat transfer coefficient.

The numerical solutions $(\mathrm{d}\theta/\mathrm{d}\eta)_{\eta=0}$ of the governing equations (4.18), (4.25), and (4.33) for some monatomic and diatomic gases, air, and water vapor are obtained. Some solutions $(\mathrm{d}\theta/\mathrm{d}\eta)_{\eta=0}$ are listed in Table 4.2 and plotted in Fig. 4.11 for laminar free convection of different gases. By using

Table 4.2. Calculated results of $(-\mathrm{d}\theta/\mathrm{d}\eta)_{\eta=0}$ obtained from A, numerical solution, B, (4.54) with (4.55) and (4.56), and C (4.54) with (4.55) and (4.57), cited from Shang and Wang [6]

T_w/T_∞		Ar	H_2	Air	N_2	CO	O_2	water vapor
		Pr =0.622	Pr =0.68	Pr =0.7	Pr =0.71	Pr =0.72	Pr =0.733	Pr =1
		n_μ =0.72	n_μ =0.68	n_μ =0.68	n_μ =0.67	n_μ =0.71	n_μ =0.694	n_μ =1.04
		n_λ =0.73	n_λ =0.8	n_λ =0.81	n_λ =0.76	n_λ =0.83	n_λ =0.86	n_λ =1.185
		$n_{\mu\lambda}$ = 0.7255	$n_{\mu\lambda}$ = 0.746	$n_{\mu\lambda}$ = 0.7515	$n_{\mu\lambda}$ = 0.72	$n_{\mu\lambda}$ = 0.776	$n_{\mu\lambda}$ = 0.785	$n_{\mu\lambda}$ = 1.12
3	A	0.1940	0.1974	0.1987	0.2043	0.1973	0.1973	0.1738
	B	0.1935	0.1975	0.1988	0.2044	0.1975	0.1975	0.1738
5/2	A	0.2256	0.2300	0.2316	0.2374	0.2306	0.2307	0.2110
	B	0.2249	0.2300	0.2318	0.2374	0.2308	0.2311	0.2115
2	A	0.2714	0.2772	0.2794	0.2852	0.2792	0.2796	0.2670
	B	0.2703	0.2772	0.2796	0.2850	0.2794	0.2801	0.2689
3/2	A	0.3438	0.3526	0.3557	0.3609	0.3570	0.3582	0.3651
	B	0.3427	0.3527	0.3561	0.3609	0.3575	0.3590	0.3665
5/4	A	0.3990	0.4105	0.4144	0.4188	0.4172	0.4193	0.4448
	B	0.3983	0.4109	0.4151	0.4191	0.4179	0.4201	0.4459
$\to 1$	A	0.4784	0.4943	0.4995	0.5021	0.5046	0.5079	0.5671
	B	0.4787	0.4953	0.5007	0.5033	0.5059	0.5092	0.5670
3/4	A	0.6035	0.6276	0.6351	0.6336	0.6446	0.6507	0.7775
	C	0.6011	0.6247	0.6333	0.6312	0.6423	0.6479	0.7761
1/2	A	0.8344	0.8774	0.8898	0.8776	0.9093	0.9225	1.2181
	C	0.8285	0.8666	0.8786	0.8684	0.8993	0.9098	1.2081
1/3	A	1.1492	1.2247	1.2448	1.2124	1.2812	1.3075	1.9198
	C	1.1419	1.2022	1.2209	1.1949	1.2591	1.2774	1.8805

curve-fitting method, Shang and Wang [6] obtained the following formulated equations for simple and reliable prediction of the values $(\mathrm{d}\theta/\mathrm{d}\eta)_{\eta=0}$ for laminar free convection of monatomic and diatomic gases, air, and water vapor:

$$-\left(\frac{\mathrm{d}\theta}{\mathrm{d}\eta}\right)_{\eta=0} = \psi(Pr)\left(\frac{T_w}{T_\infty}\right)^{-m}, \tag{4.54}$$

where

$$\psi(Pr) = 0.567 + 0.186 \times \ln(Pr) \qquad (0.6 \le Pr \le 1), \tag{4.55}$$

$$m = 0.64n_{\mu\lambda} + 0.36 = 0.35n_\lambda + 0.29n_\mu + 0.36 \qquad (T_w/T_\infty > 1), \tag{4.56}$$

$$m = 0.76n_{\mu\lambda} + 0.24 = 0.42n_\lambda + 0.34n_\mu + 0.24 \qquad (T_w/T_\infty < 1). \tag{4.57}$$

The predicted results, $(-\mathrm{d}\theta/\mathrm{d}\eta)_{\eta=0}$, of (4.54) are compared with those of the numerical results from (4.18), (4.25), and (4.33), as shown in Table 4.2. The agreement is quite good.

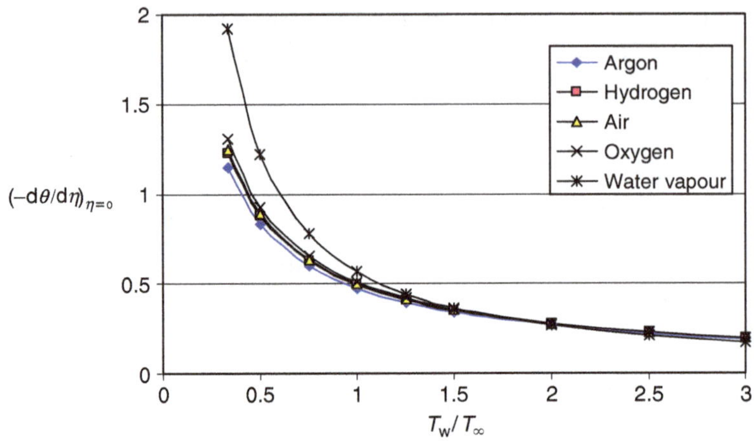

Fig. 4.11. Numerical solutions of temperature gradient $-\left(\frac{d\theta}{d\eta}\right)_{\eta=0}$ for laminar free convection of argon, hydrogen, air, oxygen, and water vapor

In addition, (4.54) with (4.55)–(4.57) shows clearly the following effects of variable thermophysical properties on heat transfer of gas laminar free convection: For same boundary temperature ratio T_w/T_∞, the temperature gradient $(-d\theta/d\eta)_{\eta=0}$ will increase with increasing Pr. For $T_w/T_\infty < 1$ the temperature gradient $(-d\theta/d\eta)_{\eta=0}$ will increase with increasing the value n_μ or n_λ. While, for $T_w/T_\infty > 1$ the temperature gradient $(d\theta/d\eta)_{\eta=0}$ will decrease with increasing the value n_μ or n_λ. With increasing the temperature boundary ratio T_w/T_∞, the temperature gradient $(-d\theta/d\eta)_{\eta=0}$ will decrease. The effect of Pr, n_μ, n_λ, and T_w/T_∞ on temperature gradient $-\left(\frac{d\theta}{d\eta}\right)_{\eta=0}$ can be briefly summarized in Table 4.3 for laminar free convection of monatomic and diatomic gases, air, and water vapor.

Only when boundary temperature ratio T_w/T_∞ is very close to unity, the free convection is corresponding to Boussinesq approximation, and then the temperature gradient $\left(-\frac{d\theta}{d\eta}\right)_{\eta=0}$ only depends on Pr, i.e.,

$$-\left(\frac{d\theta}{d\eta}\right)_{\eta=0} = \psi(Pr), \tag{4.58}$$

where $\psi(Pr)$ expresses the well-known Boussinesq solution.

4.8 Summary

So far, the governing equations for laminar free convection of monatomic and diatomic gases, air, and water vapor and expressions related to heat transfer can be summarized in Table 4.4 with consideration of variable thermophysical properties.

Table 4.3. Effects of Pr, n_μ, n_λ, and T_w/T_∞ on temperature gradient $-\left(\frac{d\theta}{d\eta}\right)_{\eta=0}$ for laminar free convection of monatomic and diatomic gases, air, and water vapor

term	$-\left(\frac{d\theta}{d\eta}\right)_{\eta=0}$	
	for $T_w - T_\infty > 1$	for $T_w/T_\infty < 1$
for effect of Pr	$-\left(\frac{d\theta}{d\eta}\right)_{\eta=0}$ increases with increase of Pr	$-\left(\frac{d\theta}{d\eta}\right)_{\eta=0}$ increases with increase of Pr
for effect of n_λ	$-\left(\frac{d\theta}{d\eta}\right)_{\eta=0}$ decreases with increasing n_λ	$\left(\frac{d\theta}{d\eta}\right)_{\eta=0}$ increases with increasing n_λ
for effect of n_μ	$-\left(\frac{d\theta}{d\eta}\right)_{\eta=0}$ decreases with increasing n_μ	$-\left(\frac{d\theta}{d\eta}\right)_{\eta=0}$ increases with increasing n_μ
for effect of T_w/T_∞	$-\left(\frac{d\theta}{d\eta}\right)_{\eta=0}$ decrease with increasing T_w/T_∞	$-\left(\frac{d\theta}{d\eta}\right)_{\eta=0}$ decrease with increasing T_w/T_∞

4.9 Remarks

The following points can be concluded from the earlier presentation:

In this chapter a novel system of analysis and transformation models is introduced for transformation of governing equations of laminar free convection, instead of traditional Falkner–Skan type transformation. Meanwhile, velocity component method is induced to directly transform velocity components w_x and w_y into the corresponding dimensionless velocity components W_x and W_y. However, for Falkner–Skan transformations, the flow function ψ has to be induced, and at last, the transformed governing ordinary equations are dominated by the f function. Compared with the f function, the dimensionless velocity components W_x and W_y have definite physical meanings. In the later chapters you can find that the dimensionless velocity components W_x and W_y also provide a convenience for investigation of mass transfer at the interface for film boiling or condensation. In addition, the novel analysis and transformation models also provide a convenience to treat the variable thermophysical property problem for free convection and film flows.

An advanced method, viz. the temperature parameter method is introduced for treatment of variable thermophysical properties of gases. With this method a simple power law is applied to treat the variations of density, viscosity, and thermal conductivity of gases with temperature. The temperature parameters are used to express the extent of the property variations with temperature. A series of values of the temperature parameters n_μ and n_λ are obtained from the typical experimental results. For monatomic and diatomic gases, air, and water vapor the value of n_μ varies from 0.64 to 1.04, while the value of n_λ varies from 0.71 to 1.185. It follows that with the temperature parameter method the variations of μ and λ of gases with temperature can be described more accurately than that with the traditional point of view summarized in [12] in which the variations of μ and λ values of gases with temperature are usually described by about 0.8 power of the absolute temperature.

Table 4.4. Governing equations for laminar free convection of monatomic and diatomic gases, air, and water vapor and expressions related to heat transfer

term	expression
	governing partial differential equations
mass equation	$\frac{\partial}{\partial x}(\rho w_x) + \frac{\partial}{\partial y}(\rho w_y) = 0$
momentum equation	$\rho(w_x \frac{\partial w_x}{\partial x} + w_y \frac{\partial w_x}{\partial y}) = \frac{\partial}{\partial y}(\mu \frac{\partial w_x}{\partial y}) + \rho g \left\lvert \frac{T-T_\infty}{T_\infty} \right\rvert$
energy equation	$\rho c_p(w_x \frac{\partial T}{\partial x} + w_y \frac{\partial T}{\partial y}) = \frac{\partial}{\partial y}(\lambda \frac{\partial T}{\partial y})$
boundary conditions	$y = 0$: $w_x = 0$, $w_y = 0$, $T = T_{\mathrm{w}}$ $y \to \infty$: $w_x \to 0$, $T = T_\infty$
	assumed similarity variables
η	$\frac{y}{x}(\frac{1}{4}Gr_{x,\infty})^{1/4}$
$Gr_{x,\infty}$	$(Gr_{x,\infty})_v = \frac{g\lvert T_{\mathrm{w}}/T_\infty - 1\rvert x^3}{\nu_\infty^2}$
θ	$\theta = \frac{T-T_\infty}{T_{\mathrm{w}}-T_\infty}$
W_x	$(2\sqrt{gx}\,\lvert T_{\mathrm{w}}/T_\infty - 1\rvert^{1/2})^{-1} w_x$
W_y	$(2\sqrt{gx} T_{\mathrm{w}}/T_\infty - 1^{1/2}(\frac{1}{4}Gr_{x,\infty})^{-1/4})^{-1} w_y$
	governing ordinary differential equations
mass equations	$2W_x - \eta \frac{\mathrm{d}W_x}{\mathrm{d}\eta} + 4\frac{\mathrm{d}W_y}{\mathrm{d}\eta} - \frac{1}{\rho}\frac{\mathrm{d}\rho}{\mathrm{d}\eta}(\eta W_x - 4W_y) = 0$
momentum equation	$\frac{\nu_\infty}{\nu}(W_x(2W_x - \eta \frac{\mathrm{d}W_x}{\mathrm{d}\eta}) + 4W_y \frac{\mathrm{d}W_x}{\mathrm{d}\eta}) =$ $\frac{\mathrm{d}^2 W_x}{\mathrm{d}\eta^2} + \frac{1}{\mu}\frac{\mathrm{d}\mu}{\mathrm{d}\eta}\frac{\mathrm{d}W_x}{\mathrm{d}\eta} + \frac{\nu_\infty}{\nu}\theta$
energy equation	$Pr\frac{\nu_\infty}{\nu}(-\eta W_x + 4W_y)\frac{\mathrm{d}\theta}{\mathrm{d}\eta} = \frac{1}{\lambda}\frac{\mathrm{d}\lambda}{\mathrm{d}\eta}\frac{\mathrm{d}\theta}{\mathrm{d}\eta} + \frac{\mathrm{d}^2\theta}{\mathrm{d}\eta^2}$
boundary conditions	$\eta = 0:$ $W_x = 0,$ $W_y = 0,$ $\theta = 0;$ $\eta \to 0:$ $W_x = 0,$ $\theta = 0$
	equations related to heat transfer
q_x (defined as $-\lambda_{\mathrm{w}}(\frac{\partial T}{\partial y})_{y=0}$)	$-\lambda_{\mathrm{w}}(T_{\mathrm{w}} - T_\infty)(\frac{1}{4}Gr_{x,\infty})^{1/4} x^{-1}(\frac{\mathrm{d}\theta}{\mathrm{d}\eta})_{\eta=0}$
α_x (defined as $\frac{q_x}{(T_{\mathrm{w}}-T_\infty)}$)	$-\lambda_{\mathrm{w}}(\frac{1}{4}Gr_{x,\infty})^{1/4} x^{-1}(\frac{\mathrm{d}\theta}{\mathrm{d}\eta})_{\eta=0}$
Q_x (defined as $\int_0^x q_x b\mathrm{d}x$)	$-\frac{4}{3}b\lambda_{\mathrm{w}}(T_{\mathrm{w}} - T_\infty)(\frac{1}{4}Gr_{x,\infty})^{1/4}(\frac{\mathrm{d}\theta}{\mathrm{d}\eta})_{\eta=0}$
$\overline{\alpha}_x$ (defined as $\frac{Q_x}{(T_{\mathrm{w}}-T_\infty)bx}$	$-\frac{4}{3}\lambda_{\mathrm{w}}(\frac{1}{4}Gr_{x,\infty})^{1/4} x^{-1}(\frac{\mathrm{d}\theta}{\mathrm{d}\eta})_{\eta=0}$
$Nu_{x,w}$ (defined as $\frac{\alpha_x x}{\lambda_{\mathrm{w}}}$)	$\frac{Nu_{x,w}}{(\frac{1}{4}Gr_{x,\infty})^{1/4}} = -(\frac{\mathrm{d}\theta}{\mathrm{d}\eta})_{\eta=0}$
$\overline{Nu}_{x,w}$ (defined as $\frac{\overline{\alpha}_x x}{\lambda_{\mathrm{w}}}$)	$\frac{\overline{Nu}_{x,\mathrm{w}}}{(\frac{1}{4}Gr_{x,\infty})^{1/4}} = -\frac{4}{3}\left(\frac{\mathrm{d}\theta}{\mathrm{d}\eta}\right)_{\eta=0}$
$-(\frac{\mathrm{d}\theta}{\mathrm{d}\eta})_{\eta=0}$	$\psi(Pr)(\frac{T_{\mathrm{w}}}{T_\infty})^{-m}$ $\psi(Pr)$ is Boussinesq solution
$\psi(Pr)$	$0.567 + 0.186 \times \ln(Pr)$ $(0.6 \le Pr \le 1)$
m	$0.35n_\lambda + 0.29n_\mu + 0.36$ for $T_{\mathrm{w}}/T_\infty > 1$ $0.42n_\lambda + 0.34n_\mu + 0.24$ for $T_{\mathrm{w}}/T_\infty < 1$

Furthermore, with the temperature parameter method the treatment of variable thermophysical properties becomes very simple and convenient.

The method proposed in this chapter for analyzing the laminar free convection of monatomic and diatomic gases, air, and water vapor along vertical isothermal flat plates with considerations of variable thermophysical properties can yield reliable results.

The analysis presented here extends the former ones reported in literatures such as in [3], [4], and [13] for gas laminar free convection.

The well-known relation, (4.58), holds true only for the case $T_w/T_\infty \to 1$ such that the Boussinesq approximation applies.

4.10 Calculation Example

Question: A flat plate with $b = 2$ m in width and $x = 0.25$ m in length is suspended vertically in air. The ambient temperature is $t_\infty = 20°$C. Calculate the free convection heat transfer of the plate for boundary temperature ratio $T_w/T_\infty = 1.1, 1.2, 1.4, 1.7$, and 2.1.

Solution. From $t_\infty = 20°$C and $T_w/T_\infty = 1.1, 1.2, 1.4, 1.7, 2.1$, we obtain $T_w = 322.3, 351.6, 410.2, 498.1, 615.3$ K or $T_w = 49.3, 78.6, 137.2, 225.1, 342.3°$C. The air physical properties are as follows: $\nu_\infty = 15.06 \times 10^{-6}$ m^2 s^{-1} for air $t_\infty = 20°$C; $\lambda_w = 2.825 \times 10^{-2}, 3.037 \times 10^{-2}, 3.649 \times 10^{-2}, 4.86 \times 10^{-2}$, and 4.8622×10^{-2} W (m°C)$^{-1}$ for air at $t_w = 49.3, 78.6, 137.2, 225.1$, and $342.3°$C, respectively. From Tables 4.1 and 4.3, we obtain $n_\mu = 0.68, n_\lambda = 0.81$ and $Pr = 0.7$ for air.

Then,

$$\psi(Pr) = 0.567 + 0.186 \times \ln\ 0.7 = 0.50066,$$

m is evaluated as later with $T_w/T_\infty > 1$

$$\begin{aligned} m &= 0.35n_\lambda + 0.29n_\mu + 0.36 \\ &= 0.35 \times 0.81 + 0.29 \times 0.68 + 0.36 = 0.8407. \end{aligned}$$

In this case, the dimensionless temperature gradient $\left(\frac{d\theta}{d\eta}\right)_{\eta=0}$ can be evaluated as

$$\begin{aligned} -\left(\frac{d\theta}{d\eta}\right)_{\eta=0} &= \psi(Pr)\left(\frac{T_w}{T_\infty}\right)^{-m} \\ &= 0.50066 \times \left(\frac{T_w}{T_\infty}\right)^{-0.8407}. \end{aligned}$$

The evaluated values of $-\left(\frac{d\theta}{d\eta}\right)_{\eta=0}$ are plotted in Table 4.5 for different temperature rations.

Table 4.5. Calculated results

$(T_{\mathrm{w}}/T_\infty)$	1.1	1.2	1.4	1.7	2.1
$t_{\mathrm{w}}(^\circ\mathrm{C})$	49.3	78.6	137.2	225.1	342.3
$t_\infty(^\circ\mathrm{C})$	20	20	20	20	20
$\lambda_{\mathrm{w}}(\mathrm{W\ (m\ ^\circ C)^{-1}})$	2.825×10^{-2}	3.037×10^{-2}	3.4675×10^{-2}	4.1007×10^{-2}	4.8622×10^{-2}
$-\left(\frac{\mathrm{d}\theta}{\mathrm{d}\eta}\right)_{\eta=0}$	0.4621	0.4295	0.3773	0.3205	0.2683
$Gr_{x,\infty}$	6.75×10^{7}	1.35×10^{8}	2.7×10^{8}	4.73×10^{8}	7.43×10^{8}
$\overline{Nu_{x,\mathrm{w}}}$	39.4927	43.652	45.602	44.550	41.761
$\overline{\alpha_x}$ $(\mathrm{W\ (m^2\ K)^{-1}})$	4.46237	5.3026	6.3247	7.3074	8.1218
Q (W)	65.3782	154.37	370.63	749.38	1308.83

Also

$$Gr_{x,\infty} = \frac{g\,|T_{\mathrm{w}}/T_\infty - 1|\,x^3}{\nu_\infty^2} = \frac{9.8\times|T_{\mathrm{w}}/T_\infty - 1|\times 0.25^3}{(15.06\times10^{-6})^2}.$$

The calculated values of $Gr_{x,\infty}$ are plotted in Table 4.5.

With (4.53) the average Nusselt number can be expressed as

$$\overline{Nu_{x,\mathrm{w}}} = -\frac{4}{3}\left(\frac{1}{4}Gr_{x,\infty}\right)^{1/4}\left(\frac{\mathrm{d}\theta}{\mathrm{d}\eta}\right)_{\eta=0}.$$

Then, the average Nusselt number $\overline{Nu_{x,\mathrm{w}}}$ are evaluated with the calculated values of $Gr_{x,\infty}$ and $-(\mathrm{d}\theta/\mathrm{d}\eta)_{\eta=0}$, and then, plotted in Table 4.5.

From the definition of the average Nusselt number $\overline{Nu_{x,\mathrm{w}}} = \overline{\alpha_x}x/\lambda_{\mathrm{w}}$, the average heat transfer coefficient can be calculated as

$$\overline{\alpha_x} = \frac{\lambda_{\mathrm{w}}}{x}\overline{Nu_{x,\mathrm{w}}} = \frac{\lambda_{\mathrm{w}}}{0.25}\overline{Nu_{x,\mathrm{w}}}$$

The average heat transfer coefficients $\overline{\alpha_x}$ are calculated with the related $\overline{Nu_{x,\mathrm{w}}}$ and λ_{w}, and plotted in Table 4.5 also.

Finally, heat transfer Q_x is calculated as

$$Q_x = \overline{\alpha_x}(T_{\mathrm{w}} - T_\infty)x\times b = \overline{\alpha_x}(T_{\mathrm{w}} - T_\infty)\times 0.25\times 2.$$

The values of Q_x is calculated with the related values of $\overline{\alpha_x}$ and $t_{\mathrm{w}} - t_\infty$, and plotted in Table 4.5.

References

1. T. Hara, The free-convection flow about a heated vertical plate in air, Trans. Jpn Soc. Mech. Eng 20, pp. 517–520, 1954

2. A.A. Tataev, Heat exchange in condition of free laminar movement of gas with variable viscosity at a vertical wall, Zh. Tekh. Fiz. 26, pp. 2714–2719, 1956
3. E.M. Sparrow and J.L. Gregg, The variable fluid-property problem in free convection, Trans. ASME 80, pp. 879–886, 1958
4. D.D. Gray and A. Giogini, The validity of the Boussinesq approximation for liquids and gases, Int. J. Heat Mass Transfer 19, pp. 546–577, 1977
5. A.M. Clausing and S.N. Kempka, The influences of property variations on natural convection from vertical surfaces, J. Heat Transfer 103, pp. 609–612, 1981
6. D.Y. Shang and B.X. Wang, Effect of variable thermophysical properties on laminar free convection of gas, Int. J. Heat Mass Transfer 33, No. 7, pp. 1387–1395, 1990
7. D.Y. Shang and B.X. Wang, Effect of variable thermophysical properties on laminar free convection of polyatomic gas, Int. J. Heat Mass Transfer 34, No. 3, pp. 749–755, 1991
8. E.R.G. Eckert and R.M. Drake, Analysis of Heat and Mass Transfer, McGraw-Hill, New York, 1972
9. A.G. Hansen, Similarity Analysis of Boundary Value Problems in Engineering, Prentice-Hall, Englewood Cliffs, NJ, 1964
10. T.Y. Na, Computational Methods in Engineering Boundary Value Problems, Academic, New York, 1979
11. J. Hisenrath et al., Tables of Thermodynamic and Transport Properties. National Bureau of Standards, Washington, 1955
12. S. Kakac et al., Natural Convection: Fundamentals and Applications, Hemisphere Publishing Corporation, pp. 729, 1985
13. A. Brown, The effect on laminar free convection heat transfer of temperature dependence of the coefficient of volumetric expansion, Trans. ASME, Ser. C, J. Heat Transfer 97, pp. 133–135, 1975

5

Laminar Free Convection of Polyatomic Gas

Nomenclature

b	width of plate, m
c_p	specific heat at constant pressure, J (kg K)$^{-1}$
$Gr_{x,\infty}$	local Grashof number for the free convection of gas on isothermal vertical flat plate, $Gr_{x,\infty} = g\,\lvert T_w/T_\infty - 1\rvert\, x^3/\nu_\infty^2$
$Nu_{x,w}$	local Nusselt number, $\alpha_x x/\lambda_w$
$\overline{Nu}_{x,w}$	average Nusselt number, $\overline{\alpha}_x x/\lambda_w$
n_{c_p}	specific heat parameter of gas
n_λ	thermal conductivity parameter of gas
n_μ	viscosity parameter of gas
$n_{\mu\lambda}$	overall temperature parameter of gas
Pr	Prandtl number
q_x	local heat transfer rate at position x per unit area on plate, W m^{-2}
Q_x	total heat transfer rate for position $x = 0$ to x with width of b on plate, W
t	temperature, °C
T	absolute temperature, K
T_w/T_∞	boundary temperature ratio
w_x, w_y	velocity components in the x- and y-directions, respectively, m s^{-1}
W_x, W_y	dimensionless velocity components in the x- and y-directions, respectively

Greek symbols

α_x	local heat transfer coefficient, W (m^2 K)$^{-1}$
$\overline{\alpha}_x$	average heat transfer coefficient, W (m^2 K)$^{-1}$

δ	boundary layer thickness, m
η	dimensionless coordinate variable for boundary layer
θ	dimensionless temperature, $\theta = (t - t_\infty)/(t_\mathrm{w} - t_\infty)$
λ	thermal conductivity, W $(\mathrm{m\ K})^{-1}$
μ	absolute viscosity, kg $(\mathrm{m\ s})^{-1}$
ν	kinetic viscosity, $\mathrm{m}^2\ \mathrm{s}^{-1}$
ρ	density, kg m^{-3}
$\left(\frac{\mathrm{d}\theta}{\mathrm{d}\eta}\right)_{\eta=0}$	dimensionless temperature gradient on the plate
$\psi(Pr)$	Boussinesq solution
$\frac{1}{\rho}\frac{\mathrm{d}p}{\mathrm{d}x}$	density factor
$\frac{1}{\mu}\frac{\mathrm{d}\mu}{\mathrm{d}\eta}$	viscosity factor
$\frac{1}{\lambda}\frac{\mathrm{d}\lambda}{\mathrm{d}\eta}$	thermal conductivity factor

Subscripts

w	at wall
δ	boundary layer
∞	far from the wall surface
sub	subcooling state

5.1 Introduction

In Chap. 4 the velocity component method and temperature parameter method were presented for the similarity transformation of the governing partial differential equations of laminar free convection of monatomic and diatomic gases, air and water vapour with consideration of variable thermophysical properties. Meanwhile, density, thermal conductivity and absolute viscosity of the gases are assumed to vary with absolute temperature according to a simple power law. The temperature parameters of thermal conductivity and the absolute viscosity are proposed and the corresponding values are provided according to the typical experimental data of the corresponding thermophysical properties. The density is taken as inversely proportional to absolute temperature at constant pressure, while the Prandtl number is assumed constant. Since the variation of specific heat for these gases with temperature is very small, it is taken as constant for the treatment of variable thermophysical properties.

However, for polyatomic gases, the variation of specific heat is not so small, and then it cannot be taken as constant. In this chapter, I focus on the presentation of free convection of polyatomic gases along an isothermal vertical flat plate with large temperature difference [1]. For this purpose the governing equations for laminar free convection of a gas are also transformed to dimensionless ordinary equations by the velocity component method. For treatment of variable thermophysical properties the temperature parameter method is used to further treat variation of specific heat with temperature. Not only the density, thermal conductivity, and dynamic viscosity but also specific heat is assumed to vary with absolute temperature according to the simple power law. The temperature parameters n_λ, n_μ and n_{c_p} are further introduced and the corresponding values are proposed according to the typical experimental results for polyatomic gases. The density is taken as inversely proportional to absolute temperature at constant pressure, while the Prandtl number is assumed constant. The numerical calculation results for the free convection of a series of polyatomic gases with various values of boundary temperature ratio T_w/T_∞ are presented, and the formulated correlations for heat transfer are derived which agree closely with the numerical solutions.

5.2 Variable Thermophysical Properties

The effect of the variable thermophysical properties on laminar free convection and heat transfer of monatomic and diatomic gases, air and water vapour along an isothermal vertical flat plate has been reported in Chap 4. However, for the polyatomic gases the variation of specific heat with temperature is more obvious. Hence, it is necessary to consider the effect of the variable specific heat of polyatomic gases on the laminar free convection. Employing the temperature parameter method proposed in Chap. 4, the specific heat parameter is taken as n_{c_p} for the polyatomic gas. Thus, for polyatomic gas, the equations of viscosity, thermal conductivity, density and specific heat with temperature are described as follows:

$$\mu/\mu_\infty = (T/T_\infty)^{n_\mu}, \tag{5.1}$$
$$\lambda/\lambda_\infty = (T/T_\infty)^{n_\lambda}, \tag{5.2}$$
$$\rho/\rho_\infty = (T/T_\infty)^{-1}, \tag{5.3}$$
$$\nu/\nu_\infty = (T/T_\infty)^{n_\mu+1}, \tag{5.4}$$
$$c_p/c_{p_\infty} = (T/T_\infty)^{n_{c_p}}. \tag{5.5}$$

According to the summarized experimental values of μ, λ and c_p for several polyatomic gases reported in [2–5], the temperature parameters n_λ, n_μ and n_{c_p} and the deviation of μ, λ and c_p arising from the corresponding experimental data are listed in Table 5.1.

5.3 Governing Partial Differential Equations and their Similarity Transformations

The physical analytical model and coordinate system used for laminar free convection of polyatomic gas on an isothermal vertical flat plate is shown in Fig. 4.1 also. Consulting the governing equations in Chap. 4 we can express the following governing partial differential equations for continuity, momentum and energy conservations and their boundary conditions for laminar free

Table 5.1. The values of temperature parameters n_λ, n_μ and n_{c_p} and the deviations of μ, λ and c_p from experimental data, cited from Shang and Wang [1]

gas	T_0 (K)	n_μ	temperature range (K)	deviation (%)	n_λ	temperature range (K)	deviation (%)	n_{c_p}	temperature range (K)	deviation (%)
gas mixture	273	0.75	273–1,173	±3	1.02	273–1,473	±1	0.134	273–1,473	±2.5
CO_2	273	0.88	220–700	±2	1.3	220–720	±2	0.34	230–810	±1
C_2H_4	273	0.853	210–900	±3	1.71	220–600	±3	0.73	273–922	±1.5
CH_4	273	0.78	220–1,000	±3	1.29	230–1,000	±3.6	0.534	273–1,033	±4.2
CCl_4	273	0.912	270–500	±1	1.29	260–400	±1	0.28	270–700	±2
SO_2	273	0.91	200–1,250	±4	1.323	250–900	±4.5	0.257	273–1,200	±3
H_2S	273	1	270–500	±1	1.29	260–400	±1	0.18	230–700	±2
NH_3	273	1.04	220–1,000	±1	1.375	250–900	±2	0.34	230–1,033	±0.36
C_2H_5OH	273	0.925	270–600	±0.6	1.95	260–430	±4	0.59	250–922	±0.25

Data of gas mixture come from [5], and data of CO_2 to C_2H_5 OH from [2–4] Components of the gas mixture: $CO_2 = 0.13$, water vapour $= 0.11$ and $N_2 = 0.76$

convection of polyatomic gases along an isothermal vertical plate:

$$\frac{\partial}{\partial x}(\rho w_x) + \frac{\partial}{\partial y}(\rho w_y) = 0, \tag{5.6}$$

$$\rho\left(w_x\frac{\partial w_x}{\partial x} + w_y\frac{\partial w_x}{\partial y}\right) = \frac{\partial}{\partial y}\left(\mu\frac{\partial w_x}{\partial y}\right) + g\rho\left|\frac{T}{T_\infty} - 1\right| \tag{5.7}$$

$$\rho\left(w_x\frac{\partial(c_pT)}{\partial x} + w_y\frac{\partial(c_pT)}{\partial y}\right) = \frac{\partial}{\partial y}\left(\lambda\frac{\partial T}{\partial y}\right), \tag{5.8}$$

$$y = 0:\ w_x = 0,\ w_y = 0, \qquad T = T_{\mathrm{w}}, \tag{5.9}$$

$$y \to \infty: \qquad w_x \to 0, \quad T = T_\infty, \tag{5.10}$$

where the temperature-dependent specific heat of polyatomic gas is considered in the energy equation.

For similarity transformation of the governing equations (5.6)–(5.8) we use the velocity component method and assume the following dimensionless variables, which are same as those for laminar free convection of monatomic and diatomic gases, air and water vapour in Chap. 4

$$\eta = \frac{x}{y}\left(\frac{1}{4}Gr_{x,\infty}\right)^{1/4}, \tag{5.11}$$

$$Gr_{x,\infty} = \frac{g\left|T_{\mathrm{w}}/T_\infty - 1\right| x^3}{\nu_\infty^2}, \tag{5.12}$$

$$\theta = \frac{T - T_\infty}{T_{\mathrm{w}} - T_\infty}, \tag{5.13}$$

$$W_x = \left[2\sqrt{gx}\left|T_{\mathrm{w}}/T_\infty - 1\right|^{1/2}\right]^{-1} w_x, \tag{5.14}$$

$$W_y = \left[2\sqrt{gx}\left|T_{\mathrm{w}}/T_\infty - 1\right|^{1/2}\left(\frac{1}{4}Gr_{x,\infty}\right)^{-1/4}\right]^{-1} w_y. \tag{5.15}$$

According to the derivation same as that in Chap. 4 the corresponding governing dimensionless equations of (5.6)–(5.7) should be

$$2W_x - \eta\frac{\mathrm{d}W_x}{\mathrm{d}\eta} + 4\frac{\mathrm{d}W_y}{\mathrm{d}\eta} - \frac{1}{\rho}\frac{\mathrm{d}\rho}{\mathrm{d}\eta}(\eta W_x - 4W_y) = 0, \tag{5.16}$$

$$\frac{\nu_\infty}{\nu}\left(W_x\left(2W_x - \eta\frac{\mathrm{d}W_x}{\mathrm{d}\eta}\right) + 4W_y\frac{\mathrm{d}W_x}{\mathrm{d}\eta}\right) = \frac{\mathrm{d}^2W_x}{\mathrm{d}\eta^2} + \frac{1}{\mu}\frac{\mathrm{d}\mu}{\mathrm{d}\eta}\frac{\mathrm{d}W_x}{\mathrm{d}\eta} + \frac{\nu_\infty}{\nu}\theta. \tag{5.17}$$

However, because the variation of specific heat of the polyatomic gases with temperature must be considered, the similarity transformation for (5.8) should be separately done.

At first, (5.8) can be further expressed as

$$\rho c_p\left(w_x\frac{\partial T}{\partial x} + w_y\frac{\partial T}{\partial y}\right) + \rho T\left(w_x\frac{\partial(c_p)}{\partial x} + w_y\frac{\partial(c_p)}{\partial y}\right) = \frac{\partial\lambda}{\partial y}\frac{\partial T}{\partial y} + \lambda\frac{\partial^2 T}{\partial y^2},$$

or

$$\left[\rho c_p w_x \frac{\partial T}{\partial x} + \rho T w_x \frac{\partial (c_p)}{\partial x}\right] + \left[\rho c_p w_y \frac{\partial T}{\partial y} + \rho T w_y \frac{\partial (c_p)}{\partial y}\right] = \frac{\partial \lambda}{\partial y}\frac{\partial T}{\partial y} + \lambda \frac{\partial^2 T}{\partial y^2}. \tag{5.18}$$

According to Chap. 4, we have

$$\frac{\partial \lambda}{\partial y} = \frac{\mathrm{d}\lambda}{\mathrm{d}\eta}\left(\frac{1}{4}Gr_{x,\infty}\right)^{1/4} x^{-1}, \tag{5.19}$$

$$\frac{\partial T}{\partial x} = -(T_{\mathrm{w}} - T_\infty)\frac{\mathrm{d}\theta}{\mathrm{d}\eta}\left(\frac{1}{4}\right)\eta x^{-1}, \tag{5.20}$$

$$\frac{\partial T}{\partial y} = (T_{\mathrm{w}} - T_\infty)\frac{\mathrm{d}\theta}{\mathrm{d}\eta}\left(\frac{1}{4}Gr_{x,\infty}\right)^{1/4} x^{-1}, \tag{5.21}$$

$$\frac{\partial^2 T}{\partial y^2} = (T_{\mathrm{w}} - T_\infty)\frac{\mathrm{d}^2\theta}{\mathrm{d}\eta^2}\left(\frac{1}{4}Gr_{x,\infty}\right)^{1/2} x^{-2}. \tag{5.22}$$

Similar to (4.14) we have

$$\frac{\partial c_p}{\partial x} = -\frac{1}{4}x^{-1}\eta\frac{\mathrm{d}c_p}{\mathrm{d}\eta}. \tag{5.23}$$

Similar to (5.22) we have

$$\frac{\partial c_p}{\partial y} = \frac{\mathrm{d}c_p}{\mathrm{d}\eta}\left(\frac{1}{4}Gr_{x,\infty}\right)^{1/4} x^{-1}, \tag{5.24}$$

where

$$T = (T_{\mathrm{w}} - T_\infty)\theta + T_\infty. \tag{5.25}$$

Then, (5.18) is changed to

$$\begin{aligned}
&2\rho\sqrt{gx}\,|T_{\mathrm{w}}/T_\infty - 1|^{1/2}\, W_x\left[T\left(-\frac{1}{4}\eta x^{-1}\frac{\mathrm{d}c_p}{\mathrm{d}\eta}\right) + c_p\left(-(T_w - T_\infty)\frac{\mathrm{d}\theta}{\mathrm{d}\eta}\left(\frac{1}{4}\right)\eta x^{-1}\right)\right]\\
&+2\rho\sqrt{gx}\,|T_{\mathrm{w}}/T_\infty - 1|^{1/2}\left(\frac{1}{4}Gr_{x,\infty}\right)^{-1/4} W_y\left[T\frac{\mathrm{d}c_p}{\mathrm{d}\eta}\left(\frac{1}{4}Gr_{x,\infty}\right)^{1/4} x^{-1}\right.\\
&\left.+c_p(T_{\mathrm{w}} - T_\infty)\frac{\mathrm{d}\theta}{\mathrm{d}\eta}\left(\frac{1}{4}Gr_{x,\infty}\right)^{1/4} x^{-1}\right]\\
&=\frac{\mathrm{d}\lambda}{\mathrm{d}\eta}\left(\frac{1}{4}Gr_{x,\infty}\right)^{1/4} x^{-1}(T_{\mathrm{w}} - T_\infty)\frac{\mathrm{d}\theta}{\mathrm{d}\eta}\left(\frac{1}{4}Gr_{x,\infty}\right)^{1/4} x^{-1}\\
&\quad+\lambda(T_{\mathrm{w}} - T_\infty)\frac{\mathrm{d}^2\theta}{\mathrm{d}\eta^2}\left(\frac{1}{4}Gr_{x,\infty}\right)^{1/2} x^{-2}.
\end{aligned} \tag{5.26}$$

With the definition of Local Grashof number, the earlier equation is simplified to

$$
\begin{aligned}
&2\rho\sqrt{gx}\,|T_\mathrm{w}/T_\infty-1|^{1/2}\,W_x\left[T\left(-\frac{1}{4}\eta x^{-1}\frac{\mathrm{d}c_p}{\mathrm{d}\eta}\right)+c_p\left(-(T_\mathrm{w}-T_\infty)\frac{\mathrm{d}\theta}{\mathrm{d}\eta}\left(\frac{1}{4}\right)\eta x^{-1}\right)\right]\\
&+2\rho\sqrt{gx}\,|T_\mathrm{w}/T_\infty-1|^{1/2}\,W_y\left[T\frac{\mathrm{d}c_p}{\mathrm{d}\eta}x^{-1}+c_p(T_\mathrm{w}-T_\infty)\frac{\mathrm{d}\theta}{\mathrm{d}\eta}x^{-1}\right]\\
&=\frac{\mathrm{d}\lambda}{\mathrm{d}\eta}x^{-1}(T_\mathrm{w}-T_\infty)\frac{\mathrm{d}\theta}{\mathrm{d}\eta}\left(\frac{g\,|T_\mathrm{w}/T_\infty-1|\,x^3}{4\nu_\infty^2}\right)^{1/2}x^{-1}+\lambda(T_\mathrm{w}-T_\infty)\\
&\quad\times\frac{\mathrm{d}^2\theta}{\mathrm{d}\eta^2}\left(\frac{g\,|T_\mathrm{w}/T_\infty-1|\,x^3}{4\nu_\infty^2}\right)^{1/2}x^{-2}.
\end{aligned}
\tag{5.27}
$$

The earlier equation is divided by

$$
\sqrt{\frac{g}{x}}\,|T_\mathrm{w}/T_\infty-1|^{1/2}\,(T_\mathrm{w}-T_\infty)
$$

and this leads to

$$
\begin{aligned}
&\frac{1}{2}\rho W_\mathrm{x}\left[\frac{T}{T_\mathrm{w}-T_\infty}\left(-\frac{1}{4}\eta\frac{\mathrm{d}c_p}{\mathrm{d}\eta}\right)-c_p\left(\frac{\mathrm{d}\theta}{\mathrm{d}\eta}\left(\frac{1}{4}\right)\eta\right)\right]+2\rho W_y\left[\frac{T}{T_\mathrm{w}-T_\infty}\frac{\mathrm{d}c_p}{\mathrm{d}\eta}+c_p\frac{\mathrm{d}\theta}{\mathrm{d}\eta}\right]\\
&=\frac{\mathrm{d}\lambda}{\mathrm{d}\eta}\frac{\mathrm{d}\theta}{\mathrm{d}\eta}\left(\frac{1}{4\nu_\infty^2}\right)^{1/2}+\lambda\frac{\mathrm{d}^2\theta}{\mathrm{d}\eta^2}\left(\frac{1}{4\nu_\infty^2}\right)^{1/2}.
\end{aligned}
\tag{5.28}
$$

The earlier equation is multiplied by $2\nu_\infty/\lambda$ and is simplified, then

$$
\begin{aligned}
&\frac{\nu_\infty}{\lambda}\rho W_x\left[\frac{T}{T_\mathrm{w}-T_\infty}\left(-\eta\frac{\mathrm{d}c_p}{\mathrm{d}\eta}\right)-c_p\left(\frac{\mathrm{d}\theta}{\mathrm{d}\eta}\eta\right)\right]\\
&\quad+4\frac{\nu_\infty}{\lambda}\rho W_y\left[\frac{T}{T_\mathrm{w}-T_\infty}\frac{\mathrm{d}c_p}{\mathrm{d}\eta}+c_p\frac{\mathrm{d}\theta}{\mathrm{d}\eta}\right]=\frac{1}{\lambda}\frac{\mathrm{d}\lambda}{\mathrm{d}\eta}\frac{\mathrm{d}\theta}{\mathrm{d}\eta}+\frac{\mathrm{d}^2\theta}{\mathrm{d}\eta^2}.
\end{aligned}
$$

Since $(\nu_\infty/\lambda)c_p\rho=Pr(\nu_\infty/\nu)$, The earlier equation can be simplified to

$$
\begin{aligned}
&-Pr\frac{\nu_\infty}{\nu}\eta W_x\left[\frac{T}{T_\mathrm{w}-T_\infty}\left(\frac{1}{c_p}\frac{\mathrm{d}c_p}{\mathrm{d}\eta}\right)+\left(\frac{\mathrm{d}\theta}{\mathrm{d}\eta}\right)\right]\\
&\quad+4Pr\frac{\nu_\infty}{\nu}W_y\left(\frac{T}{T_\mathrm{w}-T_\infty}\frac{1}{c_p}\frac{\mathrm{d}c_p}{\mathrm{d}\eta}+\frac{\mathrm{d}\theta}{\mathrm{d}\eta}\right)=\frac{1}{\lambda}\frac{\mathrm{d}\lambda}{\mathrm{d}\eta}\frac{\mathrm{d}\theta}{\mathrm{d}\eta}+\frac{\mathrm{d}^2\theta}{\mathrm{d}\eta^2},
\end{aligned}
$$

or

$$
Pr\frac{\nu_\infty}{\nu}(-\eta W_x+4W_y)\left(\frac{\mathrm{d}\theta}{\mathrm{d}\eta}+\frac{T}{T_\mathrm{w}-T_\infty}\frac{1}{c_p}\frac{\mathrm{d}c_p}{\mathrm{d}\eta}\right)=\frac{1}{\lambda}\frac{\mathrm{d}\lambda}{\mathrm{d}\eta}\frac{\mathrm{d}\theta}{\mathrm{d}\eta}+\frac{\mathrm{d}^2\theta}{\mathrm{d}\eta^2}.
\tag{5.29}
$$

Now we simplify

$$
\frac{T}{T_\mathrm{w}-T_\infty}\frac{1}{c_p}\frac{\mathrm{d}c_p}{\mathrm{d}\eta}.
$$

From (5.5) and (5.25) we obtain the following equation:

$$\frac{\mathrm{d}c_p}{\mathrm{d}\eta} = c_{p\infty}\frac{\mathrm{d}\left(\frac{(T_\mathrm{w}-T_\infty)\theta+T_\infty}{T_\infty}\right)^{n_{c_p}}}{\mathrm{d}\eta} = c_{p\infty}n_{c_p}\left(\frac{(T_\mathrm{w}-T_\infty)\theta+T_\infty}{T_\infty}\right)^{n_{c_p}-1}$$

$$\frac{T_\mathrm{w}-T_\infty}{T_\infty}\frac{\mathrm{d}\theta}{\mathrm{d}\eta} = c_{p\infty}n_{c_p}\left(\frac{T}{T_\infty}\right)^{n_{c_p}-1}\frac{T_\mathrm{w}-T_\infty}{T_\infty}\frac{\mathrm{d}\theta}{\mathrm{d}\eta}.$$

Then

$$\frac{T}{T_\mathrm{w}-T_\infty}\frac{1}{c_p}\frac{\mathrm{d}c_p}{\mathrm{d}\eta} = \frac{c_{p\infty}}{c_p}n_{c_p}\left(\frac{T}{T_\infty}\right)^{n_{c_p}-1}\frac{T}{T_\infty}\frac{\mathrm{d}\theta}{\mathrm{d}\eta}$$

$$= \left(\frac{T}{T_\infty}\right)^{-n_{c_p}}n_{c_p}\left(\frac{T}{T_\infty}\right)^{n_{c_p}-1}\frac{T}{T_\infty}\frac{\mathrm{d}\theta}{\mathrm{d}\eta} = n_{c_p}\frac{\mathrm{d}\theta}{\mathrm{d}\eta}.$$

Consequently (5.29) is changed as

$$Pr\frac{\nu_\infty}{\nu}(-\eta W_x + 4W_y)\left(\frac{\mathrm{d}\theta}{\mathrm{d}\eta} + n_{c_p}\frac{\mathrm{d}\theta}{\mathrm{d}\eta}\right) = \frac{1}{\lambda}\frac{\mathrm{d}\lambda}{\mathrm{d}\eta}\frac{\mathrm{d}\theta}{\mathrm{d}\eta} + \frac{\mathrm{d}^2\theta}{\mathrm{d}\eta^2},$$

i.e.

$$(1+n_{c_p})Pr\frac{\nu_\infty}{\nu}(-\eta W_x + 4W_y)\frac{\mathrm{d}\theta}{\mathrm{d}\eta} = \frac{1}{\lambda}\frac{\mathrm{d}\lambda}{\mathrm{d}\eta}\frac{\mathrm{d}\theta}{\mathrm{d}\eta} + \frac{\mathrm{d}^2\theta}{\mathrm{d}\eta^2}. \tag{5.30}$$

Now we summarize the transformed dimensionless equations for the laminar free convection of polyatomic gases as follows:

$$2W_x - \eta\frac{\mathrm{d}W_x}{\mathrm{d}\eta} + 4\frac{\mathrm{d}W_y}{\mathrm{d}\eta} - \frac{1}{\rho}\frac{\mathrm{d}\rho}{\mathrm{d}\eta}(\eta W_x - 4W_y) = 0, \tag{5.16}$$

$$\frac{\nu_\infty}{\nu}\left(W_x\left(2W_x - \eta\frac{\mathrm{d}W_x}{\mathrm{d}\eta}\right) + 4W_y\frac{\mathrm{d}W_x}{\mathrm{d}\eta}\right) = \frac{\mathrm{d}^2W_x}{\mathrm{d}\eta^2} + \frac{1}{\mu}\frac{\mathrm{d}\mu}{\mathrm{d}\eta}\frac{\mathrm{d}W_x}{\mathrm{d}\eta} + \frac{\nu_\infty}{\nu}\theta, \tag{5.17}$$

$$(1+n_{c_p})Pr\frac{\nu_\infty}{\nu}(-\eta W_x + 4W_y)\frac{\mathrm{d}\theta}{\mathrm{d}\eta} = \frac{1}{\lambda}\frac{\mathrm{d}\lambda}{\mathrm{d}\eta}\frac{\mathrm{d}\theta}{\mathrm{d}\eta} + \frac{\mathrm{d}^2\theta}{\mathrm{d}\eta^2}, \tag{5.30}$$

with boundary conditions

$$\eta = 0,\ W_x = 0,\quad W_y = 0,\quad \theta = 1, \tag{5.31}$$

$$\eta \to \infty,\ W_x \to 0,\quad \theta \to 1. \tag{5.32}$$

It is obvious that when $n_{c_p} = 0$, (5.30) will turn back to that for laminar free convection of monatomic and diatomic gases, air and water vapour presented in Chap. 4. Therefore, the energy equation of laminar free convection of monatomic and diatomic gases, air and water vapour is a special case of that of the polyatomic gas laminar free convection when $n_{c_p} = 0$.

Similar to the derivation in Chap. 4, the corresponding expressions for the thermophysical property factors in the governing equations (5.16), (5.17) and (5.31) become forms for treatment of variable thermophysical properties of gases.

$$\frac{1}{\rho}\frac{\mathrm{d}\rho}{\mathrm{d}\eta} = -\frac{\left(\frac{T_\mathrm{w}}{T_\infty}-1\right)\frac{\mathrm{d}\theta}{\mathrm{d}\eta}}{\left(\frac{T_\mathrm{w}}{T_\infty}-1\right)\theta+1}, \tag{5.33}$$

$$\frac{1}{\mu}\frac{\mathrm{d}\mu}{\mathrm{d}\eta} = \frac{n_\mu\left(\frac{T_\mathrm{w}}{T_\infty}-1\right)\frac{\mathrm{d}\theta}{\mathrm{d}\eta}}{\left(\frac{T_\mathrm{w}}{T_\infty}-1\right)\theta+1}, \tag{5.34}$$

$$\frac{1}{\lambda}\frac{\mathrm{d}\lambda}{\mathrm{d}\eta} = \frac{n_\lambda\left(\frac{T_\mathrm{w}}{T_\infty}-1\right)\frac{\mathrm{d}\theta}{\mathrm{d}\eta}}{\left(\frac{T_\mathrm{w}}{T_\infty}-1\right)\theta+1}, \tag{5.35}$$

$$\frac{\nu_\infty}{\nu} = \left(\left(\frac{T_\mathrm{w}}{T_\infty}-1\right)\theta+1\right)^{-(n_\mu+1)}. \tag{5.36}$$

The velocity and temperature fields can be solved from (5.16), (5.17) and (5.30) with boundary conditions, (5.31) and (5.32), combined with (5.33)–(5.36). It will be expected that, for the case of variable properties, the dimensionless velocity field and the dimensionless temperature field depends, not only on $T_\mathrm{w}/T_\infty, Pr, n_\lambda, n_\mu$ and c_{p_∞} but also on n_{c_p} for the laminar free convection of polyatomic gases.

5.4 Heat Transfer Analysis

Same as the heat transfer analysis in Chap. 4 the analytical expressions on heat transfer for laminar free convection of polyatomic gases are as follows.

The local heat transfer rate q_x at position x per unit area on the plate, the corresponding local heat transfer coefficient α_x, and the local Nusselt number $Nu_{x,\mathrm{w}}$ are expressed as follows:

$$q_x = -\lambda_\mathrm{w}(T_\mathrm{w}-T_\infty)\left(\frac{1}{4}Gr_{x,\infty}\right)^{1/4} x^{-1}\left(\frac{\mathrm{d}\theta}{\mathrm{d}\eta}\right)_{\eta=0}, \tag{5.37}$$

$$\alpha_x = -\lambda_\mathrm{w}\left(\frac{1}{4}Gr_{x,\infty}\right)^{1/4} x^{-1}\left(\frac{\mathrm{d}\theta}{\mathrm{d}\eta}\right)_{\eta=0}, \tag{5.38}$$

$$\frac{Nu_{x,\mathrm{w}}}{\left(\frac{1}{4}Gr_{x,\infty}\right)^{1/4}} = \left(\frac{\mathrm{d}\theta}{\mathrm{d}\eta}\right)_{\eta=0}, \tag{5.39}$$

where q_x, α_x and $Nu_{x,\mathrm{w}}$ are defined as $q_x = -\lambda_\mathrm{w}(\partial T/\partial y)_{y=0}, q_x = \alpha_x(T_\mathrm{w} - T_\infty)$ and $Nu_{x,\mathrm{w}} = \alpha_x x/\lambda_\mathrm{w}$, respectively. Also, total heat transfer rate Q_x for position $x=0$ to x with width of b on the plate, defined as $Q_x = \int_0^x bq_x\,\mathrm{d}x$, is

$$Q_x = -\frac{4}{3}b\lambda_\mathrm{w}(T_\mathrm{w}-T_\infty)\left(\frac{1}{4}Gr_{x,\infty}\right)^{1/4}\left(\frac{\mathrm{d}\theta}{\mathrm{d}\eta}\right)_{\eta=0}. \tag{5.40}$$

The corresponding average heat transfer coefficient $\overline{\alpha}_x$ defined as $Q_x = \overline{\alpha_x}(T_w - T_\infty) \times b \times x$ is expressed as

$$\overline{\alpha}_x = -\frac{4}{3}\lambda_w \left(\frac{1}{4}Gr_{x,\infty}\right)^{1/4} x^{-1} \left(\frac{d\theta}{d\eta}\right)_{\eta=0}. \tag{5.41}$$

The mean Nusselt number defined as $\overline{Nu}_{x,w} = \overline{\alpha}_x x/\lambda_w$ is expressed as

$$\frac{\overline{Nu}_{x,w}}{\left(\frac{1}{4}Gr_{x,\infty}\right)^{1/4}} = \frac{4}{3}\left(\frac{d\theta}{d\eta}\right)_{\eta=0}. \tag{5.42}$$

Same as that reported in Chap. 4, it is seen that, for practical calculation of heat transfer, only $(d\theta/d\eta)_{\eta=0}$ dependent on numerical solution is no-given variable.

5.5 Numerical Solutions

The governing dimensionless equations (5.16), (5.17) and (5.30) were calculated numerically with the boundary condition equations (5.31) and (5.32) as well as (5.33)–(5.36) for the thermophysical property factors. The calculations were carried out by the shooting method presented in Chap. 4. The typical results for the velocity and temperature field were obtained with different Pr, n_λ, n_μ and n_{c_p} at different boundary temperature ratios T_w/T_∞. Some of the solutions were plotted in Figs. 5.1–5.4. Meanwhile, the solutions of temperature gradient on the wall, $-(d\theta/d\eta)_{\eta=0}$, for several polyatomic gases

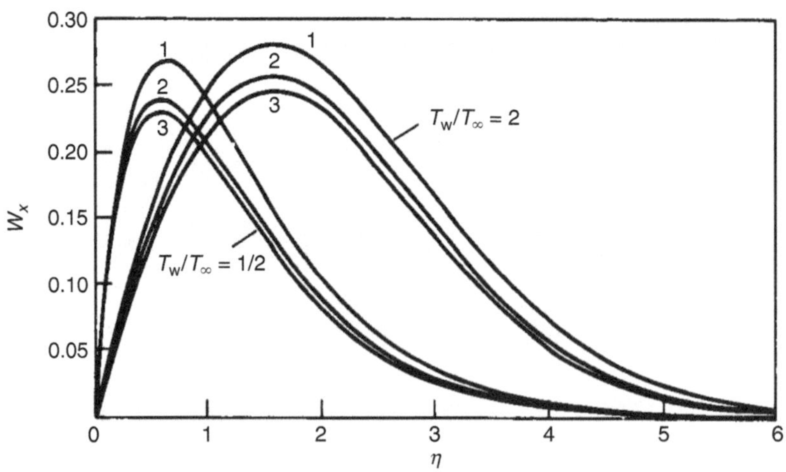

Fig. 5.1. Velocity profiles for free convection of different polyatomic gases, cited from Shang and Wang [1]. (1) Gas mixture ($CO_2 = 0.13, H_2O = 0.11, N_2 = 0.76$); (2) SO_2; (3) NH_3

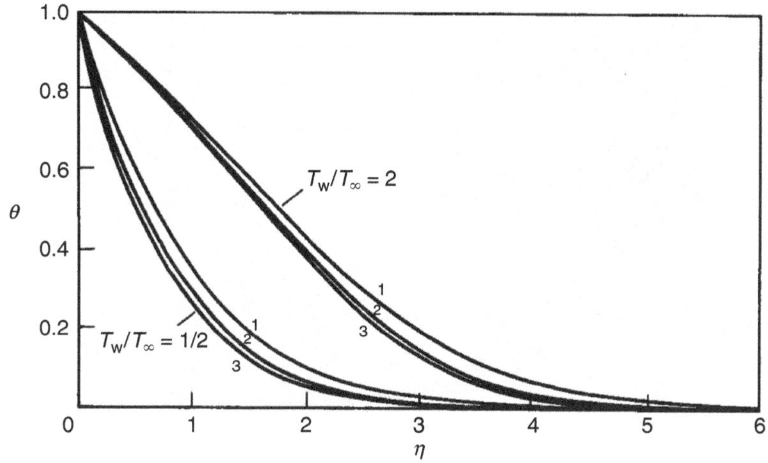

Fig. 5.2. Temperature profiles for free convection of different polyatomic gases, cited from Shang and Wang [1]. (1) Gas mixture ($CO_2 = 0.13, H_2O = 0.11, N_2 = 0.76$); (2) SO_2; (3) NH_3

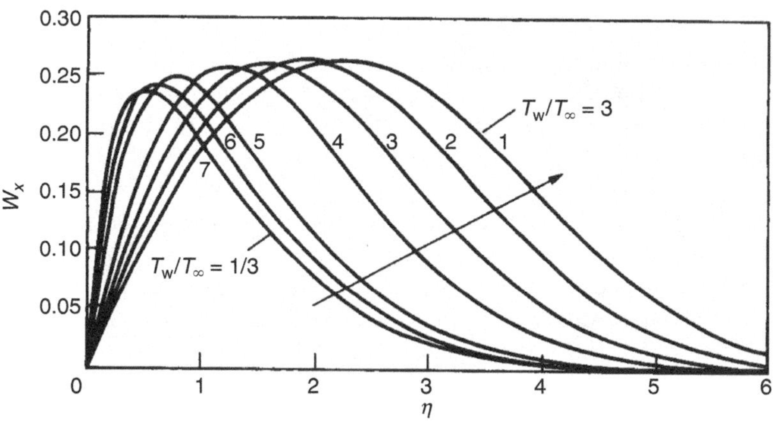

Fig. 5.3. Velocity profiles for free convection of CO_2 with different boundary temperature rations, cited from Shang and Wang [1]. (1) $T_w/T_\infty = 3$; (2) $T_w/T_\infty = 5/2$; (3) $T_w/T_\infty = 2$; (4) $T_w/T_\infty = 3/2$; (5) $T_w/T_\infty = 3/4$; (6) $T_w/T_\infty = 1/2$; (7) $T_w/T_\infty = 1/3$

with the related Pr, n_λ, n_μ and n_{c_p} at various boundary temperature ratios T_w/T_∞ are shown in Table 5.2 and plotted in Fig. 5.5. These solutions shown in Figs. 5.1–5.4 and Table 5.2 describe the effects of Pr, n_λ, n_μ and n_{c_p} on velocity and temperature Fields as well as heat transfer of polyatomic gas laminar free convection. It is found that the effects of Pr, n_λ and n_μ with the boundary temperature ratios T_w/T_∞ on the velocity and temperature fields of polyatomic laminar free convection are same as those on the velocity and temperature fields of laminar free convection of monatomic and diatomic gases, air and water vapour.

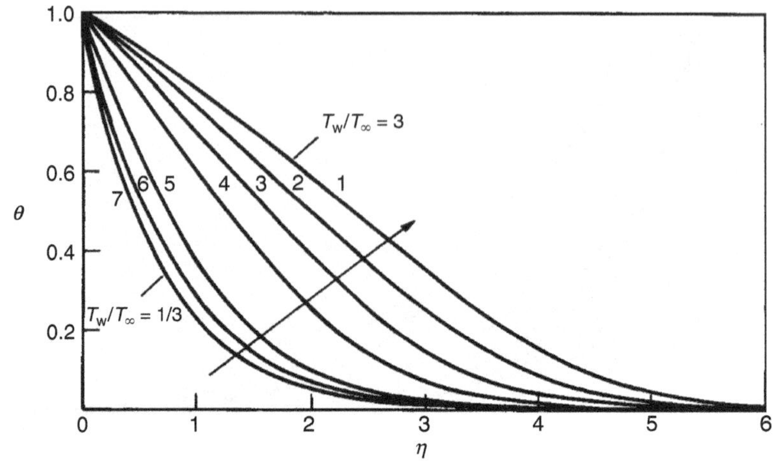

Fig. 5.4. Temperature profiles for free convection of CO_2 with different boundary temperature rations, cited from Shang and Wang [1]. (1) $T_w/T_\infty = 3$; (2) $T_w/T_\infty = 5/2$; (3) $T_w/T_\infty = 2$; (4) $T_w/T_\infty = 3/2$; (5) $T_w/T_\infty = 3/4$; (6) $T_w/T_\infty = 1/2$; (7) $T_w/T_\infty = 1/3$

For consideration of variation of specific heat of the monatomic and diatomic gases, air and water vapour with temperature, the corresponding numerical solutions are calculated by (5.16), (5.17) and (5.30) with the boundary condition equations (5.31) and (5.32) as well as (5.33)–(5.36) for the thermophysical property factors, and listed in Table 5.3. It is found that these numerical solutions are very identical to the related numerical solutions presented in Chap. 4 without consideration of variation of specific heat. Then it follows that it is acceptable to neglect the effect of the specific heat for calculation of the free convection heat transfer coefficient of the monatomic and diatomic gases, air and water vapour.

5.6 Curve-Fit Formulas for Heat Transfer

According to the numerical solutions shown in Table 5.2 for temperature gradient $-(\mathrm{d}\theta/\mathrm{d}\eta)_{\eta=0}$, a curve-fit formula of $-(\mathrm{d}\theta/\mathrm{d}\eta)_{\eta=0}$ was obtained by Shang and Wang [1] as follows for laminar free convection of polyatomic gases with consideration of variable thermophysical properties:

$$-\left(\frac{\mathrm{d}\theta}{\mathrm{d}\eta}\right)_{\eta=0} = (1+0.3n_{c_p})\psi(Pr)\left(\frac{T_w}{T_\infty}\right)^{-m}, \tag{5.43}$$

where $\psi(Pr)$ is Boussinesq solution in the range of gas Prandtl number, and is expressed as follows according to Chap. 4:

$$\psi(Pr) = 0.567 + 0.186 \times \ln Pr. \qquad (0.6 \le Pr \le 1) \tag{5.44}$$

Table 5.2. Calculated results of $(-\mathrm{d}\theta/\mathrm{d}\eta)_{\eta=0}$ for consideration of temperature-dependent specific heat, cited from Shang and Wang [1]

T_w/T_∞		gas mixture $Pr = 0.63$ $n_\mu = 0.75$ $n_\lambda = 1.02$ $n_{c_p} = 0.134$	CO_2 $Pr = 0.73$ $n_\mu = 0.88$ $n_\lambda = 1.3$ $n_{c_p} = 0.34$	CH_4 $Pr = 0.74$ $n_\mu = 0.78$ $n_\lambda = 1.29$ $n_{c_p} = 0.534$	CCl_4 $Pr = 0.8$ $n_\mu = 0.912$ $n_\lambda = 1.29$ $n_{c_p} = 0.28$	SO_2 $Pr = 0.81$ $n_\mu = 0.91$ $n_\lambda = 1.323$ $n_{c_p} = 0.257$	H_2S $Pr = 0.85$ $n_\mu = 1$ $n_\lambda = 1.29$ $n_{c_p} = 0.18$	NH_3 $Pr = 0.87$ $n_\mu = 1.04$ $n_\lambda = 1.375$ $n_{c_p} = 0.34$
3	A	0.1805	0.1744	0.1908	0.1760	0.1736	0.1696	0.1709
	B	0.1792	0.1729	0.1896	0.1747	0.1722	0.1687	0.1699
5/2	A	0.2136	0.2107	0.2289	0.2130	0.2104	0.2064	0.2093
	B	0.2125	0.2102	0.2291	0.2125	0.2100	0.2062	0.2093
2	A	0.2626	0.2664	0.2870	0.2696	0.2669	0.2630	0.2690
	B	0.2618	0.2669	0.2880	0.2702	0.2676	0.2637	0.2700
3/2	A	0.3435	0.3619	0.3860	0.3670	0.3644	0.3609	0.3733
	B	0.3426	0.3631	0.3892	0.3682	0.3659	0.3620	0.3752
5/4	A	0.4076	0.4408	0.4674	0.4475	0.4452	0.4419	0.4608
	B	0.4062	0.4413	0.4703	0.4480	0.4461	0.4425	0.4621
3/4	A	0.6613	0.7751	0.8095	0.7783	0.7902	0.7873	0.8415
	C	0.6602	0.7745	0.8105	0.7889	0.7903	0.7903	0.8448
1/2	A	0.9752	1.2291	1.2700	1.2504	1.2625	1.2582	1.3751
	C	0.9757	1.2223	1.2594	1.2483	1.2573	1.2658	1.3805
1/3	A	1.4434	1.9722	2.0185	2.0274	2.0429	2.0308	2.2802
	C	1.4420	1.9289	1.9569	1.9753	2.0001	2.0274	2.2558

Note: component of the gas mixture: $CO_2 = 0.13$, water vapour $= 0.11$ and $N_2 = 0.76$.
A. The numerical solution of governing equations (5.16), (5.17) and (5.30) with boundary condition equations (5.31) and (5.32) coupled with (5.33)–(5.36); B. from (5.43) with (5.44) and (5.46); C. from (5.43) with (5.44) and (5.46)

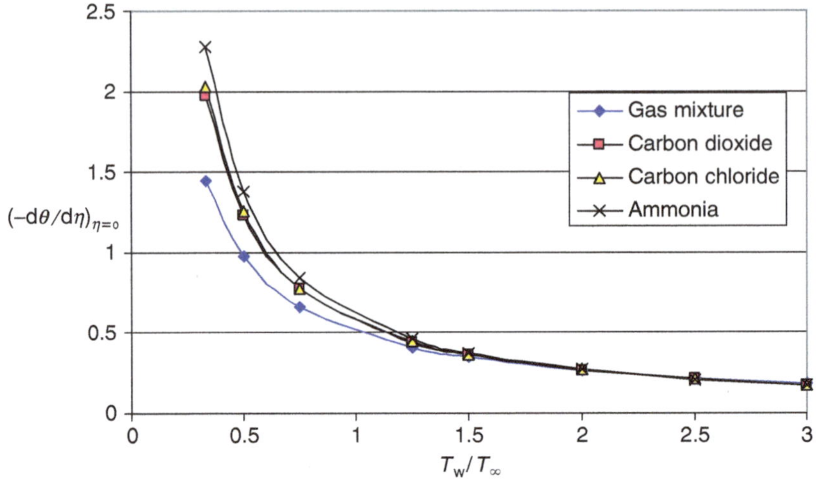

Fig. 5.5. Numerical solutions $-(\mathrm{d}\theta/\mathrm{d}\eta)_{\eta=0}$ for the free convection of gas mixture, CO_2, CCl_4 and NH_3

While,

$$m = 0.35n_\lambda + 0.29n_\mu + 0.36, \qquad (T_\mathrm{w}/T_\infty > 1) \tag{5.45}$$

$$m = 0.42n_\lambda + 0.34n_\mu + 0.28. \qquad (T_\mathrm{w}/T_\infty < 1) \tag{5.46}$$

Some results of temperature gradient $-(\mathrm{d}\theta/\mathrm{d}\eta)_{\eta=0}$ for laminar free convection of different polyatomic gases predicted by using (5.43)–(5.46) are listed in Table 5.2 also. From Table 5.2 it is found that these predicted results are very well identical to the related numerical solutions.

On the other hand, it is clear from the curve-fitting formulae that the temperature gradient $-(\mathrm{d}\theta/\mathrm{d}\eta)_{\eta=0}$ will increase with increasing the specific heat parameter n_{c_p}. Then, it is clearly known that heat transfer will increase with increasing specific heat parameter n_{c_p} for gas laminar free convection. If n_{c_p} is very small and regarded as zero, (5.43) becomes

$$-\left(\frac{\mathrm{d}\theta}{\mathrm{d}\eta}\right)_{\eta=0} = \psi(Pr)\left(\frac{T_\mathrm{w}}{T_\infty}\right)^{-m}, \tag{5.47}$$

which is identical to equation for laminar free convection of monatomic gases, diatomic gases, air and water vapour.

If boundary temperature ratio T_w/T_∞ tends to unity, the specific heat parameter n_{c_p} is regarded as zero, (5.43) is transformed to

$$-\left(\frac{\mathrm{d}\theta}{\mathrm{d}\eta}\right)_{\eta=0} = \psi(Pr), \tag{5.48}$$

where is identical to Boussinesq approximation, while $\psi(Pr)$ expresses well-known Boussinesq solution.

Table 5.3. Calculated results of $-(\mathrm{d}\theta/\mathrm{d}\eta)_{\eta=0}$ for consideration of temperature-dependent specific heat, cited from Shang and Wang [1]

T_w/T_∞		H_2 $Pr = 0.68$ $n_\mu = 0.68$ $n_\lambda = 0.8$ $n_{c_p} = 0.042$	air $Pr = 0.7$ $n_\mu = 0.68$ $n_\lambda = 0.81$ $n_{c_p} = 0.078$	N_2 $Pr = 0.71$ $n_\mu = 0.67$ $n_\lambda = 0.76$ $n_{c_p} = 0.07$	CO $Pr = 0.72$ $n_\mu = 0.71$ $n_\lambda = 0.83$ $n_{c_p} = 0.068$	O_2 $Pr = 0.733$ $n_\mu = 0.694$ $n_\lambda = 0.86$ $n_{c_p} = 0.108$	water vapour $Pr = 1$ $n_\mu = 1.04$ $n_\lambda = 1.185$ $n_{c_p} = 0.003$
3	A	0.2004	0.2042	0.2094	0.2020	0.2047	0.1740
	B	0.1999	0.2035	0.2087	0.2015	0.2039	0.1739
5/2	A	0.2334	0.2380	0.2433	0.2362	0.2394	0.2113
	B	0.2329	0.2372	0.2323	0.2355	0.2386	0.2117
2	A	0.2814	0.2871	0.2923	0.2859	0.2901	0.2682
	B	0.2807	0.2861	0.2910	0.2851	0.2892	0.2691
3/2	A	0.3579	0.3655	0.3699	0.3556	0.3716	0.3655
	B	0.3572	0.3644	0.3685	0.3648	0.3706	0.3668
5/4	A	0.4167	0.4258	0.4292	0.4272	0.4384	0.4453
	B	0.4167	0.4274	0.4279	0.4262	0.4337	0.4464
3/4	A	0.6370	0.6525	0.6492	0.6601	0.6751	0.7783
	C	0.6399	0.6546	0.6519	0.6630	0.6766	0.7858
1/2	A	0.8906	0.9141	0.8992	0.9311	0.9569	1.2194
	C	0.9022	0.9245	0.9116	0.9435	0.9656	1.2432
1/3	A	1.2431	1.2789	1.2420	1.3116	1.3560	1.9208
	C	1.2720	1.3056	1.2748	1.3426	1.3781	1.9667

A: The numerical solution of governing (5.16), (5.17) and (5.30) with boundary condition equations (5.31) and (5.32) coupled with (5.33)–(5.36); B: from (5.43) with (5.44) and (5.46); and C: from (5.43) with (5.44) and (5.46)

Table 5.4. Effects of $Pr, n_\mu, n_\lambda, n_{cp}$ and T_w/T_∞ on temperature gradient $-(d\theta/d\eta)_{\eta=0}$ for laminar free convection of polyatomic gases

term	$-\left(\frac{d\theta}{d\eta}\right)_{\eta=0}$	
	$T_w/T_\infty > 1$	$T_w/T_\infty < 1$
effect of Pr	$-\left(\frac{d\theta}{d\eta}\right)_{\eta=0}$ increases with the increase of Pr	$-\left(\frac{d\theta}{d\eta}\right)_{\eta=0}$ increases with the increase of Pr
effect of n_λ	$-\left(\frac{d\theta}{d\eta}\right)_{\eta=0}$ decreases with the increase of n_λ	$-\left(\frac{d\theta}{d\eta}\right)_{\eta=0}$ increases with the increase of n_λ
effect of n_μ	$-\left(\frac{d\theta}{d\eta}\right)_{\eta=0}$ decreases with the increase of n_μ	$-\left(\frac{d\theta}{d\eta}\right)_{\eta=0}$ increases with the increase of n_μ
effect of n_{c_p}	$-\left(\frac{d\theta}{d\eta}\right)_{\eta=0}$ increases with the increase of n_{c_p}	$-\left(\frac{d\theta}{d\eta}\right)_{\eta=0}$ increases with the increase n_{c_p}

So far, the effect of $Pr, n_\mu, n_\lambda, n_{cp}$ and T_w/T_∞ on temperature gradient $-(d\theta/d\eta)_{\eta=0}$ can be briefly summarized in Table 5.4 for laminar free convection of polyatomic gases.

5.7 Summary

Comparing the analyses and results in Chap. 4 with that in this chapter, it is obvious to find that laminar free convection and heat transfer of the monatomic and diatomic gases, air and water vapour can be regarded as a special case of that of polyatomic gases. In fact, the results of free convection heat transfer of polyatomic gases are very well identical to those of the monatomic and diatomic gases, air and water vapour. So far, the governing equations of laminar free convection for general gases and expressions related to heat transfer can be summarized in Table 5.5.

5.8 Remarks

From the earlier presentations in this chapter the following remarks can be concluded: The method proposed in this chapter, for analysing the laminar free convection of polyatomic gases along a vertical isothermal flat plate can be taken as a general one for all gases, and it could yield reliable results.

For the monatomic, diatomic gases, air and water vapour, specific heat parameter n_{c_p} is generally below 0.1, especially for monatomic gases with the value of n_{c_p} tending to zero, n_{c_p} could be taken as zero when the variable fluid properties are considered.

In considering the variable fluid properties, as $T_w/T_\infty \to 1$ the classical Bousinesq approximation holds true for monoatomic gases, and nearly

Table 5.5. Summary of the governing equations of laminar free convection for general gases and the expressions related to heat transfer

term	expression	
	governing partial differential equations	
mass equation	$\frac{\partial}{\partial x}(\rho w_x) + \frac{\partial}{\partial y}(\rho w_y) = 0$	
momentum equation	$\rho\left(w_x \frac{\partial w_x}{\partial x} + w_y \frac{\partial w_x}{\partial y}\right) = \frac{\partial}{\partial y}\left(\mu \frac{\partial w_x}{\partial y}\right) + \rho g \left\lvert \frac{T-T_\infty}{T_\infty} \right\rvert$	
energy equation	$\rho c_p \left(w_x \frac{\partial (c_p T)}{\partial x} + w_y \frac{\partial (c_p T)}{\partial y}\right) = \frac{\partial}{\partial y}\left(\lambda \frac{\partial T}{\partial y}\right)$	
boundary conditions	$y = 0$: $w_x = 0, w_y = 0, T = T_\mathrm{w}$ $y \to \infty$: $w_x \to 0, T = T_\infty$	
	assumed similarity variables	
η	$\frac{y}{x}\left(\frac{1}{4}Gr_{x,\infty}\right)^{1/4}$	
$Gr_{x,\infty}$	$\frac{g\lvert T_\mathrm{w}/T_\infty - 1\rvert x^3}{\nu_\infty^2}$	
θ	$\frac{T-T_\infty}{T_\mathrm{w}-T_\infty}$	
W_x	$(2\sqrt{gx}\,\lvert T_\mathrm{w}/T_\infty - 1\rvert^{1/2})^{-1} w_x$	
W_y	$(2\sqrt{gx}T_\mathrm{w}/T_\infty - 1^{1/2}\left(\frac{1}{4}Gr_{x,\infty}\right)^{-1/4})^{-1} w_y$	
	governing ordinary differential equations	
mass equations	$2W_x - \eta \frac{dW_x}{d\eta} + 4\frac{dW_y}{d\eta} - \frac{1}{\rho}\frac{d\rho}{d\eta}(\eta W_x - 4W_y) = 0$	
momentum equation	$\frac{\nu_\infty}{\nu}\left(W_x\left(2W_x - \eta\frac{dW_x}{d\eta}\right) + 4W_y\frac{dW_x}{d\eta}\right) = \frac{d^2 W_x}{d\eta^2} + \frac{1}{\mu}\frac{d\mu}{d\eta}\frac{dW_x}{d\eta} + \frac{\nu_\infty}{\nu}\theta$	
energy equation	$(1 + n_{c_p})Pr\frac{\nu_\infty}{\nu}(-\eta W_x + 4W_y)\frac{d\theta}{d\eta} = \frac{1}{\lambda}\frac{d\lambda}{d\eta}\frac{d\theta}{d\eta} + \frac{d^2\theta}{d\eta^2}$	
boundary condition	$\eta = 0$: $W_x = 0,\ W_y = 0,\ \theta = 0$ $\eta \to 0$: $W_x = 0,\ \theta = 0$	
	equations related to heat transfer	
q_x $\left(\text{defined as } -\lambda_\mathrm{w}\left(\frac{\partial T}{\partial y}\right)_{y=0}\right)$	$-\lambda_\mathrm{w}(T_\mathrm{w} - T_\infty)\left(\frac{1}{4}Gr_{x,\infty}\right)^{1/4} x^{-1}\left(\frac{d\theta}{d\eta}\right)_{\eta=0}$	
α_x $\left(\text{defined as } \frac{q_x}{(T_w - T_\infty)}\right)$	$-\lambda_w\left(\frac{1}{4}Gr_{x,\infty}\right)^{1/4} x^{-1}\left(\frac{d\theta}{d\eta}\right)_{\eta=0}$	
Q_x $\left(\text{defined as } \int_0^x q_x b\, dx\right)$	$-\frac{4}{3}b\lambda_\mathrm{w}(T_\mathrm{w} - T_\infty)\left(\frac{1}{4}Gr_{x,\infty}\right)^{1/4}\left(\frac{d\theta}{d\eta}\right)_{\eta=0}$	
$\overline{\alpha}_x$ $\left(\text{defined as} \frac{Q_x}{(T_\mathrm{w} - T_\infty)bx}\right)$	$-\frac{4}{3}\lambda_\mathrm{w}\left(\frac{1}{4}Gr_{x,\infty}\right)^{1/4} x^{-1}\left(\frac{d\theta}{d\eta}\right)_{\eta=0}$	
$Nu_{x,\mathrm{w}}$ $\left(\text{defined as} \frac{\alpha_x x}{\lambda_\mathrm{w}}\right)$	$\frac{Nu_{x,\mathrm{w}}}{\left(\frac{1}{4}Gr_{x,\infty}\right)^{1/4}} = -\left(\frac{d\theta}{d\eta}\right)_{\eta=0}$	
$\overline{Nu}_{x,\mathrm{w}}$ $\left(\text{defined as} \frac{\overline{\alpha}_x x}{\lambda_\mathrm{w}}\right)$	$\frac{\overline{Nu}_{x,\mathrm{w}}}{\left(\frac{1}{4}Gr_{x,\infty}\right)^{1/4}} = -\frac{4}{3}\left(\frac{d\theta}{d\eta}\right)_{\eta=0}$	
$-\left(\frac{d\theta}{d\eta}\right)_{\eta=0}$	$(1 + 0.3n_{c_p})\psi(Pr)\left(\frac{T_\mathrm{w}}{T_\infty}\right)^{-m}$	
$\psi(Pr)$	$0.567 + 0.186 \times \ln Pr$	$(0.6 \le Pr \le 1)$
	$0.35n_\lambda + 0.29n_\mu + 0.36$	for $T_\mathrm{w}/T_\infty > 1$
m	$0.42n_\lambda + 0.34n_\mu + 0.28$	for $T_\mathrm{w}/T_\infty < 1$

true for the biatomic gases, air and water vapour, but the classic Boussinesq approximation does not hold true for polyatomic gases, except n_{c_p} is taken as zero.

The analysis, presented in this chapter extends the one reported in Chap. 4.

5.9 Calculation Example

Question: An plate with uniform temperature $t_{\mathrm{w}} = 0°\mathrm{C}$, width $b = 2$ m and height $x = 0.9$ m is suspended in a gas mixture with temperature $t_\infty = 500°\mathrm{C}$. The kinetic viscosity of the gas mixture is $\nu_\infty = 7.63 \times 10^{-5}\mathrm{m}^2\ \mathrm{s}^{-1}$ at $t_\infty = 500°\mathrm{C}$, and the thermal conductivity of the gas mixture is $\lambda_{\mathrm{w}} = 0.0228\ \mathrm{W}\ (\mathrm{m}°\mathrm{C})^{-1}$ at $t_{\mathrm{w}} = 0°\mathrm{C}$. The temperature parameters of the gas mixture are $n_\mu = 0.75$, $n_\lambda = 1.02$, and $n_{c_p} = 0.134$, respectively. The gas mixture Prandtl number is $Pr = 0.63$. Suppose the free convection is laminar. Please calculate the average heat transfer coefficients and free convection heat transfer on the plate.

Solution. The temperature ratio T_{w}/T_∞ of the gas laminar free convection is

$$T_{\mathrm{w}}/T_\infty = 273/(500 + 273) = 0.35317.$$

Since $T_{\mathrm{w}}/T_\infty < 1$, from (5.46) we have

$$\begin{aligned} m &= 0.42n_\lambda + 0.34n_\mu + 0.28 \\ &= 0.42 \times 1.02 + 0.34 \times 0.75 + 0.28 \\ &= 0.9634. \end{aligned}$$

Also

$$\begin{aligned} \psi(Pr) &= 0.567 + 0.186 \times \ln Pr \\ &= 0.567 + 0.186 \times \ln(0.63). \\ &= 0.48106 \end{aligned}$$

Then, temperature gradient $-(\mathrm{d}\theta/\mathrm{d}\eta)_{\eta=0}$ is evaluated as

$$\begin{aligned} -\left(\frac{\mathrm{d}\theta}{\mathrm{d}\eta}\right)_{\eta=0} &= (1 + 0.3n_{c_p})\psi(Pr)\left(\frac{T_{\mathrm{w}}}{T_\infty}\right)^{-m} \\ &= (1 + 0.3 \times 0.134) \times 0.48106 \times 0.35317^{-0.9634} \\ &= 1.3639. \end{aligned}$$

Then, local Grashof number is calculated as

$$\begin{aligned} Gr_{x,\infty} &= \frac{g\,|T_{\mathrm{w}}/T_\infty - 1|\,x^3}{\nu_\infty^2} \\ &= \frac{9.8 \times |0.35317 - 1| \times 0.9^3}{(7.63 \times 10^{-5})^2} \\ &= 793770003 \\ &= 0.79377 \times 10^9. \end{aligned}$$

In this case, average Nusselt number$\overline{Nu_{x,\mathrm{w}}}$ can be calculated as follows by using (5.42):

$$\begin{aligned}\overline{Nu_{x,\mathrm{w}}} &= -\frac{4}{3}\left(\frac{\mathrm{d}\theta}{\mathrm{d}\eta}\right)_{\eta=0}\left(\frac{1}{4}Gr_{x,\infty}\right)^{1/4}\\ &= \frac{4}{3}\times 1.3639\times\left(\frac{1}{4}\times 79377003\right)^{1/4}\\ &= 215.84.\end{aligned}$$

With the definition of $\overline{Nu_{x,\mathrm{w}}}$, $\overline{Nu_{x,\mathrm{w}}} = \overline{\alpha_x}x/\lambda_\mathrm{w}$, the average heat transfer coefficient $\overline{\alpha_x}$ is expressed as

$$\overline{\alpha_x} = \frac{\overline{Nu_{x,\mathrm{w}}}\cdot\lambda_\mathrm{w}}{x} = \frac{215.84\times 0.0228}{0.9} = 5.468\ \mathrm{W}\ (\mathrm{m}^2\ {}^\circ\mathrm{C})^{-1}.$$

Then, total heat transfer rate on the plate is

$$\begin{aligned}Q_x &= \overline{\alpha_x}(T_\mathrm{w} - T_\infty)x\times b\\ &= 5.468\times(0-500)\times 0.9\times 2\\ &= -4921.2\ \mathrm{W},\end{aligned}$$

where the negative sign implies that the total heat transfer rate Q_x is from gas mixture to the plate.

References

1. D.Y. Shang and B.X. Wang, Effect of variable thermophysical properties on laminar free convection of polyatomic gas, Int. J. Heat Mass Transfer 34, No. 3, pp. 749–755, 1991
2. Y.S. Touloukian, S.C. Saxena and P. Hestermans, Thermophysical Properties of Matter, Vol. 2; Viscosity, IFI/Plenum, New York, 1970
3. Y.S. Touloukian, P.E. Liley and S.C. Saxena, Thermophysical properties of matter, Vol. 3; Thermal Conductivity, Non-Metallic Liquids and Gases, IFI/Plenum, New York, 1970
4. Y.S. Touloukian and T. Makita, Thermophysical properties of matter, Vol. 6; Specific Heat, Non-Metallic Liquids and Gases, IFI/Plenum, New York, 1970
5. S.M. Yang, Heat Transfer, 2nd Edn, pp. 443, Higher Education Press, in Chinese 1987

6

Laminar Free Convection of Liquid

Nomenclature

b	width of plate, m
c_p	specific heat at constant pressure, J $(\text{kg K})^{-1}$
g	gravitation acceleration, m $(\text{s}^2)^{-1}$
$Gr_{x,\infty}$	local Grashof number for liquid laminar free convection on isothermal vertical flat plate, $\frac{g\lvert\rho_\infty/\rho_\text{w}-1\rvert x^3}{\nu_\infty^2}$
$Nu_{x,\text{w}}$	local Nusselt number, $\alpha_x x/\lambda_\text{w}$
$\overline{Nu}_{x,\text{w}}$	average Nusselt number defined as $\overline{\alpha}_x x/\lambda_\text{w}$
Pr	Prandtl number
q_x	local heat transfer rate at position x per unit area on the plate, W m^{-2}
Q_x	total heat transfer rate for position $x = 0$ to x with width of b on the plate, W
t	temperature, °C
T	absolute temperature, K
T_r	a reference temperature, $T_\text{w} - (T_\text{w} - T_\infty)/4$
w_x, w_y	velocity components in the x- and y- directions, respectively, m s^{-1}
W_x, W_y	dimensionless velocity components in the x- and y- directions, respectively

Greek symbols

α_x	local heat transfer coefficient, W $(\text{m}^2\ \text{K})^{-1}$
$\overline{\alpha}_x$	average heat transfer coefficient, W $(\text{m}^2\ \text{K})^{-1}$
δ	boundary layer thickness, m
η	dimensionless coordinate variable for boundary layer
θ	dimensionless temperature
λ	thermal conductivity, W $(\text{m K})^{-1}$
μ	absolute viscosity, kg $(\text{m s})^{-1}$
ν	kinetic viscosity, $\text{m}^2\ \text{s}^{-1}$
ρ	density, kg m^{-3}

$\dfrac{\frac{\rho_\infty}{\rho}-1}{\frac{\rho_\infty}{\rho_{\mathrm{w}}}-1}$	buoyancy factor
$\left(\dfrac{d\theta}{d\eta}\right)_{\eta=0}$	dimensionless temperature gradient on the plate
$\dfrac{1}{\rho}\dfrac{\mathrm{dp}}{\mathrm{dx}}$	density factor
$\dfrac{1}{\mu}\dfrac{\mathrm{d}\mu}{d\eta}$	viscosity factor
$\dfrac{1}{\lambda}\dfrac{\mathrm{d}\lambda}{\mathrm{d}\eta}$	thermal conductivity factor

Subscripts

w	at wall
δ	boundary layer
∞	far from the wall surface

6.1 Introduction

The theoretical analysis of laminar free convection of liquid along an isothermal vertical flat plate was also started by means of Boussinesq approximation. For the case of larger temperature difference, the effects of variable thermophysical properties should be taken into consideration, as those in [1–9]. In [1] Fujii, et al. used two methods of correlating to examine the effects of variable thermophysical properties on heat transfer for free convection from vertical surfaces in liquids. The first method of correlating the data consisted of using the constant property correlations for Nusselt number and evaluating all properties at a reference temperature, $T\mathrm{r} = T_{\mathrm{w}} - (T_{\mathrm{w}} - T_\infty)/4$. They noted that the choice of the reference temperature agrees with the solution provided by two previous studies of Fujii [2] and Akagi [3]. The second method that they used to correlate their data in oils was first proposed by Akagi [3] and applies only to liquids for which viscosity variation is dominant. Piau [4] treated the similarity analysis of variable property effects in free convection from vertical surfaces in high Prandtl number liquids. It was indicated that the main property variations in water at moderate temperature levels are in the viscosity, μ, and the volumetric coefficient of thermal expansion, β, and that for higher Prandtl number liquids, the variation of β is often negligible. In [5] Piau included the effect of thermal stratification of the ambient fluid in an analysis, which also includes variables, μ and β for water. Brown [6] used an integral method and studied the effect of the coefficient of volumetric expansion on laminar free convection heat transfer. Carey and Mollendorf [7] have shown the mathematical forms of viscosity variation with temperature, which results in similarity solutions for laminar free convection from a vertical isothermal surface in liquids with temperature-dependent viscosity. Sabhapathy and Cheng [8] studied the effects of temperature-dependent viscosity and coefficient of thermal expansion on the stability of laminar free convection boundary-layer flow of a liquid along an isothermal, vertical surface, employing linear stability theory for Prandtl numbers between 7 and 10. Qureshi and

Gebhart [9] studied the stability of vertical thermal buoyancy-induced flow in cold and saline water. They indicated that the anomalous density behaviour of cold water, for example, a density extremum at about 4°C in pure water at atmospheric pressure, commonly has very large effects on flow and transport. However, the results reported so far are not convenient for heat transfer prediction due to difficulty of treating the variable thermophysical properties in governing equations.

In this chapter an advanced development [10] of laminar free convection of liquid with large temperature difference is introduced. Velocity component method for similarity transformation presented in Chaps. 4 and 5 is used here for similarity transformation of the governing partial differential equations of liquid free convection. Meanwhile, the polynomial equations are suggested to express the variable thermophysical properties of a liquid. For example, polynomial equations of the density and thermal conductivity of water are proposed, and expression of absolute viscosity of water is also based on a polynomial. A typical example of the laminar free convection of water was provided. It is concluded that the Nusselt number could be predicted by local Grashof number and the dimensionless temperature gradient on the wall. Furthermore, a reliable curve-fit formula of the dimensionless temperature gradient is presented for simple and accurate prediction of water free convection with large temperature difference.

6.2 Governing Partial Deferential Equations and their Similarity Transformation

6.2.1 Governing Partial Differential Equations

The physical analytical model and coordinate system used for laminar free convection of liquid on an isothermal vertical flat plate is shown in Fig. 4.1 also. According to the presentation in Chap.2, the conservation equation for mass, momentum and energy of steady laminar free convection of liquid in the boundary layer are

$$\frac{\partial}{\partial x}(\rho w_x) + \frac{\partial}{\partial y}(\rho w_y) = 0, \tag{6.1}$$

$$\rho\left(w_x \frac{\partial w_x}{\partial x} + w_y \frac{\partial w_x}{\partial y}\right) = \frac{\partial}{\partial y}\left(\mu \frac{\partial w_x}{\partial y}\right) + g\left|\rho_\infty - \rho\right|, \tag{6.2}$$

$$\rho c_p\left(w_x \frac{\partial t}{\partial x} + w_y \frac{\partial t}{\partial y}\right) = \frac{\partial}{\partial y}\left(\lambda \frac{\partial t}{\partial y}\right). \tag{6.3}$$

The absolute value of buoyancy term $g|\rho_\infty - \rho|$ shows that it has always positive sign no matter which one is larger between ρ and ρ_∞. In this case, the buoyancy term $g|\rho_\infty - \rho|$ and the velocity component w_x have same sign. The boundary conditions are

$$y = 0: \quad W_x = 0, \quad W_y = 0, \quad t = t_\mathrm{w}, \tag{6.4}$$

$$y \to \infty: \quad W_x \to 0, \quad t = t_\infty, \tag{6.5}$$

where the variable thermophysical properties are considered except the specific heat. In fact, such treatment for physical properties is suitable for a lot of liquids.

6.2.2 Dimensionless Transformation Variables

For similarity transformation of the governing partial deferential equations for the laminar free convection of liquid, the velocity component method is also used. Consulting the assumed dimensionless variables in Chap. 4 for the similarity transformation of the governing partial differential equations of gas laminar free convection, the following dimensionless transformation variables can be assumed for the transformation of governing equations of liquid laminar free convection:

$$\eta = \frac{y}{x}\left(\frac{1}{4}Gr_{x,\infty}\right)^{1/4}, \tag{6.6}$$

$$\theta = \frac{t - t_\infty}{t_\mathrm{w} - t_\infty}, \tag{6.7}$$

$$W_x = \left[2\sqrt{gx}\left|\frac{\rho_\infty}{\rho_\mathrm{w}} - 1\right|^{1/2}\right]^{-1} w_x, \tag{6.8}$$

$$W_y = \left[2\sqrt{gx}\left|\frac{\rho_\infty}{\rho_\mathrm{w}} - 1\right|^{1/2}\left(\frac{1}{4}Gr_{x,\infty}\right)^{-1/4}\right]^{-1} w_y, \tag{6.9}$$

$$Gr_{x,\infty} = \frac{g|\rho_\infty/\rho_\mathrm{w} - 1|x^3}{{\nu_\infty}^2}. \tag{6.10}$$

6.2.3 Similarity Transformation

For convenience of similarity transformation it is necessary to rewrite the governing equations (6.1)–(6.3) into the following form, respectively:

$$\rho\left(\frac{\partial w_x}{\partial x} + \frac{\partial w_y}{\partial y}\right) + w_x\frac{\partial \rho}{\partial x} + w_y\frac{\partial \rho}{\partial y} = 0, \tag{6.11}$$

$$\rho\left(w_x\frac{\partial w_x}{\partial x} + w_y\frac{\partial w_x}{\partial y}\right) = \mu\frac{\partial^2 w_x}{\partial y^2} + \frac{\partial w_x}{\partial y}\frac{\partial \mu}{\partial y} + g\left|\rho_\infty - \rho\right|, \tag{6.12}$$

$$\rho c_p\left(w_x\frac{\partial t}{\partial x} + w_y\frac{\partial t}{\partial y}\right) = \lambda\frac{\partial^2 t}{\partial y^2} + \frac{\partial \lambda}{\partial y}\frac{\partial t}{\partial y}. \tag{6.13}$$

Similar to the derivation of the partial differential equations in Chap.4, the related partial differential equations (6.11)–(6.13) are obtained and expressed as follows:

$$\frac{\partial w_x}{\partial x} = \sqrt{\frac{g}{x}}\left|\frac{\rho_\infty}{\rho_\mathrm{w}} - 1\right|^{1/2}\left(W_x - \frac{1}{2}\eta\frac{\mathrm{d}W_x}{\mathrm{d}\eta}\right), \tag{6.14}$$

$$\frac{\partial w_y}{\partial y} = 2\sqrt{\frac{g}{x}} \left|\frac{\rho_\infty}{\rho_\mathrm{w}} - 1\right|^{1/2} \frac{\mathrm{d}W_y}{\mathrm{d}\eta}, \tag{6.15}$$

$$\frac{\partial \rho}{\partial x} = -\frac{1}{4}\eta x^{-1} \frac{\mathrm{d}\rho}{\mathrm{d}\eta}, \tag{6.16}$$

$$\frac{\partial \rho}{\partial y} = \frac{\mathrm{d}\rho}{\mathrm{d}\eta} \left(\frac{1}{4} Gr_{x,\infty}\right)^{1/4} x^{-1}, \tag{6.17}$$

$$\frac{\partial w_x}{\partial y} = 2\sqrt{gx} \left|\frac{\rho_\infty}{\rho_\mathrm{w}} - 1\right|^{1/2} \frac{\mathrm{d}W_x}{\mathrm{d}\eta} x^{-1} \left(\frac{1}{4} Gr_{x,\infty}\right)^{1/4}, \tag{6.18}$$

$$\frac{\partial^2 w_x}{\partial y^2} = 2\sqrt{gx} \left|\frac{\rho_\infty}{\rho_\mathrm{w}} - 1\right|^{1/2} \frac{\mathrm{d}^2 W_x}{\mathrm{d}\eta^2} x^{-2} \left(\frac{1}{4} Gr_{x,\infty}\right)^{1/2}, \tag{6.19}$$

$$\frac{\partial \mu}{\partial y} = \frac{\mathrm{d}\mu}{\mathrm{d}\eta} x^{-1} \left(\frac{1}{4} Gr_{x,\infty}\right)^{1/4}, \tag{6.20}$$

$$\frac{\partial t}{\partial x} = -(t_\mathrm{w} - t_\infty) \frac{\mathrm{d}\theta}{\mathrm{d}\eta} \left(\frac{1}{4}\right) \eta x^{-1}, \tag{6.21}$$

$$\frac{\partial t}{\partial y} = (t_\mathrm{w} - t_\infty) \frac{\mathrm{d}\theta}{d\eta} \left(\frac{1}{4} Gr_{x,\infty}\right)^{1/4} x^{-1}, \tag{6.22}$$

$$\frac{\partial^2 T}{\partial y^2} = (t_w - t_\infty) \frac{\mathrm{d}^2\theta}{\mathrm{d}\eta^2} \left(\frac{1}{4} Gr_{x,\infty}\right)^{1/2} x^{-2}, \tag{6.23}$$

$$\frac{\partial \lambda}{\partial y} = \frac{\mathrm{d}\lambda}{\mathrm{d}\eta} \left(\frac{1}{4} Gr_{x,\infty}\right)^{1/4} x^{-1}, \tag{6.24}$$

where

$$t = (t_\mathrm{w} - t_\infty)\theta + t_\infty. \tag{6.25}$$

Similar to the derivations in Chap. 4, by using (6.14)–(6.25), the following governing ordinary differential equations can be obtained from (6.11)–(6.13):

$$2W_x - \eta \frac{\mathrm{d}W_x}{\mathrm{d}\eta} + 4\frac{\mathrm{d}W_y}{\mathrm{d}\eta} - \frac{1}{\rho}\frac{\mathrm{d}\rho}{\mathrm{d}\eta}(\eta W_x - 4W_y) = 0, \tag{6.26}$$

$$\frac{\nu_\infty}{\nu}\left(W_x\left(2W_x - \eta\frac{\mathrm{d}W_x}{\mathrm{d}\eta}\right) + 4W_y\frac{\mathrm{d}W_x}{\mathrm{d}\eta}\right) = \frac{\mathrm{d}^2 W_x}{\mathrm{d}\eta^2} + \frac{1}{\mu}\frac{\mathrm{d}\mu}{\mathrm{d}\eta}\frac{\mathrm{d}W_x}{\mathrm{d}\eta} + \frac{\nu_\infty}{\nu}\frac{\frac{\rho_\infty}{\rho} - 1}{\frac{\rho_\infty}{\rho_\mathrm{w}} - 1}, \tag{6.27}$$

$$Pr\frac{\nu_\infty}{\nu}(-\eta W_x + 4W_y)\frac{\mathrm{d}\theta}{\mathrm{d}\eta} = \frac{1}{\lambda}\frac{\mathrm{d}\lambda}{\mathrm{d}\eta}\frac{\mathrm{d}\theta}{\mathrm{d}\eta} + \frac{\mathrm{d}^2\theta}{\mathrm{d}\eta^2}, \tag{6.28}$$

with boundary conditions

$$\eta = 0, \quad W_x = 0, \quad W_y = 0, \quad \theta = 0, \tag{6.29}$$

$$\eta \to 0, \quad W_x \to 0, \quad \theta \to 0. \tag{6.30}$$

Equations (6.26)–(6.30) are dimensionless governing equations and the boundary conditions of laminar free convection of liquid.

6.2.4 Identical Buoyancy Factor

In fact, the buoyancy factor

$$\frac{\frac{\rho_\infty}{\rho}-1}{\frac{\rho_\infty}{\rho_\mathrm{w}}-1}$$

in (6.27) is suitable for all fluids, i.e. both liquid and gas. For gas the buoyancy factor can be rewritten as follows by using the simple power law of gas:

$$\frac{\frac{\rho_\infty}{\rho}-1}{\frac{\rho_\infty}{\rho_\mathrm{w}}-1}=\frac{\frac{T}{T_\infty}-1}{\frac{T_\mathrm{w}}{T_\infty}-1}=\frac{T-T_\infty}{T_\mathrm{w}-T_\infty}=\theta. \tag{6.31}$$

By using (6.31), (6.26)–(6.28) are completely identical to the dimensionless governing equations (4.18), (4.23) and (4.25) for laminar free convection of gas presented in Chap. 4. Therefore, (4.18), (4.23), and (4.25) can be regarded as the special case of (6.26)–(6.28).

6.3 Treatment of Variable Thermophysical Properties

We take water as an example to introduce the treatment of variable thermophysical properties of liquid. In fact, the variation of the physical properties of water can represent that of most of liquids.

The specific heat c_p of water and a lot of liquids changes little with temperature, and so, it is possible to regard the specific heat c_p as the value c_{p_∞} in a special temperature range for engineering applications. This implies

$$\begin{aligned}
Pr\frac{\nu_\infty}{\nu}&=\frac{\mu c_p}{\lambda}\frac{\nu_\infty}{\nu}\\
&=Pr_\infty\frac{\lambda_\infty}{\mu_\infty c_{p_\infty}}\frac{\mu c_p}{\lambda}\frac{\nu_\infty}{\nu}\\
&=Pr_\infty\frac{\frac{\mu}{\nu}}{\frac{\mu_\infty}{\nu_\infty}}\frac{\lambda_\infty}{\lambda}\\
&=Pr_\infty\frac{\rho}{\rho_\infty}\frac{\lambda_\infty}{\lambda}.
\end{aligned} \tag{6.32}$$

With (6.32), (6.31) can be rewritten as follows for water and many other liquids in laminar free convection:

$$Pr_\infty\frac{\rho}{\rho_\infty}\frac{\lambda_\infty}{\lambda}(-\eta W_x+4W_y)\frac{\mathrm{d}\theta}{\mathrm{d}\eta}=\frac{1}{\lambda}\frac{\mathrm{d}\lambda}{\mathrm{d}\eta}\frac{\mathrm{d}\theta}{\mathrm{d}\eta}+\frac{\mathrm{d}^2\theta}{\mathrm{d}\eta^2}. \tag{6.33}$$

The dimensionless thermophysical property factors

$$\frac{\rho}{\rho_\infty},\ \frac{\lambda_\infty}{\lambda},\ \frac{\mu_\infty}{\mu},\ \frac{\nu_\infty}{\nu},\ \frac{1}{\rho}\frac{\mathrm{d}\rho}{\mathrm{d}\eta},\ \frac{1}{\mu}\frac{\mathrm{d}\mu}{\mathrm{d}\eta}$$

and

$$\frac{1}{\lambda}\frac{\mathrm{d}\lambda}{\mathrm{d}\eta}$$

with the variable thermophysical properties in the transformed dimensionless governing (6.26)–(6.28) have to be dealt with for obtaining their numerical solutions.

From [10], the temperature-dependent expressions of density and thermal conductivity of water with the temperature range between 0 and 100°C are respectively expressed by Shang et al. with polynomials as

$$\rho = -4.48 \times 10^{-3}\, t^2 + 999.9, \tag{6.34}$$

$$\lambda = -8.01 \times 10^{-6}\, t^2 + 1.94 \times 10^{-3} t + 0.563. \tag{6.35}$$

The deviation predicted is less than 0.35% by (6.34), and less than 0.18% by (6.35), compared with the experimental data [11]. For the absolute viscosity of water, the following expression described in [12] is applied:

$$\mu = \exp\left[-1.6 - \frac{1,150}{T} + \left(\frac{690}{T}\right)^2\right] \times 10^{-3}. \tag{6.36}$$

The deviation predicted by (6.36) is less than 1.8%, as compared with the experimental data [11]. With the earlier expressions of ρ, λ and μ as well as (6.25), the earlier thermophysical property factors can be described as follows:

$$\frac{1}{\rho}\frac{\mathrm{d}\rho}{\mathrm{d}\eta} = \frac{-2 \times 4.48 \times 10^{-3}\, t(t_{\mathrm{w}} - t_\infty)\frac{\mathrm{d}\theta}{\mathrm{d}\eta}}{-4.48 \times 10^{-3}\, t^2 + 999.9}, \tag{6.37}$$

$$\frac{1}{\mu}\frac{\mathrm{d}\mu}{\mathrm{d}\eta} = \left(\frac{1,150}{T^2} - 2 \times \frac{690^2}{T^3}\right)(t_{\mathrm{w}} - t_\infty)\frac{\mathrm{d}\theta}{\mathrm{d}\eta}, \tag{6.38}$$

$$\frac{1}{\lambda}\frac{\mathrm{d}\lambda}{\mathrm{d}\eta} = \frac{(-2 \times 8.01 \times 10^{-6}\, t + 1.94 \times 10^{-3})(t_{\mathrm{w}} - t_\infty)\frac{\mathrm{d}\theta}{\mathrm{d}\eta}}{-8.01 \times 10^{-6}\, t^2 + 1.94 \times 10^{-3} t + 0.563}. \tag{6.39}$$

In addition,

$$\frac{\rho}{\rho_\infty} = \frac{-4.48 \times 10^{-3}\, t^2 + 999.9}{-4.48 \times 10^{-3}\, t_\infty^2 + 999.9}, \tag{6.40}$$

$$\frac{\lambda_\infty}{\lambda} = \frac{-8.01 \times 10^{-6}\, t_\infty^2 + 1.94 \times 10^{-3}\, t_\infty + 0.563}{-8.01 \times 10^{-6}\, t^2 + 1.94 \times 10^{-3}\, t + 0.563}, \tag{6.41}$$

$$\frac{\mu_\infty}{\mu} = \frac{\exp\left[-1.6 - \frac{1,150}{T_\infty} + \left(\frac{690}{T_\infty}\right)^2\right]}{\exp\left[-1.6 - \frac{1,150}{T} + \left(\frac{690}{T}\right)^2\right]}, \tag{6.42}$$

While,

$$\frac{\nu_\infty}{\nu} = \frac{\mu_\infty}{\mu}\frac{\rho}{\rho_\infty}. \tag{6.43}$$

6.4 Heat Transfer Analysis

Consulting the heat transfer analysis in Chap. 4 for gas laminar film free convection, the analytic expressions related to heat transfer of liquid laminar free convection are obtained as follows:

$$q_x = -\lambda_{\rm w}(t_{\rm w} - t_\infty)\left(\frac{1}{4}Gr_{x,\infty}\right)^{1/4} x^{-1}\left(\frac{{\rm d}\theta}{{\rm d}\eta}\right)_{\eta=0}, \tag{6.44}$$

$$\alpha_x = -\lambda_{\rm w}\left(\frac{1}{4}Gr_{x,\infty}\right)^{1/4} x^{-1}\left(\frac{{\rm d}\theta}{{\rm d}\eta}\right)_{\eta=0}, \tag{6.45}$$

$$Nu_{x,{\rm w}} = -\left(\frac{1}{4}Gr_{x,\infty}\right)^{1/4}\left(\frac{{\rm d}\theta}{{\rm d}\eta}\right)_{\eta=0}, \tag{6.46}$$

$$Q_x = -\frac{4}{3}b\lambda_{\rm w}(t_{\rm w} - t_\infty)\left(\frac{1}{4}Gr_{x,\infty}\right)^{1/4}\left(\frac{{\rm d}\theta}{{\rm d}\eta}\right)_{\eta=0}, \tag{6.47}$$

$$\overline{\alpha}_x = -\frac{4}{3}\lambda_{\rm w}\left(\frac{1}{4}Gr_{x,\infty}\right)^{1/4} x^{-1}\left(\frac{{\rm d}\theta}{{\rm d}\eta}\right)_{\eta=0}, \tag{6.48}$$

$$\overline{Nu}_x = -\frac{4}{3}\left(\frac{1}{4}Gr_{x,\infty}\right)^{1/4}\left(\frac{{\rm d}\theta}{{\rm d}\eta}\right)_{\eta=0}. \tag{6.49}$$

Here, q_x is local heat transfer rate at position x per unit area on the plate, defined as $q_x = -\lambda_{\rm w}\left(\partial T/\partial y\right)_{y=0}$, α_x is local heat transfer coefficient, defined as $q_x = \alpha_x(t_{\rm w} - t_\infty)$, $Nu_{x,{\rm w}}$ is local Nusselt number, defined as $Nu_{x,{\rm w}} = \alpha_x x/\lambda_{\rm w}$, Q_x is total heat transfer rate for position $x = 0$ to x with width of b on the plate, defined as $Q_x = \int_0^x q_x b\,{\rm d}x$, $\overline{\alpha}_x$ is average heat transfer coefficient, defined as $Q_x = \overline{\alpha}_x(T_{\rm w} - T_\infty) \times b \times x$, and $\overline{Nu}_x$ is average Nusselt number, defined as $\overline{Nu}_{x,{\rm w}} = \overline{\alpha_x}x/\lambda_{\rm w}$.

Same as that reported in Chap. 4, it is seen that, for practical calculation of heat transfer, only $({\rm d}\theta/{\rm d}\eta)_{\eta=0}$ dependent on numerical solution is no-given variable.

6.5 Numerical Solutions

As a typical liquid laminar free convection, the water laminar free convection can be taken as an example for presentation of the numerical calculation. The shooting method described in Chap. 4 has been adopted to solve numerically the nonlinear governing equations (6.26)–(6.28) with the boundary conditions (6.29) and (6.30) at different temperature conditions $t_{\rm w}$ and t_∞. The water thermophysical property values of ρ_∞, $\rho_{\rm w}$, ν_∞, λ_∞ and Pr_∞ at different temperature are taken directly from [11], as quoted in Table 6.1. The typical results for velocity and temperature fields of the boundary layer are plotted as Figs. 6.1–6.4, respectively. The corresponding solutions for the dimensionless

Table 6.1. The values of thermophysical properties of water

t (°C)	0	10	20	30	40	50	60	70	80	90	100
ρ (kg m^{-3})	999.8	999.8	998.3	995.8	992.3	988.1	983.2	977.7	971.4	965.1	958.4
ν $\times 10^{-6}$ m^2 s^{-1}	1.792	1.308	1.004	0.798	0.658	0.554	0.457	0.414	0.365	0.326	0.296
λ (W (K m)$^{-1}$)	0.562	0.582	0.5996	0.6151	0.6287	0.6405	0.6507	0.6595	0.6668	0.6728	0.6773
Pr	13.44	9.42	6.99	5.42	4.34	3.57	3.00	2.57	2.23	1.97	1.76

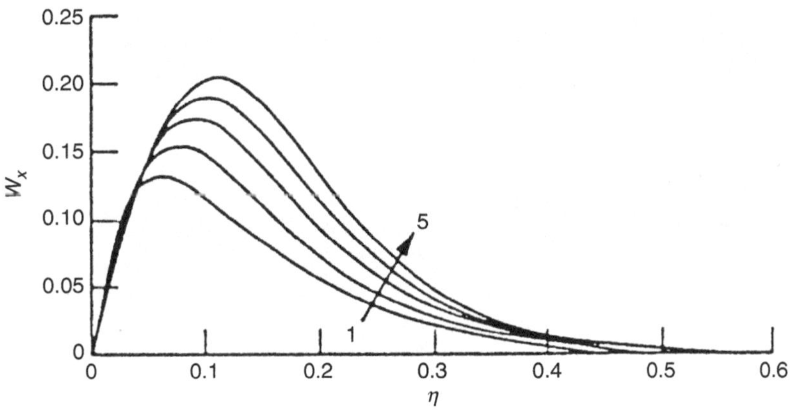

Fig. 6.1. The velocity profiles at $t_w = 40°C$ with different t_∞ ($1 \rightarrow 5$: $t_\infty = 20, 39.99, 60, 80, 100°C$), cited from Shang, Wang and Quan [10]

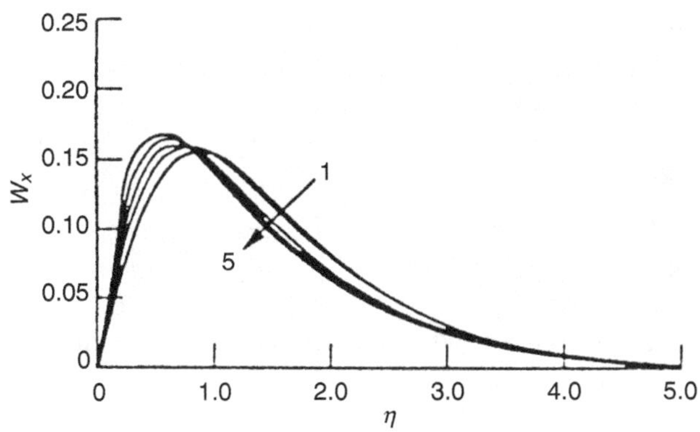

Fig. 6.2. The velocity profiles at $t_\infty = 40°C$ and different surface temperatures t_w ($1 \rightarrow 5$: $t_w = 20, 39.99, 60, 80, 100°C$), cited from Shang, Wang and Quan [10]

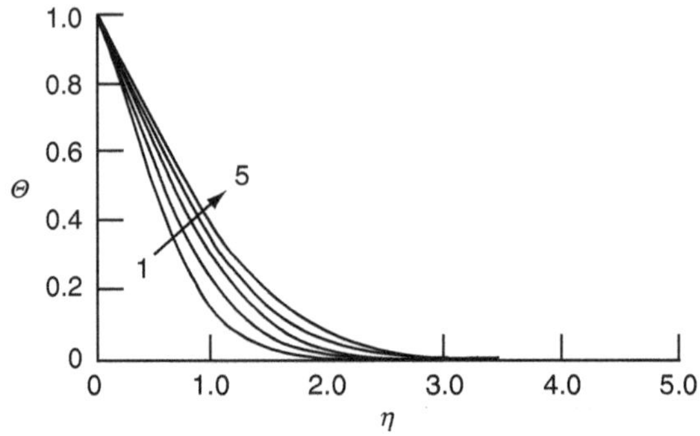

Fig. 6.3. The temperature profiles at $t_\mathrm{w} = 40°\mathrm{C}$ and different t_∞ $(1 \rightarrow 5 : t_\infty = 20, 39.99, 60, 80, 100°\mathrm{C})$, cited from Shang, Wang and Quan [10]

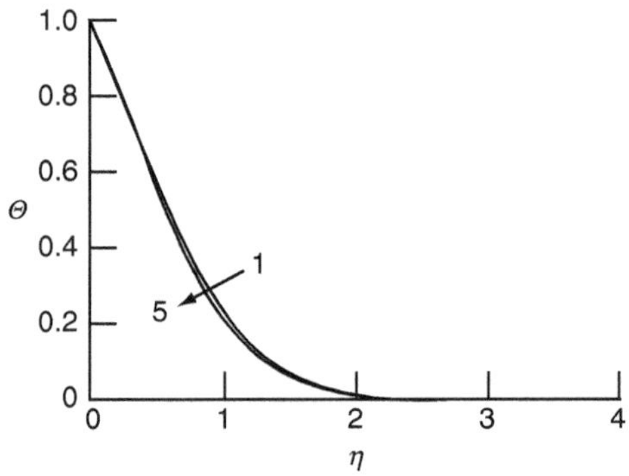

Fig. 6.4. The temperature profiles at $t_\infty = 40°\mathrm{C}$ and different surface temperatures t_w $(1 \rightarrow 5 : t_\mathrm{w} = 20, 39.99, 60, 80, 100°\mathrm{C})$, cited from Shang, Wang and Quan [10]

temperature gradient $(-\mathrm{d}\theta/\mathrm{d}\eta)_{\eta=0}$ are described in Table 6.2 and plotted in Fig. 6.5. The velocity and temperature profiles show clearly the effects of the variable thermophysical properties on velocity and temperature distributions as well as heat transfer of the water free convection. The related influences are presented as follows:

Effects of t_∞. The bulk temperature t_∞ causes a great effect on the velocity and temperature profiles. With increase of t_∞, the velocity W_x and the temperature θ obviously increase, meanwhile, the maximum of w_x shifts further from the plate. While, with increase of t_∞ the temperature gradient $-(\mathrm{d}\theta/\mathrm{d}\eta)_{\eta=0}$ decreases obviously.

Effects of t_w. The effects of t_w on the velocity W_x and temperature θ are much less than those of t_∞. With the increase of t_w, the maximum velocity of W_x increases and shifts slightly close to the plate. Generally, the effects of t_w on the temperature θ and temperature gradient $-(\mathrm{d}\theta/\mathrm{d}\eta)_{\eta=0}$ are slightly. Together with increasing t_∞, the effects of t_w on the temperature field θ and temperature gradient become even smaller and smaller.

Table 6.2. The typical numerical solutions of dimensionless temperature gradient $-(\mathrm{d}\theta/\mathrm{d}\eta)_{\eta=0}$ for water laminar free convection along a vertical plate, cited from Shang, Wang, and Quan [10]

t_w	t_∞	Pr_∞	$-\left(\frac{\mathrm{d}\theta}{\mathrm{d}\eta}\right)_{\eta=0}$	t_w	t_∞	Pr_∞	$-\left(\frac{\mathrm{d}\theta}{\mathrm{d}\eta}\right)_{\eta=0}$
4.99	5	11.16	1.21	5	10	9.42	1.153
10	5	11.16	1.169	9.99	10	9.42	1.137
15	5	11.16	1.158	15	10	9.42	1.133
20	5	11.16	1.156	20	10	9.42	1.13
30	5	11.16	1.164	30	10	9.42	1.231
40	5	11.16	1.179	40	10	9.42	1.139
50	5	11.16	1.196	50	10	9.42	1.15
60	5	11.16	1.212	60	10	9.42	1.162
70	5	11.16	1.229	70	10	9.42	1.175
80	5	11.16	1.245	80	10	9.42	1.187
90	5	11.16	1.26	90	10	9.42	1.199
100	5	11.16	1.275	100	10	9.42	1.211
5	20	6.99	1.076	5	30	5.42	0.989
10	20	6.99	1.063	10	30	5.42	0.983
15	20	6.99	1.056	15	30	5.42	0.979
19.99	20	6.99	1.05	20	30	5.42	0.977
30	20	6.99	1.051	29.99	30	5.42	0.971
40	20	6.99	1.054	40	30	5.42	0.977
50	20	6.99	1.06	50	30	5.42	0.983
60	20	6.99	1.068	60	30	5.42	0.988
70	20	6.99	1.075	70	30	5.42	0.994
80	20	6.99	1.083	80	30	5.42	0.999
90	20	6.99	1.092	90	30	5.42	1.005
100	20	6.99	1.1	100	30	5.42	1.012
5	40	4.34	0.917	5	50	3.57	0.858
10	40	4.34	0.913	10	50	3.57	0.856
15	40	4.34	0.911	15	50	3.57	0.855
20	40	4.34	0.91	20	50	3.57	0.855
30	40	4.34	0.91	30	50	3.57	0.856
39.99	40	4.34	0.914	40	50	3.57	0.858
50	40	4.34	0.916	49.99	50	3.57	0.861
60	40	4.34	0.92	60	50	3.57	0.864
70	40	4.34	0.924	70	50	3.57	0.867
80	40	4.34	0.929	80	50	3.57	0.87
90	40	4.34	0.933	90	50	3.57	0.874
100	40	4.34	0.938	100	50	3.57	0.878
5	60	3	0.809	5	70	2.57	0.768
10	60	3	0.808	10	70	2.57	0.767
15	60	3	0.807	15	70	2.57	0.767
20	60	3	0.807	20	70	2.57	0.767
30	60	3	0.809	30	70	2.57	0.769
40	60	3	0.81	40	70	2.57	0.77
50	60	3	0.813	50	70	2.57	0.772

Table 6.2. *Continued*

t_w	t_∞	Pr_∞	$-\left(\frac{d\theta}{d\eta}\right)_{\eta=0}$	t_w	t_∞	Pr_∞	$-\left(\frac{d\theta}{d\eta}\right)_{\eta=0}$
59.99	60	3	0.814	60	70	2.57	0.774
70	60	3	0.818	69.99	70	2.57	0.779
80	60	3	0.821	80	70	2.57	0.779
90	60	3	0.824	90	70	2.57	0.781
100	60	3	0.827	100	70	2.57	0.783
5	80	2.23	0.733	5	90	1.97	0.704
10	80	2.23	0.732	10	90	1.97	0.704
15	80	2.23	0.733	15	90	1.97	0.704
20	80	2.23	0.733	20	90	1.97	0.704
30	80	2.23	0.734	30	90	1.97	0.705
40	80	2.23	0.735	40	90	1.97	0.706
50	80	2.23	0.737	50	90	1.97	0.707
60	80	2.23	0.739	60	90	1.97	0.709
70	80	2.23	0.74	70	90	1.97	0.71
79.99	80	2.23	0.74	80	90	1.97	0.711
90	80	2.23	0.744	89.99	90	1.97	0.714
100	80	2.23	0.746	100	90	1.97	0.714
5	100	1.76	0.679				
10	100	1.76	0.679				
15	100	1.76	0.679				
20	100	1.76	0.679				
30	100	1.76	0.680				
40	100	1.76	0.681				
50	100	1.76	0.682				
60	100	1.76	0.683				
70	100	1.76	0.684				
80	100	1.76	0.685				
90	100	1.76	0.686				
99.99	100	1.76	0.686				

The velocity and temperature profiles show the large differences between the momentum and temperature boundary layer thicknesses for laminar free convection of liquid due to $Pr \gg 1$. Therefore, it is very difficult to make a solution of the governing equations for liquid free convection, especially for consideration of variable thermophysical properties. An accurate solution could be converged only when a special boundary layer thicknesses η is taken for a special t_∞, or special Pr_∞. Such special boundary layer thicknesses η for obtaining convergence solution only depends on t_∞ or Pr_∞. Actually, the range of such special boundary layer thicknesses η, which can be taken, is very narrow for the convergence of solution. It is found that the solution is only converged in the very narrow range of the special boundary layer thicknesses η for laminar free convection of liquid. Otherwise the convergence solution will never be obtained.

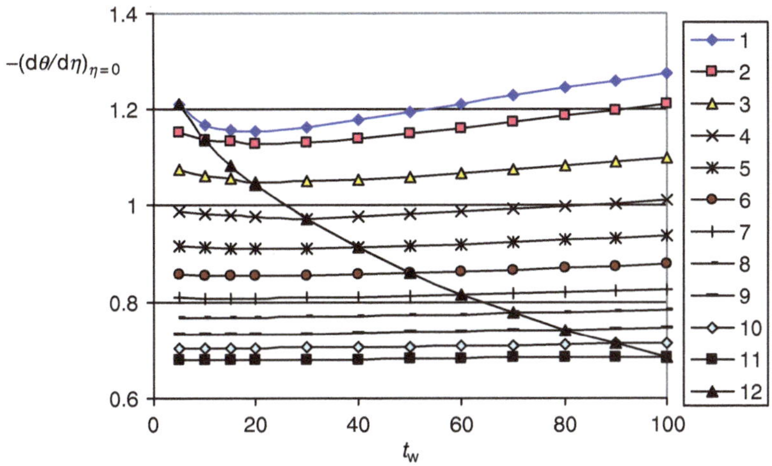

Fig. 6.5. Dimensionless temperature gradient for water laminar free convection along vertical plate 1 → 11: for $t_\infty = 5, 10, 20, 30, 40, 50, 60, 70, 80, 90, 100°\mathrm{C}$ 12: for related solution under Boussinesq approximation

6.6 A Curve-Fit Formula for Heat Transfer

Accurate solutions $-(\mathrm{d}\theta/\mathrm{d}\eta)_{\eta=0}*$ for Boussinesq approximation could be obtained from Table 6.2 in which the plate temperature t_w is very close to the bulk temperature t_∞. These Boussinesq solutions are listed in Table 6.3 for laminar free convection at the water Prandtl number range.

Based on the Boussinesq solutions listed in Table 6.3, a curve-fit formula (6.50) is obtained by Shang et al. [10] for prediction of the solutions of liquid laminar free convection under the Boussinesq approximation.

$$-\left(\frac{\mathrm{d}\theta}{\mathrm{d}\eta}\right)^*_{\eta=0} = 0.5764 + 0.1797 \times Ln(Pr) + 0.0331 \times Ln^2(Pr) \quad (1.7 < Pr_\infty < 13.5). \tag{6.50}$$

The values predicted by using (6.50) are obtained and listed in Table 6.3. It is seen from Table 6.3 that by using (6.50) the predicted deviation of Boussinesq solutions $-(\mathrm{d}\theta/\mathrm{d}\eta)_{\eta=0}*$ is less than 0.618%.

From the typical solutions for temperature gradient $-(\mathrm{d}\theta/\mathrm{d}\eta)_{\eta=0}$ in Table 6.2 it is found that the effect of t_w on temperature gradient $-(\mathrm{d}\theta/\mathrm{d}\eta)_{\eta=0}$ is not obvious generally, but the temperature t_∞ dominates the effect on the temperature gradient $-(\mathrm{d}\theta/\mathrm{d}\eta)_{\eta=0}$. On this basis, Shang et al. [10] found if the Prandtl number Pr in (6.50) is replaced by a reference Prandtl number Pr_∞ as (6.51), the maximal deviation for prediction of the temperature gradient $-(\mathrm{d}\theta/\mathrm{d}\eta)_{\eta=0}$ of water laminar free convection is less than 6% for t_∞

Table 6.3. Boussinesq solutions for laminar free convection in the Prandtl number rage of water The superscript * denotes the case of Boussinesq approximation.

Pr	1.76	1.97	2.23	2.57	3	3.57	4.34	5.42	6.99	9.42	11.16
$-\left(\frac{d\theta}{d\eta}\right)_{\eta=0} *$ (1)	0.686	0.714	0.740	0.779	0.814	0.861	0.914	0.971	1.050	1.137	1.210
$-\left(\frac{d\theta}{d\eta}\right)_{\eta=0} *$ (2)	0.6886	0.7135	0.7418	0.7756	0.8138	0.8587	0.9115	0.9747	1.0510	1.1459	1.2025
$E(\%)$	−0.374	0.076	−0.245	0.448	0.028	0.269	0.274	−0.377	−0.093	−0.786	0.618

$-\left(\frac{d\theta}{d\eta}\right)_{\eta=0} *$ (1): numerical solutions

$-\left(\frac{d\theta}{d\eta}\right)_{\eta=0} *$ (2): evaluated by (6.50)

$$E = \frac{-\left(\frac{d\theta}{d\eta}\right)_{\eta=0}^{*(1)} - \left[-\left(\frac{d\theta}{d\eta}\right)_{\eta=0}^{*(2)}\right]}{-\left(\frac{d\theta}{d\eta}\right)_{\eta=0}^{*(1)}}$$

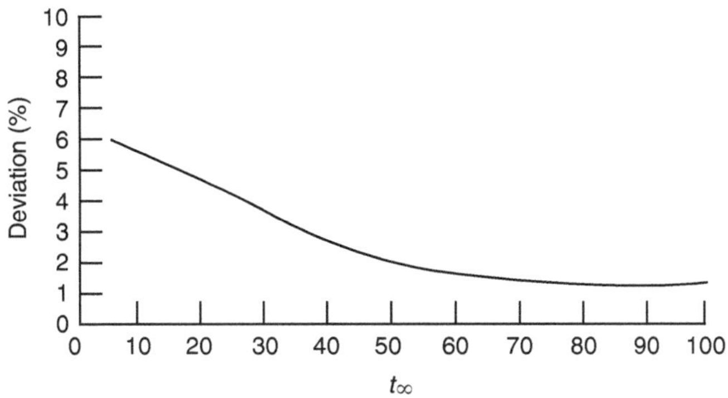

Fig. 6.6. Maximum calculated deviation of $-(\mathrm{d}\theta/\mathrm{d}\eta)_{\eta=0}$ by (6.51), cited from Shang, Wang and Quan [10]

range from 5 to 100°C with consideration of variable thermophysical properties. However, the maximal predicted deviation will be less than 2% for t_∞ range from 50 to 100°C, as shown in Fig. 6.6.

$$-\left(\frac{\mathrm{d}\theta}{\mathrm{d}\eta}\right)_{\eta=0} = 0.5764 + 0.1797 \times Ln(Pr_\infty) + 0.0331 \times Ln^2(Pr_\infty), \quad (6.51)$$

where Pr_∞ is reference Prandtl number at bulk temperature t_∞.

Therefore, combining (6.51) with (6.44)–(6.49) we can predict heat transfer of water laminar free convection conveniently for meeting engineering requirements.

6.7 Summary

So far, the governing equations of liquid laminar free convection and the equations related to heat transfer can be summarized in Table 6.4.

6.8 Remarks

From the contents introduced in this chapter the following remarks are pertinent: The analyses focus on liquid laminar free convection along an isothermal vertical flat plate with consideration of variable fluid thermophysical properties. The velocity component method can be used conveniently to transform the governing partial differential equations of laminar free convection of liquid into the corresponding dimensionless system.

Taking water as an example, the temperature-dependent expressions of the density, thermal conductivity and viscous are introduced, while the specific heat at constant pressure is taken as constant with maximum possible

Table 6.4. Summary of the governing equations of liquid laminar free convection and the equations related to heat transfer

term	expression
	governing partial deferential equations
mass equation	$\frac{\partial}{\partial x}(\rho w_x) + \frac{\partial}{\partial y}(\rho w_y) = 0$
momentum equation	$\rho\left(w_x \frac{\partial w_x}{\partial x} + w_y \frac{\partial w_x}{\partial y}\right) = \frac{\partial}{\partial y}\left(\mu \frac{\partial w_x}{\partial y}\right) + g\left\|\rho_\infty - \rho\right\|$
energy equation	$\rho c_p \left(w_x \frac{\partial T}{\partial x} + w_y \frac{\partial T}{\partial y}\right) = \frac{\partial}{\partial y}\left(\lambda \frac{\partial T}{\partial y}\right)$
boundary conditions	$y = 0: \quad w_x = 0, \quad w_y = 0, \quad T = T_w$ $y \to \infty: \quad w_x \to 0, \quad T = T_\infty$
	assumed similarity variables
η	$\frac{y}{x}\left(\frac{1}{4}Gr_{x,\infty}\right)^{1/4}$
$Gr_{x,\infty}$	$\frac{g\left\|\rho_\infty/\rho_w - 1\right\| x^3}{\nu_\infty^2}$
θ	$\frac{T - T_\infty}{T_w - T_\infty}$
W_x	$\left[2\sqrt{gx}\left\|\frac{\rho_\infty}{\rho} - 1\right\|^{1/2}\right]^{-1} w_x$
W_y	$\left[2\sqrt{gx}\left\|\frac{\rho_\infty}{\rho} - 1\right\|^{1/2}\left(\frac{1}{4}Gr_{x,\infty}\right)^{-1/4}\right]^{-1} w_y$
	governing ordinary differential equations
mass equations	$2W_x - \eta\frac{dW_x}{d\eta} + 4\frac{dW_y}{d\eta} - \frac{1}{\rho}\frac{d\rho}{d\eta}(\eta W_x - 4W_y) = 0$
momentum equation	$\frac{\nu_\infty}{\nu}\left(W_x\left(2W_x - \eta\frac{dW_x}{d\eta}\right) + 4W_y\frac{dW_x}{d\eta}\right) = \frac{d^2W_x}{d\eta^2} + \frac{1}{\mu}\frac{d\mu}{d\eta}\frac{dW_x}{d\eta} + \frac{\nu_\infty}{\nu}\frac{\frac{\rho_\infty}{\rho} - 1}{\frac{\rho_\infty}{\rho_w} - 1}$
energy equation	$Pr\frac{\nu_\infty}{\nu}(-\eta W_x + 4W_y)\frac{d\theta}{d\eta} = \frac{1}{\lambda}\frac{d\lambda}{d\eta}\frac{d\theta}{d\eta} + \frac{d^2\theta}{d\eta^2}$
boundary condition	$\eta = 0: \quad w_x = 0, \quad w_y = 0, \quad \theta = 1$ $\eta \to \infty: \quad w_x = 0, \quad \theta = 0$
	equations related to heat transfer
$q_x \left(\text{defined as} - \lambda_w\left(\frac{\partial T}{\partial y}\right)_{y=0}\right)$	$-\lambda_w(T_w - T_\infty)\left(\frac{1}{4}Gr_{x,\infty}\right)^{1/4} x^{-1}\left(\frac{d\theta}{d\eta}\right)_{\eta=0}$
$\alpha_x \left(\text{defined as} \frac{q_x}{(T_w - T_\infty)}\right)$	$-\lambda_w\left(\frac{1}{4}Gr_{x,\infty}\right)^{1/4} x^{-1}\left(\frac{d\theta}{d\eta}\right)_{\eta=0}$
$Q_x \left(\text{defined as } \int_0^x q_x b\, dx\right)$	$-\frac{4}{3}b\lambda_w(T_w - T_\infty)\left(\frac{1}{4}Gr_{x,\infty}\right)^{1/4}\left(\frac{d\theta}{d\eta}\right)_{\eta=0}$
$\overline{\alpha}_x \left(\text{defined as} \frac{Q_x}{(T_w - T_\infty)bx}\right)$	$-\frac{4}{3}\lambda_w\left(\frac{1}{4}Gr_{x,\infty}\right)^{1/4} x^{-1}\left(\frac{d\theta}{d\eta}\right)_{\eta=0}$
$Nu_{x,w} \left(\text{defined as} \frac{\alpha_x x}{\lambda_w}\right)$	$\frac{Nu_{x,w}}{\left(\frac{1}{4}Gr_{x,\infty}\right)^{1/4}} = -\left(\frac{d\theta}{d\eta}\right)_{\eta=0}$
$\overline{Nu}_{x,w} \left(\text{defined as} \frac{\overline{\alpha}_x x}{\lambda_w}\right)$	$\frac{\overline{Nu}_{x,w}}{\left(\frac{1}{4}Gr_{x,\infty}\right)^{1/4}} = -\frac{4}{3}\left(\frac{d\theta}{d\eta}\right)_{\eta=0}$
$-\left(\frac{d\theta}{d\eta}\right)_{\eta=0}$	$0.5764 + 0.1797 \times Ln(Pr_\infty) + 0.0331 \times Ln^2(Pr_\infty)$ ($1.7 < Pr < 13.5$) for laminar free convection of water

deviation of 0.45% only. The polynomial is suggested as the appropriate form for expressions of the variations of ρ and λ with temperature. While, absolute viscosity μ is expressed as a power function, and a polynomial is taken as its exponent. These expressions are reliable according to the typical experimental results.

The non-linear governing equations with corresponding boundary conditions are solved numerically by the shooting method. The expressions related to heat transfer dependent on the temperature gradient $-(\mathrm{d}\theta/\mathrm{d}\eta)_{\eta=0}$ are presented. The effect of variable thermophysical properties on water laminar free convection along an isothermal vertical plate was investigated. The method proposed to predict heat transfer for liquid laminar free convection in coordination of variable thermophysical properties is reliable.

The reported expression, (6.50), for prediction of Boussinesq solution could be used to conveniently predict heat transfer of water laminar free convection, if the Prandtl number Pr is replaced by a reference Prandtl number Pr_∞. The precision of such prediction can meet engineering application.

6.9 Calculation Examples

Question 1: A flat plate with $b = 1\,\mathrm{m}$ in width and $x = 0.25\,\mathrm{m}$ in length is suspended vertically in the space of water. The ambient temperature is $t_\infty = 5°\mathrm{C}$, and the plate temperature is $t_\mathrm{w} = 60°\mathrm{C}$. The water properties are as follows:

$\nu_\infty = 1.5475 \times 10^{-6}\,\mathrm{m}^2\ \mathrm{s}^{-1}, Pr_\infty = 11.16$ and $\rho_\infty = 999.8\ \mathrm{kg\ m}^{-3}$ at $t_\infty = 5°\mathrm{C}$; $\lambda_\mathrm{w} 0.659\,\mathrm{W\ (m° C)}^{-1}$ and $\rho_\mathrm{w} = 983.8\,\mathrm{kg\,m}^{-3}$ at $t_\mathrm{w} = 60°\mathrm{C}$.

Suppose the free convection is laminar, please calculate the free convection heat transfer on the plate.

Solution. With the definition of local Grashof number shown in (6.10) we have

$$\begin{aligned} Gr_{x,\infty} &= \frac{g\,|\rho_\infty/\rho_\mathrm{w} - 1|\,x^3}{\nu_\infty^2} \\ &= \frac{9.8 \times |999.8/983.8 - 1| \times 0.25^3}{(1.5475 \times 10^{-6})^2} \\ &= 1.04 \times 10^9. \end{aligned}$$

The flow of free convection can be regarded as laminar flow. With (6.51) the dimensionless temperature gradient $-(\mathrm{d}\theta/\mathrm{d}\eta)_{\eta=0}$ for water laminar free convection can be calculated by the following equation:

$$\begin{aligned} -\left(\frac{\mathrm{d}\theta}{\mathrm{d}\eta}\right)_{\eta=0} &= 0.5764 + 0.1797 \times Ln(Pr_\infty) + 0.0331 \times Ln^2(Pr_\infty) \\ &= 0.5764 + 0.1797 \times Ln(11.16) + 0.0331 \times Ln^2(11.16) \\ &= 1.2025. \end{aligned}$$

(The evaluated value $-(\mathrm{d}\theta/\mathrm{d}\eta)_{\eta=0}$ is very close to the related typical value 1.212 got from Table 6.2). On this basis, with (6.49) the following average

Nusselt number $\overline{Nu}_{x,\mathrm{w}}$ can be obtained as

$$\begin{aligned}\overline{Nu}_{x,\mathrm{w}} &= -\frac{4}{3}\left(\frac{1}{4}Gr_{x,\infty}\right)^{1/4}\left(\frac{\mathrm{d}\theta}{\mathrm{d}\eta}\right)_{\eta=0}\\ &= (4/3)\times(0.25\times 1.04\times 10^9)^{1/4}\times 1.2025\\ &= 203.6.\end{aligned}$$

With the definition of average Nusselt number, $\overline{Nu}_{x,\mathrm{w}} = \overline{\alpha}_x x/\lambda_{\mathrm{w}}$, the following average heat transfer coefficient can be calculated as

$$\overline{\alpha}_x = \overline{Nu}_{x,\mathrm{w}}\frac{\lambda_{\mathrm{w}}}{x} = 203.6\times\frac{0.659}{0.25} = 536.7\,\mathrm{W}\ (\mathrm{m}^2\,{}^\circ\mathrm{C})^{-1}.$$

With the definition of the average heat transfer coefficient, $Q_x = \overline{\alpha}_x(t_{\mathrm{w}} - t_\infty)\times x\times b$ we obtain the following total free convection heat transfer on the plate

$$\begin{aligned}Q_x &= \overline{\alpha}_x(t_{\mathrm{w}} - t_\infty)\times x\times b\\ &= 536.7\times(60-5)\times 0.25\times 1\\ &= 7379.6\,\mathrm{W}\\ &= 7.3796\,\mathrm{kW}.\end{aligned}$$

Question 2: For the flat plate of question 1, if the plate height is changed to 0.11 m, and temperatures are changed into $t_\infty = 60^\circ\mathrm{C}$ and $t_{\mathrm{w}} = 5^\circ\mathrm{C}$, please calculate the corresponding heat transfer.

The water physical properties are as follows: kinematic viscosity $\nu_\infty = 0.478\times 10^{-6}\,\mathrm{m}^2\ \mathrm{s}^{-1}$, Prandtl number $Pr_\infty = 3$ and the density $\rho_\infty = 983.1\ \mathrm{kg\ m}^{-3}$ at $t_\infty = 60^\circ\mathrm{C}$; thermal conductivity $\lambda_{\mathrm{w}} = 0.5625$ and density $\rho_{\mathrm{w}} = 999.8\ \mathrm{kg\ m}^{-3}$ at $t_{\mathrm{w}} = 5^\circ\mathrm{C}$.

Solution. With the definition of local Grashof number shown in (6.10) we get

$$\begin{aligned}Gr_{x,\infty} &= \frac{g\,|\rho_\infty/\rho_{\mathrm{w}} - 1|\,x^3}{\nu_\infty^2}\\ &= \frac{9.8\times|983.1/999.8 - 1|\times 0.11^3}{(0.478\times 10^{-6})^2}\\ &= 0.95357\times 10^9.\end{aligned}$$

The free convection can be regarded as laminar flow.

With (6.52) the dimensionless temperature gradient $-(\mathrm{d}\theta/\mathrm{d}\eta)_{\eta=0}$ will be

$$\begin{aligned}-\left(\frac{\mathrm{d}\theta}{\mathrm{d}\eta}\right)_{\eta=0} &= 0.5764 + 0.1797\times Ln(Pr_\infty) + 0.0331\times Ln^2(Pr_\infty)\\ &= 0.5764 + 0.1797\times Ln(3) + 0.0331\times Ln^2(3)\\ &= 0.5764 + 0.1797\ Ln(3) + 0.0331\ (Ln(3))^2\\ &= 0.81387.\end{aligned}$$

On this basis, with (6.49) the following average Nusselt number $\overline{Nu}_{x,\mathrm{w}}$ can be obtained as

$$\begin{aligned}\overline{Nu}_{x,\mathrm{w}} &= -\frac{4}{3}\left(\frac{1}{4}Gr_{x,\infty}\right)^{1/4}\left(\frac{\mathrm{d}\theta}{\mathrm{d}\eta}\right)_{\eta=0} \\ &= (4/3) \times (0.25 \times 0.95357 \times 10^9)^{1/4} \times 0.81387 \\ &= 134.84.\end{aligned}$$

With the definition of average Nusselt number, $\overline{Nu}_{x,\mathrm{w}} = \overline{\alpha}_x x/\lambda_\mathrm{w}$, the following mean heat transfer coefficient can be calculated as

$$\overline{\alpha}_x = \overline{Nu}_{x,\mathrm{w}}\frac{\lambda_\mathrm{w}}{x} = 134.84 \times \frac{0.5625}{0.11} = 689.5\ \mathrm{W\ (m^2\ {}^\circ C)^{-1}}$$

With the definition of the average heat transfer coefficient $\overline{\alpha}_x, Q_x = \overline{\alpha}_x(t_\mathrm{w} - t_\infty) \times x \times b$ we have the following free convection heat transfer on the plate

$$\begin{aligned}Q_x &= \overline{\alpha}_x(t_\mathrm{w} - t_\infty) \times x \times b \\ &= 689.5 \times (5 - 60) \times 0.11 \times 1 \\ &= -4171.5\,\mathrm{W} \\ &= -4.1715\,\mathrm{kW}.\end{aligned}$$

The negative sign denotes the heat flux is to the plate from the liquid.

References

1. T. Fujii et al., Experiments on natural convection heat transfer from the outer surface of a vertical cylinder to liquids, Int. J. Heat Mass Transfer 13, pp. 753–787, 1970
2. T.N. Fujii, Heat transfer from a vertical flat surface by laminar free convection – the case where the physical constants of fluids depend on the temperature and the surface has an arbitrary temperature distribution in the vertical direction, Trans. Jpn Soc. Mech. Eng. 24, pp. 964–972, 1958
3. S. Akagi, Free convection heat transfer in viscous oil, Trans. Jpn Soc. Mech. Eng. 30, pp. 624–635, 1964
4. J.M. Piau, Convection Natural Laminar en Regime Permanent dans les Liquids, Influence des Variations des Properties Physique avec la Temperature, C.R. Hebd, Seanc, Acad. Sci., Paris, Vol. 271, pp. 935–956, 1970
5. J.M. Piau, , Influence des variations des properties physiques et la stratification en convection naturelle, Int. J. Heat Mass Transfer 17, pp. 465–476, 1974
6. A. Brown, The effect on laminar free convection heat transfer of temperature dependence of the coefficient of volumetric expansion, Trans. ASME, Ser. C, J. Heat Transfer 97, pp. 133–135, 1975
7. V.P. Carey and J.C. Mollendorf, Natural convection in liquids with temperature dependence viscosity, In Proceedings of the Sixth International Heat Transfer Conference, Toronto, NC-5, Vol. 2, pp. 211–217, Hemisphere, Washington, DC, 1978

8. P. Sabhapathy and K.C. Cheng, The effect of temperature-dependent viscosity and coefficient of thermal expansion on the stability of laminar, natural convective flow along an isothermal, vertical surface, Int. J. Heat Mass Transfer 29, pp. 1521–1529, 1986
9. Z.H. Qureshi and B. Gebhart, The stability of vertical thermal buoyancy induced flow in cold pure and saline water, Int. J. Heat Mass Transfer 29, pp. 1383–1392, 1986
10. D.Y. Shang, B.X. Wang, Y. Wang and Y. Quan, Study on liquid laminar free convection with consideration of variable thermophysical properties, Int. J. Heat Mass Transfer 36, No. 14, pp. 3411–3419, 1993
11. WVDI – Warmeatlas, Berechnungsblatter fur den Warmeübertragung, 5, erweiterte Auflage, VDI Verlage GmbH, Düsseldorf, 1988
12. J.Q. Chang, Real Fluid Mechanics, Tsinghua University Press, Beijing, 1986

7

Heat Transfer Deviation of Laminar Free Convection Caused by Boussinesq Approximation

Nomenclature

c_p	specific heat at constant pressure, $\mathrm{J\,(kg\,K)^{-1}}$
$E^*_{\alpha_x}$	relative predicted deviation of heat transfer coefficient caused by Boussinesq approximation, $1-(\alpha^*_x/\alpha_x)$
g	gravitation acceleration, $\mathrm{m\,s^{-2}}$
$Gr_{x,\mathrm{f}}$	Grashof number related to average temperature T_f, $\frac{g\vert\rho_\infty/\rho_\mathrm{w}-1\vert x^3}{{\nu_\mathrm{f}}^2}$
$Gr_{x,\infty}$	local Grashof number for the free convection of gas on isothermal vertical flat plate, $\frac{g\vert T_\mathrm{w}/T_\infty-1\vert x^3}{\nu^2_\infty}$
	local Grashof number for the free convection of liquid on isothermal vertical flat plate, $\frac{g\vert\rho_\infty/\rho_\mathrm{w}-1\vert x^3}{\nu^2_\infty}$
$Nu_{x,\mathrm{w}}$	local Nusselt number, $\alpha_x x/\lambda_\mathrm{w}$
$\overline{Nu}_{x,\mathrm{w}}$	average Nusselt number, $\overline{\alpha}_x x/\lambda_\mathrm{w}$
n_{c_p}	specific heat parameter of gas
n_λ	thermal conductivity parameter of gas
n_μ	viscosity parameter of gas
$n_{\mu\lambda}$	overall temperature parameter
P	pressure, $\mathrm{N\ m^{-2}}$
Pr	Prandtl number
ΔPr	relative difference of Prandtl number, ΔPr, $\frac{Pr_\mathrm{w}-Pr_\infty}{Pr_\mathrm{max}-Pr_\mathrm{min}}$
q_x	local heat transfer rate at position x per unit area on the plate, $\mathrm{W\ m^{-2}}$
Q_x	total heat transfer rate for position $x=0$ to x with width of b on the plate, W
t	temperature, °C
t_f	average temperature, $(t_\mathrm{w}+t_\infty)/2$, °C
T	absolute temperature, K

w_x, w_y	velocity components in the x- and y- directions, respectively, m s^{-1}
W_x, W_y	dimensionless velocity components in the x- and y- directions, respectively

Greek symbols

α_x	local heat transfer coefficient, W $(\text{m K})^{-1}$
$\overline{\alpha}_x$	average heat transfer coefficient, W $(\text{m}^2\,\text{K})^{-1}$
δ	boundary layer thickness, m
η	dimensionless coordinate variable for boundary layer
θ	dimensionless temperature
λ	thermal conductivity, W $(\text{m K})^{-1}$
μ	absolute viscosity, kg $(\text{m s})^{-1}$
ν	kinetic viscosity, $\text{m}^2\ \text{s}^{-1}$
ρ	density, kg m^{-3}
$\dfrac{\dfrac{\rho_\infty}{\rho}-1}{\dfrac{\rho_\infty}{\rho_w}-1}$	buoyancy term
$(\mathrm{d}\theta/\mathrm{d}\eta)_{\eta=0}$	dimensionless temperature gradient on the plate
$\dfrac{1}{\rho}\dfrac{\mathrm{d}\rho}{\mathrm{d}x}$	density factor
$\dfrac{1}{\mu}\dfrac{\mathrm{d}\mu}{\mathrm{d}\eta}$	viscosity factor
$\dfrac{1}{\lambda}\dfrac{\mathrm{d}\lambda}{\mathrm{d}\eta}$	thermal conductivity factor

Superscripts

*	denotes the case under Boussinesq approximation

Subscripts

max	maximum
min	minimum
w	at wall
δ	thickness of boundary layer
$\propto$	far from the wall surface

7.1 Introduction

In Chaps. 4 and 5 we have presented effects of variable thermophysical properties on laminar free convection of monatomic and diatomic gases, air and water vapour as well as polyatomic gases along isothermal flat plate. In Chap.6 the effects of variable thermophysical properties on laminar free convection of liquids were presented. For liquid, the density, thermal conductivity and dynamic viscosity are assumed to vary with temperature according to polynomial or power with the exponent of polynomial. While specific heat is regarded as constant. For gases the thermal conductivity, dynamic viscosity and specific heat are assumed to vary with absolute temperature according to a simple power law, i.e. $\lambda \propto T^{n_\lambda}$, $\mu \propto T^{n_\mu}$ and $C_p \propto T^{n_{c_p}}$. While, the density is taken as inversely proportional to absolute temperature at constant pressure, i.e. $\rho \propto 1/T$, and Prandtl number is assumed being constant. Since the variation of c_p for monatomic and diatomic gases, air and water vapour is very small, it is taken as constant. Consequently, the rigorous numerical solutions are obtained and, on this basis, the curve-fitting formulations for prediction of temperature gradient $(\mathrm{d}\theta/\mathrm{d}\eta)_{\eta=0}$ are derived, which is the key work for simple and reliable calculation of heat transfer of fluid laminar free convection. These analyses originally presented in [1–4] extend the former ones reported in the literatures [5–7].

In this chapter we will further present the validity of conventional heat transfer prediction under Boussinesq approximation for fluid laminar free convection.

7.2 Governing Equations of Fluid Laminar Free Convection under Boussinesq Approximation

7.2.1 For Fluid Laminar Free Convection

Fluid thermophysical properties, such as thermal conductivity, dynamic viscosity and density vary obviously with its temperature. Strictly speaking, the Boussinesq approximation for fluid laminar free convection is valid only in a special case that the wall temperature is very close to that of fluid bulk, i.e. $T_\mathrm{w} \to T_\infty$. As we know, with Boussinesq approximation the fluid thermophysical properties are regarded as constant, except for density in the buoyancy term of momentum equation. With Boussinesq approximation the basis governing partial differential equations of the laminar free convection of fluid along a vertical plate will be

$$\frac{\partial}{\partial x}(w_x) + \frac{\partial}{\partial y}(w_y) = 0, \tag{7.1}$$

$$\left(w_x\frac{\partial w_x}{\partial x} + w_y\frac{\partial w_x}{\partial y}\right) = \nu\frac{\partial^2 w_x}{\partial y^2} + g\left|\frac{\rho_\infty - \rho}{\rho}\right|, \tag{7.2}$$

$$w_x\frac{\partial t}{\partial x} + w_y\frac{\partial t}{\partial y} = a\frac{\partial^2 t}{\partial y^2}, \tag{7.3}$$

with boundary conditions

$$y = 0: \ w_x = 0, \qquad w_y = 0, \quad t = t_\mathrm{w}, \tag{7.4}$$

$$y \to \infty: \qquad w_x \to 0, \quad t = t_\infty. \tag{7.5}$$

Applying the velocity component method, the following transformation variables can be applied for similarity transformation of the governing partial differential (7.1)–(7.3) with the boundary conditions (7.4) and (7.5):

$$\eta = \frac{x}{y}\left(\frac{1}{4}Gr_{x,\mathrm{f}}\right)^{1/4}, \tag{7.6}$$

$$Gr_{x\mathrm{f}} = \frac{g|\rho_\infty/\rho\mathrm{w} - 1|x^3}{\mathrm{v_f}^2}, \tag{7.7}$$

$$\theta = \frac{t - t_\infty}{t_\mathrm{w} - t_\infty}, \tag{7.8}$$

$$W_x = \left[2\sqrt{gx}|\rho_\infty/\rho_\mathrm{w} - 1|^{1/2}\right]^{-1} w_x, \tag{7.9}$$

$$W_y = \left[2\sqrt{gx}|\rho_\infty/\rho_\mathrm{w} - 1|^{1/2}\left(\frac{1}{4}Gr_{x,\mathrm{f}}\right)^{-1/4}\right]^{-1} w_y, \tag{7.10}$$

where subscript f denotes a case for average temperature, i.e.

$$t_\mathrm{f} = (t_\mathrm{w} + t_\infty)/2. \tag{7.11}$$

Here, $Gr_{x,\mathrm{f}}$ denotes the Grashof number at the average temperature t_f, while, η, θ, W_x and W_y are dimensionless coordinate variable, dimensionless temperature and dimensionless velocities components, respectively, related to Boussinesq approximation.

With the earlier assumed transformation variables from (7.6) to (7.11), the governing partial (7.1)–(7.3) with the boundary conditions (7.4) and (7.5) of the fluid laminar free convection are changed into the following dimensionless ones:

$$(2\mathrm{W}_x - \eta\frac{\mathrm{d}W_x}{\mathrm{d}\eta} + 4\frac{\mathrm{d}W_y}{\mathrm{d}\eta} = 0, \tag{7.12}$$

$$W_x\left(2W_x - \eta\frac{\mathrm{d}W_x}{\mathrm{d}\eta}\right) + 4W_y\frac{\mathrm{d}W_x}{\mathrm{d}\eta} = \frac{\mathrm{d}^2W_x}{\mathrm{d}\eta^2} + \frac{\frac{\rho_\infty}{\rho} - 1}{\frac{\rho_\infty}{\rho_\mathrm{w}} - 1}, \tag{7.13}$$

$$(-\eta W_x + 4W_y)\frac{\mathrm{d}\theta}{\mathrm{d}\eta} = \frac{1}{Pr}\frac{\mathrm{d}^2\theta}{\mathrm{d}\eta^2}, \tag{7.14}$$

with boundary conditions

$$\eta = 0,\ W_x = 0,\ W_y = 0, \theta = 0, \tag{7.15}$$

$$\eta \to 0,\ W_x \to 0, \theta \to 0. \tag{7.16}$$

Equations (7.12)–(7.14) with the boundary condition equations (7.15) and (7.16) can be regarded as typical dimensionless governing equations of fluid laminar free convection for Boussinesq approximation.

7.2.2 For Gas Laminar Free Convection

In principle, the governing equations (7.1)–(7.3) with their boundary condition equations (7.4) and (7.5) are valid both for Newtonian liquid and gas laminar free convections under Boussinesq approximation. However, for gas laminar free convection, the following further simplifications are necessary. For simplification of the gas laminar free convection for Boussinesq approximation, it is convenient to express the temperature by absolute temperature T, and then, the temperature t in (7.3)–(7.5), (7.8) and (7.11) is replaced by the absolute temperature T to still keep the identification of these equations. By using the simple power law $\rho_\infty/\rho = T/T_\infty$ of gas shown in Chap. 4, theoretical models in Sect. 7.2.1 can be rewritten as follows for gas laminar free convection. With Boussinesq approximation the basic fluid governing partial differential equations of the laminar free convection of gas along a vertical plate will be

$$\frac{\partial}{\partial x}(w_x) + \frac{\partial}{\partial y}(w_y) = 0, \tag{7.1}$$

$$\left(w_x\frac{\partial w_x}{\partial x} + w_y\frac{\partial w_x}{\partial y}\right) = \nu\frac{\partial^2 w_x}{\partial y^2} + g\left|\frac{T - T_\infty}{T_\infty}\right|, \tag{7.2a}$$

$$w_x\frac{\partial T}{\partial x} + w_y\frac{\partial T}{\partial y} = a\frac{\partial^2 T}{\partial y^2}, \tag{7.3a}$$

with boundary conditions

$$y = 0: \ w_x = 0. \quad w_y = 0, \quad T = T_\mathrm{w}, \tag{7.4a}$$

$$y \to \infty: \quad w_x \to 0, \quad T = T_\infty. \tag{7.5a}$$

It is obvious that the earlier expressions for transformation variables, (7.6)–(7.10), and the transformed ordinary equations (7.12)–(7.14) with the bounadary conditions, (7.15) and (7.16) for fluid laminar free convection under Boussinesq approximation still keep for the gas laminar free convection, only the related density ratios are changed to the absolute temperature ratios by means of the simple power law. Obviously, the theoretical models in this section for gas laminar free convection under Boussinesq approximation is only special case of those presented in Sect. 7.2.1 for fluid laminar free convection under Boussinesq approximation.

7.3 Heat Transfer Deviation of Liquid Laminar Free Convection Caused by Boussinesq Approximation

7.3.1 Boussinesq Solutions for Laminar Free Convection

It can be seen from (7.12)–(7.14) that the Boussinesq approximation solutions of the liquid free convection depend on Pr in the case of $\rho_\infty \to \rho_\mathrm{w}$,

Table 7.1. Boussinesq solutions $(\mathrm{d}\theta/\mathrm{d}\eta)^*_{\eta=0}$ related to Pr range for water

Pr	1.76	1.97	2.23	2.57	3	3.57	4.34	5.42	6.99	9.42	11.16
$-\left(\frac{\mathrm{d}\theta}{\mathrm{d}\eta}\right)^*_{\eta=0}$ (1)	0.686	0.714	0.740	0.779	0.814	0.861	0.914	0.971	1.050	1.137	1.210
$-\left(\frac{\mathrm{d}\theta}{\mathrm{d}\eta}\right)^*_{\eta=0}$ (2)	0.6886	0.7135	0.7418	0.7756	0.8138	0.8587	0.9115	0.9747	1.0510	1.1459	1.2025

$-\left(\frac{\mathrm{d}\theta}{\mathrm{d}\eta}\right)^*_{\eta=0}$ (1): Numerical solutions
$-\left(\frac{\mathrm{d}\theta}{\mathrm{d}\eta}\right)^*_{\eta=0}$ (2): Evaluated results by (7.17)

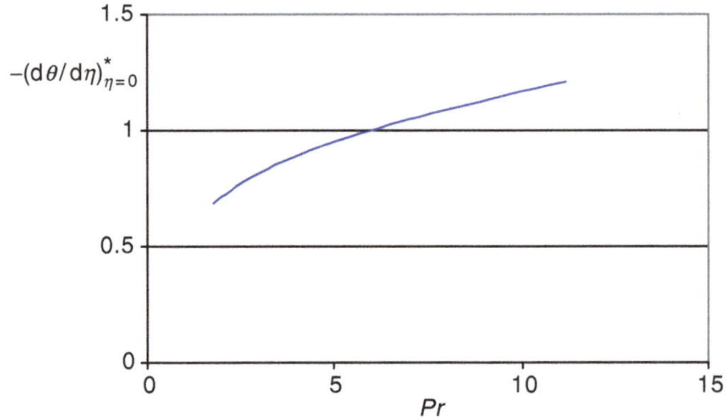

Fig. 7.1. The Boussinesq solutions $-(\mathrm{d}\theta/\mathrm{d}\eta)^*_{\eta=0}$ of laminar free convection in range of water Prandtl number

i.e. $t_\infty \to t_\mathrm{w}$. Let us take water free convection as an example. Based on Table 6.3 the Boussinesq solutions $(\mathrm{d}\theta/\mathrm{d}\eta)^*_{\eta=0}$ are shown in Table 7.1 and plotted here in Fig. 7.1.

According to the curve-fit equation (6.50) the Boussinesq solution for laminar free convection can be expressed as follows for the range of water Prandtl number:

$$-\left(\frac{\mathrm{d}\theta}{\mathrm{d}\eta}\right)^*_{\eta=0} = 0.5764 + 0.1797\,\ln(Pr) + 0.0331\,\ln^2(Pr). \tag{7.17}$$

The results of Boussinesq solutions predicted by using (7.17) are also listed in Table 7.1 for the range of water Prandtl number. It is seen that the predicted results are very identical to the numerical results.

7.3.2 Models for Predicted Deviation on Heat Transfer Caused by Boussinesq Approximation

Consulting the heat transfer analysis described in Chap. 6, the following equations can be obtained for heat transfer coefficient of liquid laminar free convection, respectively, with considerations of Boussinesq approximation and variable thermophysical properties:

$$\alpha_x^* = -\lambda_\mathrm{w}\left(\frac{1}{4}Gr_{x,\mathrm{f}}\right)^{1/4} x^{-1}\left(\frac{\mathrm{d}\theta}{\mathrm{d}\eta}\right)_{\eta=0}^*, \tag{7.18}$$

$$\alpha_x = -\lambda_\mathrm{w}\left(\frac{1}{4}Gr_{x,\infty}\right)^{1/4} x^{-1}\left(\frac{\mathrm{d}\theta}{\mathrm{d}\eta}\right)_{\eta=0}. \tag{7.19}$$

Here, the superscript asterisk * denotes the case for consideration of Boussinesq approximation in order to distinguish it from the case for consideration of variable thermophysical properties. λ_w is defined as the fluid thermal conductivity at the wall temperature t_w. While, $Gr_{x,\mathrm{f}}$ and $Gr_{x,\infty}$ denote average Grashof number, and local Grashof number, respectively.

If $E_{\alpha_x}*$ is taken as relative predicted deviation of heat transfer coefficient due to the Boussinesq approximation, it can be expressed as

$$E_{\alpha_x}* = 1 - \frac{\alpha_x *}{\alpha_x}. \tag{7.20}$$

With (7.18) and (7.19), (7.20) is changed into

$$E_{\alpha_x}* = 1 - \left(\frac{Gr_{x,\mathrm{f}}}{Gr_{x,\infty}}\right)^{1/4} \frac{\left(\frac{\mathrm{d}\theta}{\mathrm{d}\eta}\right)_{\eta=0}^*}{\left(\frac{\mathrm{d}\theta}{\mathrm{d}\eta}\right)_{\eta=0}}. \tag{7.21}$$

With (7.7) and (6.10), (7.21) becomes

$$E_{\alpha_x}* = 1 - \left(\frac{\nu_\infty}{\nu_\mathrm{f}}\right)^{1/2} \frac{\left(\frac{\mathrm{d}\theta}{\mathrm{d}\eta}\right)_{\eta=0}^*}{\left(\frac{\mathrm{d}\theta}{\mathrm{d}\eta}\right)_{\eta=0}}, \tag{7.22}$$

which can be used for prediction of the relative deviation of heat transfer coefficient for fluid laminar free convection caused by Boussinesq approaximation. It is seen here that the multiplication of the ratios $(\nu_\infty/\nu_\mathrm{f})^{1/2}$ and

$$\frac{\left(\frac{\mathrm{d}\theta}{\mathrm{d}\eta}\right)_{\eta=0}^*}{\left(\frac{\mathrm{d}\theta}{\mathrm{d}\eta}\right)_{\eta=0}}$$

dominate the deviation.

For prediction of heat transfer deviation $E_{\alpha x}^*$ of liquid laminar free convection, here we induce relative difference of Prandtl number, ΔPr, defined as

$$\Delta Pr = \frac{Pr_\mathrm{w} - Pr_\infty}{Pr_\mathrm{max} - Pr_\mathrm{min}}, \tag{7.23}$$

where Pr_w and Pr_∞ denote the liquid Prandtl numbers at wall and bulk temperatures, respectively, while, Pr_max and Pr_min denote the liquid maximum and minimum Prandtl numbers, respectively. The relative difference of Prandtl number ΔPr will be used to investigate $E_{\alpha_x}*$ variation late.

7.3.3 Prediction of Heat Transfer Deviation $E^*_{\alpha_x}$ for Water Laminar Free Convection

Now, water laminar free convection is taken here as an example for prediction of deviation of heat transfer of fluid laminar free convection caused by Boussinesq approximation. A system of values of ratios $(\nu_\infty/\nu_\mathrm{f})^{1/2}$ for water laminar free convection are evaluated here at the given T_w and T_∞, and the results are listed in Table 7.2 and plotted in Fig. 7.2, respectively, varying with the relative difference of Prandtl number, ΔPr. While, according to Chap. 6, the related temperature gradients $(\mathrm{d}\theta/\mathrm{d}\eta)_{\eta=0}$ and $(\mathrm{d}\theta/\mathrm{d}\eta)^*_{\eta=0}$ are obtained and listed in Table 7.2 where the reference temperature of $(\mathrm{d}\theta/\mathrm{d}\eta)^*_{\eta=0}$ is taken as $t_\mathrm{f} = (t_\mathrm{w} + t_\infty)/2$. On this basis, the predicted values of

$$\frac{\left(\frac{\mathrm{d}\theta}{\mathrm{d}\eta}\right)_{\eta=0}{}^*}{\left(\frac{\mathrm{d}\theta}{\mathrm{d}\eta}\right)_{\eta=0}}$$

are evaluated, listed in Table 7.2, and plotted in Fig. 7.3 varying with the relative difference of Prandtl number, ΔPr. Here Pr_max and Pr_min are taken as 13.44 at 0°C and 1.76 at 100°C, respectively. It is seen in Fig. 7.2 that with increase of ΔPr, the value of $(\nu_\infty/\nu_\mathrm{f})^{1/2}$ decreases. If relative difference of Prandtl number, ΔPr, tends to zero, the value $(\nu_\infty/\nu_\mathrm{f})^{1/2}$ will tend to unity.

It is seen in Fig. 7.3 that with increase of ΔPr, the ratio $(\mathrm{d}\theta/\mathrm{d}\eta)_{\eta=0} * /(\mathrm{d}\theta/\mathrm{d}\eta)_{\eta=0}$ will increase. On the other hand, if ΔPr tends to zero, the ratio $(\mathrm{d}\theta/\mathrm{d}\eta)_{\eta=0} * /(\mathrm{d}\theta/\mathrm{d}\eta)_{\eta=0}$ will tend to unity.

Finally, for water laminar free convection the relative deviations of heat transfer coefficient, $E_{\alpha_x}*$, are predicted by using (7.22) and the results are listed in Table 7.2 and plotted in Fig 7.4 with relative Prandtl number ΔPr. It is founds that with increase of relative Prandtl number ΔPr, the value

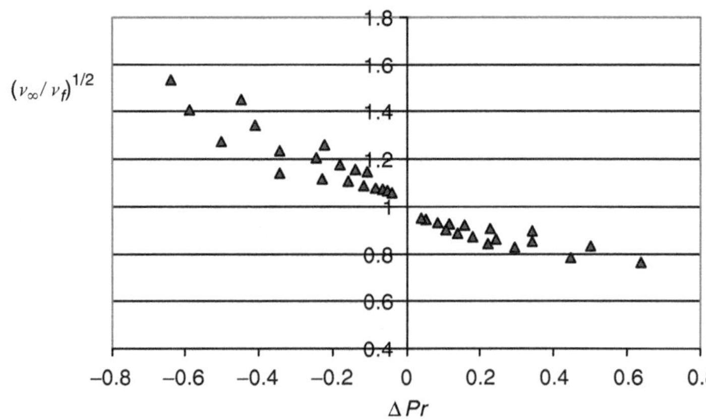

Fig. 7.2. The ratio $(\nu_\infty/\nu_\mathrm{f})^{1/2}$ varies with ΔPr for water laminar free convection

Table 7.2 (1). Predicted heat transfer deviation $E_{\alpha_x}*$ with ratio $(\nu_\infty/\nu_f)^{1/2}$ and $(d\theta/d\eta)_{\eta=0}*/(d\theta/d\eta)_{\eta=0}$ for water laminar free convection

t_w (°C)	Pr_w	t_∞ (°C)	Pr_∞	ΔPr	t_f (°C)	ν_∞ $\times 10^6$ $(m^2 s^{-1})$	ν_f $\times 10^6$ $(m^2 s^{-1})$	$(\nu_\infty/\nu_f)^{1/2}$	Pr_f
90	1.97	10	9.42	−0.638	50	1.306	0.556	1.5326	3.57
70	2.57	10	9.42	−0.586	40	1.306	0.659	1.4078	4.34
50	3.57	10	9.42	−0.501	30	1.306	0.805	1.2737	5.42
30	5.42	10	9.42	−0.342	20	1.306	1.006	1.1394	6.99
100	1.76	20	6.99	−0.448	60	1.006	0.478	1.4507	3
80	2.23	20	6.99	−0.408	50	1.006	0.556	1.3451	3.57
60	3	20	6.99	−0.342	40	1.006	0.659	1.2355	4.34
40	4.34	20	6.99	−0.227	30	1.006	0.805	1.1179	5.42
70	2.57	30	5.42	−0.244	50	0.805	0.556	1.2033	3.57
50	3.57	30	5.42	−0.158	40	0.805	0.659	1.1052	4.34
10	9.42	30	5.42	0.342	20	0.805	1.006	0.8945	6.99
100	1.76	40	4.34	−0.221	70	0.659	0.415	1.2601	2.57
80	2.23	40	4.34	−0.181	60	0.659	0.478	1.1742	3
60	3	40	4.34	−0.115	50	0.659	0.556	1.0887	3.57
20	6.99	40	4.34	0.227	30	0.659	0.805	0.9048	5.42
90	1.97	50	3.57	−0.137	70	0.556	0.415	1.1575	2.57
70	2.57	50	3.57	−0.086	60	0.556	0.478	1.0785	3
30	5.42	50	3.57	0.158	40	0.556	0.659	0.9185	4.34
10	9.42	50	3.57	0.501	30	0.556	0.805	0.8311	5.42
100	1.76	60	3	−0.106	80	0.478	0.365	1.1444	2.23
80	2.23	60	3	−0.066	70	0.478	0.415	1.0732	2.57
40	4.34	60	3	0.115	50	0.478	0.556	0.9272	3.57
20	6.99	60	3	0.342	40	0.478	0.659	0.8517	4.34
90	1.97	70	2.57	−0.051	80	0.415	0.365	1.0663	2.23
50	3.57	70	2.57	0.0856	60	0.415	0.478	0.9318	3
30	5.42	70	2.57	0.2440	50	0.415	0.556	0.8640	3.57
100	1.76	80	2.23	−0.040	90	0.365	0.326	1.0581	1.97
40	4.34	80	2.23	0.1807	60	0.365	0.478	0.8738	3
70	2.57	90	1.97	0.0514	80	0.326	0.365	0.9451	2.23
50	3.57	90	1.97	0.137	70	0.326	0.415	0.8863	2.57
30	5.42	90	1.97	0.295	60	0.326	0.478	0.8258	3
10	9.42	90	1.97	0.638	50	0.326	0.556	0.7657	3.57
80	2.23	100	1.76	0.040	90	0.295	0.326	0.9513	1.97
60	3	100	1.76	0.106	80	0.295	0.365	0.8990	2.23
40	4.34	100	1.76	0.221	70	0.295	0.415	0.8431	2.57
20	6.99	100	1.76	0.448	60	0.295	0.478	0.7856	3

Table 7.2 (2). Predicted heat transfer deviation E_{ax} with ratio $(\nu_\infty/\nu_f)^{1/2}$ and $(d\theta/d\eta)_{\eta=0} * /(d\theta/d\eta)_{\eta=0}$ for water laminar free convection

t_w (°C)	Pr_w	t_∞ (°C)	Pr_∞	ΔPr	$\left(\frac{d\theta}{d\eta}\right)^*_{\eta=0}$	$\left(\frac{d\theta}{d\eta}\right)_{\eta=0}$	$\frac{(\frac{d\theta}{d\eta})_{\eta=0}*}{(\frac{d\theta}{d\eta})_{\eta=0}}$	$E^*_{\alpha_x}$ (%)
90	1.97	10	9.42	−0.638	0.861	1.199	0.7182	−10.06
70	2.57	10	9.42	−0.586	0.914	1.175	0.7779	−9.51
50	3.57	10	9.42	−0.501	0.971	1.150	0.8443	−7.55
30	5.42	10	9.42	−0.342	1.050	1.131	0.9284	−5.78
100	1.76	20	6.99	−0.448	0.814	1.100	0.74	−7.35
80	2.23	20	6.99	−0.408	0.861	1.083	0.7950	−6.94
60	3	20	6.99	−0.342	0.914	1.068	0.8558	−5.74
40	4.34	20	6.99	−0.227	0.971	1.054	0.9213	−2.99
70	2.57	30	5.42	−0.244	0.861	0.994	0.8662	−4.23
50	3.57	30	5.42	−0.158	0.914	0.983	0.9298	−2.77
10	9.42	30	5.42	0.342	1.050	0.983	1.0682	4.45
100	1.76	40	4.34	−0.221	0.779	0.938	0.8305	−4.65
80	2.23	40	4.34	−0.181	0.814	0.929	0.8762	−2.88
60	3	40	4.34	−0.115	0.861	0.920	0.9359	−1.89
20	6.99	40	4.34	0.227	0.971	0.911	1.0659	3.56
90	1.97	50	3.57	−0.137	0.779	0.874	0.8913	−3.17
70	2.57	50	3.57	−0.086	0.814	0.867	0.9389	−1.26
30	5.42	50	3.57	0.158	0.914	0.856	1.0678	1.92
10	9.42	50	3.57	0.501	0.971	0.856	1.1343	5.73
100	1.76	60	3	−0.106	0.740	0.827	0.8948	−2.40
80	2.23	60	3	−0.066	0.779	0.821	0.9488	−1.83
40	4.34	60	3	0.115	0.861	0.810	1.0630	1.44
20	6.99	60	3	0.342	0.914	0.807	1.1326	3.54
90	1.97	70	2.57	−0.051	0.740	0.781	0.9475	−1.03
50	3.57	70	2.57	0.0856	0.814	0.772	1.0544	1.75
30	5.42	70	2.57	0.2440	0.861	0.769	1.1196	3.27
100	1.76	80	2.23	−0.040	0.714	0.746	0.9571	−1.27
40	4.34	80	2.23	0.1807	0.814	0.735	1.1075	3.22
70	2.57	90	1.97	0.0514	0.740	0.710	1.0423	1.50
50	3.57	90	1.97	0.137	0.779	0.707	1.1018	2.34
30	5.42	90	1.97	0.295	0.814	0.705	1.1546	4.65
10	9.42	90	1.97	0.638	0.861	0.704	1.2230	6.35
80	2.23	100	1.76	0.040	0.714	0.685	1.0423	0.85
60	3	100	1.76	0.106	0.740	0.683	1.0835	2.60
40	4.34	100	1.76	0.221	0.779	0.681	1.1439	3.56
20	6.99	100	1.76	0.448	0.814	0.679	1.1988	5.82

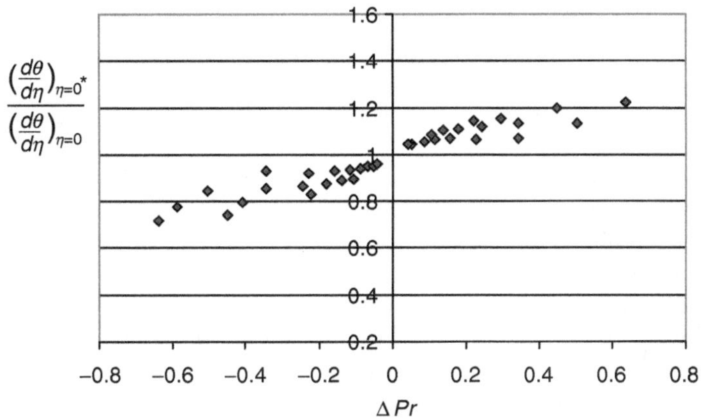

Fig. 7.3. The ratio $\frac{(d\theta/d\eta)_{\eta=0^*}}{(d\theta/d\eta)_{\eta=0}}$ varies with ΔPr for water laminar free convection

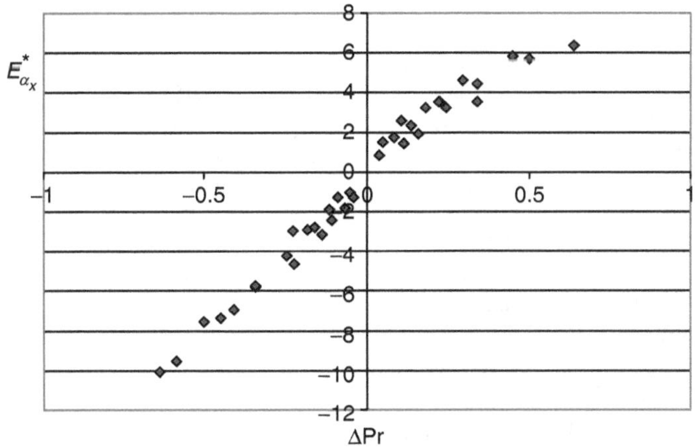

Fig. 7.4. The predicted ratio $E_{\alpha_x}*$ varies with ΔPr for water laminar free convection

of $E_{\alpha_x}*$ increases. On the other hand, with increasing the absolute value $|\Delta Pr|$, the absolute value $|E_{\alpha_x}*|$ will increase. If the value of ΔPr decreases from 0, −0.2, −0.3, −0.5 to −0.6, the value of $E_{\alpha_x}*$ will vary from 0, −3.5%, −6%, −8.5% to −10.5%. While, if the value of ΔPr increases from 0, 0.2, 0.3, 0.5 to 0.6, the value of $E_{\alpha_x}*$ will vary from 0, 3, 5, 7 to 8%, respectively.

It follows that the predicted heat transfer deviation $E_{\alpha_x}*$ for liquid laminar free convection caused by Boussinesq approximation are obvious, due to ignoring variable thermophysical properties.

7.4 Heat Transfer Deviation of Gas Laminar Free Convection Caused by Boussinesq Approximation

7.4.1 Boussinesq Solutions for Gas Laminar Free Convection

Equations (7.12), (7.13) and (7.14) with the boundary condition equations (7.15) and (7.16) are solved for gas laminar free convection in the range of gas Prandtl number, from 0.61 to 1 under Boussinesq approximation. The numerical solutions for the Boussinesq approximation are listed and plotted in Table 7.3 and Fig. 7.5, respectively.

By using a curve matching method, an equation for prediction of Boussinesq solutions $(\mathrm{d}\theta/\mathrm{d}\eta)_{\eta=0}*$ of gas laminar free convection in the range of gas Prandtl number is obtained as follows:

Table 7.3. Boussinesq solutions related to gas laminar free convection

Pr	$\left(\frac{\mathrm{d}\theta}{\mathrm{d}\eta}\right)_{\eta=0}*$ (1)	$\left(\frac{\mathrm{d}\theta}{\mathrm{d}\eta}\right)_{\eta=0}*$ (2)	Pr	$\left(\frac{\mathrm{d}\theta}{\mathrm{d}\eta}\right)_{\eta=0}*$ (1)	$(\frac{\mathrm{d}\theta}{\mathrm{d}\eta})_{\eta=0}*$ (2)
0.61	0.475	0.47511	0.74	0.50964	0.51099
0.622	0.47841	0.47868	0.75	0.51211	0.51349
0.63	0.48066	0.48106	0.8	0.52411	0.52549
0.65	0.48619	0.48687	0.81	0.52644	0.52781
0.68	0.49426	0.49527	0.83	0.53105	0.53234
0.7	0.49949	0.50066	0.85	0.53557	0.53677
0.71	0.50207	0.50330	0.87	0.54001	0.54110
0.715	0.50334	0.50460	0.88	0.54220	0.54322
0.72	0.50461	0.50590	0.9	0.54653	0.54740
0.73	0.50714	0.50846	0.95	0.55704	0.55746
0.733	0.50789	0.50923	1	0.56714	0.567

$(\mathrm{d}\theta/\mathrm{d}\eta)_{\eta=0}*$ (1): numerical solution
$(\mathrm{d}\theta/\mathrm{d}\eta)_{\eta=0}*$ (2): predicted value by using (7.24)

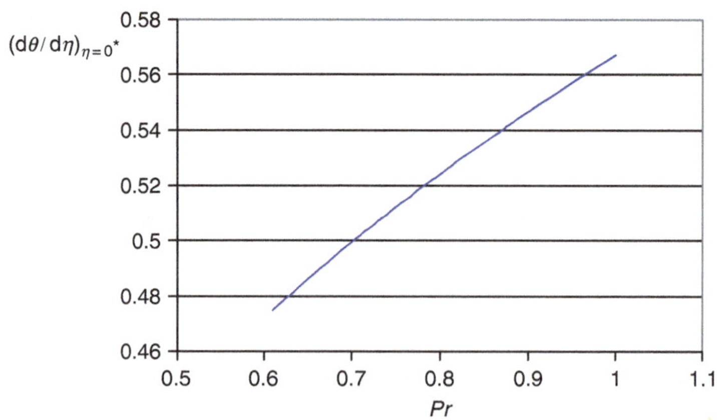

Fig. 7.5. Boussinesq solutions related to gas laminar free convection

$$-\left(\frac{\mathrm{d}\theta}{\mathrm{d}\eta}\right)^*_{\eta=0} = \psi(Pr) = 0.567 + 0.186 \times \ln Pr \qquad (0.6 \le Pr \le 1). \quad (7.24)$$

The Boussinesq approximation solutions $-(\mathrm{d}\theta/\mathrm{d}\eta)_{\eta=0}*$ related to gas laminar free convection are listed in Table 7.3 also. It is seen from Table 7.3 that the evaluated Boussinesq solutions agree very well with the related numerical solutions.

7.4.2 Models on Predicted Deviation of Heat Transfer of Gas Laminar Free Convection Caused by Boussinesq Approximation

For gas laminar free convection, according to simple power law provided in (4.39), the ratio $(\nu_\infty/\nu_\mathrm{f})^{1/2}$ can be expressed as

$$\left(\frac{\nu_\infty}{\nu_\mathrm{f}}\right)^{1/2} = \left(\frac{T_\infty}{T_\mathrm{f}}\right)^{\frac{n_\mu+1}{2}}. \quad (7.25)$$

With (7.11) for the average temperature T_f, the ratio $(\nu_\infty/\nu_\mathrm{f})^{1/2}$ can be further expressed as:

$$\left(\frac{\nu_\infty}{\nu_\mathrm{f}}\right)^{1/2} = \left(\frac{T_\infty}{\frac{T_\mathrm{w}+T_\infty}{2}}\right)^{(n_\mu+1)/2} = \left(\frac{2}{\frac{T_\mathrm{w}}{T_\infty}+1}\right)^{(n_\mu+1)/2}. \quad (7.26)$$

According to Chap. 4, the temperature gradient $(\mathrm{d}\theta/\mathrm{d}\eta)_{\eta=0}$ for gas laminar free convection with consideration of variable thermophysical properties can be described by the following very accurate curve-fitting equations

$$-\left(\frac{\mathrm{d}\theta}{\mathrm{d}\eta}\right)_{\eta=0} = \psi(Pr)\left(\frac{T_\mathrm{w}}{T_\infty}\right)^{-m}, \quad (7.27)$$

where

$$\psi(Pr) = 0.567 + 0.186 \times \ln Pr (0.6 \leqslant Pr \leqslant 1), \quad (7.28)$$

$$m = 0.35 n_\lambda + 0.29 n_\mu + 0.36 \quad (T_\mathrm{w}/T_\infty > 1), \quad (7.29)$$

$$m = 0.42 n_\lambda + 0.34 n_\mu + 0.24 \quad (T_\mathrm{w}/T_\infty < 1). \quad (7.30)$$

With (7.24) and (7.27), the ratio

$$\frac{\left(\frac{\mathrm{d}\theta}{\mathrm{d}\eta}\right)_{\eta=0}^*}{\left(\frac{\mathrm{d}\theta}{\mathrm{d}\eta}\right)_{\eta=0}}$$

is expressed as

$$\frac{\left(\frac{\mathrm{d}\theta}{\mathrm{d}\eta}\right)_{\eta=0}^{*}}{\left(\frac{\mathrm{d}\theta}{\mathrm{d}\eta}\right)_{\eta=0}} = \left(\frac{T_{\mathrm{w}}}{T_{\infty}}\right)^{m} . \tag{7.31}$$

With (7.26) and (7.31), (7.22) is changed into the following equation for gas laminar free convection:

$$E_{\alpha_x}* = 1 - \left(\frac{2}{\frac{T_{\mathrm{w}}}{T_{\infty}}+1}\right)^{\frac{n_\mu+1}{2}} \cdot \left(\frac{T_{\mathrm{w}}}{T_{\infty}}\right)^{m} . \tag{7.32}$$

It is clear that (7.32) is suitable for gas laminar free convection. Equation (7.32), (7.29) and (7.30) are exact equations for prediction of the relative deviation $E_{\alpha_x}*$ of heat transfer coefficient for gas laminar free convection caused by Boussinesq approximation. It is clear that $E_{\alpha_x}*$ is dominated by n_μ, n_λ and boundary temperature ratio $(T_{\mathrm{w}}/T_{\infty})$.

7.4.3 Prediction Results of Deviation $E_{\alpha x}^*$ for Gas Laminar Free Convection

Some gases such as air, CO and water vapour are taken as examples for prediction of the deviation $E_{\alpha_x}*$ for gas laminar free convection. According to [1] and [2], the Prandtl numbers Pr are taken as 0.7, 0.72 and 1, viscous parameters n_μ are taken as 0.68, 0.71 and 1.04, and the thermal conductivity parameters n_λ are taken as 0.81, 0.83 and 1.185 for air, CO and water vapour, respectively. According to (7.29), (7.30) and (7.32), the values

$$\left(\frac{2}{\frac{T_{\mathrm{w}}}{T_{\infty}}+1}\right)^{\frac{n_\mu+1}{2}},$$

$(T_{\mathrm{w}}/T_{\infty})^m$ and $E_{\alpha_x}*$ are calculated for the range of temperature ratio $T_{\mathrm{w}}/T_{\infty}$ from 0.7 to 1.9, listed in Table 7.4 in detail and plotted in Figs. 7.6–7.8, respectively. It is seen from Table 7.4 and Fig. 7.6 that the value

$$\left(\frac{2}{\frac{T_{\mathrm{w}}}{T_{\infty}}+1}\right)^{\frac{n_\mu+1}{2}}$$

depends on n_μ. With increasing the temperature ratio $T_{\mathrm{w}}/T_{\infty}$, the value

$$\left(\frac{2}{\frac{T_{\mathrm{w}}}{T_{\infty}}+1}\right)^{\frac{n_\mu+1}{2}}$$

will decrease. If the temperature ratio $T_{\mathrm{w}}/T_{\infty}$ tends to unity, the value

$$\left(\frac{2}{\frac{T_w}{T_\infty}+1}\right)^{\frac{n_\mu+1}{2}}$$

will tend to unity also.

It is seen from Table 7.4 and Fig. 7.7 that the value $(T_w/T_\infty)^m$ depends on n_μ and n_λ. With increasing the temperature ratio T_w/T_∞, the value

Table 7.4. Predicted deviation $E_{\alpha_x}*$ with factor $\left(\frac{2}{(T_w/T_\infty)+1}\right)^{(n_\mu+1)/2}$ and $(T_w/T_\infty)^m$ for air laminar free convection

air $n_\mu = 0.68, n_\lambda = 0.81, Pr = 0.7$				
T_w/T_∞	m	$\left(\frac{2}{\frac{T_w}{T_\infty}+1}\right)^{(n_\mu+1)/2}$	$\left(\frac{T_w}{T_\infty}\right)^m$	$E_{\alpha_x}*$
1.9	0.8329	0.73189854	1.706766	−0.24918
1.8	0.8329	0.75379361	1.63161	−0.2299
1.7	0.8329	0.77717645	1.555754	−0.2091
1.5	0.8329	0.82907838	1.401737	−0.16215
1.4	0.8329	0.85800092	1.323458	−0.13553
1.3	0.8329	0.8892294	1.244238	−0.10641
1.2	0.8329	0.92306045	1.163992	−0.07444
1.1	0.8329	0.95984474	1.08262	−0.03915
1	0.8329	1	1	0
0.9	0.801	1.04402806	0.919069	0.040466
0.8	0.801	1.09253735	0.836325	0.086284
0.7	0.801	1.1462731	0.751491	0.138587
0.6	0.801	1.20615858	0.6642	0.198869
0.5	0.801	1.27335219	0.573951	0.269158
CO $n_\mu = 0.71$ $n_\lambda = 0.83$, $Pr = 0.72$				
T_w/T_∞	m	$\left(\frac{2}{\frac{T_w}{T_\infty}+1}\right)^{(n_\mu+1)/2}$	$\left(\frac{T_w}{T_\infty}\right)^m$	$E_{\alpha_x}*$
1.9	0.8492	0.72783068	1.72471586	−0.25530112
1.8	0.8492	0.749998733	1.64731788	−0.23548632
1.7	0.8492	0.773685803	1.56926838	−0.21412067
1.5	0.8492	0.826307965	1.41103146	−0.16594653
1.4	0.8492	0.855657644	1.33073607	−0.13865449
1.3	0.8492	0.887367148	1.2495703	−0.10882764
1.2	0.8492	0.921741739	1.16745652	−0.0760934
1.1	0.8492	0.959142531	1.08430302	−0.04000115
1	0.8492	1	1	0
0.9	0.8204	1.044831645	0.91719263	0.04168812
0.8	0.8204	1.094265368	0.83271239	0.088791673
0.7	0.8204	1.149070879	0.74630858	0.142438549
0.6	0.8204	1.210202543	0.6576507	0.204109452
0.5	0.8204	1.278858869	0.56628491	0.275801516

Table 7.4. *Continued*

		water vapour $n_\mu = 1.04$ $n_\lambda = 1.185$, $Pr = 1$		
T_w/T_∞	m	$\left(\frac{2}{\frac{T_w}{T_\infty}+1}\right)^{(n_\mu+1)/2}$	$\left(\frac{T_w}{T_\infty}\right)^m$	$E_{\alpha_x}*$
1.9	1.07635	0.68454915	1.99543	−0.3660
1.8	1.07635	0.70949511	1.88262	−0.3357
1.7	1.07635	0.73630806	1.770287	−0.3035
1.5	1.07635	0.79643766	1.547162	−0.2322
1.4	1.07635	0.83030017	1.436431	−0.1927
1.3	1.07635	0.86713797	1.326304	−0.1501
1.2	1.07635	0.90735965	1.216821	−0.1041
1.1	1.07635	0.95145207	1.108034	−0.0542
1	1.07635	1	1	0
0.9	1.0913	1.05371199	0.891384	0.0607
0.8	1.0913	1.11345492	0.783866	0.1272
0.7	1.0913	1.18030078	0.677572	0.2003
0.6	1.0913	1.25559106	0.572659	0.28010
0.5	1.0913	1.34102697	0.469338	0.3706

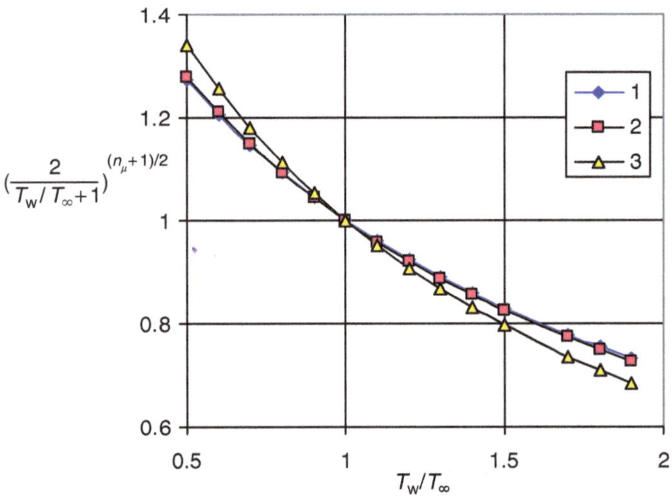

Fig. 7.6. The value $\left(\frac{2}{T_w/T_\infty+1}\right)^{(n_\mu+1)/2}$ (i.e. ratio $(\nu_\infty/\nu_f)^{1/2}$) varies with T_w/T_∞ for laminar free convection of some gases. (1) Air, (2) CO, and (3) water vapour

$(T_w/T_\infty)^m$ will increase. If the temperature ratio T_w/T_∞ tends to unity, the value $(T_w/T_\infty)^m$ will tend to unity also.

It is seen from Table 7.4 and Fig. 7.8 that the predicted deviation $E_{\alpha_x}*$ is very sensitive due to Boussinesq approximation. With increasing the absolute

$$\left|1 - \frac{T_w}{T_\infty}\right|,$$

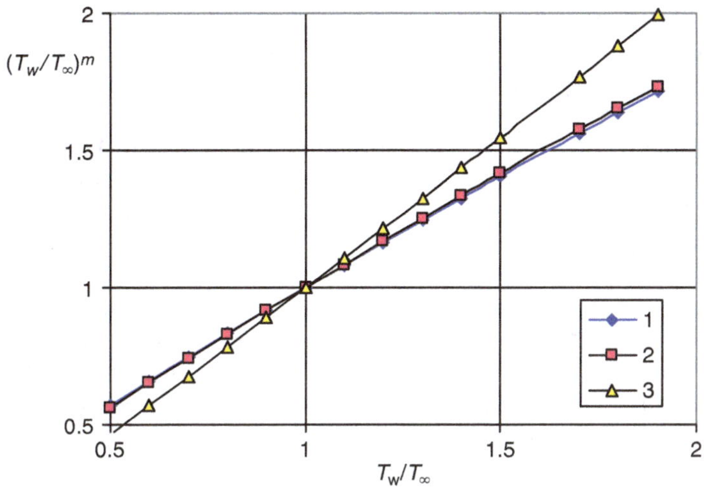

Fig. 7.7. The value $(T_w/T_\infty)^m$ $\left(\text{i.e. ratio} \frac{\left(\frac{d\theta}{d\eta}\right)_{\eta=0}^{*}}{\left(\frac{d\theta}{d\eta}\right)_{\eta=0}}\right)$ varies with T_w/T_∞ for laminar free convection of some gases. (1) Air, (2) CO, and (3) water vapour

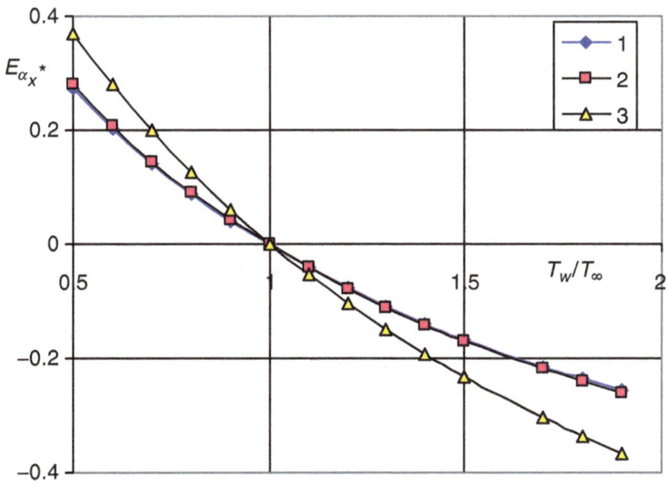

Fig. 7.8. Predicted relative deviation $E_{\alpha_x}*$ varying with the temperature ratio T_w/T_∞ for laminar free convection of some gases. (1) Air, (2) CO, and (3) water vapour. (1) for air, (2) for water vapour and (3) for NH_3

the value of $E^*_{\alpha_x}$ will increase rapidly. If the temperature ratio T_w/T_∞ tends to unity, the value of $E^*_{\alpha_x}$ will tend to zero.

It is seen that the absolute value of deviation $E_{\alpha_x}*$ of water vapour laminar free convection is larger than that for CO free convection, while, the absolute value of deviation $E^*_{\alpha_x}$ of the free convection with CO is little bit larger than that with air. Observing the viscosity parameter n_μ or the thermal conductivity parameter n_λ, it is found that water vapour has much larger n_μ and n_λ than those CO has, while, CO has little bit larger n_μ and n_λ than those air has. The value of n_μ or n_λ reflects the degree of variation of gas absolute viscosity or thermal conductivity with temperature. Therefore, with increase in the value n_μ or n_λ, the absolute value of deviation $E^*_{\alpha_x}$ increases.

From the $E^*_{\alpha_x}$ prediction, it is possible to investigate the validity of gas laminar free convection under Boussinesq approximation. For example, with temperature ratio T_w/T_∞ between 0.9 and 1.1, a small range of the value T_w/T_∞, the predicted deviations $E^*_{\alpha_x}$ are between 4 and −3.9% for air laminar free convection, between 4.17 and −4% , for CO laminar free convection, and between 6.1 and −5.4% for water vapour laminar free convection. It can be seen that the valid range of temperature ratio T_w/T_∞ is very limited for Boussinesq approximation of gas laminar free convection,

In principle, only when $T_w/T_\infty \to 1$, the deviation $E^*_{\alpha_x}$ tents to zero, which means that only the situation for $T_w/T_\infty \to 1$ is strictly suitable for Boussinesq approximation.

7.5 Summary

For prediction of heat transfer deviation for fluid laminar free convection caused by Boussinesq approximation, the related models can be summarized in Table 7.5.

7.6 Remarks

From the analysis and prediction of heat transfer coefficient deviation for fluid laminar free convection caused by Boussinesq approximation, the following remarks are obtained:

It is clear that the predicted deviation $E^*_{\alpha_x}$ of heat transfer coefficient for fluid laminar free convection is dominated by the square root of a kinetic ratio, $(v_\infty/v_f)^{1/2}$, and a temperature gradient ratio, $(\mathrm{d}\theta/\mathrm{d}\eta)^*_{\eta=0} / (\mathrm{d}\theta/\mathrm{d}\eta)_{\eta=0}$.

For liquid laminar free convection it is significant to transform the temperature diference into a relative difference of Prandtl number in order to investigate $E^*_{\alpha_x}$ variation. It is found that with increasing the absolute value $|\Delta\ Pr|$ the absolute value $\left|E^*_{\alpha_x}\right|$ will increase. For water laminar free convection, for example, if the value $\Delta\ Pr$ varies from 0 to −0.2, −0.3, −0.5 and −0.6, the predicted deviation of heat transfer coefficient will vary from 0 to −3.5, −6, −8.5 to −10.5%. While, if the value of $\Delta\ Pr$ varies from 0 to 0.2, 0.3, 0.5 and 0.6,

Table 7.5. Summary of prediction on heat transfer deviation for fluid laminar free convection caused by Boussinesq approximation

term	symbol and equation	
	consideration of variable thermophysical properties	under Boussinesq approximation
	for liquid laminar free convection	
heat transfer coefficient	$\alpha_x = -\lambda_{\mathrm{w}}\left(\frac{1}{4}Gr_{x,\infty}\right)^{1/4} x^{-1}\left(\frac{\mathrm{d}\theta}{\mathrm{d}\eta}\right)_{\eta=0}$	$\alpha_x^* = -\lambda_{\mathrm{w}}\left(\frac{1}{4}Gr_{x,\mathrm{f}}\right)^{1/4} x^{-1}\left(\frac{\mathrm{d}\theta}{\mathrm{d}\eta}\right)_{\eta=0}$
grashof number	$Gr_{x,\infty} = \frac{g\lvert\rho_\infty/\rho_{\mathrm{w}}-1\rvert x^3}{\nu_\infty^2}$	$Gr_{x,\infty} = \frac{g\lvert\rho_\infty/\rho_{\mathrm{w}}-1\rvert x^3}{\nu_{\mathrm{f}}^2}$
Temperature gradient	$-\left(\frac{\mathrm{d}\theta}{\mathrm{d}\eta}\right)_{\eta=0} =$ $0.5764 + 0.1797\ln Pr_\infty + 0.0331\ln^2 Pr_\infty$ (for water laminar free convection)	$-\left(\frac{\mathrm{d}\theta}{\mathrm{d}\eta}\right)_{\eta=0}$ $= 0.5674 + 0.1797\ln Pr_{\mathrm{f}} + 0.0331\ln^2 Pr_{\mathrm{f}}$ (in the range of water Prandtl number)
$E_{\alpha_x}^*$	$E_{\alpha_x}^* = 1 - \left(\frac{v_\infty}{v_{\mathrm{f}}}\right)^{1/2}\frac{\left(\frac{\mathrm{d}\theta}{\mathrm{d}\eta}\right)_{\eta=0}^*}{\left(\frac{\mathrm{d}\theta}{\mathrm{d}\eta}\right)_{\eta=0}}$	
	for gas laminar free convection	
heat transfer coefficient	$\alpha_x = -\lambda_{\mathrm{w}}\left(\frac{1}{4}Gr_{x,\infty}\right)^{1/4} x^{-1}\left(\frac{\mathrm{d}\theta}{\mathrm{d}\eta}\right)_{\eta=0}$	$\alpha_x^* = -\lambda_{\mathrm{w}}\left(\frac{1}{4}Gr_{x,f}\right)^{1/4} x^{-1}\left(\frac{\mathrm{d}\theta}{\mathrm{d}\eta}\right)_{\eta=0}$
grashof number	$Gr_{x,\infty} = \frac{g\lvert T_{\mathrm{w}}/T_\infty-1\rvert x^3}{\nu_\infty^2}$	$Gr_{x,f} = \frac{g\lvert T_{\mathrm{w}}/T_\infty-1\rvert x^3}{\nu_f^2}$
temperature gradient	$-\left(\frac{\mathrm{d}\theta}{\mathrm{d}\eta}\right)_{\eta=0} = \psi(Pr)\left(\frac{T_{\mathrm{w}}}{T_\infty}\right)^{-m}$	$-\left(\frac{\mathrm{d}\theta}{\mathrm{d}\eta}\right)_{\eta=0}^* = \psi(Pr)$
	$\psi(Pr) = 0.567 + 0.186\times\ln Pr$	$(0.6 \le Pr \le 1)$
	$m = 0.35n_\lambda + 0.29n_\mu + 0.36$	$(T_{\mathrm{w}}/T_\infty > 1)$
	$m = 0.42n_\lambda + 0.34n_\mu + 0.24$	$(T_{\mathrm{w}}/T_\infty < 1)$
$E_{\alpha_x}^*$	$E_{\alpha_x}^* = 1 - \left(\frac{2}{(T_{\mathrm{w}}/T_\infty)+1}\right)^{(n_\mu+1)/2}\left(\frac{T_{\mathrm{w}}}{T_\infty}\right)^m$	

the value of $E^*_{\alpha_x}$ will vary from 0, 3, 5, 7 to 8%, respectively. If Boussinesq approximation is applied to predict water laminar free convection, the relative Prandtl number $\Delta\ Pr$ is suggested to be limited between – 0.3 and 0.3. Only with such condition, it could be sure that the predicted relative deviation $E^*_{\alpha_x}$ of heat transfer coefficient is between −6 and 5%.

By means of the theoretical models for treatment of variable thermophysical properties provided in [1], for gas laminar free convection, the kinematic ratio $(v_\infty/v_\mathrm{f})^{1/2}$ and temperature gradient ratio $(\mathrm{d}\theta/\mathrm{d}\eta)^*_{\eta=0} / (\mathrm{d}\theta/\mathrm{d}\eta)_{\eta=0}$ are transformed into the factors $\left(\frac{2}{(T_\mathrm{w}/T_\infty)+1}\right)^{(n_\mu+1)/2}$ and $(T_\mathrm{w}/T_\infty)^m$, respectively. Then, it is found that $E^*_{\alpha_x}$ depends on n_μ, n_λ and T_w/T_∞ for laminar free convection of monatomic and diatomic gases, air and water vapour.

The predicted deviation $E^*_{\alpha_x}$ is great also for gas laminar free convection under Boussinesq approximation. For temperature ratios T_w/T_∞ between 0.9 and 1.1, the predicted deviations $E^*_{\alpha_x}$ is between 4 and −3.9% for air laminar free convection, between 4.17 and −4% for CO laminar free convection, and between 6.1 and −5.4% for water vapour laminar free convection. It follows that the validity of the Boussinesq approximation is also very limited for gas laminar free convection.

7.7 Calculation example

Example 1: A flat plate with $b = 1\,\mathrm{m}$ in width and $x = 0.175\,\mathrm{m}$ in length is suspended vertically in the space of water. The ambient temperature is at $t_\infty = 10°\,\mathrm{C}$, and the plate temperature is $t_\mathrm{w} = 50°\,\mathrm{C}$. The water physical properties are as follows: water kinetic viscosity $\nu_\infty = 1.306 \times 10^{-6}\,\mathrm{m^2\,s^{-1}}$, density $\rho_\infty = 999.7\,\mathrm{kg\,m^{-3}}$ and $Pr_\infty = 9.52$ at the temperature $t_\infty = 10°\,\mathrm{C}$; $\rho_\mathrm{w} = 988.1\,\mathrm{kg\,m^{-3}}$ at wall temperature and $t_\mathrm{w} = 50°\,\mathrm{C}$. Suppose the free convection is laminar, please calculate the heat transfer deviation $E^*_{\alpha_x}$.

Solution:

$$t_\mathrm{f} = (t_\mathrm{w} + t_\infty)/2 = (50 + 10)/2 = 30°\,\mathrm{C}$$

Then, $Pr_\mathrm{f} = 5.42$ and $\nu_\mathrm{f} = 0.805 \times 10^{-6}\,\mathrm{m^2\,s^{-1}}$ at $t_\mathrm{f} = 30°\,\mathrm{C}$ for water. According to (7.7), the local Grashof number is evaluated as

$$\begin{aligned} Gr_{x,\mathrm{f}} &= \frac{g\,|\rho_\infty/\rho_\mathrm{w} - 1|\,x^3}{\nu_\mathrm{f}^2} \\ &= \frac{9.8 \times |999.7/988.1 - 1| \times 0.175^3}{(0.805 \times 10^{-6})^2} \\ &= 951492898 \\ &= 0.951492898 \times 10^9. \end{aligned}$$

According to (7.17), the Boussinesq approximation for $(\mathrm{d}\theta/\mathrm{d}\eta)^*_{\eta=0}$ can be predicted as

$$
\begin{aligned}
-\left(\frac{\mathrm{d}\theta}{\mathrm{d}\eta}\right)^*_{\eta=0} &= 0.5764 + 0.1797\ln\left(Pr_{\mathrm{f}}\right) + 0.0331\ln^2\left(Pr_{\mathrm{f}}\right)\\
&= 0.5764 + 0.1797\ln(5.42) + 0.0331\ln^2(5.42)\\
&= 0.9746.
\end{aligned}
$$

While, the temperature gradient $-(\mathrm{d}\theta/\mathrm{d}\eta)_{\eta=0}$ at the reference temperature $t_\infty = 10°\,\mathrm{C}\,(Pr_\infty = 9.42)$ for water laminar free convection can be predicted as

$$
\begin{aligned}
-\left(\frac{\mathrm{d}\theta}{\mathrm{d}\eta}\right)_{\eta=0} &= 0.5764 + 0.1797\ln\left(Pr_\infty\right) + 0.0331\ln^2\left(Pr_\infty\right)\\
&= 0.5764 + 0.1797\ln(9.42) + 0.0331\ln^2(9.42)\\
&= 1.1459.
\end{aligned}
$$

According to (7.22) the predicted deviation of heat transfer for the water laminar free convection under Boussinesq approximation could be evaluated as

$$
\begin{aligned}
E^*_{\alpha_x} &= 1 - \left(\frac{\nu_\infty}{\nu_{\mathrm{f}}}\right)^{1/2} \frac{\left(\frac{\mathrm{d}\theta}{\mathrm{d}\eta}\right)^*_{\eta=0}}{\left(\frac{\mathrm{d}\theta}{\mathrm{d}\eta}\right)_{\eta=0}}\\
&= 1 - \left(\frac{1.306}{0.805}\right)^{1/2} \frac{0.9746}{1.1459}\\
&= -0.08331\\
&= -8.33\%.
\end{aligned}
\tag{7.33}
$$

Example 2: A flat plate with $b = 2\,\mathrm{m}$ in width and $x = 0.25\,\mathrm{m}$ in length is suspended vertically in air. The ambient temperature is $t_\infty = 20°$ C. Suppose the free convection is laminar, calculate the heat transfer deviation $E^*_{\alpha_x}$ for laminar free convection at the temperature ratio $T_{\mathrm{w}}/T_\infty = 1.1, 1.2, 1.4, 1.7$ and 2.1.

Solution: According to (7.32) the predicted deviation of heat transfer for gas free convection under Boussinesq approximation could be evaluated as

$$
E^*_{\alpha_x} = 1 - \left(\frac{2}{\frac{T_{\mathrm{w}}}{T_\infty}+1}\right)^{(n_\mu+1)/2} \left(\frac{T_{\mathrm{w}}}{T_\infty}\right)^m, \tag{7.34}
$$

where

$$
m = 0.35n_\lambda + 0.29n_\mu + 0.36 \quad (T_{\mathrm{w}}/T_\infty > 1) \tag{7.35}
$$

Table 7.6. Calculated results for question 2

T_w/T_∞	m	$\left(\frac{2}{\frac{T_w}{T_\infty}+1}\right)^{(n_\mu+1)/2}$	$\left(\frac{T_w}{T_\infty}\right)^m$	$E^*_{\alpha_x}$ (%)
1.1	0.8407	0.959845	1.083425	−3.99
1.2	0.8407	0.92306	1.165649	−7.596
1.4	0.8407	0.858001	1.326936	−13.85
1.7	0.8407	0.777176	1.562206	−21.41
2.1	0.8407	0.692024	1.865906	−29.125

According to Table 4.1, it is found that $n_\mu = 0.68$ and $n_\lambda = 0.81$ for air. Then, $m = 0.35n_\lambda + 0.29n_\mu + 0.36 = 0.35 + 0.81 + 0.29 \times 0.68 + 0.36 = 0.8407$. On this basis, the heat transfer deviations $E^*_{\alpha_x}$ for laminar free convection at the temperature ratio $T_w/T_\infty = 1.1, 1.2, 1.4, 1.7$ and 2.1 are predicted and listed in Table 7.6.

References

1. D.Y. Shang and B.X. Wang, Effect of variable thermophysical properties on laminar free convection of gas, Int. J. Heat Mass Transfer 33, No. 7, pp. 1387–1395, 1990
2. D.Y. Shang and B.X. Wang, Effect of variable thermophysical properties on laminar free convection of polyatomic gas, Int. J. Heat Mass Transfer 34, No. 3, pp. 749–755, 1991
3. D.Y. Shang and B.X. Wang, The deviation of heat transfer calculation for laminar free convection of gas due to igonoring the variable thermophysical properties, Warme- und Stoffubertragong 28, 33–36, 1993
4. D.Y. Shang, B.X. Wang, Y, Wang, and Y. Quan, Study on liquid laminar free convection with consideration of variable thermophysical properties, Int. J. Heat Mass Transfer 36, No. 14, pp. 3411–3419, 1993
5. E.M. Sparrow and J.L. Gregg, The variable fluid-property problem in free convection, Trans. ASME 80, pp. 879–886, 1958
6. A. Brown, The effect on laminar free convection heat transfer of temperature dependence of the coefficient of volumetric expension, Trans. ASME, Ser. C, J. Heat Transfer 97, pp. 133–135, 1975
7. D.D. Gray and A. Giogini, The validity of the Boussinesq approximation for liquids and gases, Int. J. Heat Mass Transfer 19, pp. 546–577, 1977

Experimental Measurements of Free Convection with Large Temperature Difference

Nomenclature

c_p	specific heat at constant pressure, J (kg K)$^{-1}$
g	gravitation acceleration, m s^{-2}
$Gr_{x,\infty}$	local Grashof number for laminar free convection of gas on isothermal vertical flat plate, $\frac{g\lvert T_w/T_\infty - 1\rvert x^3}{\nu_\infty^2}$
	local Grashof number for laminar free convection of liquid on isothermal vertical flat plate, $\frac{g\lvert \rho_\infty/\rho_w - 1\rvert x^3}{\nu_\infty^2}$
LDV	Laser Doppler Velocimeter
n_λ	thermal conductivity parameter of gas
n_μ	viscosity parameter of gas
Pr	Prandtl number
t	temperature, °C
T	absolute temperature, K
w_x, w_y	velocity components in the x- and y-directions, respectively, m s^{-1}
W_x, W_y	dimensionless velocity components in the x- and y-directions, respectively

Greek symbols

δ	boundary layer thickness, m
η	dimensionless coordinate variable for boundary layer
θ	dimensionless temperature
λ	thermal conductivity, W (m K)$^{-1}$
μ	absolute viscosity, kg (m s)$^{-1}$
ν	kinetic viscosity, m^2 s^{-1}
ρ	density, kg m^{-3}
$\left(\frac{d\theta}{d\eta}\right)_{\eta=0}$	dimensionless temperature gradient on the plate
$\frac{\frac{\rho_\infty}{\rho} - 1}{\frac{\rho_\infty}{\rho_w} - 1}$	buoyancy factor

$\frac{1}{\rho}\frac{\mathrm{d}p}{\mathrm{d}x}$	density factor
$\frac{1}{\mu}\frac{\mathrm{d}\mu}{\mathrm{d}\eta}$	viscosity factor
$\frac{1}{\lambda}\frac{\mathrm{d}\lambda}{\mathrm{d}\eta}$	thermal conductivity factor

Subscripts

w	at wall
δ	boundary layer
$\propto$	far from the wall surface

8.1 Introduction

The classical measurement of the velocity field for free convection of air along an isothermal vertical plate was originally made by Schmidt and Beckman [1]. Their results showed excellent agreement with the corresponding numerical results for the Boussinesq approximation calculated by Pohlhausen [2] shown in Fig. 8.1. It is further seen that the velocity and thermal boundary-layer thicknesses are proportional to $x^{1/4}$.

However, in their experimental measurements only small differences between the surface and the ambient temperatures were considered. Since then,

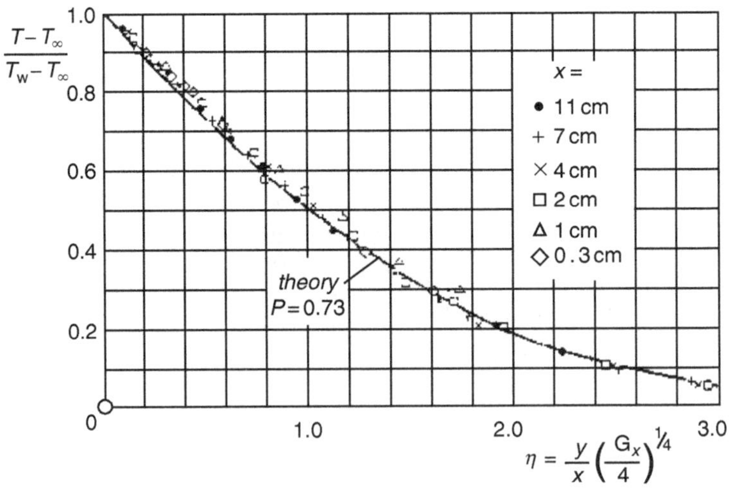

Fig. 8.1. Velocity distribution in the laminar boundary layer on a hot vertical flat plate in natural convection of air, as measured by Schmidt and Beckman [1] (From H. Schlichting, Boundary-Layer Theory, 6th ed., McGraw-Hill, New York, 1968.)

there has been a shortage of accurate measuremental results for consideration of the larger temperature differences. The reasons of this shortage are twofold: (1) the lower velocity of free convection and (2) the restriction of the measuring devices. First of all, the fluid velocity in free convection is typically much slower. In consequence of this, the experimental measurements become more difficult and less reliable due to the increasing influence of various interferences. In addition, due to the weak flow of free convection, the pressure gradients are also quite small and the measuremental techniques based on pressure differences, such as the Pitot tube, cannot be used very accurately. The hot-wire anemometer has been used in velocity measurements, but its basic principle is heat transfer from a heated wire. The heat transfer from the wire is dependent on the flow velocity. However, the major problem that easily produces deviation for the measurement in free convection is the small magnitude of the velocity. Additionally, since the velocity boundary layer for the free convection is very thin, with the above instrument, the interference, which cannot be negligible, will be manifest in the measurement.

Fortunately, the laser doppler velocimeter (LDV) has been developed in recent years. The LDV demonstrates higher accuracy for the measurement of fluid velocity. An instrument, which does not contact the flow field, will not produce any interference in the velocity field. It can measure very low velocity flow. All these features give LDV great advantage over hot-wire anemometers. With consideration of variable thermophysical properties, two experimental results of the laminar free convection for air and water for larger temperature differences, which were provided by Shang, et al. [3–5], are introduced in this chapter. The experimental results were verified by the calculation methods with consideration of variable thermophysical properties introduced in Chaps. 4 and 6 [4, 6], respectively. In this chapter we discuss the measurements of the velocity fields in the laminar boundary layer for free convection of air and water studied with the LDV. Large temperature differences were considered in the experimental measurements for the free convection. The experimental results were verified by the corresponding numerical solutions, and it is shown that the experimental results agreed well with the corresponding numerical solutions for variable thermophysical properties.

8.2 Experimental Measurements of Velocity Field for Air Laminar Free Convection

8.2.1 Experimental Devices and Instruments

The experimental device established is shown schematically in Fig. 8.2. It consists essentially of three parts: an isothermal vertical testing plate, a LDV and a particulate generator.

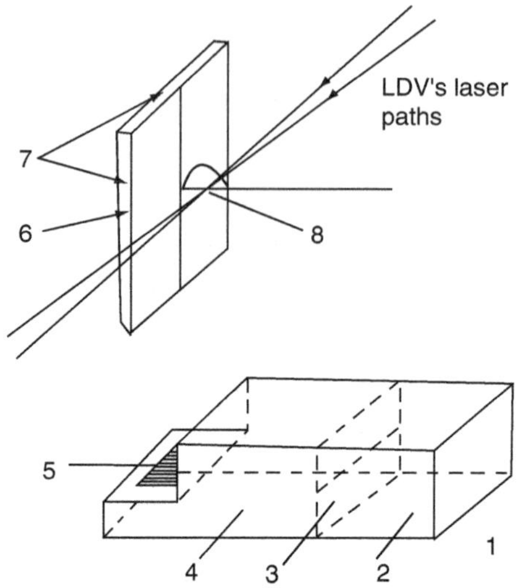

Fig. 8.2. Schematic diagram of experimental device, cited from Shang and Wang [3] (1) Particulate generator, (2) Chamber of mosquito-repellent incense, (3) Spacer, (4) Storage smoke chamber, (5) Wire net of copper, (6) Isothermal flat plate, (7) Thermocouples, (8) Focus of lasers. Note: A dotted oblique line at the left end of the Storage smoke chamber, should be real line

Isothermal testing plate. This is a flat copper plate with a polished surface, 300 mm in height, 195 mm in width and 7 mm in thickness. A sharp angle is made at the bottom of the plate to minimize the possible distortion of the measured velocity field for air free convection. A thin film heater is embedded in the testing plate, the electric power supplied to the heater being adjusted by a current transformer. ϕ 0.1 mm Cu–Constantan thermocouples are installed in the plate to monitor and measure the temperature.

Laser doppler velocimeter (LDV). The short wavelength LDV at Northeastern University, China was used to measure the velocity field of air free convection. The velocity measured with this LDV is so small that it is suitable for detecting the air velocity field being studied.

Particulate generator. The experimental measurement of the velocity field by the LDV requires a particulate generator with an ability to track the air convection. The particulate generator, as shown in Fig. 8.2, consists of a chamber for burning mosquito-repellent incense, a storage chamber of smoke, and a net made of copper wire. The mosquito-repellent incense is burnt in the burning chamber and the smoke produced enters into the storage chamber through the upper gap of the spacer. The smoke cools down in a storage chamber, and then, diffuses through a copper-wire net into the air stream. The velocity of the smoke through the net is very small, and consequently it will disturb the velocity field only to a very small extent.

8.2.2 Measurement Results

Experiments were conducted at three temperature conditions: $T_w/T_\infty = 1.1$ and $T_\infty = 291$ K; $T_w/T_\infty = 1.5$ and $T_\infty = 293$ K; and $T_w/T_\infty = 1.8$ and $T_\infty = 287$ K. For each case, the measurements were made at four heights counted from the bottom edge of the testing plate, i.e., $x = 25$ mm, $x = 50$ mm, $x = 100$ mm, and $x = 150$ mm. The measurement conditions with the related thermophysical properties for air are listed in Table 8.1. Measured velocities w_x are plotted in Figs. 8.3–8.5. It is clear from each of the figures that, w_x would increase along x, and simultaneously, the position for maximum w_x shifts far away from the surface. Comparing the results shown in Figs. 8.3–8.5, it is also seen that, for the same height, x, the larger the boundary temperature ratio T_w/T_∞, the thinner the thickness of boundary layer would be, and so, the position of maximum w_x will be closer to the plate surface with an increased value of maximum w_x. Additionally, the dimensionless velocity component w_x transformed by using (8.1) and (8.2) are plotted in Figs. 8.6–8.8, respectively.

8.2.3 Governing Equations

The governing partial differential equations of gas laminar free convection and their boundary conditions are shown as (4.1)–(4.5) in Chap. 4. According to Chap. 4, the related defined similarity variables are shown as

$$\eta = \frac{x}{y}\left(\frac{1}{4}Gr_{x,\infty}\right)^{1/4} \quad Gr_{x,\infty} = \frac{g\,|T_w/T_\infty - 1|\,x^3}{\nu_\infty^2}, \tag{8.1}$$

Table 8.1. The measurement conditions with the related thermophysical properties for air free convection

heights	temperature conditions	$\nu_\infty \times 10^6$ ($m^2\ s^{-1}$)
$x = 0.025$ m	$T_w/T_\infty = 1.1$ and $T_\infty = 291$ K	14.88
	$T_w/T_\infty = 1.5$ and $T_\infty = 293$ K	15.06
	$T_w/T_\infty = 1.8$ and $T_\infty = 287$ K	14.52
$x = 0.05$ m	$T_w/T_\infty = 1.1$ and $T_\infty = 291$ K	14.88
	$T_w/T_\infty = 1.5$ and $T_\infty = 293$ K	15.06
	$T_w/T_\infty = 1.8$ and $T_\infty = 287$ K	14.52
$x = 0.1$ m	$T_w/T_\infty = 1.1$ and $T_\infty = 291$ K	14.88
	$T_w/T_\infty = 1.5$ and $T_\infty = 293$ K	15.06
	$T_w/T_\infty = 1.8$ and $T_\infty = 287$ K	14.52
$x = 0.15$ m	$T_w/T_\infty = 1.1$ and $T_\infty = 291$ K	14.88
	$T_w/T_\infty = 1.5$ and $T_\infty = 293$ K	15.06
	$T_w/T_\infty = 1.8$ and $T_\infty = 287$ K	14.52

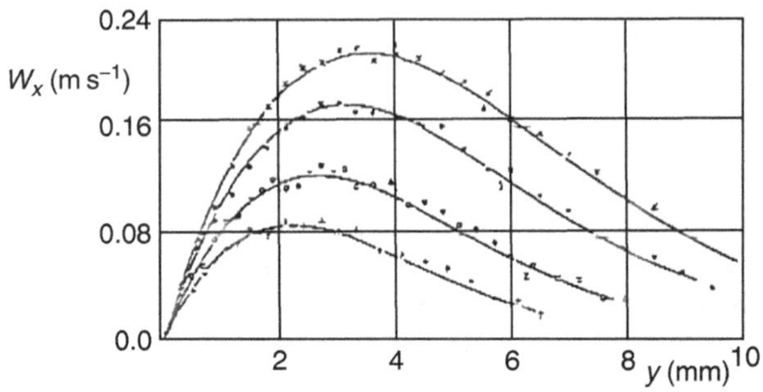

Fig. 8.3. Measured and calculated values for the dimensional velocity of air free convection for conditions $T_\mathrm{w}/T_\infty = 1.1$ and $T_\infty = 291\,\mathrm{K}$, cited from Shang and Wang [3] $+, x = 0025\,\mathrm{m}$; ■, $x = 0.050\,\mathrm{m}$; □, $x = 0.100\,\mathrm{m}$; $\times, x = 0.150\,\mathrm{m}$; — numerical prediction

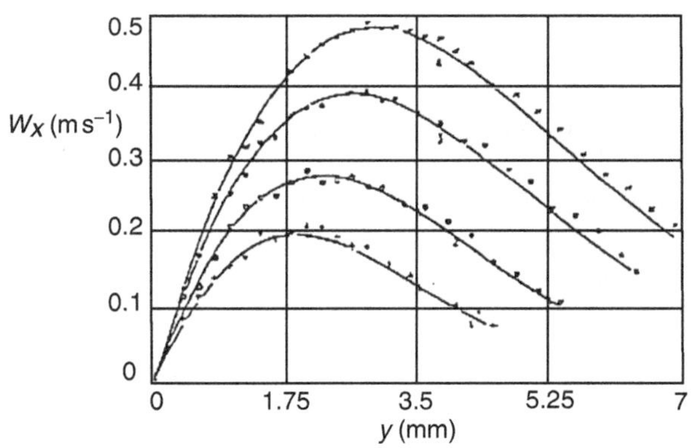

Fig. 8.4. Measured and calculated values for dimensional velocity of air free convection for conditions $T_\mathrm{w}/T_\infty = 1.5$ and $T_\infty = 293\,\mathrm{K}$, cited from Shang and Wang [3] $+, x = 0025\,\mathrm{m}$; ■, $x = 0.050\,\mathrm{m}$; □, $x = 0.100\,\mathrm{m}$; $\times, x = 0.150\,\mathrm{m}$; — numerical prediction

$$W_x = \left[2\sqrt{gx}(T_\mathrm{w}/T_\infty - 1)^{1/2}\right]^{-1} w_x, \tag{8.2}$$

$$W_y = \left[2\sqrt{gx}\,|T_\mathrm{w}/T_\infty - 1|^{1/2}\,(\frac{1}{4}Gr_{x,\infty})^{-1/4}\right]^{-1} w_y, \tag{8.3}$$

and the transformed dimensionless governing equations and boundary conditions are

$$\left(2W_x - \eta\frac{\mathrm{d}W_x}{\mathrm{d}\eta} + 4\frac{\mathrm{d}W_y}{\mathrm{d}\eta}\right) - \frac{1}{\rho}\frac{\mathrm{d}\rho}{\mathrm{d}\eta}(\eta W_x - 4W_y) = 0, \tag{8.4}$$

$$\frac{\nu_\infty}{\nu}\left[W_x\left(2W_x - \eta\frac{\mathrm{d}W_x}{\mathrm{d}\eta}\right) + 4W_y\frac{\mathrm{d}W_x}{\mathrm{d}\eta}\right] = \frac{\mathrm{d}^2W_x}{\mathrm{d}\eta^2} + \frac{1}{\mu}\frac{\mathrm{d}\mu}{\mathrm{d}\eta}\frac{\mathrm{d}W_x}{\mathrm{d}\eta} + \frac{\nu_\infty}{\nu}\theta, \tag{8.5}$$

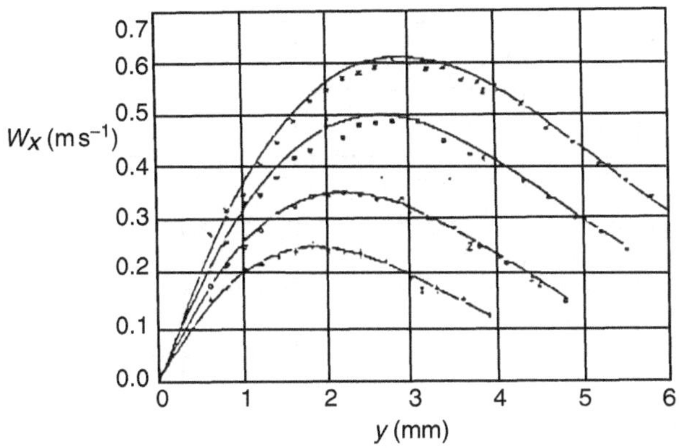

Fig. 8.5. Measured and calculated values for dimensional velocity of air free convection for condition $T_w/T_\infty = 1.8$ and $T_\infty = 287$ K, cited from Shang and Wang [3] $+, x = 0025$ m; ■, $x = 0.050$ m; □, $x = 0.100$ m; $\times, x = 0.150$ m; — numerical prediction

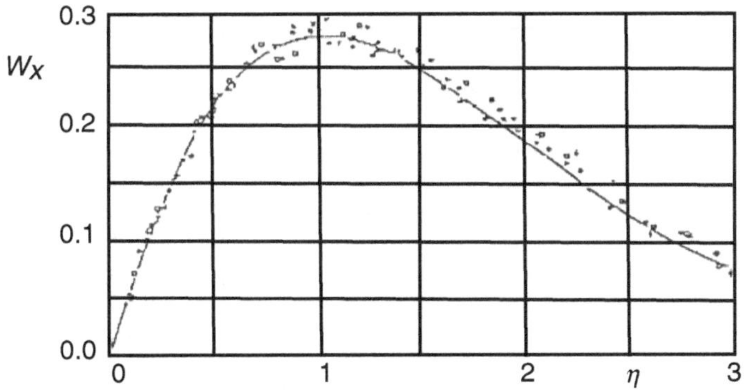

Fig. 8.6. Measured and calculated values for dimensionless velocity of air free convection for condition $T_w/T_\infty = 1.1$ and $T_\infty = 291$ K, cited from Shang and Wang [3] $+, x = 0\ 025$ m; ■, $x = 0.050$ m; □, $x = 0.100$ m; $\times, x = 0.150$ m; — numerical prediction

$$Pr\frac{\nu_\infty}{\nu}\left(-\eta W_x + 4W_y)\frac{\mathrm{d}\theta}{\mathrm{d}\eta}\right) = \frac{\mathrm{d}^2\theta}{\mathrm{d}\eta^2} + \frac{1}{\lambda}\frac{\mathrm{d}\lambda}{\mathrm{d}\eta}\frac{\mathrm{d}\theta}{\mathrm{d}\eta}, \tag{8.6}$$

$$\eta = 0,\ W_x = 0,\ W_y = 0, \theta = 0, \tag{8.7}$$

$$\eta \to 0,\ W_x \to 0,\ \theta \to 0, \tag{8.8}$$

for gas laminar free convection.

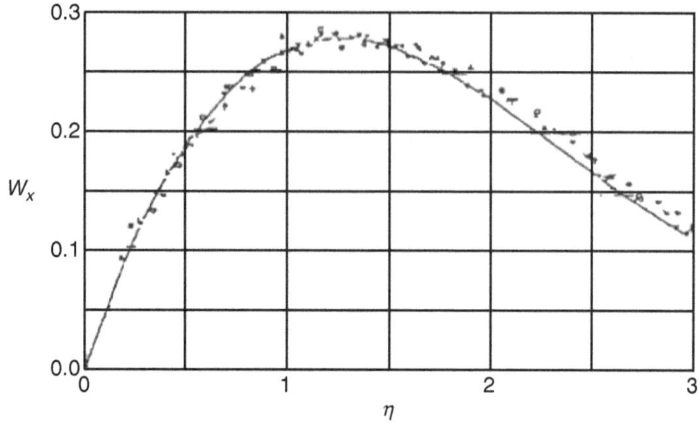

Fig. 8.7. Measured and calculated values for dimensionless velocity of air free convection for condition $T_w/T_\infty = 1.5$ and $T_\infty = 293\,K$, cited from Shang and Wang [3] $+, x = 0025\,m$; ■, $x = 0.050\,m$; □, $x = 0.100\,m$; $\times, x = 0.150\,m$; — numerical prediction

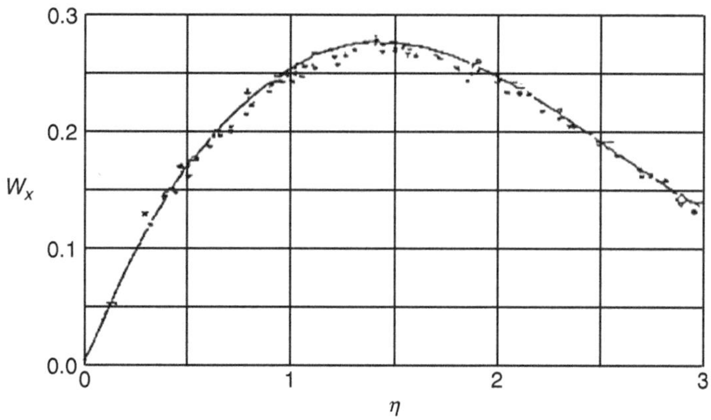

Fig. 8.8. Measured and calculated values for dimensionless velocity of air free convection for condition $T_w/T_\infty = 1.8$ and $T_\infty = 287$ K, cited from Shang and Wang [3] $+, x = 0025\,m$; ■, $x = 0.050\,m$; □, $x = 0.100\,m$; $\times, x = 0.150\,m$; — numerical prediction

8.2.4 The Numerical Solutions

By using the shooting method, the governing dimensionless differential equations (8.4)–(8.6) with their boundary conditions are solved for $T_w/T_\infty = 1.1$ and $T_\infty = 291$ K; $T_w/T_\infty = 1.5$ and $T_\infty = 293$ K; and $T_w/T_\infty = 1.8$ and $T_\infty = 287$ K, respectively, for the air free convection ($n_\mu = 0.68, n_\lambda = 0.81$).

The numerical solutions for dimensionless velocity components W_x are plotted also in Figs. 8.3–8.5. While, the dimensionless numerical solutions transformed by using (8.2) and (8.3) are plotted in Figs. 8.6–8.8, respectively. It can be seen that the measured results agree very well with the predicted results.

8.3 Experimental Measurements of Velocity Field for Water Laminar Free Convection

8.3.1 Main Experimental Apparatus

An isothermal vertical flat plate, a LDV and a water tank (see Fig. 8.9) constitute the main experimental apparatus.

Isothermal vertical flat plate. The isothermal vertical flat plate (called here the plate) made of copper, is 250 mm in length, 140 mm in width, and 7 mm in thickness. The surface of the plate is well polished. In the plate, a nickel–chromium wire of 0.5 mm in diameter and 389 m in length is uniformly inserted. The nickel–chromium wire serves as an electrical heat source, and it is insulated. A sharp angle is made in the bottom of the plate so that the velocity field would not be influenced by the free convection near to the bottom surface. Thermocouples are installed in the plate and are very close to the surface. By controlling the electric current passing through the nickel–chromium wire, the temperature at the surface of the plate will be maintained at a certain level. On the top of the plate, two metal plates with 150 mm in length and 3 mm in thickness are welded. The upper part of both metal plates is drilled so that the plate can be suspended on the frame.

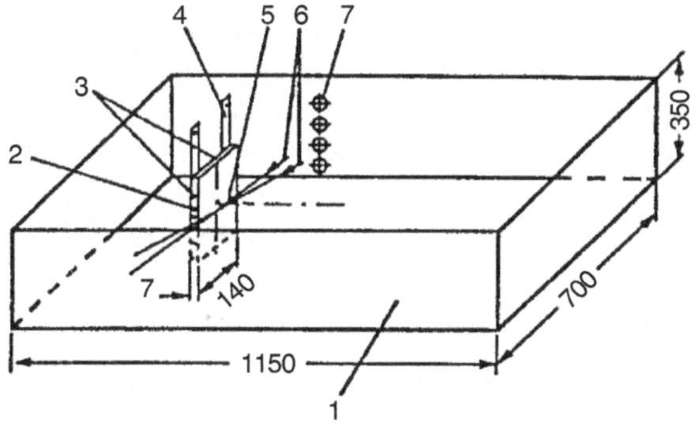

Fig. 8.9. Schematic diagram of the device used in the experiment of water free convection, cited from Shang, Wang, Wang, and Quan [4]: (1) water tank; (2) isothermal vertical flat plate; (3) thermocouples; (4) metal plate; (5) focus of laser; (6) laser paths; (7) drilled hole for laser path

LDV. The equipment used to measure the velocity field of the water free convection on the plate is the LDV of the 606 Institute in Shenyang. In order to measure very small velocities such as that of water free convection, the technique of frequency-deviation-shift is applied to the LDV.

Water tank. The water tank is rectangular in shape. It is made of organic glass plate with 8 mm thickness. The top of the water tank is open. The tank is 1.1 m in length, 0.7 m in width, and 0.35 m in height. With such a large space the water tank can efficiently keep away the free convection near to the surface of the plate face from any disturbing influences. In the side of the tank are drilled four drill ways of 20 mm diameter each. The drilled ways serve as paths of laser light. Through the drill ways the laser will reach the surface of the plate to measure the velocity field of the water free convection. The distance between each two centers of the drilled ways is 50 mm, which just matches the measured heights. The locations of the drilled ways are covered with the thin organic glass, 1 mm in thickness, so that the laser power wasted though the organic glass can be minimized.

8.3.2 The Results of Experiment

At the start of this experiment the surface of the plate should be heated slowly so that the temperature of the plate rises slowly. For this purpose the voltage and electricity current through the nickel–chromium wire is increased slowly by means of a voltage regulator. After the temperature of the measured surface is raised to given level, the temperature is stabilized for 3 min, and then the measurements commenced.

In this experiment, the measurements are carried out under three temperature conditions: $t_{\mathrm{w}} = 40°\mathrm{C}$ and $t_\infty = 20°\mathrm{C}, t_{\mathrm{w}} = 50°\mathrm{C}$ and $t_\infty = 20°\mathrm{C}, t_{\mathrm{w}} = 60°\mathrm{C}$ and $t_\infty = 20°\mathrm{C}$, respectively. For each condition the measurements are made at four heights from the bottom of the plate, i.e., $x = 0.05\mathrm{m}$, $x = 0.10\,\mathrm{m}$, $x = 0.15\,\mathrm{m}$ and $x = 0.20\,\mathrm{m}$. The measurement conditions are listed in Table 8.2 in detail. The measured values of the velocity components w_x are described in Tables 8.4–8.6 and plotted in Figs. 8.10–8.12 respectively. The measured values w_x and the corresponding coordinate variable x are further transformed to the dimensionless values by using the expressions (8.9)–(8.11), described in Tables 8.4–8.6 and Figs. 8.13–8.15, respectively.

8.3.3 Governing Equations

The governing partial differential equations of liquid laminar free convection and their boundary conditions are shown as (6.1)–(6.5) in Chap. 6. According to Chap. 6, the related defined similarity variables for liquid laminar free convection are shown as

$$\eta = \left(\frac{1}{4}Gr_{x,\infty}\right)^{1/4}\frac{y}{x}, \qquad Gr_{x,\infty} = \frac{g\,|\rho_\infty/\rho_{\mathrm{w}} - 1|\,x^3}{\nu_\infty^2}, \tag{8.9}$$

Table 8.2. The measurement conditions

heights	temperature
$x = 0.05\,\mathrm{m}$	$t_\mathrm{w} = 40°\mathrm{C}$ and $t_\infty = 20°\mathrm{C}$
	$t_\mathrm{w} = 50°\mathrm{C}$ and $t_\infty = 20°\mathrm{C}$
	$t_\mathrm{w} = 60°\mathrm{C}$ and $t_\infty = 20°\mathrm{C}$
$x = 0.10\,\mathrm{m}$	$t_\mathrm{w} = 40°\mathrm{C}$ and $t_\infty = 20°\mathrm{C}$
	$t_\mathrm{w} = 50°\mathrm{C}$ and $t_\infty = 20°\mathrm{C}$
	$t_\mathrm{w} = 60°\mathrm{C}$ and $t_\infty = 20°\mathrm{C}$
$x = 0.15\,\mathrm{m}$	$t_\mathrm{w} = 40°\mathrm{C}$ and $t_\infty = 20°\mathrm{C}$
	$t_\mathrm{w} = 50°\mathrm{C}$ and $t_\infty = 20°\mathrm{C}$
	$t_\mathrm{w} = 60°\mathrm{C}$ and $t_\infty = 20°\mathrm{C}$
$x = 0.20\,\mathrm{m}$	$t_\mathrm{w} = 40°\mathrm{C}$ and $t_\infty = 20°\mathrm{C}$
	$t_\mathrm{w} = 50°\mathrm{C}$ and $t_\infty = 20°\mathrm{C}$
	$t_\mathrm{w} = 60°\mathrm{C}$ and $t_\infty = 20°\mathrm{C}$

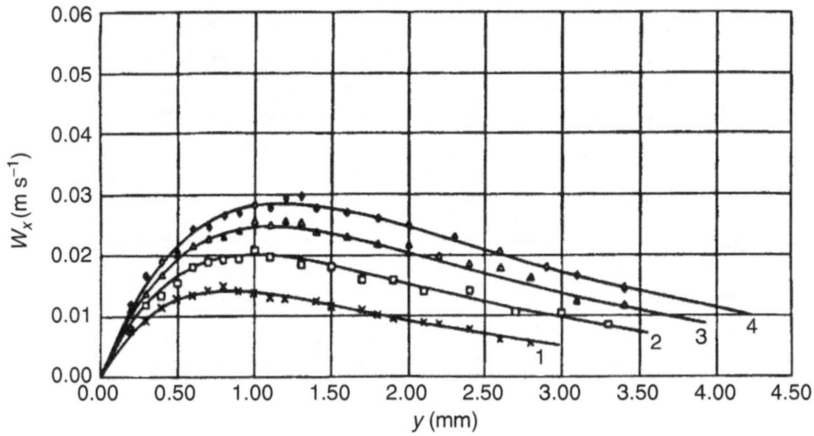

Fig. 8.10. Measured and numerical values of velocity w_x of water laminar free convection for $t_\mathrm{w} = 40°\mathrm{C}$ and $t_\infty = 20°\mathrm{C}$ [4], cited from Shang, Wang, and Takhar [5]. Full line: numerical solution, symbol: corresponding measured value. (1) $\times, x = 0.05\,\mathrm{m}$; (2) $\square, x = 0.10\,\mathrm{m}$; (3) $\Delta, x = 0.15\,\mathrm{m}$, and (4) $\diamondsuit, x = 0.20\,\mathrm{m}$

$$W_x = (2\sqrt{gx}\,|\rho_\infty/\rho_\mathrm{w} - 1|^{1/2})^{-1} w_x, \tag{8.10}$$

$$W_y = \left[2\sqrt{gx}\left|\frac{\rho_\infty}{\rho} - 1\right|^{-1/2}\left(\frac{1}{4}Gr_{x,\infty}\right)^{-1/4}\right]^{-1} w_y. \tag{8.11}$$

The transformed dimensionless governing equations and boundary conditions for liquid laminar free convection are

$$2W_x - \eta\frac{\mathrm{d}W_x}{\mathrm{d}\eta} + 4\frac{\mathrm{d}W_y}{\mathrm{d}\eta} - \frac{1}{\rho}\frac{\mathrm{d}\rho}{\mathrm{d}\eta}(\eta W_x - 4W_y) = 0, \tag{8.12}$$

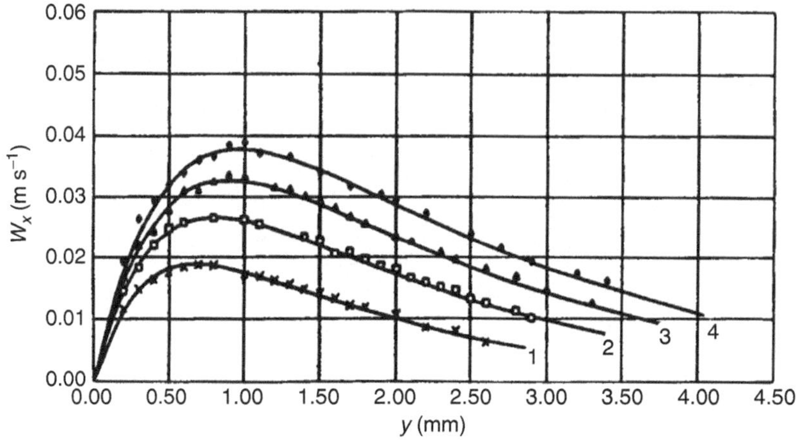

Fig. 8.11. Measured and numerical values of velocity w_x of water laminar free convection for $t_w = 50°C$ and $t_\infty = 20°C$, cited from Shang, Wang, and Takhar [5]. Full line: numerical solution, symbol: corresponding measured value. (1) $\times, x = 0.05\,\mathrm{m}$; (2) $\square, x = 0.10\,\mathrm{m}$; (3) $\Delta, x = 0.15\,\mathrm{m}$, and (4) $\diamondsuit, x = 0.20\,\mathrm{m}$

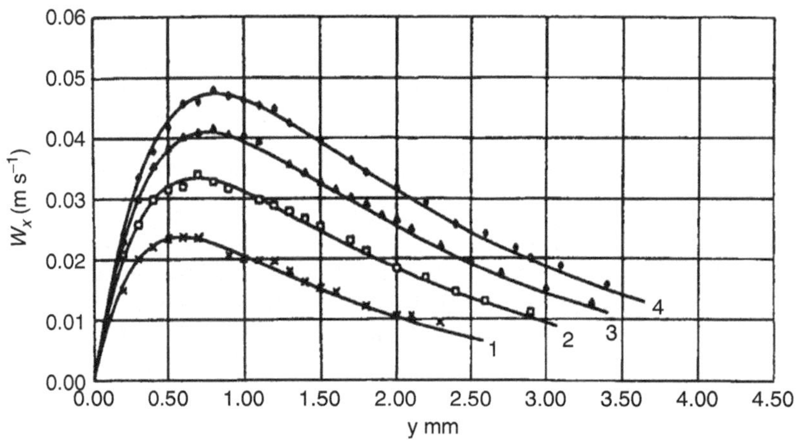

Fig. 8.12. Measured and numerical values for velocity w_x of water laminar free convection for $t_w = 60°C$ and $t_\infty = 20°C$, cited from Shang, Wang, and Takhar [5]. Full line: numerical solution, symbol: corresponding measured value. (1) $\times, x = 0.05\,\mathrm{m}$; (2) $\square, x = 0.10\,\mathrm{m}$; (3) $\Delta, x = 0.15\,\mathrm{m}$, and (4) $\diamondsuit, x = 0.20\,\mathrm{m}$

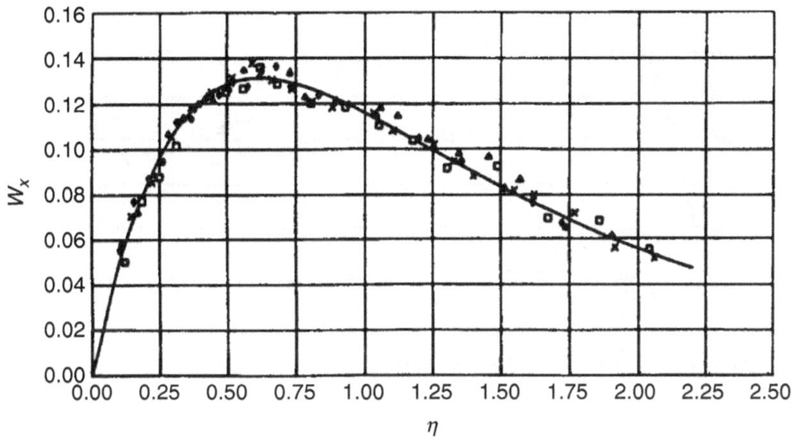

Fig. 8.13. Measured and calculated values of the dimensional velocity W_x of water laminar free Convection, for $t_\mathrm{w} = 40°\mathrm{C}$ and $t_\infty = 20°\mathrm{C}$, cited from Shang, Wang, and Takhar [5]. Full line: numerical solution, symbol: corresponding measured value. (1) $\times, x = 0.05\,\mathrm{m}$; (2) $\square, x = 0.10\,\mathrm{m}$; (3) $\Delta, x = 0.15\,\mathrm{m}$, and (4) $\diamondsuit, x = 0.20\,\mathrm{m}$

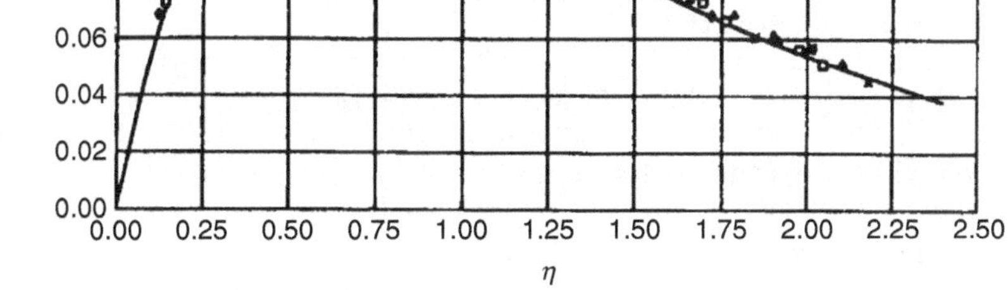

Fig. 8.14. Measured and calculated values for dimensional velocity W_x of water laminar free convection, for $t_\mathrm{w} = 50°\mathrm{C}$ and $t_\infty = 20°\mathrm{C}$, cited from Shang, Wang, and Takhar [5]. Full line: numerical solution, symbol: corresponding measured value. (1) $\times, x = 0.05$ m; (2) $\square, x = 0.10$ m; (3) $\Delta, x = 0.15$ m, and (4) $\diamondsuit, x = 0.20$ m

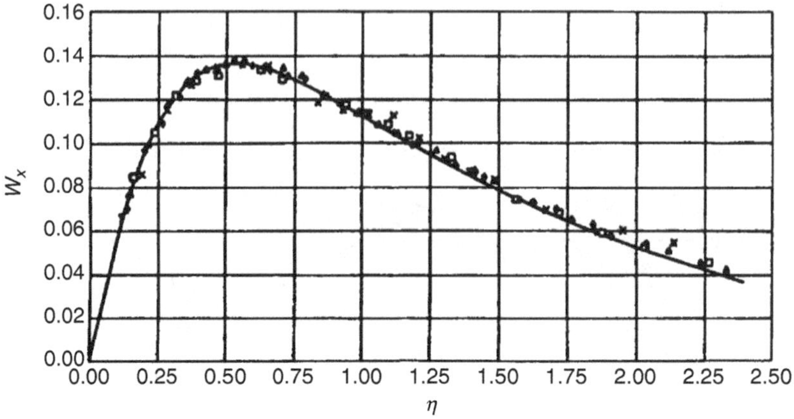

Fig. 8.15. Measured and calculated values for dimensional velocity W_x of water laminar free convection, for condition $t_w = 60°C$ and $t_\infty = 20°C$, cited from Shang, Wang, and Takhar [5]. Full line: numerical solution, symbol: corresponding measured value. (1) $\times, x = 0.05$ m; (2) $\square, x = 0.10$ m; (3) $\Delta, x = 0.15$ m, and (4) $\diamondsuit, x = 0.20$ m

$$\frac{\nu_\infty}{\nu}\left(W_x\left(2W_x - \eta\frac{\mathrm{d}W_x}{\mathrm{d}\eta}\right) + 4W_y\frac{\mathrm{d}W_x}{\mathrm{d}\eta}\right) = -\frac{\mathrm{d}^2W_x}{\mathrm{d}\eta^2} + \frac{1}{\mu}\frac{\mathrm{d}\mu}{\mathrm{d}\eta}\frac{\mathrm{d}W_x}{\mathrm{d}\eta} + \frac{\nu_\infty}{\nu}\frac{\frac{\rho_\infty}{\rho} - 1}{\frac{\rho_\infty}{\rho_w} - 1}, \tag{8.13}$$

$$Pr\frac{\nu_\infty}{\nu}(-\eta W_x + 4W_y)\frac{\mathrm{d}\theta}{\mathrm{d}\eta} = \frac{1}{\lambda}\frac{\mathrm{d}\lambda}{\mathrm{d}\eta}\frac{\mathrm{d}\theta}{\mathrm{d}\eta} + \frac{\mathrm{d}^2\theta}{\mathrm{d}\eta^2}, \tag{8.14}$$

with boundary conditions

$$\eta = 0,\ W_x = 0, W_y = 0, \theta = 0, \tag{8.15}$$

$$\eta \to 0,\ W_x \to 0,\ \theta \to 0. \tag{8.16}$$

8.3.4 Numerical Solutions

As the analysis in Chap. 6, if the specific heat c_p of water is substituted by c_{p_∞}, i.e., its value at the temperature at infinity, the maximum predicted deviation will be less than 0.455 % for the temperature range from 0 to 100°C according to typical experiment values [7]. Such small deviation is allowed for the treatment of variable thermophysical properties. Consequently, property factor $Pr(\nu_\infty/\nu)$ of (8.14) can be changed to the following form for water laminar free convection:

$$Pr\frac{\nu_\infty}{\nu} = Pr_\infty\frac{\lambda_\infty}{\lambda}\frac{\rho}{\rho}. \tag{8.17}$$

The water thermophysical property values of ρ, μ, ν, λ and Pr are taken from Chap. 6. For convenience some specimen values of the water thermophysical properties for the experiment are listed in Table 8.3.

Table 8.3. Water property values

t (°C)	20	40	50	60
ρ (kg m^{-3})	998.3	992.3	988.1	983.2
λ (W (m K)$^{-1}$)	0.5996			
$\nu \times 10^{-6}$ (kg (m s)$^{-1}$)	1.004			
Pr	6.99			

According to the approach of the numerical calculation of Chap. 6, the solutions for water laminar free convection are obtained from the governing ordinary differential equation (8.12)–(8.14) with (8.17) and the boundary conditions (8.15) and (8.16) by shooting method, respectively, for $t_{\mathrm{w}} = 40°\mathrm{C}$ and $t_\infty = 20°\mathrm{C}, t_{\mathrm{w}} = 50°\mathrm{C}$ and $t_\infty = 20°\mathrm{C}, t_{\mathrm{w}} = 60°\mathrm{C}$, and $t_\infty = 20°\mathrm{C}$. While, the water thermophysical properties such as ρ, λ and ν are described respectively, by (8.18)–(8.20). Meanwhile (6.37)–(6.43) are applied for describing the related physical property factors of the governing equations. The numerical solutions for velocity component w_x obtained for the water laminar free convection are listed in Tables 8.4–8.6 and plotted in Figs. 8.10–8.12, respectively. In addition, these numerical solutions w_x are transformed into the corresponding dimensional ones W_x by means of the (8.10). The transformed dimensionless solutions are described in Tables 8.4–8.6 and plotted in Figs. 8.13–8.15, respectively. It can be seen that the measured results agree very well with the predicted results Tables 8.7–8.9.

$$\rho = -4.48 \times 10^{-3} t^2 + 999.9, \tag{8.18}$$

$$\lambda = -8.01 \times 10^{-6} t^2 + 1.94 \times 10^{-3} t + 0.563, \tag{8.19}$$

$$\mu = \exp\left[-1.6 - \frac{1150}{T} + \left(\frac{690}{T}\right)^2\right] \times 10^{-3}. \tag{8.20}$$

8.4 Remarks

Experimental investigations were carried out to study effects of variable thermophysical properties on laminar free convection of air and water and to further verify the results of the previous chapters, Chaps. 4 and 6. The following points are made:

By increasing the temperature t_{w} for the liquid laminar free convection or with increasing the boundary temperature ratio T_{w}/T_∞ for gas laminar free convection of gas, the velocity component w_x of the free convection increases,

Table 8.4. The measurement results for velocities w_x and the transformed values of W_x for water laminar free convection at $t_w = 40°C$ and $t_\infty = 20°C$, cited from Shang, Wang, and Takhar [5]

$x = 0.05$m				$x = 010$m			
y(mm)	η	w_x(m s^{-1})	W_x	y(mm)	η	w_x(m s^{-1})	W_x
0.2	0.147	0.007	0.0707	0.2	0.124	0.007	0.0500
0.3	0.221	0.0093	0.0854	0.3	0.186	0.0119	0.0773
0.4	0.294	0.0115	0.1056	0.4	0.248	0.0135	0.0877
0.5	0.368	0.0129	0.1185	0.5	0.309	0.0156	0.1013
0.6	0.441	0.0136	0.1249	0.6	0.371	0.0182	0.1182
0.7	0.515	0.0143	0.1313	0.7	0.433	0.0190	0.1234
0.8	0.589	0.0150	0.1377	0.8	0.495	0.0193	0.1253
0.9	0.662	0.0142	0.1304	0.9	0.557	0.0195	0.1266
1.0	0.736	0.0139	0.1276	1.0	0.619	0.0209	0.1357
1.2	0.883	0.0129	0.1185	1.3	0.804	0.0185	0.1201
1.4	1.030	0.0126	0.1157	1.5	0.928	0.0182	0.1182
1.5	1.104	0.0118	0.1084	1.7	1.052	0.0161	0.1104
1.7	1.251	0.0111	0.1019	1.9	1.176	0.0160	0.1039
1.9	1.398	0.0096	0.0882	2.1	1.299	0.0141	0.0916
2.1	1.545	0.0089	0.0817	2.4	1.485	0.0142	0.0922
2.4	1.766	0.0078	0.0716	2.7	1.671	0.0107	0.0695
2.6	1.913	0.0061	0.0560	3.0	1.856	0.0105	0.0682
2.8	2.060	0.0056	0.0514	3.3	2.042	0.0085	0.0552

$x = 0.15$m				$x = 0.20$ m			
y(mm)	η	w_x(m s^{-1})	W_x	y(mm)	η	w_x (m s^{-1})	W_x
0.2	0.112	0.0109	0.0578	0.2	0.104	0.0120	0.0551
0.3	0.168	0.0136	0.0721	0.3	0.156	0.0168	0.0771
0.4	0.224	0.0168	0.0890	0.4	0.208	0.0190	0.0872
0.5	0.280	0.0201	0.1065	0.5	0.260	0.0206	0.0946
0.7	0.391	0.0226	0.1198	0.7	0.364	0.0247	0.1134
0.9	0.503	0.0239	0.1267	0.9	0.468	0.0270	0.1240
1.1	0.615	0.0249	0.1320	1.1	0.572	0.0278	0.1276
1.3	0.727	0.0252	0.1335	1.2	0.624	0.0294	0.1350
1.4	0.783	0.0237	0.1229	1.3	0.676	0.0297	0.1364
1.6	0.894	0.0229	0.1214	1.4	0.728	0.0278	0.1276
1.8	1.056	0.0217	0.1130	1.6	0.832	0.0270	0.1240
2.0	1.118	0.0216	0.1145	1.8	0.937	0.0260	0.1194
2.2	1.230	0.0197	0.1044	2.0	1.041	0.0250	0.1148
2.4	1.342	0.0185	0.0980	2.3	1.197	0.0229	0.1051
2.6	1.453	0.0182	0.0964	2.6	1.353	0.0206	0.0946
2.8	1.565	0.0163	0.0864	2.9	1.509	0.0180	0.0826
3.1	1.733	0.0123	0.0650	3.1	1.613	0.0167	0.0767
3.4	1.901	0.0116	0.0615	3.4	1.719	0.0146	0.0670

Table 8.5. The measurement results for velocities w_x and the transformed value W_x for water laminar free convection at $t_w = 50°C$ and $t_\infty = 20°C$, cited from Shang, Wang, and Takhar [5]

$x = 0.05\,m$				$x = 0.10\,m$			
y(mm)	η	$w_x(m\,s^{-1})$	W_x	y(mm)	η	$w_x(m\,s^{-1})$	W_x
0.2	0.168	0.0116	0.0819	0.2	0.141	0.0146	0.0729
0.3	0.252	0.0148	0.1045	0.3	0.212	0.0184	0.0919
0.4	0.336	0.0162	0.1144	0.4	0.282	0.0221	0.1103
0.5	0.419	0.0174	0.1229	0.5	0.353	0.0249	0.1243
0.6	0.503	0.0184	0.1299	0.6	0.423	0.0259	0.1293
0.7	0.587	0.0188	0.1328	0.8	0.565	0.0266	0.1328
0.8	0.671	0.0187	0.1321	1.0	0.706	0.0263	0.1313
1.0	0.839	0.0172	0.1215	1.1	0.776	0.0256	0.1278
1.1	0.923	0.0168	0.1186	1.4	0.988	0.0234	0.1168
1.3	1.091	0.0155	0.1095	1.6	1.129	0.0206	0.1028
1.4	1.174	0.0148	0.1045	1.7	1.200	0.0209	0.1043
1.5	1.258	0.0141	0.0996	1.8	1.270	0.0196	0.0979
1.7	1.426	0.0122	0.0862	2.0	1.411	0.0180	0.0899
1.8	1.510	0.0118	0.0833	2.1	1.482	0.0167	0.0834
2.0	1.678	0.0107	0.0756	2.3	1.623	0.0151	0.0754
2.2	1.846	0.0086	0.0607	2.5	1.764	0.0133	0.0664
2.4	2.013	0.0081	0.0572	2.6	1.835	0.0126	0.0629
2.6	2.181	0.0064	0.0452	2.9	2.047	0.0102	0.0509

$x = 0.15\,m$				$x = 0.20\,m$			
y(mm)	η	$w_x(m\,s^{-1})$	W_x	y(mm)	η	$w_x(m\,s^{-1})$	W_x
0.2	0.127	0.0167	0.0681	0.2	0.119	0.0194	0.0685
0.3	0.191	0.0219	0.0893	0.3	0.178	0.0264	0.0932
0.4	0.255	0.0242	0.0987	0.4	0.237	0.0294	0.1028
0.5	0.319	0.0277	0.1129	0.5	0.297	0.0321	0.1133
0.6	0.382	0.0309	0.1260	0.6	0.356	0.0340	0.1201
0.7	0.446	0.0310	0.1264	0.7	0.415	0.0360	0.1271
0.9	0.574	0.0334	0.1362	0.9	0.534	0.0384	0.1356
1.0	0.637	0.0332	0.1353	1.0	0.593	0.0389	0.1374
1.2	0.765	0.0315	0.1284	1.1	0.652	0.0372	0.1314
1.3	0.828	0.0312	0.1272	1.3	0.771	0.0365	0.1289
1.4	0.892	0.0301	0.1227	1.5	0.890	0.0342	0.1208
1.5	0.956	0.0293	0.1194	1.7	1.008	0.0317	0.1119
1.6	1.020	0.0281	0.1151	2.9	1.127	0.0304	0.1073
1.8	1.147	0.0255	0.1040	2.2	1.305	0.0272	0.0960
2.0	1.275	0.0233	0.0950	2.5	1.483	0.0239	0.0844
2.3	1.466	0.0208	0.0848	2.9	1.720	0.0193	0.0682
2.8	1.784	0.0168	0.0685	3.2	1.898	0.0174	0.0614
3.3	2.103	0.0125	0.0510	3.4	2.017	0.0161	0.0569

Table 8.6. The measurement results for velocities w_x and the transformed values W_x for water laminar free convection at $t_w = 60°C$ and $t_\infty = 20°C$, cited from Shang, Wang, and Takhar [5]

$x = 0.05$ m				$x = 0.10$ m			
y(mm)	η	w_x(m s^{-1})	W_x	y(mm)	η	w_x(m s^{-1})	W_x
0.2	0.186	0.0149	0.0858	0.2	0.156	0.0208	0.0847
0.3	0.279	0.0200	0.1152	0.3	0.234	0.0258	0.1051
0.4	0.371	0.0220	0.1267	0.4	0.313	0.0299	0.1218
0.5	0.464	0.0234	0.1348	0.5	0.391	0.0315	0.1283
0.6	0.557	0.0236	0.1359	0.6	0.469	0.0321	0.1308
0.7	0.650	0.0235	0.1354	0.7	0.547	0.0340	0.1385
0.9	0.836	0.0206	0.1187	0.8	0.625	0.0328	0.1336
1.0	0.929	0.0200	0.1152	0.9	0.703	0.0317	0.1291
1.1	1.021	0.0197	0.1135	1.1	0.859	0.0298	0.1214
1.2	1.114	0.0196	0.1129	1.2	0.938	0.0289	0.1177
1.3	1.207	0.0178	0.1025	1.3	1.016	0.0279	0.1136
1.4	1.300	0.0161	0.0927	1.4	1.094	0.0267	0.1088
1.5	1.393	0.0151	0.0870	1.5	1.172	0.0254	0.1035
1.6	1.486	0.0145	0.0835	1.7	1.328	0.0230	0.0937
1.8	1.671	0.0121	0.0697	1.8	1.406	0.0213	0.0868
2.0	1.857	0.0104	0.0599	2.0	1.563	0.0183	0.0745
2.1	1.950	0.0105	0.0605	2.4	1.875	0.0145	0.0591
2.3	2.136	0.0095	0.0547	2.9	2.266	0.0112	0.0456

$x = 0.15$m				$x = 0.20$m			
y(mm)	η	w_x(m s^{-1})	W_x	y(mm)	η	w_x(m s^{-1})	W_x
0.2	0.141	0.0230	0.0765	0.2	0.131	0.0243	0.0700
0.3	0.212	0.0299	0.0994	0.3	0.197	0.0336	0.0968
0.4	0.282	0.0353	0.1174	0.4	0.263	0.0379	0.1092
0.5	0.353	0.0384	0.1277	0.5	0.329	0.0420	0.1210
0.7	0.494	0.0408	0.1357	0.7	0.460	0.0460	0.1325
0.9	0.635	0.0405	0.1347	0.9	0.591	0.0470	0.1354
1.1	0.776	0.0393	0.1307	1.1	0.723	0.0455	0.1310
1.3	0.917	0.0355	0.1181	1.2	0.788	0.0449	0.1293
1.4	0.988	0.0342	0.1137	1.3	0.854	0.0425	0.1224
1.6	1.129	0.0314	0.1044	1.7	1.117	0.0364	0.1048
1.8	1.270	0.0290	0.0964	2.0	1.314	0.0319	0.0919
2.0	1.411	0.0264	0.0878	2.2	1.445	0.0294	0.0847
2.1	1.482	0.0249	0.0828	2.4	1.577	0.0258	0.0743
2.3	1.623	0.0220	0.0732	2.6	1.708	0.0243	0.0700
2.5	1.764	0.0196	0.0652	2.8	1.840	0.0219	0.0631
2.7	1.905	0.0175	0.0582	2.9	1.905	0.0201	0.0579
3.0	2.117	0.0152	0.0505	3.1	2.037	0.0188	0.0541
3.3	2.329	0.0126	0.0419	3.4	2.234	0.0157	0.0452

Table 8.7. The numerical solutions of velocity components w_x and W_x at $t_w = 40°C$ and $t_\infty = 20°C$, cited from Shang, Wang, and Takhar [5]

$x = 0.05$ m				$x = 0.10$ m			
η	y (mm)	W_x	w_x (m s^{-1})	η	y (mm)	W_x	w_x (m s^{-1})
0	0	0	0	0	0	0	0
0.075	0.102	0.0378	0.0041	0.075	0.121	0.0378	0.0058
0.150	0.204	0.0674	0.0073	0.150	0.242	0.0674	0.0104
0.225	0.306	0.0898	0.0098	0.225	0.364	0.0898	0.0138
0.300	0.408	0.1063	0.0116	0.300	0.485	0.1063	0.0164
0.375	0.510	0.1178	0.0128	0.375	0.606	0.1178	0.0181
0.450	0.612	0.1253	0.0136	0.450	0.727	0.1253	0.0193
0.525	0.713	0.1297	0.0141	0.525	0.848	0.1297	0.0200
0.600	0.815	0.1314	0.0143	0.600	0.970	0.1314	0.0202
0.700	0.951	0.1308	0.0142	0.700	1.131	0.1308	0.0201
0.800	1.087	0.1277	0.0139	0.800	1.293	0.1277	0.0197
0.900	1.223	0.1231	0.0134	0.900	1.454	0.1231	0.0190
1.050	1.427	0.1143	0.0124	1.050	1.700	0.1143	0.0176
1.200	1.631	0.1046	0.0114	1.200	1.939	0.1046	0.0161
1.350	1.835	0.0948	0.0103	1.350	2.182	0.0948	0.0146
1.500	2.039	0.0853	0.0093	1.500	2.424	0.0853	0.0131
1.800	2.446	0.0681	0.0074	1.800	2.909	0.0681	0.0105
2.100	2.854	0.0536	0.0058	2.100	3.394	0.0536	0.0083

$x = 0.15$ m				$x = 0.20$ m			
η	y (mm)	W_x	w_x (m s^{-1})	η	y (mm)	W_x	w_x (m s^{-1})
0	0	0	0	0	0	0	0
0.075	0.134	0.0378	0.0071	0.075	0.144	0.0378	0.0082
0.150	0.268	0.0674	0.0127	0.150	0.288	0.0674	0.0147
0.225	0.403	0.0898	0.0169	0.225	0.432	0.0898	0.0196
0.300	0.537	0.1063	0.0201	0.300	0.577	0.1063	0.0232
0.375	0.671	0.1178	0.0222	0.375	0.721	0.1178	0.0257
0.450	0.805	0.1253	0.0236	0.450	0.865	0.1253	0.0273
0.525	0.939	0.1297	0.0245	0.525	1.009	0.1297	0.0282
0.600	1.073	0.1314	0.0248	0.600	1.153	0.1314	0.0286
0.700	1.252	0.1308	0.0247	0.700	1.345	0.1308	0.0285
0.800	1.431	0.1277	0.0241	0.800	1.538	0.1277	0.0278
0.900	1.610	0.1231	0.0232	0.900	1.730	0.1231	0.0268
1.050	1.878	0.1143	0.0216	1.050	2.018	0.1143	0.0249
1.200	2.147	0.1046	0.0197	1.200	2.306	0.1046	0.0228
1.350	2.415	0.0948	0.0179	1.350	2.595	0.0948	0.0206
1.500	2.684	0.0853	0.0161	1.500	2.883	0.0853	0.0186
1.800	3.220	0.0681	0.0129	1.800	2.460	0.0681	0.0148
2.100	3.757	0.0536	0.0101	2.100	4.036	0.0536	0.0117

Table 8.8. The numerical solutions of velocity components w_x and W_x at $t_w = 50°C$ and $t_\infty = 20°C$, cited from Shang, Wang, and Takhar [5]

$x = 0.05\,m$				$x = 0.10\,m$			
η	y (mm)	W_x	w_x (m s^{-1})	η	y (mm)	W_x	w_x (m s^{-1})
0	0	0	0	0	0	0	0
0.075	0.089	0.0416	0.0059	0.075	0.106	0.0416	0.0083
0.150	0.179	0.0732	0.0104	0.150	0.213	0.0732	0.0147
0.225	0.268	0.0964	0.0137	0.225	0.319	0.0964	0.0193
0.300	0.358	0.1127	0.0160	0.300	0.425	0.1127	0.0226
0.375	0.447	0.1235	0.0175	0.375	0.531	0.1235	0.0247
0.450	0.536	0.1300	0.0184	0.450	0.638	0.1300	0.0260
0.525	0.626	0.1331	0.0188	0.525	0.744	0.1331	0.0267
0.600	0.715	0.1337	0.0189	0.600	0.850	0.1337	0.0268
0.700	0.834	0.1317	0.0186	0.700	0.992	0.1317	0.0264
0.800	0.954	0.1274	0.0180	0.800	1.134	0.1274	0.0255
0.900	1.073	0.1219	0.0173	0.900	1.275	0.1219	0.0244
1.050	1.252	0.1123	0.0159	1.050	1.488	0.1123	0.0225
1.200	1.430	0.1022	0.0145	1.200	1.700	0.1022	0.0205
1.350	1.609	0.0922	0.0131	1.350	1.913	0.0922	0.0185
1.650	1.967	0.0740	0.0105	1.650	2.338	0.0740	0.0148
1.950	2.324	0.0586	0.0083	1.950	2.763	0.0586	0.0117
2.250	2.682	0.0459	0.0065	2.250	3.188	0.0459	0.0092

$x = 0.15\,m$				$x = 0.20\,m$			
η	y (mm)	W_x	w_x (m s^{-1})	η	y (mm)	W_x	w_x (m s^{-1})
0	0	0	0	0	0	0	0
0.075	0.118	0.0416	0.0102	0.075	0.126	0.0416	0.0118
0.150	0.235	0.0732	0.0180	0.150	0.253	0.0732	0.0207
0.225	0.353	0.0964	0.0236	0.225	0.379	0.0964	0.0273
0.300	0.471	0.1127	0.0276	0.300	0.506	0.1127	0.0319
0.375	0.588	0.1235	0.0303	0.375	0.632	0.1235	0.0350
0.450	0.706	0.1300	0.0319	0.450	0.759	0.1300	0.0368
0.525	0.824	0.1331	0.0326	0.525	0.885	0.1331	0.0377
0.600	0.941	0.1337	0.0328	0.600	1.012	0.1337	0.0379
0.700	1.098	0.1317	0.0323	0.700	1.180	0.1317	0.0373
0.800	1.255	0.1274	0.0313	0.800	1.349	0.1274	0.0361
0.900	1.412	0.1219	0.0299	0.900	1.517	0.1219	0.0345
1.050	1.647	0.1123	0.0275	1.050	1.770	0.1123	0.0318
1.200	1.883	0.1022	0.0251	1.200	2.023	0.1022	0.0289
1.350	2.118	0.0922	0.0226	1.350	2.276	0.0922	0.0261
1.650	2.589	0.0740	0.0182	1.650	2.782	0.0740	0.0210
1.950	3.064	0.0586	0.0144	1.950	3.288	0.0586	0.0166
2.250	3.530	0.0459	0.0113	2.250	3.794	0.0459	0.0130

Table 8.9. The numerical solutions of velocities components w_x and W_x at $t_w = 60°C$ and $t_\infty = 20°C$, cited from Shang, Wang, and Takhar [5]

$x = 0.05$ m				$x = 0.10$ m			
η	y (mm)	W_x	w_x (m s^{-1})	η	y (mm)	W_x	w_x (m s^{-1})
0	0	0	0	0	0	0	0
0.075	0.081	0.0454	0.0079	0.075	0.096	0.0454	0.0111
0.150	0.162	0.0789	0.0137	0.150	0.192	0.0789	0.0194
0.225	0.242	0.1028	0.0178	0.225	0.288	0.1028	0.0252
0.300	0.323	0.1189	0.0206	0.300	0.384	0.1189	0.0292
0.375	0.404	0.1290	0.0224	0.375	0.480	0.1290	0.0317
0.450	0.485	0.1345	0.0233	0.450	0.576	0.1345	0.0330
0.525	0.565	0.1365	0.0237	0.525	0.672	0.1365	0.0335
0.600	0.646	0.1360	0.0236	0.600	0.768	0.1360	0.0334
0.700	0.754	0.1326	0.0230	0.700	0.896	0.1326	0.0326
0.800	0.862	0.1273	0.0221	0.800	1.024	0.1273	0.0313
0.900	0.969	0.1209	0.0210	0.900	1.152	0.1209	0.0297
1.050	1.131	0.1106	0.0192	1.050	1.344	0.1106	0.0272
1.200	1.292	0.1001	0.0174	1.200	1.536	0.1001	0.0246
1.350	1.454	0.0901	0.0156	1.350	1.728	0.0901	0.0221
1.650	1.777	0.0721	0.0125	1.650	2.112	0.0721	0.0177
1.950	2.100	0.0570	0.0099	1.950	2.496	0.0570	0.0140
2.250	2.423	0.0446	0.0077	2.250	2.880	0.0446	0.0109

$x = 0.15$ m				$x = 0.20$ m			
η	y (mm)	W_x	w_x (m s^{-1})	η	y (mm)	W_x	w_x (m s^{-1})
0	0	0	0	0	0	0	0
0.075	0.106	0.0454	0.0137	0.075	0.114	0.0454	0.0158
0.150	0.213	0.0789	0.0237	0.150	0.228	0.0789	0.0274
0.225	0.319	0.1028	0.0309	0.225	0.342	0.1028	0.0357
0.300	0.425	0.1189	0.0358	0.300	0.457	0.1189	0.0413
0.375	0.531	0.1290	0.0388	0.375	0.571	0.1290	0.0448
0.450	0.638	0.1345	0.0404	0.450	0.685	0.1345	0.0467
0.525	0.744	0.1365	0.0410	0.525	0.799	0.1365	0.0474
0.600	0.850	0.1360	0.0409	0.600	0.913	0.1360	0.0472
0.700	0.992	0.1326	0.0399	0.700	1.065	0.1326	0.0464
0.800	1.134	0.1273	0.0383	0.800	1.218	0.1273	0.0442
0.900	1.275	0.1209	0.0364	0.900	1.370	0.1209	0.0420
1.050	1.488	0.1106	0.0333	1.050	1.598	0.1106	0.0384
1.200	1.700	0.1001	0.0301	1.200	1.827	0.1001	0.0348
1.350	1.913	0.0901	0.0271	1.350	2.055	0.0901	0.0313
1.650	2.338	0.0721	0.0217	1.650	2.511	0.0721	0.0250
1.950	2.763	0.0570	0.0171	1.950	2.968	0.0570	0.0198
2.250	3.188	0.0446	0.0134	2.250	3.425	0.0446	0.0155

and the velocity profile moves to the direction of the flat plate. Consequently, the thickness of the velocity boundary layer decreases.

With an increase of the height x, the velocity component w_x of water or air free convection increases, and the maximum velocity of the velocity profile moves towards the fluid bulk. As a result the thickness of velocity boundary layer increases.

It is found that the agreement between the measured and calculated velocity fields is good, thus confirming that the results in Chaps. 4–6 are reliable.

With regard to the study on the condition of stability for the laminar film free convection of fluid, I emphasize the need for this kind of measurement in an extended experiment.

References

1. E. Schmidt and W. Beckman, Ads Temperature- und Geschwindigkeitsfeld von einer Wärme Abgebenden Senkrechten Plätte bei Naturlicher Konvection, Forsch. Ing. -Wes. 1, pp. 391, 1930
2. E. Pohlhausen, Der Wärmesustausch zwischen festen Körpern und Flüssigkeiten mit kleiner Reibung und kleiner Wärmeleitung, Zeitschrift für angewandte Mathematik und Mechanik 1, pp. 115–121, 1921
3. D.Y. Shang and B.X. Wang, Measurement on velocity of laminar boundary layer for gas free convection along an isothermal vertical flat plat, Anon(Ed.), Inst. Chem. Eng. Symp. Ser, Vol. 1, No. 12, 3rd UK National Conf. incorporating 1st European Conf. on Thermal Sciences, Birmingham Engl., Sep. 16–18, Hemisphere Publishing Corporation, pp. 484–489, 1992
4. D.Y. Shang, B.X. Wang, Y. Wang, and Y. Quan, Study on liquid laminar free convection with consideration of variable thermophysical properties, Int. J. Heat Mass Transfer, Vol. 36, No. 14, pp. 3411–3419, 1993
5. D.Y. Shang, B.X. Wang and H.S. Takhar, Measurements of the velocity field of the laminar boundary layer for water free convection along an isothermal vertical flat plate, Appl. Mech. Eng. 3, No. 4, pp. 553–570, 1998
6. D.Y. Shang and B.X. Wang, Effect of variable thermophysical properties on laminar free convection of gas, Int. J. Heat Mass Transfer 33, No. 7, pp. 1387–1395, 1990
7. VDI - Wärmeatlas, Berechnungsblätter für den Wärmeübertragung, 5, erweiterte Auflage, VDI Verlage GmbH, Düsseldorf, 1988

9

Relationship on Laminar Free Convection and Heat Transfer Between Inclined and Vertical Cases

Nomenclature

a	thermal diffusive coefficient, $\mathrm{m}^2\ \mathrm{s}^{-1}$
b	width of plate, m
c_p	specific heat at constant pressure, J $(\mathrm{kg\,K})^{-1}$
g	gravitation acceleration, m s^{-2}
$(Gr_{x,\infty})_{\mathrm{i}}$	local Grashof number for fluid laminar free convection on isothermal inclined flat plate, $Gr_{x,\infty} = \frac{g\cos\gamma\lvert\rho_\infty/\rho_{\mathrm{w}}-1\rvert x^3}{\nu_\infty{}^2}$
	local Grashof number for gas laminar free convection on isothermal inclined flat plate, $\frac{g\cos\gamma\lvert T_{\mathrm{w}}/T_\infty-1\rvert x^3}{\nu_\infty^2}$
$(Gr_{x,\infty})_{\mathrm{v}}$	local Grashof number for fluid laminar free convection on isothermal vertical flat plate, $\frac{g\lvert\rho_\infty/\rho_{\mathrm{w}}-1\rvert x^3}{\nu_\infty{}^2}$
	local Grashof number for gas laminar free convection on isothermal vertical flat plate, $\frac{g\lvert T_{\mathrm{w}}/T_\infty-1\rvert x^3}{\nu_\infty^2}$
$(g_x)_{\mathrm{i}}$	local mass flow rate entering the boundary layer for inclined case at position x per unit area of the plate, kg $(\mathrm{m}^2\,\mathrm{s})^{-1}$
$(g_x)_{\mathrm{v}}$	local mass flow rate entering the boundary layer for vertical case at position x per unit area of the plate, kg $(\mathrm{m}^2\,\mathrm{s})^{-1}$
$(G_x)_{\mathrm{i}}$	total mass flow rate entering the boundary layer for inclined case for position $x = 0$ to x with width of b of the plate, kg s^{-1}
$(G_x)_{\mathrm{v}}$	total mass flow rate entering the boundary layer for vertical case for position $x = 0$ to x with width of b of the plate, kg s^{-1}

$(Nu_{x,\mathrm{w}})_\mathrm{i}$	local Nusselt number for inclined case, $\alpha_x x/\lambda_\mathrm{w}$
$(Nu_{x,w})_\mathrm{v}$	local Nusselt number for vertical case, $\alpha_x x/\lambda_\mathrm{w}$
$(\overline{Nu}_{x,w})_\mathrm{i}$	average Nusselt number for inclined case, $\overline{\alpha}_x x/\lambda_\mathrm{w}$
$(\overline{Nu}_{x,w})_\mathrm{v}$	average Nusselt number for vertical case, $\overline{\alpha}_x x/\lambda_\mathrm{w}$
n_{c_p}	specific heat parameter of gas
n_λ	thermal conductivity parameter of gas
n_μ	viscosity parameter of gas
Pr	Prandtl number
$(q_x)_\mathrm{i}$	local heat transfer rate for inclined case at position x per unit area on the plate, W m^{-2}
$(Q_x)_\mathrm{i}$	total heat transfer rate for inclined case for position $x = 0$ to x with width of b on the plate, W
$(q_x)_\mathrm{v}$	local heat transfer rate for vertical case at position x per unit area on the plate, W m^{-2}
$(Q_x)_\mathrm{v}$	total heat transfer rate for vertical case for position $x = 0$ to x with width of b on the plate, W
t	temperature, °C
T	absolute temperature, K
$(w_x)_\mathrm{i}, (w_y)_\mathrm{i}$	velocity components for inclined case in the x- and y-directions, respectively, m s^{-1}
$(W_x)_\mathrm{i}, (W_y)_\mathrm{i}$	dimensionless velocity components for inclined case in the x- and y-directions, respectively
$(w_x)_\mathrm{v}, (w_y)_\mathrm{v}$	velocity components for vertical case in the x- and y-directions, respectively, m s^{-1}
$(W_x)_\mathrm{v}, (w_y)_\mathrm{v}$	dimensionless velocity components for vertical case in the x- and y-directions, respectively

Greek symbols

α_x	local heat transfer coefficient, W (m^2 K)$^{-1}$
$\overline{\alpha}_x$	average heat transfer coefficient, W (m^2 K)$^{-1}$
δ	boundary layer thickness, m
η_i	dimensionless coordinate variable for boundary layer for inclined case
η_v	dimensionless coordinate variable for boundary layer for vertical case
γ	inclined angle of surface
θ	dimensionless temperature,
λ	thermal conductivity, W (m K)$^{-1}$
μ	absolute viscosity, kg (m s)$^{-1}$
ν	kinematic viscosity, m^2 s^{-1}
ρ	density, kg m^{-3}
$\left[\left(\frac{\mathrm{d}\theta}{\mathrm{d}\eta}\right)_{\eta=0}\right]_\mathrm{i}$	dimensionless temperature gradient on plate for inclined case

$\left[\left(\frac{d\theta}{d\eta}\right)_{\eta=0}\right]_{\mathrm{v}}$	dimensionless temperature gradient on plate for vertical case
$(\eta_\delta)_{\mathrm{i}}(W_{x,\delta})_{\mathrm{i}} - 4(W_{y,\delta})_{\mathrm{i}}$	mass flow rate parameter entering in the boundary layer for inclined case
$(\eta_\delta)_{\mathrm{v}}(W_{x,\delta})_{\mathrm{v}} - 4(W_{y,\delta})_{\mathrm{v}}$	mass flow rate parameter entering in the boundary layer for vertical case
$\dfrac{\frac{\rho_\infty}{\rho}-1}{\frac{\rho_\infty}{\rho_{\mathrm{w}}}-1}$	buoyancy factor

Subscripts

i	inclined case
v	vertical case
w	at wall
∞	far from the wall surface

9.1 Introduction

It was Rich [1] who first suggested theoretically the procedure for obtaining the heat transfer rate from an inclined surface. According to his procedure, the problem of free convection on an inclined surface is identical to that of flow over a vertical surface except that g is replaced by $g\cos\gamma$, and therefore, a replacement of g by $g\cos\gamma$ in all the relationships derived earlier. This implies using $Gr_x\cos\gamma$ for Gr_x. As a result, his experimental data are in general agreement with the anticipated values. The data obtained by Vliet [2] for a uniform-flux, heated surface in air and in water indicate the validity of the procedure mentioned earlier up to inclination angles as large as 60°C. Detailed experimental results on this were obtained by Fujii and Imura [3]. They also discuss the separation of the boundary layer for the inclined surface facing upward.

However, so far, there has been a shortage of a theoretically rigorous derivations to support the earlier conclusions by means of a replacement of g with $g\cos\gamma$ for all the relationships, and there is a shortage of the clear correlations in describing the transformations of heat transfer, momentum transfer, and mass flow rate drawn from the vertical case to the inclined case for the free convection. As we know, the traditional method for the treatment of similarity transformation of the governing equations for laminar free convection is the Falkner–Skan transformation [4–6]. It is difficult with the traditional Falkner–Skan transformation to realize such a derivation.

Fortunately, the velocity component method presented in Chap. 4–6 [7–9] for similarity transformation of the governing partial deferential equations of fluid laminar boundary layer has provided the possibility to realize such derivation. It is shown that in these studies the velocity component method has its advantages over the Falkner–Skan transformation for the treatment of variable thermophysical properties and other various physical factors. On this basis, Shang and Takhar [10] clarified the relationships of heat, momentum, and mass transfer of laminar fluid free convection between inclined and vertical cases for consideration of variable thermophysical properties.

In this present chapter, I will introduce the exact relationships of heat, momentum, and mass transfer between inclined and vertical cases with consideration of variable thermophysical properties in order to satisfy the requirement in industrial applications. To this end, the governing equations of laminar free convection of fluid in the inclined case are transformed by means of a developed similarity transformation approach, viz, the velocity component method, instead of traditional Falkner–Skan type of transformation. Meanwhile, the suitable forms of some dimensionless variables such as an appropriate suitable local Grashof number $Gr_{x,\infty}$ and suitable dimensionless velocity components for the free convection are proposed. It will be found that the formation of the transformed dimensionless governing equations for the inclined case is fully same as those for the corresponding vertical cases. Then, it is obvious that, except the different assumption of local Grashof number and dimensionless velocity components, the prediction correlations of heat transfer, momentum transfer, and mass flow rate for fluid laminar free convection for the vertical case presented in the previous chapters can be completely taken as those for the related inclined case.

9.2 Fluid Free Convection on inclined plate

9.2.1 Physical Model and Basic Equations

The physical model and coordinate system of fluid free convection on inclined plate are shown schematically in Fig. 9.1. An isothermal inclined flat plate is suspended in a quiescent fluid. The surface temperature is t_w and the fluid bulk temperature is t_∞. It is assumed that t_w is not equal to t_∞, so that laminar free convection can be produced easily on the inclined surface in both

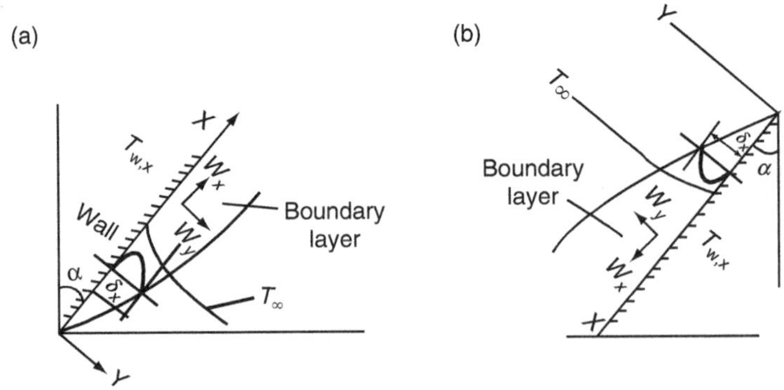

Fig. 9.1. Physical model and coordinate system. (a) Ascending flow on the inclined surface ($t_w > t_\infty$), (b) falling flow on the inclined surface ($t_w < t_\infty$)

the cases as shown in Fig. 9.1(a) and (b), respectively. The governing partial differential equations for mass, momentum, and energy for consideration of variable thermophysical properties applied to the fluid laminar free convection on the inclined surface, are

$$\frac{\partial}{\partial x}[\rho(w_x)_{\mathrm{i}}] + \frac{\partial}{\partial y}[\rho(w_y)_{\mathrm{i}}] = 0, \tag{9.1}$$

$$\rho\left[(w_x)_{\mathrm{i}}\frac{\partial(w_x)_{\mathrm{i}}}{\partial x} + (w_y)_{\mathrm{i}}\frac{\partial(w_x)_{\mathrm{i}}}{\partial y}\right] = \frac{\partial}{\partial y}\left[\mu\frac{\partial(w_x)_{\mathrm{i}}}{\partial y}\right] + g\,|\rho_\infty - \rho|\cos\gamma, \tag{9.2}$$

$$\rho c_p\left[(w_x)_{\mathrm{i}}\frac{\partial t}{\partial x} + (w_y)_{\mathrm{i}}\frac{\partial t}{\partial y}\right] = \frac{\partial}{\partial y}\left(\lambda\frac{\partial t}{\partial y}\right), \tag{9.3}$$

where γ expresses the inclined angle of the plate. Here, the buoyancy fact $|\rho_\infty - \rho|$ is taken as the absolute value because its direction is same as that of the velocity component $(w_x)_{\mathrm{i}}$. Obviously, the governing equations (9.1)–(9.3) are suitable for laminar free convection both of liquid and gas.

The boundary conditions are

$$y = 0: \quad (w_x)_{\mathrm{i}} = 0, \quad (w_y)_{\mathrm{i}} = 0, \quad t = t_{\mathrm{w}}, \tag{9.4}$$

$$y \to \infty \quad (w_x)_{\mathrm{i}} = 0, \quad t = t_\infty, \tag{9.5}$$

9.2.2 Similarity Transformation of the Basic Equations

For similarity transformation of the basic equations we use the velocity component method which was developed in [7–9] and presented in Chaps. 4–6. Taking subscript i to express the case on the inclined surfaces, we assume the following dimensionless coordinate variables for similarity transformation of the earlier governing partial differential equations of fluid laminar free convection on inclined plate:

$$\eta_{\mathrm{i}} = \frac{y}{x}\left[\frac{1}{4}(Gr_{x,\infty})_{\mathrm{i}}\right]^{1/4}, \tag{9.6}$$

where η_{i} is dimensionless coordinate variable for boundary layer. The local Grashof number $(Gr_{x,\infty})_{\mathrm{i}}$ can be assumed to be

$$(Gr_{x,\infty})_{\mathrm{i}} = \frac{g\,(\cos\gamma)\,|\rho_\infty/\rho_{\mathrm{w}} - 1|\,x^3}{{\nu_\infty}^2}. \tag{9.7}$$

The dimensionless temperature is given by

$$\theta = \frac{t - t_\infty}{t_{\mathrm{w}} - t_\infty}. \tag{9.8}$$

The dimensionless velocity components are assumed to be

$$(W_x)_{\mathrm{i}} = (2\sqrt{g\,(\cos\gamma)x}\,|\rho_\infty/\rho_{\mathrm{w}} - 1|^{1/2})^{-1}(w_x)_{\mathrm{i}}, \tag{9.9}$$

$$(W_y)_{\mathrm{i}} = \{2\sqrt{g\,(\cos\gamma)x}\,|\rho_\infty/\rho_{\mathrm{w}} - 1|^{1/2}\left[\frac{1}{4}(Gr_{x,\infty})_{\mathrm{i}}\right]^{-1/4}\}^{-1}(w_y)_{\mathrm{i}}. \tag{9.10}$$

With the earlier similarity variables defined in (9.6)–(9.10), (9.1)–(9.3) with the boundary conditions (9.4) and (9.5) can be transformed to the following governing ordinary differential equations:

$$2(W_x)_{\rm i} - \eta\frac{{\rm d}(W_x)_{\rm i}}{{\rm d}\eta_{\rm i}} + 4\frac{{\rm d}(W_y)_{\rm i}}{{\rm d}\eta_{\rm i}} - \frac{1}{\rho}\frac{{\rm d}\rho}{{\rm d}\eta_{\rm i}}[\eta_{\rm i}(W_x)_{\rm i} - 4(W_y)_{\rm i}] = 0, \tag{9.11}$$

$$\frac{\nu_\infty}{\nu}\left[(W_x)_{\rm i}\left(2(W_x)_{\rm i} - \eta_{\rm i}\frac{{\rm d}(W_x)_{\rm i}}{{\rm d}\eta_{\rm i}}\right) + 4(W_y)_{\rm i}\frac{{\rm d}(W_x)_{\rm i}}{{\rm d}\eta_{\rm i}}\right]$$
$$= \frac{{\rm d}^2(W_x)_{\rm i}}{{\rm d}\eta_{\rm i}^2} + \frac{1}{\mu}\frac{{\rm d}\mu}{{\rm d}\eta_{\rm i}}\frac{{\rm d}(W_x)_{\rm i}}{{\rm d}\eta_{\rm i}} + \frac{\nu_\infty}{\nu}\frac{\frac{\rho_\infty}{\rho} - 1}{\frac{\rho_\infty}{\rho_{\rm w}} - 1}, \tag{9.12}$$

$$Pr\frac{\nu_\infty}{\nu}[-\eta_{\rm i}(W_x)_{\rm i} + 4(W_y)_{\rm i}]\frac{{\rm d}\theta}{{\rm d}\eta_{\rm i}} = \frac{{\rm d}^2\theta}{{\rm d}\eta_{\rm i}^2} + \frac{1}{\lambda}\frac{{\rm d}\lambda}{{\rm d}\eta_{\rm i}}\frac{{\rm d}\theta}{{\rm d}\eta_{\rm i}}, \tag{9.13}$$

$$\eta_{\rm i} = 0: \qquad (W_x)_{\rm i} = 0, (W_y)_{\rm i} = 0, \theta = 0, \tag{9.14}$$

$$\eta_{\rm i} \to 0: \qquad (W_x)_{\rm i} = 0, \theta = 0. \tag{9.15}$$

The derivation processes for (9.11)–(9.13) are described in Appendix A in detail.

9.2.3 Relationships of Momentum, Heat, and Mass Transfer between Inclined and Vertical Cases

For heat transfer. (9.11)–(9.13) and their boundary conditions (9.14) and (9.15) are dimensionless forms of the equations of fluid laminar free convection in the inclined case. They are completely identical to those in the vertical case derived in Chap. 6. Therefore, for same liquid laminar free convection with same boundary temperature conditions $t_{\rm w}$ and t_∞, we have

$$-\left[\left(\frac{{\rm d}\theta}{{\rm d}\eta}\right)_{\eta=0}\right]_{\rm i} = -\left[\left(\frac{{\rm d}\theta}{{\rm d}\eta}\right)_{\eta=0}\right]_{\rm v}. \tag{9.16}$$

where the subscripts i and v denotes the related inclined and vertical cases, respectively. With the same derivation as that in Chap. 6 the correlation for $\left[\left(\frac{{\rm d}\theta}{{\rm d}\eta}\right)_{\eta=0}\right]_{\rm i}$ in the following form for water laminar free convection can be taken for consideration of variable thermophysical properties:

$$-\left[\left(\frac{{\rm d}\theta}{{\rm d}\eta}\right)_{\eta=0}\right]_{\rm i} = -\left[\left(\frac{{\rm d}\theta}{{\rm d}\eta}\right)_{\eta=0}\right]_{v} = 0.5764 + 0.1797\cdot Ln(Pr_\infty) + 0.0331\cdot Ln^2(Pr_\infty), \tag{9.17}$$

where the water bulk temperature t_∞ is defined as that of reference Prandtl number Pr_∞.

From definitions of local Grashof number for the inclined and vertical cases defined in (9.7) and (6.10), respectively, we obtain the following equation:

$$\frac{(Gr_{x,\infty})_{\mathrm{i}}}{(Gr_{x,\infty})_{\mathrm{v}}} = \cos\gamma. \tag{9.18}$$

From the related definitions of local Nusselt number of laminar free convection for vertical and inclined cases we have

$$\frac{(Nu_{x,\mathrm{w}})_{\mathrm{i}}}{(Nu_{x,\mathrm{w}})_{\mathrm{v}}} = \frac{-\left(\frac{1}{4}Gr_{x,\infty}\right)_{\mathrm{i}}^{1/4}\left[\left(\frac{\mathrm{d}\theta}{\mathrm{d}\eta}\right)_{\eta=0}\right]_{\mathrm{i}}}{-\left(\frac{1}{4}Gr_{x,\infty}\right)_{\mathrm{v}}^{1/4}\left[\left(\frac{\mathrm{d}\theta}{\mathrm{d}\eta}\right)_{\eta=0}\right]_{\mathrm{v}}} = \cos^{1/4}\gamma. \tag{9.19}$$

For momentum transfer. Since the dimensionless governing equations (9.11)–(9.13) are completely identical to (6.26)–(6.28), the solutions for dimensionless velocity components both for inclined and vertical cases for fluid laminar free convection are identical, i.e.,

$$(W_x)_{\mathrm{i}} = (W_x)_{\mathrm{v}}, \tag{9.20}$$

$$(W_y)_{\mathrm{i}} = (W_y)_{\mathrm{v}}. \tag{9.21}$$

Combining (9.20) with (9.9) and (6.8) we have

$$(2\sqrt{g\cos\gamma x}\,|\rho_\infty/\rho_{\mathrm{w}} - 1|^{1/2})^{-1}(w_x)_{\mathrm{i}} = \left[2\sqrt{gx}\left|\frac{\rho_\infty}{\rho_{\mathrm{w}}} - 1\right|^{1/2}\right]^{-1}(w_x)_{\mathrm{v}},$$

i.e.,

$$\frac{(w_x)_{\mathrm{i}}}{(w_x)_{\mathrm{v}}} = \cos^{1/2}\gamma. \tag{9.22}$$

Combining (9.21) with (9.10) and (6.9) we have

$$\{2\sqrt{g\cos\gamma x}\,|\rho_\infty/\rho_{\mathrm{w}} - 1|^{1/2}\left[\frac{1}{4}(Gr_{x,\infty})_{\mathrm{i}}\right]^{-1/4}\}^{-1}(w_y)_{\mathrm{i}}$$

$$= \left[2\sqrt{gx}\left|\frac{\rho_\infty}{\rho_{\mathrm{w}}} - 1\right|^{-1/2}\left(\frac{1}{4}(Gr_{x,\infty})_v\right)^{-1/4}\right]^{-1}(w_y)_{\mathrm{v}}$$

i.e.,

$$\frac{(w_y)_{\mathrm{i}}}{(w_y)_{\mathrm{v}}} = \cos^{1/4}\gamma. \tag{9.23}$$

For mass transfer. Let us set $(g_x)_{\mathrm{i}}$ to be a local mass flow rate entering the boundary layer at position x per unit area of the inclined plate. According to the boundary layer theory of fluid mechanics, $(g_x)_{\rm i}$ is expressed as

$$(g_x)_{\rm i} = \rho_\infty \left[(w_{x,\delta})_{\rm i} \frac{{\rm d}\delta_{\rm i}}{{\rm d}x} - (w_{y,\delta})_{\rm i} \right],$$

where $\delta_{\rm i}$ is the boundary layer thickness, and $(w_{x,\delta})_{\rm i}$ and $(w_{y,\delta})_{\rm i}$ are velocity components at $y = \delta_{\rm i}$. With (9.9) and (9.10) the earlier equation is changed into the following one:

$$\begin{aligned}(g_x)_{\rm i} &= \rho_\infty [2\sqrt{g \cos \gamma x}\,|\rho_\infty/\rho_{\rm w} - 1|^{1/2} (W_{x,\delta})_{\rm i} \frac{{\rm d}\delta_{\rm i}}{{\rm d}x} \\ &\quad -2\sqrt{g \cos \gamma x}\,|\rho_\infty/\rho_{\rm w} - 1|^{1/2} \left(\frac{1}{4} Gr_{x,\infty}\right)_{\rm i}^{-1/4} (W_{y,\delta})_{\rm i}] \\ &= 2\sqrt{g \cos \gamma x}\,|\rho_\infty/\rho_{\rm w} - 1|^{1/2} \rho_\infty [(W_{x,\delta})_{\rm i} \frac{{\rm d}\delta_{\rm i}}{{\rm d}x} - \left(\frac{1}{4} Gr_{x,\infty}\right)_{\rm i}^{-1/4} (W_{y,\delta})_{\rm i}],\end{aligned} \tag{9.24}$$

where

$$\delta_{\rm i} = (\eta_\delta)_{\rm i} \left(\frac{1}{4} Gr_{x,\infty}\right)_{\rm i}^{-1/4} x, \tag{9.25}$$

according to (9.6).
Then,

$$\frac{{\rm d}\delta_{\rm i}}{{\rm d}x} = \frac{1}{4}(\eta_\delta)_{\rm i} \left(\frac{1}{4} \frac{g \cos \gamma\,|\rho_\infty/\rho_{\rm w} - 1|}{{\nu_\infty}^2}\right)_{\rm i}^{-1/4} x^{-3/4} = \frac{1}{4}(\eta_\delta)_{\rm i} \left(\frac{1}{4} Gr_{x,\infty}\right)_{\rm i}^{-1/4}. \tag{9.26}$$

Hence (9.24) can be simplified as

$$\begin{aligned}(g_x)_{\rm i} &= 2\sqrt{g \cos \gamma x}\,|\rho_\infty/\rho_{\rm w} - 1|^{1/2} \rho_\infty \\ &\quad \times \left[\frac{1}{4}(W_{x,\delta})_{\rm i}(\eta_\delta)_{\rm i} \left(\frac{1}{4} Gr_{x,\infty}\right)_{\rm i}^{-1/4} - \left(\frac{1}{4} Gr_{x,\infty}\right)_{\rm i}^{-1/4} (W_{y,\delta})_{\rm i} \right] \\ &= 4\nu_\infty \rho_\infty \left(\frac{1}{4} Gr_{x,\infty}\right)_{\rm i}^{1/2} x^{-1} \\ &\quad \times \left[\frac{1}{4}(W_{x,\delta})_{\rm i}(\eta_\delta)_{\rm i} \left(\frac{1}{4} Gr_{x,\infty}\right)_{\rm i}^{-1/4} - \left(\frac{1}{4} Gr_{x,\infty}\right)_{\rm i}^{-1/4} (W_{y,\delta})_{\rm i} \right] \\ &= 4\mu_\infty \left(\frac{1}{4} Gr_{x,\infty}\right)_{\rm i}^{1/4} x^{-1} \left[\frac{1}{4}(W_{x,\delta})_{\rm i}(\eta_\delta)_{\rm i} - (W_{y,\delta})_{\rm i} \right] \\ &= \mu_\infty \left(\frac{1}{4} Gr_{x,\infty}\right)_{\rm i}^{1/4} [(\eta_\delta)_{\rm i}(W_{x,\delta})_{\rm i} - 4(W_{y,\delta})_{\rm i}] x^{-1}.\end{aligned} \tag{9.27}$$

If $(G_x)_{\rm i}$ is taken to express total mass flow rate entering the boundary layer for position $x = 0$ to x with width of b on the inclined plate, it should be the following integration:

$$(G_x)_{\rm i} = \iint_A (g_x)_{\rm i}\,{\rm d}A = b\int_0^x (g_x)_{\rm i}{\rm d}x. \qquad (9.28)$$

With (9.27), (9.28) becomes

$$(G_x)_i = b\int_0^x \left[\mu_\infty \left(\frac{1}{4}Gr_{x,\infty}\right)_{\rm i}^{1/4} [(\eta_\delta)_{\rm i}(W_{x,\delta})_{\rm i} - 4(W_{y,\delta})_{\rm i}]\, x^{-1}\right] {\rm d}x$$
$$= \frac{4}{3}b\cdot\mu_\infty \left(\frac{1}{4}Gr_{x,\infty}\right)_{\rm i}^{1/4} [(\eta_\delta)_{\rm i}(W_{x,\delta})_{\rm i} - 4(W_{y,\delta})_{\rm i}]\,. \qquad (9.29)$$

Set $(G_x)_{\rm v}$ is total mass flow rate entering the boundary layer for position $x = 0$ to x with width of b on the vertical plate. Obviously, $(G_x)_{\rm v}$ is the special case of $(G_x)_{\rm i}$ when the inclined angle γ equals zero. The $(G_x)_{\rm v}$ is expressed as

$$(G_x)_{\rm v} = \frac{4}{3}b\cdot\mu_\infty \left(\frac{1}{4}Gr_{x,\infty}\right)_{\rm v}^{1/4} [(\eta_\delta)_{\rm v}(W_{x,\delta})_{\rm v} - 4(W_{y,\delta})_{\rm v}]. \qquad (9.30)$$

Then, at the same temperature boundary condition the ratio of the mass flow rate drawn into the boundary layer will be

$$\frac{(G_x)_{\rm i}}{(G_x)_{\rm v}} = \frac{\frac{4}{3}b\cdot\mu_\infty \left(\frac{1}{4}Gr_{x,\infty}\right)_{\rm i}^{1/4} [(\eta_\delta)_{\rm i}(W_{x,\delta})_{\rm i} - 4(W_{y,\delta})_{\rm i}]}{\frac{4}{3}b\cdot\mu_\infty \left(\frac{1}{4}Gr_{x,\infty}\right)_{\rm v}^{1/4} [(\eta_\delta)_{\rm v}(W_{x,\delta})_{\rm v} - 4(W_{y,\delta})_{\rm v}]}. \qquad (9.31)$$

Since the dimensionless governing equations of fluid laminar free convection for the inclined case are completely identical to those for vertical case, we have

$$(\eta_\delta)_{\rm i} = (\eta_\delta)_{\rm v},$$
$$[(\eta_\delta)_{\rm i}(W_{x,\delta})_{\rm i} - 4(W_{y,\delta})_{\rm i}] = [(\eta_\delta)_{\rm v}(W_{x,\delta})_{\rm v} - 4(W_{y,\delta})_{\rm v}].$$

Then,

$$\frac{(G_x)_{\rm i}}{(G_x)_{\rm v}} = \left[\frac{(Gr_{x,\infty})_{\rm i}}{(Gr_{x,\infty})_{\rm v}}\right]^{1/4} = \cos^{1/4}\gamma. \qquad (9.32)$$

The governing basic equations and relationships of mass, momentum, and heat transfer between the inclined and vertical cases for fluid laminar free convection are summarized in Table 9.1.

Table 9.1. Governing equations of fluid laminar free convection both with the vertical and inclined cases, and the relationships for heat, momentum and mass transfer between the vertical and inclined cases

term	on vertical surface	on inclined surface
	governing partial differential equations	
mass equation	$\frac{\partial}{\partial x}(\rho w_x) + \frac{\partial}{\partial y}(\rho w_y) = 0$	
momentum equation	$\rho\left[(w_x)_\mathrm{v}\frac{\partial (w_x)_\mathrm{v}}{\partial x} + (w_y)_\mathrm{v}\frac{\partial (w_x)_\mathrm{v}}{\partial y}\right] =$ $\frac{\partial}{\partial y}\left[\mu\frac{\partial (w_x)_\mathrm{v}}{\partial y}\right] + g\,\lvert\rho_\infty - \rho\rvert$ ($g(\rho_\infty - \rho) = \rho g\frac{T-T_\infty}{T_\infty}$ for gas laminar free convection)	$\rho\left[(w_x)_\mathrm{i}\frac{\partial (w_x)_\mathrm{i}}{\partial x} + (w_y)_\mathrm{i}\frac{\partial (w_x)_\mathrm{i}}{\partial y}\right] =$ $\frac{\partial}{\partial y}\left[\mu\frac{\partial (w_x)_\mathrm{i}}{\partial y}\right] + g\,\lvert\rho_\infty - \rho\rvert\cos\gamma$ ($g(\rho_\infty - \rho)\cos\,\gamma = \rho g\frac{T-T_\infty}{T_\infty}\cos\,\gamma$ for gas laminar free convection)
energy equation	$\rho c_p\left(w_x\frac{\partial t}{\partial x} + w_y\frac{\partial t}{\partial y}\right) = \frac{\partial}{\partial y}\left(\lambda\frac{\partial t}{\partial y}\right)$	
	boundary conditions	
$y = 0$	$w_x = 0, \quad w_y = 0, \quad T = T_\mathrm{w}$	
$y \to \infty$	$w_x \to 0, \quad T = T_\infty$	
	assumed similarity variables	
η	$\eta = \frac{y}{x}\left(\frac{1}{4}Gr_{x,\infty}\right)^{1/4}$	
$Gr_{x,\infty}$	$(Gr_{x,\infty})_\mathrm{v} = \frac{g\lvert\rho_\infty/\rho_\mathrm{w} - 1\rvert x^3}{{\nu_\infty}^2}$ (for gas laminar free convection it is equivalently rewritten as $(Gr_{x,\infty})_\mathrm{v} = \frac{g\lvert T_\mathrm{w}/T_\infty - 1\rvert x^3}{{\nu_\infty}^2}$	$(Gr_{x,\infty})_\mathrm{i} = \frac{g\cos\gamma\lvert\rho_\infty/\rho_\mathrm{w} - 1\rvert x^3}{{\nu_\infty}^2}$ (for gas laminar free convection it is equivalently rewritten as $(Gr_{x,\infty})_\mathrm{i} = \frac{g\ \cos\ \gamma\lvert T_\mathrm{w}/T_\infty - 1\rvert x^3}{{\nu_\infty}^2}$
θ	$\theta = \frac{t-t_\infty}{t_\mathrm{w}-t_\infty}$ for gas laminar free convection it is equivalently rewritten as $\theta = \frac{T-T_\infty}{T_\mathrm{w}-T_\infty}$	

W_x	$(2\sqrt{gx}\,\lvert\rho_\infty/\rho_{\mathrm{w}}-1\rvert^{1/2})^{-1}(w_x)_{\mathrm{v}}$ (for gas laminar free convection it is equivalently rewritten as $(2\sqrt{gx}\,\lvert T_{\mathrm{w}}/T_\infty-1\rvert^{1/2})^{-1}(w_x)_{\mathrm{v}}$	$(2\sqrt{gx\ \cos\ \gamma}\,\lvert\rho_\infty/\rho_{\mathrm{w}}-1\rvert^{1/2})^{-1}(w_x)_{\mathrm{i}}$ (for gas laminar free convection it is equivalently rewritten as $(2\sqrt{gx\ \cos\ \gamma}\,\lvert T_{\mathrm{w}}/T_\infty-1\rvert^{1/2})^{-1}w_x)$
W_y	$(2\sqrt{gx}\,\lvert\rho_\infty/\rho_{\mathrm{w}}-1\rvert^{1/2}\left(\frac{1}{4}Gr_{x,\infty}\right)_{\mathrm{v}}^{-1/4})^{-1}(w_y)_{\mathrm{v}}$ (for gas laminar free convection it is equivalently rewritten as that similar to the expression of W_x	$(2\sqrt{gx\ \cos\ \gamma}\,\lvert\rho_\infty/\rho_{\mathrm{w}}-1\rvert^{1/2}\left(\frac{1}{4}Gr_{x,\infty}\right)_{\mathrm{i}}^{-1/4})^{-1}(w_y)_{\mathrm{i}}$ (for gas laminar free convection it is equivalently rewritten as that similar to the expression of W_x
	governing ordinary differential equations	
mass equation	$2W_x-\eta\frac{\mathrm{d}W_x}{\mathrm{d}\eta}+4\frac{\mathrm{d}W_y}{\mathrm{d}\eta}-\frac{1}{\rho}\frac{\mathrm{d}\rho}{\mathrm{d}\eta}(-\eta W_x+4W_y)=0$	
momentum equation	$\frac{\nu_\infty}{\nu}\left(W_x\left(2W_x-\eta\frac{\mathrm{d}W_x}{\mathrm{d}\eta}\right)+4W_y\frac{\mathrm{d}W_y}{\mathrm{d}\eta}\right)=\frac{\mathrm{d}^2W_x}{\mathrm{d}\eta^2}+\frac{1}{\mu}\frac{d\mu}{d\eta}\frac{dW_x}{d\eta}+\frac{\nu_\infty}{\nu}\frac{\frac{\rho_\infty}{\rho}-1}{\frac{\rho_\infty}{\rho_w}-1}$	
	(for gas laminar free convection the buoyancy factor $\frac{\frac{\rho_\infty}{\rho}-1}{\frac{\rho_\infty}{\rho_{\mathrm{w}}}-1}$ can be equivalently rewritten as the dimensionless temperature θ	
energy equations	$Pr\frac{\nu_\infty}{\nu}(-\eta W_x+4W_y)\frac{\mathrm{d}\theta}{\mathrm{d}\eta}=\frac{\mathrm{d}^2\theta}{\mathrm{d}\eta^2}+\frac{1}{\lambda}\frac{\mathrm{d}\theta}{\mathrm{d}\eta}$	
	boundary conditions	
$\eta=0$	$W_x=0,\quad W_y=0,\quad \theta=0$	
$\eta\to 0$	$W_x=0,\quad \theta=0$	
	relationships between vertical and inclined cases	
$Gr_{x,\infty}$	$\frac{(Gr_{x,\infty})_{\mathrm{i}}}{(Gr_{x,\infty})_{\mathrm{v}}}=\cos\ \gamma$	
W_x	$(W_x)_{\mathrm{i}}=(W_x)_{\mathrm{v}}$	
W_y	$(W_y)_{\mathrm{i}}=(W_y)_{\mathrm{v}}$	
w_x	$\frac{(w_x)_{\mathrm{i}}}{(w_x)_{\mathrm{v}}}=\cos^{1/2}\gamma$	
w_y	$\frac{(w_y)_{\mathrm{i}}}{(w_y)_{\mathrm{v}}}=\cos^{1/4}\gamma$	

Table 9.1. *Continued*

term	on vertical surface	on inclined surface
	for water laminar free convection:	
$\left(\frac{\mathrm{d}\theta}{\mathrm{d}\eta}\right)_{\eta=0}$	$-\left[\left(\frac{\mathrm{d}\theta}{\mathrm{d}\eta}\right)_{\eta=0}\right]_{\mathrm{i}} = -\left[\left(\frac{\mathrm{d}\theta}{\mathrm{d}\eta}\right)_{\eta=0}\right] = 0.5674 + 0.1797 \cdot Ln(Pr_{\infty}) + 0.0331 \cdot Ln^2(Pr_{\infty})$	
	for laminar free convection of monatomics or diatomic gas, air and water vapor	
	$\left[-\left(\frac{\mathrm{d}\theta}{\mathrm{d}\eta}\right)_{\eta=0}\right]_{\mathrm{i}} = -\left[\left(\frac{\mathrm{d}\theta}{\mathrm{d}\eta}\right)_{\eta=0}\right] = \psi(Pr)\left(\frac{T_{\mathrm{w}}}{T_{\infty}}\right)^{-m}$	
	$\psi(Pr) = (0.567 + 0.186 \cdot Ln(Pr))$	
	$m = 0.35n_{\lambda} + 0.29n_{\mu} + 0.36$ for $T_{\mathrm{w}}/T_{\infty} > 1$	
	$m = 0.42n_{\lambda} + 0.34n_{\mu} + 0.24$ for $T_{\mathrm{w}}/T_{\infty} < 1$	
$Nu_{x,\mathrm{w}}$		$\frac{(Nu_{x,\mathrm{w}})_{\mathrm{i}}}{(Nu_{x,\mathrm{w}})_{\mathrm{v}}} = \cos^{1/4}\gamma$
G_x		$(G_x)_{\mathrm{i}}/(G_x)_{\mathrm{v}} = \cos^{1/4}\gamma$

9.3 Gas Free Convection on Inclined Plate

In principle, the governing equations of laminar free convection of fluid are completely suitable to those of gas. Then, all the relationship equations between the inclined and vertical for fluid laminar free convection derived in Sect. 9.2 are completely suitable to those of gas laminar free convection.

However, for convenience it is necessary to use the temperature parameter method introduced in Chap. 4 for expression of gas density variation with absolute temperature, i.e.,

$$\frac{\rho}{\rho_\infty} = \frac{T_\infty}{T} \tag{9.33}$$

to rewrite the equations in Sect. 9.2 for buoyancy factor, local Grashof number, and velocity components as follows for gas laminar free convection.

For inclined case:

$$g(\rho_\infty - \rho)\cos\gamma = \rho g \frac{T - T_\infty}{T_\infty}\cos\gamma, \tag{9.34}$$

$$(Gr_{x,\infty})_\mathrm{i} = \frac{g\cos\gamma\,|T_\mathrm{w}/T_\infty - 1|\,x^3}{{\nu_\infty}^2}, \tag{9.35}$$

$$(W_x)_\mathrm{i} = (2\sqrt{g\cos\gamma x}\,|T_\mathrm{w}/T_\infty - 1|^{1/2})^{-1}(w_x)_\mathrm{i} \tag{9.36}$$

$$(W_y)_\mathrm{i} = \{2\sqrt{g\cos\gamma x}\,|T_\mathrm{w}/T_\infty - 1|^{1/2}\left[\frac{1}{4}(Gr_{x,\infty})_\mathrm{i}\right]^{-1/4}\}^{-1}(w_y)_\mathrm{i}. \tag{9.37}$$

For vertical case:

$$g(\rho_\infty - \rho) = \rho g \frac{T - T_\infty}{T_\infty}, \tag{9.38}$$

$$(Gr_{x,\infty})_\mathrm{v} = \frac{g\,|T_\mathrm{w}/T_\infty - 1|\,x^3}{{\nu_\infty}^2}, \tag{9.39}$$

$$(W_x)_\mathrm{i} = (2\sqrt{gx}\,|T_\mathrm{w}/T_\infty - 1|^{1/2})^{-1}(w_x)_\mathrm{v}, \tag{9.40}$$

$$(W_y)_\mathrm{i} = \{2\sqrt{gx}\,|T_\mathrm{w}/T_\infty - 1|^{1/2}\left[\frac{1}{4}(Gr_{x,\infty})_\mathrm{v}\right]^{-1/4}\}^{-1}(w_y)_\mathrm{v}. \tag{9.41}$$

Furthermore, with the simple power law of gas, the buoyancy factor $\rho_\infty/\rho_\mathrm{w} - 1$ in (9.7)–(9.10) for definition of local Grashof number $(Gr_{x,\infty})_\mathrm{i}$ and dimensionless velocity components W_x and W_y can be rewritten as follows:

$$\rho_\infty/\rho_\mathrm{w} - 1 = T_\mathrm{w}/T_\infty - 1. \tag{9.42}$$

In addition, the buoyancy factor

$$\frac{\frac{\rho_\infty}{\rho} - 1}{\frac{\rho_\infty}{\rho_\mathrm{w}} - 1}$$

in the transformed momentum equation (9.12) can be rewritten as dimensionless temperature θ.

Since the dimensionless governing equations (9.11)–(9.13) for fluid laminar free convection for inclined case are completely identical to those for the vertical case, the following curve-fitting formulae of dimensionless temperature gradient are suitable both to inclined and vertical cases for laminar free convection of monatomic and diatomic gases, air, and water vapor:

$$\left[\left(\frac{d\theta}{d\eta}\right)_{\eta=0}\right]_{\mathrm{i}} = \left[\left(\frac{d\theta}{d\eta}\right)_{\eta=0}\right]_{\mathrm{v}} = \psi(Pr)\left(\frac{T_{\mathrm{w}}}{T_{\infty}}\right)^{-m}, \tag{9.43}$$

where

$$\psi(Pr) = (0.567 + 0.186 \cdot Ln(Pr)), \tag{9.44}$$

$$m = 0.35n_{\lambda} + 0.29n_{\mu} + 0.36 \quad \text{for} \quad T_{\mathrm{w}}/T_{\infty} > 1, \tag{9.45}$$

$$m = 0.42n_{\lambda} + 0.34n_{\mu} + 0.24 \quad \text{for} \quad T_{\mathrm{w}}/T_{\infty} < 1. \tag{9.46}$$

9.4 Summary

So far, the governing equations of fluid laminar free convection both with the vertical and inclined cases, and the relationships for heat, momentum, and mass transfer between the vertical and inclined cases are summarized in Table9.1.

9.5 Remarks

In this chapter an advanced similarity method, the velocity component method, is applied to the similarity transformation of the governing equations of fluid laminar free convection on inclined plate, instead of the traditional Falkner–Skan transformation. This leads to the governing ordinary differential equations which are the same as those of the corresponding equations on the vertical plate. Finally, the following simple and direct correlations for describing the transformations of the velocity components, heat transfer, and mass flow rate from the vertical case to the inclined case for the free convection are derived:

$$\frac{(w_x)_{\mathrm{i}}}{(w_x)_{\mathrm{v}}} = \cos^{1/2}\gamma, \quad \frac{(w_y)_{\mathrm{i}}}{(w_y)_{\mathrm{v}}} = \cos^{1/4}\gamma, \quad \frac{(Nu_{x,\mathrm{w}})_{\mathrm{i}}}{(Nu_{x,\mathrm{w}})_{\mathrm{v}}} = \cos^{1/4}\gamma \text{ and } \frac{(G_x)_{\mathrm{i}}}{(G_x)_{\mathrm{v}}} = \cos^{1/4}\gamma.$$

In addition, $\eta_\delta W_{x,\delta} - 4W_{y,\delta}$ in (9.27) and (9.29) can be defined as mass flow rate parameter. Obviously, $W_{x,\delta} = 0$ corresponds to the case of free convection. The successful derivation for the relationships of heat, momentum, and mass transfer for laminar free convection between the inclined and vertical plates, in this paper, reveals once again the advantage of the velocity component method over the traditional Falkner–Skan transformation, for studying laminar boundary layer problems.

9.6 Calculation Example

Question: A flat plate with $b = 1$ m in width and $x = 0.3$ m in length is suspended vertically in air. The ambient temperature is $t_\infty = 20°\text{C}$. Calculate the free convection heat transfer of the plate for the temperature ratio $T_\text{w}/T_\infty = 1.7$. What is its heat transfer rate, if the plate inclined angle is 45°C.

Solution: From $t_\infty = 20°\text{C}$ and $T_\text{w}/T_\infty = 1.7$, we have $T_\text{w} = 498.1$ K or $t_\text{w} = 225.1°\text{C}$. The air physical properties are as follows: kinetic viscosity is $\nu_\infty = 15.06 \times 10^{-6}\ \text{m}^2\ \text{s}^{-1}$ at $t_\infty = 20°\text{C}$, $\lambda_\text{w} = 4.07 \times 10^{-2}\ \text{W (m°C)}^{-1}$ at $t_\text{w} = 225.1°\text{C}$. From Tables 4.1 and 4.3, we get $n_\mu = 0.68, n_\lambda = 0.81$ and $Pr = 0.7$ for air.

1. *For vertical case.* From (4.51) the local Nusselt number is expressed as

$$(Nu_{x,\text{w}})_\text{v} = -\left(\frac{1}{4}Gr_{x,\infty}\right)_\text{v}^{1/4}\left(\frac{\text{d}\theta}{\text{d}\eta}\right)_{\eta=0},$$

where $(Nu_{x,\text{w}})_\text{v}$ is defined as

$$(Nu_{x,\text{w}})_\text{v} = \frac{(\alpha_x)_\text{v} x}{\lambda_\text{w}}.$$

The local Grashof number is evaluated as

$$\begin{aligned}(Gr_{x,\infty})_\text{v} &= \frac{g\left|T_\text{w}/T_\infty - 1\right| x^3}{\nu_\infty^2} \\ &= \frac{9.8 \times |(498.1/293 - 1| \times 0.3^3}{(15.06 \times 10^{-6})^2} \\ &= 0.81665 \times 10^9 < 10^9\end{aligned}$$

Then, the flow is laminar free convection.

According to (4.54) and (4.55) the temperature gradient is expressed as

$$-\left(\frac{\text{d}\theta}{\text{d}\eta}\right)_{\eta=0} = (0.567 + 0.186 \times \ln Pr)\cdot\left(\frac{T_\text{w}}{T_\infty}\right)^{-m},$$

where parameter m is expressed as

$$\begin{aligned}m &= 0.35 n_\lambda + 0.29 n_\mu + 0.36 \\ &= 0.35 \times 0.81 + 0.29 \times 0.68 + 0.36 = 0.8407\end{aligned}$$

for $T_\text{w}/T_\infty > 1$. Then,

$$-\left(\frac{\text{d}\theta}{\text{d}\eta}\right)_{\eta=0} = (0.567 + 0.186 \times \ln 0.7)\cdot\left(\frac{498.1}{293}\right)^{-0.8407} = 0.32048.$$

On these bases, $(Nu_{x,\mathrm{w}})_\mathrm{v}$ can be evaluated as follows:

$$\begin{aligned}(Nu_{x,\mathrm{w}})_\mathrm{v} &= -\left(\frac{1}{4}Gr_{x,\infty}\right)_\mathrm{v}^{1/4}\left(\frac{\mathrm{d}\theta}{\mathrm{d}\eta}\right)_{\eta=0}\\ &= \left(\frac{1}{4}\times(0.81665\times 10^9)\right)^{1/4}\times 0.32048\\ &= 38.3085.\end{aligned}$$

With the definition of local Nussetl number for vertical case, $(Nu_{x,\mathrm{w}})_\mathrm{v} = (\alpha_x)_\mathrm{v}x/\lambda_\mathrm{w}$, the local heat transfer coefficient for vertical case can be calculated as

$$\begin{aligned}(\alpha_x)_\mathrm{v} &= (Nu_{x,\mathrm{w}})_\mathrm{v}\frac{\lambda_\mathrm{w}}{x}\\ &= 38.3085\times\frac{0.0407}{0.3}\\ &= 5.197\ \mathrm{W}\ (\mathrm{m}^2\ \mathrm{K})^{-1}.\end{aligned}$$

The average heat transfer coefficient can be calculated as

$$\begin{aligned}\overline{(\alpha_x)_\mathrm{v}} &= \frac{4}{3}(\alpha_x)_\mathrm{v}\\ &= \frac{4}{3}\times 5.197\\ &= 6.9296\ \mathrm{W}\ (\mathrm{m}^2\ \mathrm{K})^{-1}.\end{aligned}$$

The heat transfer rate of the free convection on the vertical plate is

$$\begin{aligned}(Q_x)_\mathrm{v} &= \overline{(\alpha_x)_\mathrm{v}}\cdot(T_\mathrm{w}-T_\infty)\cdot bx\\ &= 6.9296\times(498.1-293)\times 1\times 0.3\\ &= 426.38\,\mathrm{W}.\end{aligned}$$

2. For inclined case. From Table 9.1 the local Nusselt number for inclined case can be expressed as

$$\begin{aligned}(Nu_{x,\mathrm{w}})_\mathrm{i} &= (Nu_{x,\mathrm{w}})_\mathrm{v}\cdot\cos^{1/4}\gamma\\ &= 38.3085\times\cdot\cos^{1/4}(45°)\\ &= 35.129.\end{aligned}$$

With the definition of local Nusselt number for inclined case, $(Nu_{x,\mathrm{w}})_\mathrm{i} = (\alpha_x)_\mathrm{i}x/\lambda_\mathrm{w}$, the local heat transfer coefficient for inclined case can be calculated as

$$\begin{aligned}(\alpha_x)_\mathrm{i} &= (Nu_{x,\mathrm{w}})_\mathrm{i}\frac{\lambda_\mathrm{w}}{x}\\ &= 35.129\times\frac{4.07\times 10^{-2}}{0.3}\\ &= 4.7658\ \mathrm{W}\ (\mathrm{m}^2\,\mathrm{K})^{-1}.\end{aligned}$$

The average heat transfer coefficient can be calculated as

$$\begin{aligned}\overline{(\alpha_x)}_{\mathrm{i}} &= \frac{4}{3}(\alpha_x)_{\mathrm{i}}\\ &= \frac{4}{3}\times 4.7658\\ &= 6.3544\ \mathrm{W}\ (\mathrm{m}^2\,\mathrm{K})^{-1}.\end{aligned}$$

The heat transfer rate of the free convection on the inclined plate is

$$\begin{aligned}(Q_x)_{\mathrm{i}} &= \overline{(\alpha_x)}_{\mathrm{i}}\cdot(T_{\mathrm{w}}-T_\infty)\cdot bx\\ &= 6.3544\times(498.1-293)\times 1\times 0.3\\ &= 391\,\mathrm{W}.\end{aligned}$$

Appendix A. Derivation of Equations (9.1)–(9.3)

1 Derivation of equation (9.1)

Equation (9.1) can be changed to

$$\rho\left[\frac{\partial(w_x)_{\mathrm{i}}}{\partial x}+\frac{\partial(w_y)_{\mathrm{i}}}{\partial x}\right]+(w_x)_{\mathrm{i}}\frac{\partial\rho}{\partial x}+(w_y)_{\mathrm{i}}\frac{\partial\rho}{\partial y}=0. \tag{9.1a}$$

With the dimensionless variables assumed in (9.6), (9.7), (9.9), and (9.10) the following correlations are obtained:

$$\begin{aligned}\frac{\partial(w_x)_{\mathrm{i}}}{\partial x} &= \left[2\sqrt{gx}\left|\frac{\rho_\infty}{\rho_{\mathrm{w}}}-1\right|^{1/2}\right]\frac{\mathrm{d}(W_x)_{\mathrm{i}}}{\mathrm{d}\eta_{\mathrm{i}}}\frac{\partial\eta_{\mathrm{i}}}{\partial x}\cos\gamma\\ &\quad+\frac{1}{2}x^{-1/2}\left[2\sqrt{g}\left|\frac{\rho_\infty}{\rho_{\mathrm{w}}}-1\right|^{1/2}\right](W_x)_{\mathrm{i}}\cos\gamma,\end{aligned}$$

where

$$\begin{aligned}\frac{\partial\eta_{\mathrm{i}}}{\partial x} &= \frac{\partial}{\partial x}\left[\frac{y}{x}\left(\frac{1}{4}Gr_{x,\infty}\right)_{\mathrm{i}}^{1/4}\right]\\ &= \frac{\partial}{\partial x}\left[y\left(\frac{1}{4}\frac{g\left|\frac{\rho_\infty}{\rho_{\mathrm{w}}}-1\right|x^{-1}}{\nu_\infty^2}\right)^{1/4}\right]\\ &= -\frac{1}{4}\left[y\left(\frac{1}{4}\frac{g\left|\frac{\rho_\infty}{\rho_{\mathrm{w}}}-1\right|}{\nu_\infty^2}\right)^{1/4}\right]x^{-\frac{5}{4}}\\ &= -\frac{1}{4}\left[y\left(\frac{1}{4}\frac{g\left|\frac{\rho_\infty}{\rho_{\mathrm{w}}}-1\right|x^3}{\nu_\infty^2}\right)^{1/4}\right]x^{-2}\\ &= -\frac{1}{4}x^{-1}\eta_{\mathrm{i}}.\end{aligned}$$

Then,

$$\frac{\partial (w_x)_{\rm i}}{\partial x} = \left[2\sqrt{gx}\frac{\rho_\infty}{\rho_{\rm w}} - 1^{1/2}\right]\frac{{\rm d}(W_x)_{\rm i}}{{\rm d}\eta_{\rm i}}\left(-\frac{1}{4}x^{-1}\eta_{\rm i}\right)\cos^{1/2}\gamma$$
$$+\frac{1}{2}x^{-1/2}\left[2\sqrt{g}\left|\frac{\rho_\infty}{\rho_{\rm w}} - 1\right|^{1/2}\right](W_x)_{\rm i}\cos^{1/2}\gamma$$
$$= -\frac{1}{2}\left[\sqrt{\frac{g}{x}}\left|\frac{\rho_\infty}{\rho_{\rm w}} - 1\right|^{1/2}\right]\eta\frac{{\rm d}(W_x)_{\rm i}}{{\rm d}\eta_{\rm i}}\cos^{1/2}\gamma$$
$$+\left[\sqrt{\frac{g}{x}}\left|\frac{\rho_\infty}{\rho_{\rm w}} - 1\right|^{1/2}\right](W_x)_{\rm i}\cos^{1/2}\gamma$$
$$= \sqrt{\frac{g}{x}}\left|\frac{\rho_\infty}{\rho_{\rm w}} - 1\right|^{1/2}\cos^{1/2}\gamma\left((W_x)_{\rm i} - \frac{1}{2}\eta_{\rm i}\frac{{\rm d}(W_x)_{\rm i}}{{\rm d}\eta_{\rm i}}\right), \quad (9.47)$$

$$\frac{\partial (w_y)_{\rm i}}{\partial y} = 2\sqrt{\frac{g}{x}}\left|\frac{\rho_\infty}{\rho_{\rm w}} - 1\right|^{1/2}\cos^{1/2}\gamma\frac{{\rm d}(W_y)_{\rm i}}{{\rm d}\eta}, \quad (9.48)$$

$$\frac{\partial \rho}{\partial x} = \frac{{\rm d}\rho}{{\rm d}\eta_{\rm i}}\frac{\partial \eta_{\rm i}}{\partial x} = -\frac{1}{4}\eta_{\rm i}x^{-1}\frac{{\rm d}\rho}{{\rm d}\eta_{\rm i}}, \quad (9.49)$$

$$\frac{\partial \rho}{\partial y} = \frac{{\rm d}\rho}{{\rm d}\eta_{\rm i}}\frac{\partial \eta_{\rm i}}{\partial y} = \frac{{\rm d}\rho}{{\rm d}\eta_{\rm i}}\left[\frac{1}{4}(Gr_{x,\infty})_{\rm i}\right]_{\rm i}^{1/4}x^{-1} \quad (9.50)$$

With (9.47)–(9.50), (9.1a) can be transformed into

$$\rho\left[\sqrt{\frac{g}{x}}\left|\frac{\rho_\infty}{\rho_{\rm w}} - 1\right|^{1/2}\cos^{1/2}\gamma((W_x)_{\rm i} - \frac{1}{2}\eta_{\rm i}\frac{{\rm d}(W_x)_{\rm i}}{{\rm d}\eta_{\rm i}})\right.$$
$$\left.+ 2\sqrt{\frac{g}{x}}\left|\frac{\rho_\infty}{\rho_{\rm w}} - 1\right|^{1/2}\cos^{1/2}\gamma\frac{{\rm d}(W_y)_{\rm i}}{{\rm d}\eta_{\rm i}}\right] + 2\sqrt{gx\ \cos\ \gamma}\,|\rho_\infty/\rho_{\rm w} - 1|^{1/2}(W_x)_{\rm i}$$
$$\times\left(-\frac{1}{4}\eta_{\rm i}x^{-1}\frac{{\rm d}\rho}{{\rm d}\eta}\right) + 2\sqrt{gx\ \cos\ \gamma}\,|\rho_\infty/\rho_{\rm w} - 1|^{1/2}\left[\frac{1}{4}(Gr_{x,\infty})_{\rm i}\right]^{-1/4}(W_y)_{\rm i}$$
$$\times\frac{{\rm d}\rho}{{\rm d}\eta_{\rm i}}\left(\frac{1}{4}Gr_{x,\infty}\right)_{\rm i}^{1/4}x^{-1}$$
$$= 0. \quad (9.51)$$

Equation (9.51) is divided by

$$\sqrt{\frac{g\cos\gamma}{x}}\left|\frac{\rho_\infty}{\rho_{\rm w}} - 1\right|^{1/2}$$

then is simplified to

$$2(W_x)_{\rm i} - \eta\frac{{\rm d}(W_x)_{\rm i}}{{\rm d}\eta_{\rm i}} + 4\frac{{\rm d}(W_y)_{\rm i}}{{\rm d}\eta_{\rm i}} - \frac{1}{\rho}\frac{{\rm d}\rho}{{\rm d}\eta_{\rm i}}[\eta_{\rm i}(W_x)_{\rm i} - 4(W_y)_{\rm i}] = 0. \quad (9.11)$$

This is the dimensionless continuity differential equation of fluid laminar free convection for inclined case.

2 Derivation of equation (9.2)

Equation (9.2) is rewritten as

$$\rho\left[(w_x)_\mathrm{i}\frac{\partial(w_x)_\mathrm{i}}{\partial x}+(w_y)_\mathrm{i}\frac{\partial(w_x)_\mathrm{i}}{\partial y}\right]=\mu\frac{\partial(^2w_x)_\mathrm{i}}{\partial x^2}+\frac{\partial(w_x)_\mathrm{i}}{\partial y}\frac{\partial\mu}{\partial y}+g\left|\rho_\infty-\rho\right|\cos\ \gamma, \tag{9.2a}$$

where

$$\frac{\partial(w_x)_\mathrm{i}}{\partial y}=\frac{\mathrm{d}(w_x)_\mathrm{i}}{\mathrm{d}\eta_\mathrm{i}}\frac{\partial\eta_\mathrm{i}}{\partial y}=2\sqrt{gx}\left|\frac{\rho_\infty}{\rho_\mathrm{w}}-1\right|^{1/2}\frac{\mathrm{d}(W_x)_\mathrm{i}}{\mathrm{d}\eta_\mathrm{i}}\frac{\partial\eta_\mathrm{i}}{\partial y}\cos^{1/2}\gamma$$

$$\frac{\partial\eta_\mathrm{i}}{\partial y}=x^{-1}\left[\frac{1}{4}(Gr_{x,\infty})_\mathrm{i}\right]_\mathrm{i}^{1/4}.$$

Then,

$$\frac{\partial(w_x)_\mathrm{i}}{\partial y}=2\sqrt{gx}\left|\frac{\rho_\infty}{\rho_\mathrm{w}}-1\right|^{1/2}\frac{\mathrm{d}(W_x)_\mathrm{i}}{\mathrm{d}\eta_\mathrm{i}}x^{-1}\left[\frac{1}{4}(Gr_{x,\infty})_\mathrm{i}\right]_\mathrm{i}^{1/4}\cos^{1/2}\gamma, \tag{9.52}$$

$$\begin{aligned}\frac{\partial^2(w_x)_\mathrm{i}}{\partial y^2}&=2\sqrt{gx}\left|\frac{\rho_\infty}{\rho_\mathrm{w}}-1\right|^{1/2}\frac{\mathrm{d}^2(W_x)_\mathrm{i}}{\mathrm{d}\eta_\mathrm{i}^2}x^{-1}\left[\frac{1}{4}(Gr_{x,\infty})_\mathrm{i}\right]_\mathrm{i}^{1/4}\frac{\partial\eta_\mathrm{i}}{\partial y}\cos^{1/2}\gamma\\&=2\sqrt{gx}\left|\frac{\rho_\infty}{\rho_\mathrm{w}}-1\right|^{1/2}\frac{\mathrm{d}^2(W_x)_\mathrm{i}}{\mathrm{d}\eta_\mathrm{i}^2}\left[\frac{1}{4}(Gr_{x,\infty})_\mathrm{i}\right]^{1/4}x^{-1}\left[\frac{1}{4}(Gr_{x,\infty})_\mathrm{i}\right]^{1/4}\cos^{1/2}\gamma\\&=2\sqrt{gx}\left|\frac{\rho_\infty}{\rho_\mathrm{w}}-1\right|^{1/2}\frac{\mathrm{d}^2(W_x)_\mathrm{i}}{\mathrm{d}\eta_\mathrm{i}^2}\left(\frac{1}{4}Gr_{x,\infty}\right)_\mathrm{i}^{1/2}x^{-2}\cos^{1/2}\gamma.\end{aligned} \tag{9.53}$$

$$\begin{aligned}\frac{\partial\mu}{\partial y}&=\frac{\mathrm{d}\mu}{\mathrm{d}\eta_\mathrm{i}}\frac{\partial\eta_\mathrm{i}}{\partial y}\\&=\frac{\mathrm{d}\mu}{\mathrm{d}\eta_\mathrm{i}}\left(\frac{1}{4}Gr_{x,\infty}\right)^{1/4}x^{-1}.\end{aligned} \tag{9.54}$$

With (9.52)–(9.54), (9.2a) is changed into

$$\begin{aligned}&\rho[2\sqrt{gx\ \cos\ \gamma}\left|\rho_\infty/\rho_\mathrm{w}-1\right|^{1/2}(W_x)_\mathrm{i}\sqrt{\frac{g}{x}}\left|\frac{\rho_\infty}{\rho_\mathrm{w}}-1\right|^{1/2}\cos^{1/2}\gamma((W_x)_\mathrm{i}-\frac{1}{2}\eta\frac{\mathrm{d}(W_x)_\mathrm{i}}{\mathrm{d}\eta_\mathrm{i}})\\&+2\sqrt{gx\ \cos\ \gamma}\left|\rho_\infty/\rho_\mathrm{w}-1\right|^{1/2}\left[\frac{1}{4}(Gr_{x,\infty})_\mathrm{i}\right]^{-1/4}\\&\times(W_y)_\mathrm{i}2\sqrt{gx}\left|\frac{\rho_\infty}{\rho_\mathrm{w}}-1\right|^{1/2}\frac{\mathrm{d}(W_x)_\mathrm{i}}{\mathrm{d}\eta_\mathrm{i}}x^{-1}\left[\frac{1}{4}(Gr_{x,\infty})_\mathrm{i}\right]_\mathrm{i}^{1/4}\cos^{1/2}\gamma]\\&=2\mu\sqrt{gx}\left|\frac{\rho_\infty}{\rho_\mathrm{w}}-1\right|^{1/2}\frac{\mathrm{d}^2(W_x)_\mathrm{i}}{\mathrm{d}\eta^2}\left[\frac{1}{4}Gr_{x,\infty})_\mathrm{i}\right]^{1/2}x^{-2}\cos^{1/2}\gamma\\&\quad+2\sqrt{gx}\left|\frac{\rho_\infty}{\rho_\mathrm{w}}-1\right|^{1/2}\frac{\mathrm{d}(W_x)_\mathrm{i}}{\mathrm{d}\eta_\mathrm{i}}x^{-1}\left[\frac{1}{4}(Gr_{x,\infty})_\mathrm{i}\right]_\mathrm{i}^{1/4}\cos^{1/2}\gamma\\&\quad\times\frac{\mathrm{d}\mu}{\mathrm{d}\eta_\mathrm{i}}\left[\frac{1}{4}(Gr_{x,\infty})_\mathrm{i}\right]_\mathrm{i}^{1/4}x^{-1}+g\left|\rho_\infty-\rho\right|\cos\ \gamma\end{aligned} \tag{9.55}$$

With definition of $(Gr_{x,\infty})_{\mathrm{i}}$, (9.55) is rewritten as

$$\rho[2\sqrt{gx\ \cos\ \gamma}\,|\rho_\infty/\rho_{\mathrm{w}}-1|^{1/2}\,(W_x)_{\mathrm{i}}\sqrt{\frac{g}{x}}\left|\frac{\rho_\infty}{\rho_{\mathrm{w}}}-1\right|^{1/2}\cos^{1/2}\gamma((W_x)_{\mathrm{i}}-\frac{1}{2}\eta\frac{\mathrm{d}(W_x)_{\mathrm{i}}}{\mathrm{d}\eta})$$
$$+2\sqrt{gx\ \cos\ \gamma}\,|\rho_\infty/\rho_{\mathrm{w}}-1|^{1/2}\,(W_y)_{\mathrm{i}}2\sqrt{gx}\left|\frac{\rho_\infty}{\rho_{\mathrm{w}}}-1\right|^{1/2}\frac{\mathrm{d}(W_x)_{\mathrm{i}}}{\mathrm{d}\eta_{\mathrm{i}}}x^{-1}\cos^{1/2}\gamma]$$
$$=2\mu\sqrt{gx}\left|\frac{\rho_\infty}{\rho_{\mathrm{w}}}-1\right|^{1/2}\frac{\mathrm{d}^2W_x}{\mathrm{d}\eta_{\mathrm{i}}^2}\left[\frac{1}{4}\frac{g\ \cos\ \gamma\,|\rho_\infty/\rho_{\mathrm{w}}-1|\,x^3}{{\nu_\infty}^2}\right]^{1/2}x^{-2}\cos^{1/2}\gamma$$
$$+2\sqrt{gx}\left|\frac{\rho_\infty}{\rho_{\mathrm{w}}}-1\right|^{1/2}\frac{\mathrm{d}(W_x)_{\mathrm{i}}}{\mathrm{d}\eta_{\mathrm{i}}}x^{-1}\left[\frac{1}{4}(\frac{g\ \cos\ \gamma\,|\rho_\infty/\rho_{\mathrm{w}}-1|\,x^3}{{\nu_\infty}^2}\right]_{\mathrm{i}}^{1/2}\cos^{1/2}\gamma$$
$$\times\frac{\mathrm{d}\mu}{\mathrm{d}\eta_{\mathrm{i}}}x^{-1}+g\,|\rho_\infty-\rho|\cos\ \gamma. \tag{9.56}$$

Equation (9.56) is divided by

$$\rho g\left|\frac{\rho_\infty}{\rho_{\mathrm{w}}}-1\right|\cos\gamma,$$

and simplified to

$$\left[2\sqrt{x}(W_x)_{\mathrm{i}}\sqrt{\frac{1}{x}}\left((W_x)_{\mathrm{i}}-\frac{1}{2}\eta\frac{\mathrm{d}(W_x)_{\mathrm{i}}}{\mathrm{d}\eta_{\mathrm{i}}}\right)+2\sqrt{x}(W_y)_{\mathrm{i}}2\sqrt{x}\frac{\mathrm{d}(W_x)_{\mathrm{i}}}{\mathrm{d}\eta_{\mathrm{i}}}x^{-1}\right]$$
$$=2\nu\sqrt{x}\frac{\mathrm{d}^2(W_x)_{\mathrm{i}}}{\mathrm{d}\eta_{\mathrm{i}}^2}\left[\frac{1}{4}\frac{x^3}{{\nu_\infty}^2}\right]^{1/2}x^{-2}+\frac{2}{\rho}\sqrt{x}\frac{\mathrm{d}(W_x)_{\mathrm{i}}}{\mathrm{d}\eta_{\mathrm{i}}}x^{-1}$$
$$\times\left[\frac{1}{4}\left(\frac{x^3}{{\nu_\infty}^2}\right)\right]_{\mathrm{i}}^{1/2}\frac{\mathrm{d}\mu}{\mathrm{d}\eta_{\mathrm{i}}}x^{-1}+\frac{\rho_{\mathrm{w}}}{\rho}\left|\frac{\rho_\infty-\rho}{\rho_\infty-\rho_{\mathrm{w}}}\right|$$

The earlier equation is divided by ν/ν_∞, and simplified to

$$\frac{\nu_\infty}{\nu}\left[2(W_x)_{\mathrm{i}}\left((W_x)_{\mathrm{i}}-\frac{1}{2}\eta_{\mathrm{i}}\frac{\mathrm{d}(W_x)_{\mathrm{i}}}{\mathrm{d}\eta_{\mathrm{i}}}\right)+2(W_y)_{\mathrm{i}}2\frac{\mathrm{d}(W_x)_{\mathrm{i}}}{\mathrm{d}\eta_{\mathrm{i}}}\right]$$
$$=\frac{\mathrm{d}^2(W_x)_{\mathrm{i}}}{\mathrm{d}\eta_{\mathrm{i}}^2}+\frac{2}{\rho}\frac{\nu_\infty}{\nu}\frac{\mathrm{d}(W_x)_{\mathrm{i}}}{\mathrm{d}\eta_{\mathrm{i}}}\left[\frac{1}{4}\left(\frac{1}{{\nu_\infty}^2}\right]_{\mathrm{i}}^{1/2}\frac{\mathrm{d}\mu}{\mathrm{d}\eta_{\mathrm{i}}}+\frac{\nu_\infty}{\nu}\frac{\rho_{\mathrm{w}}}{\rho}\left|\frac{\rho_\infty-\rho}{\rho_\infty-\rho_{\mathrm{w}}}\right|$$

i.e.,

$$\frac{\nu_\infty}{\nu}\left[(W_x)_{\mathrm{i}}\left(2(W_x)_{\mathrm{i}}-\eta_{\mathrm{i}}\frac{\mathrm{d}(W_x)_{\mathrm{i}}}{\mathrm{d}\eta_{\mathrm{i}}}\right)+4(W_y)_{\mathrm{i}}\frac{\mathrm{d}(W_x)_{\mathrm{i}}}{\mathrm{d}\eta_{\mathrm{i}}}\right]$$
$$=\frac{\mathrm{d}^2(W_x)_{\mathrm{i}}}{\mathrm{d}\eta_{\mathrm{i}}^2}+\frac{1}{\mu}\frac{\mathrm{d}(W_x)_{\mathrm{i}}}{\mathrm{d}\eta_{\mathrm{i}}}\frac{\mathrm{d}\mu}{\mathrm{d}\eta_{\mathrm{i}}}+\frac{\nu_\infty}{\nu}\frac{\rho_{\mathrm{w}}}{\rho}\left|\frac{\rho_\infty-\rho}{\rho_\infty-\rho_{\mathrm{w}}}\right|. \tag{9.57}$$

Because $(\rho_\infty - \rho)/(\rho_\infty - \rho_\mathrm{w})$ is always positive, the earlier equation is rewritten as

$$\frac{\nu_\infty}{\nu}\left[(W_x)_\mathrm{i}\left(2(W_x)_\mathrm{i} - \eta_\mathrm{i}\frac{\mathrm{d}(W_x)_\mathrm{i}}{\mathrm{d}\eta_\mathrm{i}}\right) + 4(W_y)_\mathrm{i}\frac{\mathrm{d}(W_x)_\mathrm{i}}{\mathrm{d}\eta_\mathrm{i}}\right]$$
$$= \frac{\mathrm{d}^2(W_x)_\mathrm{i}}{\mathrm{d}\eta_\mathrm{i}^2} + \frac{1}{\mu}\frac{\mathrm{d}\mu}{\mathrm{d}\eta_\mathrm{i}}\frac{\mathrm{d}(W_x)_\mathrm{i}}{\mathrm{d}\eta_\mathrm{i}} + \frac{\nu_\infty}{\nu}\frac{\frac{\rho_\infty}{\rho} - 1}{\frac{\rho_\infty}{\rho_\mathrm{w}} - 1}. \tag{9.12}$$

This is the dimensionless momentum differential equation of fluid laminar free convection for inclined case.

3 Derivation of equation (9.3)

Equation (9.3) is rewritten as

$$\rho c_p\left[(w_x)_\mathrm{i}\frac{\partial t}{\partial x} + (w_y)_\mathrm{i}\frac{\partial t}{\partial y}\right] = \lambda\frac{\partial^2 t}{\partial y^2} + \frac{\partial \lambda}{\partial y}\frac{\partial t}{\partial y}, \tag{9.58}$$

where

$$t = (t_\mathrm{w} - t_\infty)\theta + t_\infty, \tag{9.59}$$

$$\frac{\partial t}{\partial x} = -(t_\mathrm{w} - t_\infty)\frac{\mathrm{d}\theta}{\mathrm{d}\eta_\mathrm{i}}\left(\frac{1}{4}\right)\eta x^{-1}, \tag{9.60}$$

$$\frac{\partial t}{\partial y} = (t_\mathrm{w} - t_\infty)\frac{\mathrm{d}\theta}{\mathrm{d}\eta_\mathrm{i}}\left(\frac{1}{4}Gr_{x,\infty}\right)_\mathrm{i}^{1/4} x^{-1}, \tag{9.61}$$

$$\frac{\partial^2 t}{\partial y^2} = (t_\mathrm{w} - t_\infty)\frac{\mathrm{d}^2\theta}{\mathrm{d}\eta_\mathrm{i}^2}\left(\frac{1}{4}Gr_{x,\infty}\right)_\mathrm{i}^{1/2} x^{-2}, \tag{9.62}$$

$$\frac{\partial \lambda}{\partial y} = \frac{\mathrm{d}\lambda}{\mathrm{d}\eta_\mathrm{i}}\left(\frac{1}{4}Gr_{x,\infty}\right)_\mathrm{i}^{1/4} x^{-1}. \tag{9.63}$$

With (9.59)–(9.63), (9.58) becomes:

$$\rho c_p[-2\sqrt{g\ \cos\ \gamma x}\,|\rho_\infty/\rho_\mathrm{w} - 1|^{1/2}(W_x)_\mathrm{i}(t_\mathrm{w} - t_\infty)\frac{\mathrm{d}\theta}{\mathrm{d}\eta_\mathrm{i}}\left(\frac{1}{4}\right)\eta_\mathrm{i} x^{-1}$$
$$+2\sqrt{g\ \cos\ \gamma x}\,|\rho_\infty/\rho_\mathrm{w} - 1|^{1/2}\left(\frac{1}{4}Gr_{x,\infty}\right)_\mathrm{i}^{-1/4}(W_y)_\mathrm{i}(t_\mathrm{w} - t_\infty)$$
$$\times\frac{\mathrm{d}\theta}{\mathrm{d}\eta_\mathrm{i}}\left(\frac{1}{4}Gr_{x,\infty}\right)_\mathrm{i}^{1/4} x^{-1}]$$
$$= \lambda(t_\mathrm{w} - t_\infty)\frac{\mathrm{d}^2\theta}{\mathrm{d}\eta_\mathrm{i}^2}\left(\frac{1}{4}Gr_{x,\infty}\right)_\mathrm{i}^{1/2} x^{-2} + \frac{\mathrm{d}\lambda}{\mathrm{d}\eta_\mathrm{i}}\left(\frac{1}{4}Gr_{x,\infty}\right)_\mathrm{i}^{1/4}$$
$$\times x^{-1}(t_\mathrm{w} - t_\infty)\frac{\mathrm{d}\theta}{\mathrm{d}\eta_\mathrm{i}}\left(\frac{1}{4}Gr_{x,\infty}\right)_\mathrm{i}^{1/4} x^{-1}.. \tag{9.64}$$

Equation (9.64) is divided by $(t_{\mathrm{w}} - t_\infty)$, simplified to the follow form by Consideration of definition of Grashof number, $Gr_{x,\infty}$:

$$\rho c_p[-2\sqrt{g\ \cos\ \gamma x}\,|\rho_\infty/\rho_{\mathrm{w}} - 1|^{1/2}\,(W_x)_{\mathrm{i}}\frac{\mathrm{d}\theta}{\mathrm{d}\eta_{\mathrm{i}}}\left(\frac{1}{4}\right)\eta_{\mathrm{i}}x^{-1}$$
$$+2\sqrt{g\ \cos\ \gamma x}\,|\rho_\infty/\rho_{\mathrm{w}} - 1|^{1/2}\,(W_y)_{\mathrm{i}}\frac{\mathrm{d}\theta}{\mathrm{d}\eta_{\mathrm{i}}}x^{-1}]$$
$$= \lambda\frac{\mathrm{d}^2\theta}{\mathrm{d}\eta_{\mathrm{i}}^2}\left(\frac{1}{4}\frac{g\ \cos\ \gamma\,|\rho_\infty/\rho_{\mathrm{w}} - 1|\,x^3}{{\nu_\infty}^2}\right)_{\mathrm{i}}^{1/2} x^{-2}$$
$$+\frac{\mathrm{d}\lambda}{\mathrm{d}\eta_{\mathrm{i}}}\left(\frac{1}{4}\frac{g\ \cos\ \gamma\,|\rho_\infty/\rho_{\mathrm{w}} - 1|\,x^3}{{\nu_\infty}^2}\right)_{\mathrm{i}}^{1/2} x^{-1}\frac{\mathrm{d}\theta}{\mathrm{d}\eta_{\mathrm{i}}}x^{-1}.$$

The earlier equation is divided by

$$\left|\frac{\rho_\infty}{\rho_{\mathrm{w}}} - 1\right|^{1/2}\sqrt{\frac{g\ \cos\ \gamma}{x}},$$

then, we get

$$\rho c_p\left[-2(W_x)_{\mathrm{i}}\frac{\mathrm{d}\theta}{\mathrm{d}\eta_{\mathrm{i}}}\left(\frac{1}{4}\right)\eta_{\mathrm{i}} + 2(W_y)_{\mathrm{i}}\frac{\mathrm{d}\theta}{\mathrm{d}\eta_{\mathrm{i}}}\right] = \lambda\frac{\mathrm{d}^2\theta}{\mathrm{d}\eta_{\mathrm{i}}^2}\left(\frac{1}{4}\frac{1}{{\nu_\infty}^2}\right)_{\mathrm{i}}^{1/2} + \frac{\mathrm{d}\lambda}{\mathrm{d}\eta_{\mathrm{i}}}\left(\frac{1}{4}\frac{1}{{\nu_\infty}^2}\right)_{\mathrm{i}}^{1/2}\frac{\mathrm{d}\theta}{\mathrm{d}\eta_{\mathrm{i}}} \tag{9.65}$$

This equation is multiplied by $2\nu_\infty/\lambda$ and on simplification, finally becomes:

$$\frac{\nu_\infty}{\lambda}\rho c_p\left[-(W_x)_{\mathrm{i}}\frac{\mathrm{d}\theta}{\mathrm{d}\eta_{\mathrm{i}}}\eta_{\mathrm{i}} + 4(W_y)_{\mathrm{i}}\frac{\mathrm{d}\theta}{\mathrm{d}\eta_{\mathrm{i}}}\right] = \frac{\mathrm{d}^2\theta}{\mathrm{d}\eta_{\mathrm{i}}^2} + \frac{1}{\lambda}\frac{\mathrm{d}\lambda}{\mathrm{d}\eta_{\mathrm{i}}}\frac{\mathrm{d}\theta}{\mathrm{d}\eta_{\mathrm{i}}},$$

i.e.,

$$Pr\frac{\nu_\infty}{\nu}\left[-\eta_{\mathrm{i}}(W_x)_{\mathrm{i}} + 4(W_y)_{\mathrm{i}}\right]\frac{\mathrm{d}\theta}{\mathrm{d}\eta_{\mathrm{i}}} = \frac{\mathrm{d}^2\theta}{\mathrm{d}\eta_{\mathrm{i}}^2} + \frac{1}{\lambda}\frac{\mathrm{d}\lambda}{\mathrm{d}\eta_{\mathrm{i}}}\frac{\mathrm{d}\theta}{\mathrm{d}\eta_{\mathrm{i}}}. \tag{9.13}$$

This is the dimensionless energy differential equation of fluid laminar free convection for inclined case.

References

1. B.R. Rich, An investigation of heat transfer from an inclined flat plate in free convection, Trans. ASME 75, pp. 489–499, 1953
2. G.C. Vliet, National convection local heat transfer on constant heat flux inclined surface, J. Heat Transfer 9, pp. 511–516, 1969
3. T. Fujii and H. Imura, Natural convection from a plate with arbitrary inclination, Int. J. Heat Mass Transfer 15, pp. 755–767, 1972
4. V.M. Falkner and S.W. Skan, Some approximate solutions of the boundary layer equations, Phil. Mag. 12, pp. 865, 1931

5. H. Schlichting, Boundary Layer Theory, translated by J. Kestin, pp. 316–317, 1979
6. T. Cebeci and P. Bradshaw, Physical and Computational aspects of convective heat transfer, Springer Berlin Heidelberg New York, 1984.
7. D.Y. Shang and B.X. Wang, Effect of variable thermophysical properties on laminar free convection of gas, Int. J. Heat Mass Transfer 33, No. 7, pp. 1387–1395, 1990
8. D.Y. Shang and B.X. Wang, Effect of variable thermophysical properties on laminar free convection of polyatomic gas, Int. J. Heat Mass Transfer 34, pp. 749–755, 1991
9. D.Y. Shang, B.X. Wang, Y. Wang, and Y. Quan, Study on liquid laminar free convection with consideration of variable thermophysical properties, Int. J. Heat Mass Transfer 36, No. 14, pp. 3411–3419, 1993
10. D.Y. Shang and H.S. Takhar, Extended study on relationships of heat, momentum, and mass transfer for laminar free convection between inclined and vertical plates, J. Theor. Appl. Fluid Mech. 1, No. 1, pp. 16–32, 1995

Part II

Film Boiling and Condensation

10

Laminar Film Boiling of Saturated Liquid

Nomenclature

a	thermal diffusive coefficient, $\mathrm{m^2\,s^{-1}}$
b	width of platc, m
c_p	specific heat at constant pressure, $\mathrm{J\,(kg\,K)^{-1}}$
DT_v	$\left(\frac{\mathrm{d}\theta_v}{\mathrm{d}\eta_v}\right)_\mathrm{s}$
DT_l	$-\frac{h_\mathrm{fg}\mu_\mathrm{v,s}(W_{x\mathrm{v,s}}\eta_{\mathrm{v}\delta}-4W_{y\mathrm{v,s}})}{\lambda_\mathrm{v,s}(t_\mathrm{w}-t_s)}$
$DW_{x\mathrm{v}}$	$(\mathrm{d}W_{x\mathrm{v}}/\mathrm{d}\eta_\mathrm{v})_\mathrm{s}$
$DW_{x\mathrm{l}}$	$\frac{\mu_\mathrm{v,s}}{\mu_\mathrm{v,s}}\left(\frac{\rho_\mathrm{l,s}-\rho_\mathrm{v,w}}{\rho_\mathrm{s,w}}\right)^{3/4}(\frac{\nu_{\mathrm{l},s}}{\nu_\mathrm{v,s}})^{1/2}(\frac{\mathrm{d}W_\mathrm{xv}}{\mathrm{d}\eta_\mathrm{v}})_{\eta_\mathrm{v}=\eta_{\mathrm{v}\delta}}$
g	gravitation acceleration, $\mathrm{m\,s^2}$
$Gr_{x\mathrm{l,s}}$	local Grashof number of liquid film for film boiling of saturated liquid $gx^3/\nu_\mathrm{l,s}^2$
$Gr_{x\mathrm{v,s}}$	local Grashof number of vapor film for film boiling of saturated liquid, $\frac{g(\rho_\mathrm{l,s}/\rho_\mathrm{v,w}-1)x^3}{\nu_\mathrm{v,s}^2}$
g_x	local mass flow rate entering the vapor film at position x per unit area of the plate, $\mathrm{kg\,(m^2\,s)^{-1}}$
G_x	total mass flow rate entering the vapor film for position $x=0$ to x with width of b of the plate, $\mathrm{kg\,s^{-1}}$
h_fg	latent heat of vaporization, $\mathrm{J\,kg^{-1}}$
$Nu_{x,\mathrm{w}}$	local Nusselt number for film boiling when wall temperature t_w is taken as reference temperature, $\alpha_x x/\lambda_\mathrm{w}$
$\overline{Nu}_{xv,\mathrm{w}}$	average Nusselt number for film boiling, $\overline{\alpha}_x x/\lambda_\mathrm{w}$
n_{c_p}	specific heat parameter of gas
n_λ	thermal conductivity parameter of gas
n_μ	viscosity parameter of gas

Pr	Prandtl number
q_x	local heat transfer rate at position x per unit area on the plate, $\mathrm{W\,m^{-2}}$
Q_x	total heat transfer rate for position $x = 0$ to x with width of b on the plate, W
t	temperature, $^\circ$C
T	absolute temperature, $^\circ$K
w_x, w_y	velocity components in the x- and y-directions, respectively, $\mathrm{m\,s^{-1}}$
W_x, W_y	dimensionless velocity components in the x- and y-directions, respectively
$W_{x\mathrm{v,s}}$, $W_{y\mathrm{v,s}}$	interfacial dimensionless velocity components for vapor film in the x- and y-directions, respectively

Greek symbols

α_x	local heat transfer coefficient, $\mathrm{W\,(m^2\,K)^{-1}}$
$\overline{\alpha}_x$	average heat transfer coefficient, $\mathrm{W\,(m^2\,K)^{-1}}$
δ	boundary layer thickness, m
δ_l	thickness of liquid film, m
δ_v	thickness of vapor film, m
η	dimensionless co-ordinate variable for boundary layer
θ_v	dimensionless temperature for vapor film, $\frac{T-T_\mathrm{s}}{T_\mathrm{w}-T_\mathrm{s}}$
λ	thermal conductivity, $\mathrm{W\,(m\,K)^{-1}}$
μ	absolute viscosity, $\mathrm{kg\,(m\,s)^{-1}}$
ν	kinetic viscosity, $\mathrm{m^2\,s^{-1}}$
ρ	density, $\mathrm{kg\,m^{-3}}$
Δt_w	wall superheated temperature, $t_\mathrm{w} - t_\mathrm{s}$, $^\circ$C
$\frac{\Delta t_\mathrm{w}}{t_\mathrm{s}}$	wall superheated grade, $\frac{t_\mathrm{w}-t_\mathrm{s}}{t_\mathrm{s}}$
$\left(-\left(\frac{\mathrm{d}\theta_\mathrm{v}}{\mathrm{d}\eta_\mathrm{v}}\right)_{\eta_\mathrm{v}=0}\right)_{\Delta t_\infty=0}$	dimensionless temperature gradient on the plate for film boiling of saturated liquid
$\eta_{\mathrm{v}\delta}$	dimensionless thickness of vapor film
$(\eta_{\mathrm{v}\delta} W_{x\mathrm{v,s}} - 4W_{y\mathrm{v,s}})_{\Delta t_\infty=0}$	mass flow rate parameter for film boiling of saturated liquid
$\frac{\rho_\mathrm{v,w}}{\rho_\mathrm{v}}\frac{\rho_\mathrm{l,s}-\rho_\mathrm{v}}{\rho_\mathrm{l,s}-\rho_\mathrm{v,w}}$	buoyancy factor of film boiling
$\frac{1}{\rho_\mathrm{v}}\frac{\mathrm{d}\rho_\mathrm{v}}{\mathrm{d}\eta_\mathrm{v}}$	density factor
$\frac{1}{\mu_\mathrm{v}}\frac{\mathrm{d}\mu_\mathrm{v}}{\mathrm{d}\eta_\mathrm{v}}$	viscosity factor

$\frac{1}{\lambda_v}\frac{d\lambda_v}{d\eta_v}$	thermal conductivity factor
$\frac{\nu_{v,s}}{\nu_v}$	kinetic viscosity factor

Subscripts

i	liquid film
s	saturated state, or at the vapor–liquid interface
v	vapor film
w	at wall
∞	far from the wall surface

10.1 Introduction

Bromley [1] first treated the laminar film-boiling heat transfer of saturated liquid from a horizontal cylinder, using a simple theoretical model. Later analytical investigations [2–7] were made to analyze pool film boiling from a vertical plate, in which only a few researches [5] and [7] took into account temperature-dependence of fluid thermophysical properties. McFadden and Grosh [5] developed the analysis of saturated film boiling where the temperature-dependence of density and specific heat was taken into account. Nishikawa, Ito and Matsumoto [7] made an analysis of pool film boiling as a variable property problem on the basis of the two-phase boundary layer theory, but only the effect of variation of vapor's thermophysical properties with temperature was examined in the range of lower degree of subcooling ($T_s - T_\infty = 0,\ 20,$ and $40°$ C).

However, in general film boiling, the temperature difference between the heating surface and bulk liquid is very large, where large superheated degrees on the surface and large subcooled degrees of liquid are included. In Chaps. 4–7 it is shown that the thermophysical property variations of gas and liquid with temperature have great influences on their free convection. Of course, they have definitely great effect on the film boiling of liquid. Therefore, from now on, two chapters will be devoted in this book to introducing the recent studies of Shang, Wang and Zhong [8–10] on the film boiling of saturated and subcooled liquid, respectively. In this chapter, the rigorous theoretical models for film boiling of saturated liquid along an isothermal vertical plate are established by means of the velocity component method. The temperature parameter method introduced in Chap. 4 is used for treatment of the variable thermophysical properties on the vapor film. The governing ordinary differential equations with the boundary conditions are solved by a successively iterative procedure at different wall superheated degrees for saturated water. The distributions of velocity and temperature fields of the boiling are rigorously determined. The theoretical correlations of heat transfer coefficient and mass flow rate are derived, and then the corresponding curve-fit predictive equations for the heat transfer coefficient and mass flow rate parameter are developed based on the rigorous numerical solutions.

10.2 Governing Partial Differential Equations

The analytical model and coordinating system used in this problem are shown in Fig. 10.1. The heated plate with uniform temperature T_w is submerged vertically in stagnant liquid whose temperature is higher than the liquid saturated temperature T_s. We assume that the heating surface of the plate is covered with continuous laminar vapor film, which moves upwards with the vapor. Thus a two-phase boundary layer is formed. Heat flux produced from the heating surface of the plate transfers through the two-phase boundary layer to the bulk liquid. Meanwhile, mass transfer is produced at the vapor–liquid interface due to the film boiling of liquid.

The governing conservation equations of mass, momentum, and energy conservation for steady laminar free convection of the film boiling of saturated liquid can be written as

(a.) for vapor film with consideration of variable thermophysical properties:

$$\frac{\partial}{\partial x}(\rho_v w_{xv}) + \frac{\partial}{\partial y}(\rho_v w_{yv}) = 0, \tag{10.1}$$

$$\rho_v\left(w_{xv}\frac{\partial w_{xv}}{\partial x} + w_{yv}\frac{\partial w_{xv}}{\partial y}\right) = \frac{\partial}{\partial y}\left(\mu_v\frac{\partial w_{xv}}{\partial y}\right) + g(\rho_{l,s} - \rho_v), \tag{10.2}$$

$$\rho_v c_{p_v}\left(w_{xv}\frac{\partial t_v}{\partial x} + w_{yv}\frac{\partial t_v}{\partial y}\right) = \frac{\partial}{\partial y}\left(\lambda_v\frac{\partial t_v}{\partial y}\right), \tag{10.3}$$

(b.) for liquid film with the constant physical properties:

$$\frac{\partial}{\partial x}(w_{xl}) + \frac{\partial}{\partial y}(w_{yl}) = 0, \tag{10.4}$$

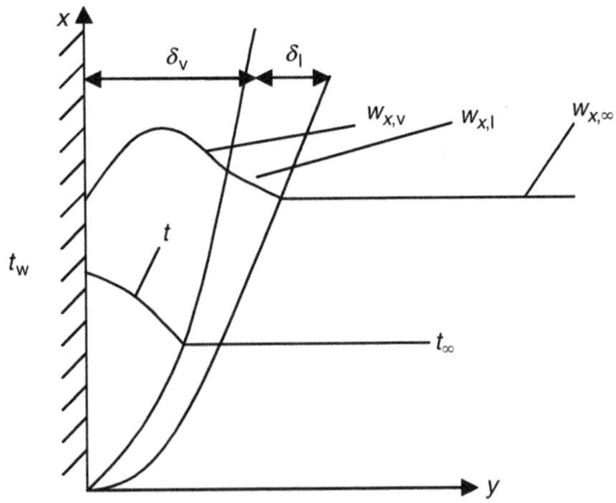

Fig. 10.1. Physical model and coordinate system of film boiling of saturated liquid

$$w_{x\mathrm{l}}\frac{\partial w_{x\mathrm{l}}}{\partial x}+w_{y\mathrm{l}}\frac{\partial w_{x\mathrm{l}}}{\partial y}=\nu_l\frac{\partial^2 w_{x\mathrm{l}}}{\partial y^2}, \tag{10.5}$$

with boundary conditions

$$y=0: w_{x\mathrm{v}}=0, w_{y\mathrm{v}}=0, \quad T_\mathrm{v}=T_\mathrm{w}, \tag{10.6}$$

$$y=\delta_\mathrm{v}: \quad w_{x\mathrm{v,s}}=w_{x\mathrm{l,s}}, \tag{10.7}$$

$$\rho_{\mathrm{v,s}}\left(w_{x\mathrm{v}}\frac{\partial\delta_\mathrm{v}}{\partial x}-w_{y\mathrm{v}}\right)_\mathrm{s}=\rho_{\mathrm{l,s}}\left(w_{x\mathrm{l}}\frac{\partial\delta_\mathrm{l}}{\partial x}-w_{y\mathrm{l}}\right)_\mathrm{s}, \tag{10.8}$$

$$\mu_{\mathrm{v,s}}\left(\frac{\partial w_{x\mathrm{v}}}{\partial y}\right)_\mathrm{s}=\mu_{\mathrm{l,s}}\left(\frac{\partial w_{x\mathrm{l}}}{\partial y}\right)_\mathrm{s}, \tag{10.9}$$

$$-\lambda_{\mathrm{v,s}}\left(\frac{\partial t_\mathrm{v}}{\partial y}\right)_{y=\delta_\mathrm{v}}= \quad h_{fg}\rho_{\mathrm{v,s}}\left(w_{x\mathrm{v}}\frac{\partial\delta_{x\mathrm{v}}}{\partial x}-w_{y\mathrm{v}}\right)_\mathrm{s}, \tag{10.10}$$

$$T_\mathrm{v}=T_\mathrm{s}, \tag{10.11}$$

$$y\to\infty: w_{x\mathrm{l}}\to 0, \tag{10.12}$$

Here, (10.7)–(10.11) express physical matching conditions of the continuities of velocity, mass flow rate, shear force, heat flux, and temperature at the vapor–liquid interface.

10.3 Similarity Transformation

In order to solve the governing equations in a suitable dimensionless form, it is necessary to transform similarly the governing partial differential equations and the boundary conditions. We still use the velocity component method to carry out this transformation. At first, we introduce the similarity variables as follows:

10.3.1 Similarity Transformation Variables

Due to the two-phase boundary layer there are two sets of the transformation variables: the transformation variables for vapor and liquid films.

For vapor film. For vapor film the dimensionless coordinate variable η_v is set up at first as follows:

$$\eta_\mathrm{v}=\left(\frac{1}{4}Gr_{x\mathrm{v,s}}\right)^{1/4}\frac{y}{x}, \tag{10.13}$$

where the local Grashof number $Gr_{x\mathrm{v,s}}$ is assumed as

$$Gr_{x\mathrm{v,s}}=\frac{g(\rho_{\mathrm{l,s}}/\rho_{\mathrm{v,w}}-1)x^3}{\nu_{\mathrm{v,s}}^2}. \tag{10.14}$$

The dimensionless temperature is given as

$$\theta_{\rm v} = \frac{T_{\rm v} - T_{\rm s}}{T_{\rm w} - T_{\rm s}}. \tag{10.15}$$

The dimensionless velocity components are assumed as

$$W_{x{\rm v}} = (2\sqrt{gx}(\rho_{\rm l,s}/\rho_{\rm v,w} - 1)^{1/2})^{-1} w_{x{\rm v}}, \tag{10.16}$$

$$W_{x{\rm v}} = \left(2\sqrt{gx}(\rho_{\rm l,s}/\rho_{\rm v,w} - 1)^{-2}\left(\frac{1}{4}Gr_{xv,s}\right)^{-1/4}\right)^{-1} w_{y{\rm v}}. \tag{10.17}$$

For liquid film. For liquid film the dimensionless coordinate variable $\eta_{\rm l}$ is set up as

$$\eta_{\rm l} = \left(\frac{1}{4}Gr_{x{\rm l,s}}\right)^{1/4}\frac{y}{x}, \tag{10.18}$$

where the local Grashof number $Gr_{x{\rm l,s}}$ is assumed as

$$Gr_{x{\rm l,s}} = \frac{gx^3}{\nu_{\rm l,s}^2}. \tag{10.19}$$

The dimensionless velocity components are assumed as

$$W_{x{\rm l}} = (2\sqrt{gx})^{-1} w_{x{\rm l}}, \tag{10.20}$$

$$W_{y{\rm l}} = \left(2\sqrt{gx}\left(\frac{1}{4}Gr_{x{\rm l,s}}\right)^{-1/4}\right)^{-1} w_{y{\rm l}}. \tag{10.21}$$

10.3.2 Similarity Transformation

The similarity transformation of the governing partial differential equations is divided into three parts: the transformations of the governing equations for the vapor film, liquid film, and the boundary conditions, respectively.

For vapor film:

Transformation of (10.1). At first, (10.1) is rewritten as

$$\rho_{\rm v}\left(\frac{\partial w_{x{\rm v}}}{\partial x} + \frac{\partial w_{y{\rm v}}}{\partial y}\right) + w_{x{\rm v}}\frac{\partial \rho_{\rm v}}{\partial x} + w_{y{\rm v}}\frac{\partial \rho_{\rm v}}{\partial y} = 0. \tag{10.22}$$

With (10.13), (10.14), (10.16), and (10.17) we can obtain the following correlations:

$$\frac{\partial w_{x\mathrm{v}}}{\partial x} = \sqrt{\frac{g}{x}}(\rho_{\mathrm{l,s}}/\rho_{\mathrm{v,w}} - 1)^{1/2}\left(W_{x\mathrm{v}} - \frac{1}{2}\eta_{\mathrm{v}}\frac{\mathrm{d}W_{x\mathrm{v}}}{\mathrm{d}\eta}\right), \tag{10.23}$$

$$\frac{\partial w_{y\mathrm{v}}}{\partial y} = 2\sqrt{\frac{g}{x}}(\rho_{\mathrm{l,s}}/\rho_{\mathrm{v,w}} - 1)^{1/2}\frac{\mathrm{d}W_{y\mathrm{v}}}{\mathrm{d}\eta_{\mathrm{v}}}, \tag{10.24}$$

$$\frac{\partial \rho_{\mathrm{v}}}{\partial x} = -\frac{1}{4}\eta_{\mathrm{v}}x^{-1}\frac{\mathrm{d}\rho_{\mathrm{v}}}{\mathrm{d}\eta_{\mathrm{v}}}, \tag{10.25}$$

$$\frac{\partial \rho_{\mathrm{v}}}{\partial y} = \frac{\mathrm{d}\rho_{\mathrm{v}}}{\mathrm{d}\eta_{\mathrm{v}}}\left(\frac{1}{4}Gr'_{x\mathrm{v,s}}\right)^{1/4} x^{-1}. \tag{10.26}$$

With (10.16), (10.17), (10.23)–(10.26), (10.22) can be changed into

$$\rho_{\mathrm{v}}\left[\sqrt{\frac{g}{x}}(\rho_{\mathrm{l,s}}/\rho_{\mathrm{v,w}} - 1)^{1/2}\left(W_{x\mathrm{v}} - \frac{1}{2}\eta_{\mathrm{v}}\frac{\mathrm{d}W_{x\mathrm{v}}}{\mathrm{d}\eta}\right) + 2\sqrt{\frac{g}{x}}(\rho_{\mathrm{l,s}}/\rho_{\mathrm{v,w}} - 1)^{1/2}\frac{\mathrm{d}W_{y\mathrm{v}}}{\mathrm{d}\eta_{\mathrm{v}}}\right]$$
$$+2\sqrt{gx}(\rho_{\mathrm{l,s}}/\rho_{\mathrm{v,w}} - 1)^{1/2}W_{x\mathrm{v}}\left(-\frac{1}{4}\eta_{\mathrm{v}}x^{-1}\frac{\mathrm{d}\rho_{\mathrm{v}}}{\mathrm{d}\eta_{\mathrm{v}}}\right)$$
$$+2\sqrt{gx}(\rho_{\mathrm{l,s}}/\rho_{\mathrm{v,w}} - 1)^{1/2}\left(\frac{1}{4}Gr_{x\mathrm{v,s}}\right)^{-1/4} W_{y\mathrm{v}}\frac{\mathrm{d}\rho_{\mathrm{v}}}{\mathrm{d}\eta_{\mathrm{v}}}\left(\frac{1}{4}Gr_{x\mathrm{v,s}}\right)^{1/4} x^{-1} = 0.$$

This earlier equation is divided by $(\rho_{\mathrm{l,s}}/\rho_{\mathrm{v,w}} - 1)^{1/2}\sqrt{\frac{g}{x}}$ and is simplified to

$$2W_{x\mathrm{v}} - \eta_{\mathrm{v}}\frac{\mathrm{d}W_{x\mathrm{v}}}{\mathrm{d}\eta_{\mathrm{v}}} + 4\frac{\mathrm{d}W_{y\mathrm{v}}}{\mathrm{d}\eta_{\mathrm{v}}} - \frac{1}{\rho_{\mathrm{v}}}\frac{\mathrm{d}\rho_{\mathrm{v}}}{\mathrm{d}\eta_{\mathrm{v}}}(\eta_{\mathrm{v}}W_{x\mathrm{v}} - 4W_{y\mathrm{v}}) = 0. \tag{10.27}$$

Transformation of (10.2). The equation (10.2) is rewritten as

$$\rho_{\mathrm{v}}\left(w_{x\mathrm{v}}\frac{\partial w_{x\mathrm{v}}}{\partial x} + w_{y\mathrm{v}}\frac{\partial w_{x\mathrm{v}}}{\partial y}\right) = \mu_{\mathrm{v}}\frac{\partial^2 w_{x\mathrm{v}}}{\partial y^2} + \frac{\partial w_{x\mathrm{v}}}{\partial y}\frac{\partial \mu_{\mathrm{v}}}{\partial y} + g(\rho_{\mathrm{l,s}} - \rho_{\mathrm{v}}) \tag{10.28}$$

With the dimensionless transformation variables assumed in (10.13), (10.14), (10.16), and (10.17) we get

$$\frac{\partial w_{x\mathrm{v}}}{\partial y} = 2\sqrt{gx}(\rho_{\mathrm{l,s}}/\rho_{\mathrm{v,w}} - 1)^{1/2}\frac{\mathrm{d}W_{x\mathrm{v}}}{\mathrm{d}\eta_{\mathrm{v}}}x^{-1}\left(\tfrac{1}{4}Gr_{x\mathrm{v,s}}\right)^{1/4}, \tag{10.29}$$

$$\frac{\partial^2 w_{x\mathrm{v}}}{\partial y^2} = 2\sqrt{gx}(\rho_{\mathrm{l,s}}/\rho_{\mathrm{v,w}} - 1)^{1/2}\frac{\mathrm{d}^2W_{x\mathrm{v}}}{\mathrm{d}\eta_{\mathrm{v}}^2}x^{-1}\left(\tfrac{1}{4}Gr_{x\mathrm{v,s}}\right)^{1/4}\left(\tfrac{1}{4}Gr_{x\mathrm{v,s}}\right)^{1/4}x^{-1}$$
$$= 2\sqrt{gx}(\rho_{\mathrm{l,s}}/\rho_{\mathrm{v,w}} - 1)^{1/2}\frac{\mathrm{d}^2W_{x\mathrm{v}}}{\mathrm{d}\eta_{\mathrm{v}}^2}\left(\tfrac{1}{4}Gr_{x\mathrm{v,s}}\right)^{1/2}x^{-2}, \tag{10.30}$$

$$\frac{\partial \mu_{\mathrm{v}}}{\partial y} = \frac{\mathrm{d}\mu_{\mathrm{v}}}{\mathrm{d}\eta_{\mathrm{v}}}\left(\tfrac{1}{4}Gr_{x\mathrm{v,s}}\right)^{1/4}x^{-1}. \tag{10.31}$$

With (10.16), (10.17), and (10.29)–(10.31), (10.28) will be changed into

$$\rho_{\rm v}\left[2\sqrt{gx}(\rho_{\rm l,s}/\rho_{\rm v,w}-1)^{1/2}W_{x{\rm v}}\sqrt{\frac{g}{x}}(\rho_{\rm l,s}/\rho_{\rm v,w-1})^{1/2}\left(W_{x{\rm v}}-\frac{1}{2}\eta_{\rm v}\frac{{\rm d}W_{x{\rm v}}}{{\rm d}\eta}\right)\right.$$
$$+2\sqrt{gx}(\rho_{\rm l,s}/\rho_{\rm v,w}-1)^{1/2}\left(\frac{1}{4}Gr_{x{\rm v,s}}\right)^{-1/4}W_{y{\rm v}}2\sqrt{gx}(\rho_{\rm l,s}/\rho_{\rm v,w}-1)^{1/2}$$
$$\left.\times\frac{{\rm d}W_{x{\rm v}}}{{\rm d}\eta_{\rm v}}x^{-1}\left(\tfrac{1}{4}Gr_{x{\rm v,s}}\right)^{1/4}\right]$$
$$=\mu_{\rm v}2\sqrt{gx}(\rho_{\rm l,s}/\rho_{\rm v,w}-1)^{1/2}\frac{{\rm d}^2W_{x{\rm v}}}{{\rm d}\eta_{\rm v}^2}\left(\tfrac{1}{4}Gr_{x{\rm v,s}}\right)^{1/2}x^{-2}$$
$$+2\sqrt{gx}(\rho_{\rm l,s}/\rho_{\rm v,w}-1)^{1/2}\frac{{\rm d}W_{x{\rm v}}}{{\rm d}\eta}x^{-1}\left(\tfrac{1}{4}Gr_{x{\rm v,s}}\right)^{1/4}$$
$$\times\frac{{\rm d}\mu_{\rm v}}{{\rm d}\eta_{\rm v}}\left(\tfrac{1}{4}Gr_{x{\rm v,s}}\right)^{1/4}x^{-1}+g(\rho_{\rm l,s}-\rho_{\rm v}).$$

The earlier equation is divided by $g(\rho_{\rm l,s}/\rho_{\rm v,w}-1)$ and with the definition of $Gr_{x{\rm v,s}}$ the equation is further simplified to

$$\rho_{\rm v}\left[2W_{x{\rm v}}\left(W_{x{\rm v}}-\frac{1}{2}\eta_{\rm v}\frac{{\rm d}W_{x{\rm v}}}{{\rm d}\eta}\right)+4W_{y{\rm v}}\frac{{\rm d}W_{x{\rm v}}}{{\rm d}\eta_{\rm v}}\right]$$
$$=\mu_{\rm v}\frac{{\rm d}^2W_{x{\rm v}}}{{\rm d}\eta_{\rm v}^2}\frac{1}{\nu_{\rm v,s}}+\frac{{\rm d}W_{x{\rm v}}}{{\rm d}\eta_{\rm v}}\frac{{\rm d}\mu_{\rm v}}{{\rm d}\eta_{\rm v}}\frac{1}{\nu_{\rm v,s}}+\frac{(\rho_{\rm l,s}-\rho_{\rm v})}{(\rho_{\rm l,s}/\rho_{\rm v,w}-1)}.$$

The earlier equation is multiplied by $(1/\rho_{\rm v})(\nu_{\rm v,s}/\nu_{\rm v})$ and further simplified to

$$\frac{\nu_{\rm v,s}}{\nu_{\rm v}}\left(W_{x{\rm v}}\left(2W_{x{\rm v}}-\eta_{\rm v}\frac{{\rm d}W_{x{\rm v}}}{{\rm d}\eta_{\rm v}}\right)+4W_{y{\rm v}}\frac{{\rm d}W_{x{\rm v}}}{{\rm d}\eta_{\rm v}}\right)=\frac{{\rm d}^2W_{x{\rm v}}}{{\rm d}\eta_{\rm v}^2}+\frac{1}{\mu_{\rm v}}\frac{{\rm d}\mu_{\rm v}}{{\rm d}\eta_{\rm v}}\frac{{\rm d}W_{x{\rm v}}}{{\rm d}\eta_{\rm v}}+\frac{\nu_{\rm v,s}}{\nu_{\rm v}}\frac{\rho_{\rm v,w}}{\rho_{\rm v}}\frac{\rho_{\rm l,s}-\rho_{\rm v}}{\rho_{\rm l,s}-\rho_{\rm v,w}}. \tag{10.32}$$

Transformation of (10.3). Equation (10.3) is firstly rewritten as

$$\rho_{\rm v}c_{p_{\rm v}}\left(w_{x{\rm v}}\frac{\partial T_{\rm v}}{\partial x}+w_{y{\rm v}}\frac{\partial T_{\rm v}}{\partial y}\right)=\lambda_{\rm v}\frac{\partial^2T_{\rm v}}{\partial y^2}+\frac{\partial\lambda_{\rm v}}{\partial y}\frac{\partial T_{\rm v}}{\partial y}, \tag{10.33}$$

where

$$T_{\rm v}=(T_{\rm w}-T_{\rm s})\theta_{\rm v}+T_{\rm s}, \tag{10.34}$$
$$\frac{\partial T_{\rm v}}{\partial x}=-(T_{\rm w}-t_{\rm s})\frac{{\rm d}\theta_{\rm v}}{{\rm d}\eta_{\rm v}}\left(\frac{1}{4}\right)\eta_{\rm v}x^{-1}, \tag{10.35}$$

$$\frac{\partial T_{\mathrm{v}}}{\partial y}=(T_{\mathrm{w}}-T_{\mathrm{s}})\frac{\mathrm{d}\theta_{\mathrm{v}}}{\mathrm{d}\eta_{\mathrm{v}}}\left(\frac{1}{4}Gr_{x\mathrm{v,s}}\right)^{1/4}x^{-1}, \tag{10.36}$$

$$\frac{\partial^2 T_{\mathrm{v}}}{\partial y^2}=(T_{\mathrm{w}}-T_{\mathrm{s}})\frac{\mathrm{d}^2\theta_{\mathrm{v}}}{\mathrm{d}\eta_{\mathrm{v}}}\left(\frac{1}{4}Gr_{x\mathrm{v,s}}\right)^{1/2}x^{-2}, \tag{10.37}$$

$$\frac{\partial \lambda_{\mathrm{v}}}{\partial y}=\frac{\mathrm{d}\lambda_{\mathrm{v}}}{\mathrm{d}\eta_{\mathrm{v}}}\left(\frac{1}{4}Gr_{x\mathrm{v,s}}\right)^{1/4}x^{-1}. \tag{10.38}$$

With (10.16), (10.17), (10.34)–(10.38), (10.33) will become

$$\rho_{\mathrm{v}}c_{p_{\mathrm{v}}}\left[2\sqrt{gx}(\rho_{\mathrm{l,s}}/\rho_{\mathrm{v,w}}-1)^{1/2}W_{x\mathrm{v}}\left(-(T_{\mathrm{w}}-T_{\mathrm{s}})\frac{\mathrm{d}\theta_{\mathrm{v}}}{\mathrm{d}\eta_{\mathrm{v}}}\left(\frac{1}{4}\right)\eta_{\mathrm{v}}x^{-1}\right)\right.$$
$$\left.+2\sqrt{gx}(\rho_{\mathrm{l,s}}/\rho_{\mathrm{v,w}})^{1/2}\left(\frac{1}{4}Gr_{x\mathrm{v,s}}\right)^{-1/4}W_{y\mathrm{v}}(T_{\mathrm{w}}-T_{\mathrm{s}})\frac{\mathrm{d}\theta_{\mathrm{v}}}{\mathrm{d}\eta_{\mathrm{v}}}\left(\frac{1}{4}Gr_{x\mathrm{v,s}}\right)^{1/4}x^{-1}\right]$$
$$=\lambda_{\mathrm{v}}(T_{\mathrm{w}}-T_{\mathrm{s}})\frac{\mathrm{d}^2\theta_{\mathrm{v}}}{\mathrm{d}\eta_{\mathrm{v}}}\left(\frac{1}{4}Gr_{x\mathrm{v,s}}\right)^{1/2}x^{-2}$$
$$+\frac{\mathrm{d}\lambda_{\mathrm{v}}}{\mathrm{d}\eta_{\mathrm{v}}}\left(\frac{1}{4}Gr_{x\mathrm{v,s}}\right)^{1/4}x^{-1}(T_{\mathrm{w}}-T_{\mathrm{s}})\frac{\mathrm{d}\theta_{\mathrm{v}}}{\mathrm{d}\nu_{\mathrm{v}}}\left(\frac{1}{4}Gr_{x\mathrm{v,s}}\right)^{1/4}x^{-1}.$$

This earlier equation is divided by $(T_{\mathrm{w}}-T_{\mathrm{s}})$ and is further simplified to

$$\rho_{\mathrm{v}}c_{p_{\mathrm{v}}}\left[-2\sqrt{gx}(\rho_{\mathrm{l,s}}/\rho_{\mathrm{v,w}}-1)^{1/2}W_{x\mathrm{v}}\frac{\mathrm{d}\theta_{\mathrm{v}}}{\mathrm{d}\eta_{\mathrm{v}}}\left(\frac{1}{4}\right)\eta_{\mathrm{v}}x^{-1}\right)$$
$$\left.+2\sqrt{gx}(\rho_{\mathrm{l,s}}/\rho_{\mathrm{v,w}})^{1/2}W_{y\mathrm{v}}\frac{\mathrm{d}\theta_{\mathrm{v}}}{\mathrm{d}\eta_{\mathrm{v}}}X^{-1}\right]$$
$$=+\lambda_{\mathrm{v}}\frac{\mathrm{d}^2\theta_{\mathrm{v}}}{\mathrm{d}\eta_{\mathrm{v}}}\left(\frac{1}{4}Gr_{x\mathrm{v,s}}\right)^{1/2}x^{-2}+\frac{\mathrm{d}\lambda_{\mathrm{v}}}{\mathrm{d}\eta_{\mathrm{v}}}\frac{\mathrm{d}\theta_{\mathrm{v}}}{\mathrm{d}\eta_{\mathrm{v}}}\left(\frac{1}{4}Gr_{x\mathrm{v,s}}\right)^{1/2}x^{-2}.$$

The earlier equation is divided by

$$\sqrt{\frac{g}{x}}(\rho_{\mathrm{l,s}}/\rho_{\mathrm{v,w}}-1)^{1/2},$$

and then is simplified to the following form with consideration of the definition of $Gr_{x\mathrm{v,s}}$:

$$\rho_{\mathrm{v}}c_{p_{\mathrm{v}}}\left[-W_{x\mathrm{v}}\frac{\mathrm{d}\theta_{\mathrm{v}}}{\mathrm{d}\eta_{\mathrm{v}}}\eta_{\mathrm{v}}+4W_{y\mathrm{v}}\frac{\mathrm{d}\theta_{\mathrm{v}}}{\mathrm{d}\eta_{\mathrm{v}}}\right]=\lambda_{\mathrm{v}}\frac{\mathrm{d}^2\theta_{\mathrm{v}}}{\mathrm{d}\eta_{\mathrm{v}}}\frac{1}{V_{\mathrm{v,s}}}+\frac{\mathrm{d}\lambda_{\mathrm{v}}}{\mathrm{d}\eta_{\mathrm{v}}}\frac{\mathrm{d}\theta_{\mathrm{v}}}{\mathrm{d}\eta_{\mathrm{v}}}\frac{1}{V_{\mathrm{v,s}}}.$$

The earlier equation is multiplied by $\nu_{\mathrm{v,s}}/\lambda_{\mathrm{v}}$ and simplified into

$$Pr_{\mathrm{v}}\frac{\nu_{\mathrm{v,s}}}{\nu_{\mathrm{v}}}(-\eta_{\mathrm{v}}W_{x\mathrm{v}}+4W_{y\mathrm{v}})\frac{\mathrm{d}\theta_{\mathrm{v}}}{\mathrm{d}\eta_{\mathrm{v}}}=\frac{\mathrm{d}^2\theta_{\mathrm{v}}}{\mathrm{d}\eta_{\mathrm{v}}}+\frac{1}{\lambda_{\mathrm{v}}}\frac{\mathrm{d}\lambda_{\mathrm{v}}}{\mathrm{d}\eta_{\mathrm{v}}}\frac{\mathrm{d}\theta_{\mathrm{v}}}{\mathrm{d}\eta_{\mathrm{v}}}, \tag{10.39}$$

where Pr_{v} is vapor Prandtl number defined as $Pr_{\mathrm{v}}=\mu_{\mathrm{v}}c_{p_{\mathrm{v}}}/\lambda_{\mathrm{v}}$.

For liquid film. Transformation of (10.4). With the (10.18)–(10.21) we get

$$\frac{\partial w_{x\mathrm{l}}}{\partial x} = \sqrt{\frac{g}{x}}\left(W_{\mathrm{lv}} - \frac{1}{2}\eta_l \frac{\mathrm{d}W_{x\mathrm{l}}}{\mathrm{d}\eta_\mathrm{l}}\right), \tag{10.40}$$

$$\frac{\partial w_{y\mathrm{v}}}{\partial y} = 2\sqrt{\frac{g}{x}}\frac{\mathrm{d}W_{y\mathrm{v}}}{\mathrm{d}\eta_\mathrm{v}}\Bigg) \tag{10.41}$$

With (10.40)–(10.41), (10.4) is changed into

$$2W_{x\mathrm{l}} - \eta_\mathrm{l}\frac{\mathrm{d}W_{x\mathrm{l}}}{\mathrm{d}\eta_\mathrm{l}} + 4\frac{\mathrm{d}W_{y\mathrm{l}}}{\mathrm{d}\eta_\mathrm{l}} = 0 \tag{10.42}$$

Transformation of (10.5). With (10.18)–(10.21) we obtain

$$\frac{\partial w_{x\mathrm{l}}}{\partial y} = 2\sqrt{gx}\frac{\mathrm{d}W_{x\mathrm{l}}}{\mathrm{d}\eta_\mathrm{l}}x^{-1}\left(\tfrac{1}{4}Gr_{x\mathrm{l,s}}\right)^{1/4} \tag{10.43}$$

$$\frac{\partial^2 w_{x\mathrm{l}}}{\partial y^2} = 2\sqrt{gx}\frac{\mathrm{d}^2W_{x\mathrm{l}}}{\mathrm{d}\eta_\mathrm{l}^2}x^{-1}\left(\tfrac{1}{4}Gr_{x\mathrm{l,s}}\right)^{1/4}\left(\tfrac{1}{4}Gr_{x\mathrm{l,s}}\right)^{1/4}x^{-1}$$

$$= 2\sqrt{gx}\frac{\mathrm{d}^2W_{x\mathrm{l}}}{\mathrm{d}\eta_\mathrm{l}^2}\left(\tfrac{1}{4}Gr_{x\mathrm{l,s}}\right)^{1/2}x^{-2}. \tag{10.44}$$

With (10.20), (10.21), (10.40), and (10.43)–(10.44), (10.5) will be simplified to

$$\left(2\sqrt{gx}W_{x\mathrm{l}}\sqrt{\frac{g}{x}}\left(W_{\mathrm{lv}} - \frac{1}{2}\eta_\mathrm{l}\frac{\mathrm{d}W_{x\mathrm{l}}}{\mathrm{d}\eta_\mathrm{l}}\right) + 4\sqrt{gx}W_{y\mathrm{l}}\sqrt{gx}\frac{\mathrm{d}W_{x\mathrm{l}}}{\mathrm{d}\eta_\mathrm{l}}x^{-1} =\right.$$

$$2\nu_\mathrm{l}\sqrt{gx}\frac{\mathrm{d}^2W_{x\mathrm{l}}}{\mathrm{d}\eta_\mathrm{l}^2}\left(\tfrac{1}{4}Gr_{x\mathrm{l,s}}\right)^{1/2}x^{-2}.$$

The earlier equation is divided by g, and then is simplified to the following form with consideration of the definition of $Gr_{x\mathrm{l,s}}$,

$$W_{x\mathrm{l}}\left(2W_{\mathrm{lv}} - \eta_\mathrm{l}\frac{\mathrm{d}W_{x\mathrm{l}}}{\mathrm{d}\eta_\mathrm{l}}\right) + 4W_{y\mathrm{l}}\frac{\mathrm{d}W_{x\mathrm{l}}}{\mathrm{d}\eta_\mathrm{l}} = \frac{\nu_\mathrm{l}}{\nu_\mathrm{l,s}}\frac{\mathrm{d}^2W_{x\mathrm{l}}}{\mathrm{d}\eta_\mathrm{l}^2}.$$

Since $\nu_\mathrm{l} = \nu_\mathrm{l,s}$ for saturated liquid film, the earlier equation is simplified to

$$W_{x\mathrm{l}}\left(2W_{\mathrm{lv}} - \eta_\mathrm{l}\frac{\mathrm{d}W_{x\mathrm{l}}}{\mathrm{d}\eta_\mathrm{l}}\right) + 4W_{y\mathrm{l}}\frac{\mathrm{d}W_{x\mathrm{l}}}{\mathrm{d}\eta_\mathrm{l}} = \frac{\mathrm{d}^2W_{x\mathrm{l}}}{\mathrm{d}\eta_\mathrm{l}^2}. \tag{10.45}$$

Then, the governing ordinary partial differential equations for vapor and liquid film obtained by the similarity transformation are rewritten as follows:

$$2W_{x\mathrm{v}} - \eta_\mathrm{v}\frac{\mathrm{d}W_{x\mathrm{v}}}{\mathrm{d}\eta_\mathrm{v}} + 4\frac{\mathrm{d}W_{y\mathrm{v}}}{\mathrm{d}\eta_\mathrm{v}} - \frac{1}{\rho_\mathrm{v}}\frac{\mathrm{d}\rho_\mathrm{v}}{\mathrm{d}\eta_\mathrm{v}}(\eta_\mathrm{v}W_{x\mathrm{v}} - 4W_{y\mathrm{v}}) = 0, \tag{10.27}$$

$$\frac{\nu_{\mathrm{v,s}}}{\nu_{\mathrm{v}}}\left(W_{x\mathrm{v}}\left(2W_{x\mathrm{v}}-\eta_{\mathrm{v}}\frac{\mathrm{d}W_{x\mathrm{v}}}{\mathrm{d}\eta_{\mathrm{v}}}\right)+4W_{y\mathrm{v}}\frac{\mathrm{d}W_{x\mathrm{v}}}{\mathrm{d}\eta_{\mathrm{v}}}\right)$$
$$=\frac{\mathrm{d}^2W_{x\mathrm{v}}}{\mathrm{d}\eta_{\mathrm{v}}^2}+\frac{1}{\mu_{\mathrm{v}}}\frac{\mathrm{d}\mu_{\mathrm{v}}}{\mathrm{d}\eta_{\mathrm{v}}}\frac{\mathrm{d}W_{x\mathrm{v}}}{\mathrm{d}\eta_{\mathrm{v}}}+\frac{\nu_{\mathrm{v,s}}}{\nu_{\mathrm{v}}}\frac{\rho_{\mathrm{v,w}}}{\rho_{\mathrm{v}}}\frac{\rho_{\mathrm{l,s}}-\rho_{\mathrm{v}}}{\rho_{l,\mathrm{s}}-\rho_{\mathrm{v,w}}}, \tag{10.32}$$

$$Pr_{\mathrm{v}}\frac{\nu_{\mathrm{v,s}}}{\nu_{\mathrm{v}}}(-\eta_{\mathrm{v}}W_{x\mathrm{v}}+4W_{y\mathrm{v}})\frac{\mathrm{d}\theta_{\mathrm{v}}}{\mathrm{d}\eta_{\mathrm{v}}}=\frac{\mathrm{d}^2\theta_{\mathrm{v}}}{\mathrm{d}\eta_{\mathrm{v}}}+\frac{1}{\lambda_{\mathrm{v}}}\frac{\mathrm{d}\lambda_{\mathrm{v}}}{\mathrm{d}\eta_{\mathrm{v}}}\frac{\mathrm{d}\theta_{\mathrm{v}}}{\mathrm{d}\eta_{\mathrm{v}}}, \tag{10.39}$$

$$2W_{x\mathrm{l}}-\eta_{\mathrm{l}}\frac{\mathrm{d}W_{x\mathrm{l}}}{\mathrm{d}\eta_{\mathrm{l}}}+4\frac{\mathrm{d}W_{y\mathrm{l}}}{\mathrm{d}\eta_{\mathrm{l}}}=0, \tag{10.42}$$

$$W_{x\mathrm{l}}\left(2W_{\mathrm{lv}}-\eta_{\mathrm{l}}\frac{\mathrm{d}W_{x\mathrm{l}}}{\mathrm{d}\eta_{\mathrm{l}}}\right)+4W_{y\mathrm{l}}\frac{\mathrm{d}W_{x\mathrm{l}}}{\mathrm{d}\eta_{\mathrm{l}}}=\frac{\mathrm{d}^2W_{x\mathrm{l}}}{\mathrm{d}\eta_{\mathrm{l}}^2}. \tag{10.45}$$

For Boundary conditions. With the assumed transformation variables the physical boundary conditions (10.6)–(10.12) are transformed equivalently to the following ones, respectively:

$$\eta_{\mathrm{v}}=0: \qquad W_{x\mathrm{v}}=0,\quad W_{y\mathrm{v}}=0,\quad \theta_{\mathrm{v}}=1, \tag{10.46}$$

$$\eta_{\mathrm{v}}=\eta_{\mathrm{v}\delta}\ \text{or}\ \eta_{\mathrm{l}}=0:$$

$$W_{x\mathrm{l,s}}=(\rho_{\mathrm{l,s}}/\rho_{\mathrm{v,w}}-1)^{1/2}W_{x\mathrm{v,s}}, \tag{10.47}$$

$$W_{y\mathrm{l,s}}=-0.25\frac{\mu_{v,\mathrm{s}}}{\mu_{\mathrm{l,s}}}\left(\frac{\nu_{\mathrm{l,s}}}{\nu_{\mathrm{v,s}}}\right)^{1/2}(\rho_{\mathrm{l,s}}/\rho_{\mathrm{v,w}}-1)^{1/4}(W_{x\mathrm{v,s}}\eta_{\mathrm{v}\delta}-4W_{y\mathrm{v,s}}), \tag{10.48}$$

$$\left(\frac{\mathrm{d}W_{x\mathrm{l}}}{\mathrm{d}\eta_{\mathrm{l}}}\right)_{\eta_{\mathrm{l}}=0}=\frac{\mu_{\mathrm{v,s}}}{\mu_{\mathrm{l,s}}}(\rho_{\mathrm{l,s}}/\rho_{\mathrm{v,w}}-1)^{3/4}\left(\frac{\nu_{\mathrm{l,s}}}{\nu_{\mathrm{v,s}}}\right)^{1/2}\left(\frac{\mathrm{d}W_{x\mathrm{v}}}{\mathrm{d}\eta_{\mathrm{v}}}\right)_{\eta_{\mathrm{v}}=\eta_{\mathrm{v}\delta}}, \tag{10.49}$$

$$h_{fg}\mu_{\mathrm{v,s}}(W_{x\mathrm{v,s}}\eta_{\mathrm{v}\delta}-4W_{y\mathrm{v,s}})+\lambda_{\mathrm{v,s}}(T_{\mathrm{w}}-T_{\mathrm{s}})\left(\frac{\mathrm{d}\theta_{\mathrm{v}}}{\mathrm{d}\eta_{\mathrm{v}}}\right)_{\eta_{\mathrm{v}}=\eta_{\mathrm{v}\delta}}=0, \tag{10.50}$$

$$\theta_{\mathrm{v}}=0. \tag{10.51}$$

$$\eta_{\mathrm{l}}\to\infty: \qquad W_{x\mathrm{l}}\to 0. \tag{10.52}$$

10.4 Numerical Calculation

10.4.1 Treatment of Variable Thermophysical Properties

Due to the larger superheated temperature of the surface, the treatment of variable thermophysical properties for vapor film must be performed. To this end, the temperature parameter method introduced in Chap. 4 [11] will be used for the treatment of variable thermophysical properties of the vapor medium. If the saturated temperature T_s is taken as reference temperature, the expressions of the thermophysical properties by means of the temperature parameter method can be taken as follows:

$$\frac{\mu_\mathrm{v}}{\mu_\mathrm{v,s}} = \left(\frac{T}{T_\mathrm{s}}\right)^{n_\mu}, \tag{10.53}$$

$$\frac{\lambda_\mathrm{v}}{\lambda_\mathrm{v,s}} = \left(\frac{T}{T_\mathrm{s}}\right)^{n_\lambda}, \tag{10.54}$$

$$\frac{\rho_\mathrm{v}}{\rho_\mathrm{v,s}} = \left(\frac{T}{T_\mathrm{s}}\right)^{-1}, \tag{10.55}$$

$$\frac{\nu_\mathrm{v}}{\nu_\mathrm{v,s}} = \left(\frac{T}{T_\mathrm{v,s}}\right)^{n_\mu+1}, \tag{10.56}$$

$$c_{p_\mathrm{v}}/c_{p_\mathrm{v,s}} = (T/T_\mathrm{s})^{n_{c_p}}. \tag{10.57}$$

In the governing ordinary differential equations (10.27), (10.32), and (10.39) for vapor film the physical property factors

$$\frac{1}{\rho_\mathrm{v}}\frac{\mathrm{d}\rho_\mathrm{v}}{\mathrm{d}\eta_\mathrm{v}}, \frac{1}{\mu_\mathrm{v}}\frac{\mathrm{d}\mu_\mathrm{v}}{\mathrm{d}\eta_\mathrm{v}}, \frac{1}{\lambda_\mathrm{v}}\frac{\mathrm{d}\lambda_\mathrm{v}}{\mathrm{d}\eta_\mathrm{v}} \text{ and } \frac{\nu_\mathrm{v,s}}{\nu_\mathrm{v}}$$

are involved. In order to solve these equations, these physical property factors must be transformed in the form of temperature and temperature gradient. From the assumed transformation variables defined in (10.13) and (10.15), as well as the expressions (10.53–10.57) for the temperature parameter method for vapor film medium, consulting the derivation in Chap. 4, we obtain the following equations for these physical property factors:

$$\frac{1}{\rho_\mathrm{v}}\frac{\mathrm{d}\rho_\mathrm{v}}{\mathrm{d}\eta_\mathrm{v}} = -\frac{(T_\mathrm{w}/T_\mathrm{s}-1)\mathrm{d}\theta_\mathrm{v}/\mathrm{d}\eta_\mathrm{v}}{(T_\mathrm{w}/T_\mathrm{s}-1)\theta_\mathrm{v}+1}, \tag{10.58}$$

$$\frac{1}{\mu_\mathrm{v}}\frac{\mathrm{d}\mu_\mathrm{v}}{\mathrm{d}\eta_\mathrm{v}} = \frac{n_\mu(T_\mathrm{w}/T_\mathrm{s}-1)\mathrm{d}\theta_\mathrm{v}/\mathrm{d}\eta_\mathrm{v}}{(T_\mathrm{w}/T_\mathrm{s}-1)\theta_\mathrm{v}+1} \tag{10.59}$$

$$\frac{1}{\lambda_\mathrm{v}}\frac{\mathrm{d}\lambda_\mathrm{v}}{\mathrm{d}\eta_\mathrm{v}} = \frac{n_\lambda(T_\mathrm{w}/T_\mathrm{s}-1)\mathrm{d}\theta_\mathrm{v}/\mathrm{d}\eta_\mathrm{v}}{(T_\mathrm{w}/T_\mathrm{s}-1)\theta_\mathrm{v}+1} \tag{10.60}$$

$$\frac{\nu_\mathrm{v,s}}{\nu_\mathrm{v}} = [(T_\mathrm{w}/T_\mathrm{s}-1)\theta_\mathrm{v}+1]^{-(n_\mu+1)}. \tag{10.61}$$

10.4.2 Numerical Calculation

The procedure chart of the calculation is shown in Fig. 10.2. The general procedure of the calculation with the theoretical model for the film boiling free convection of saturated vapor is described as follows: first the values of $\eta_{\mathrm{v}\delta}$ and $W_{x\mathrm{v,s}}$ of the vapor film at the vapor–liquid interface are guessed. These two values combined with (10.46) and (10.51) allow us to solve the governing

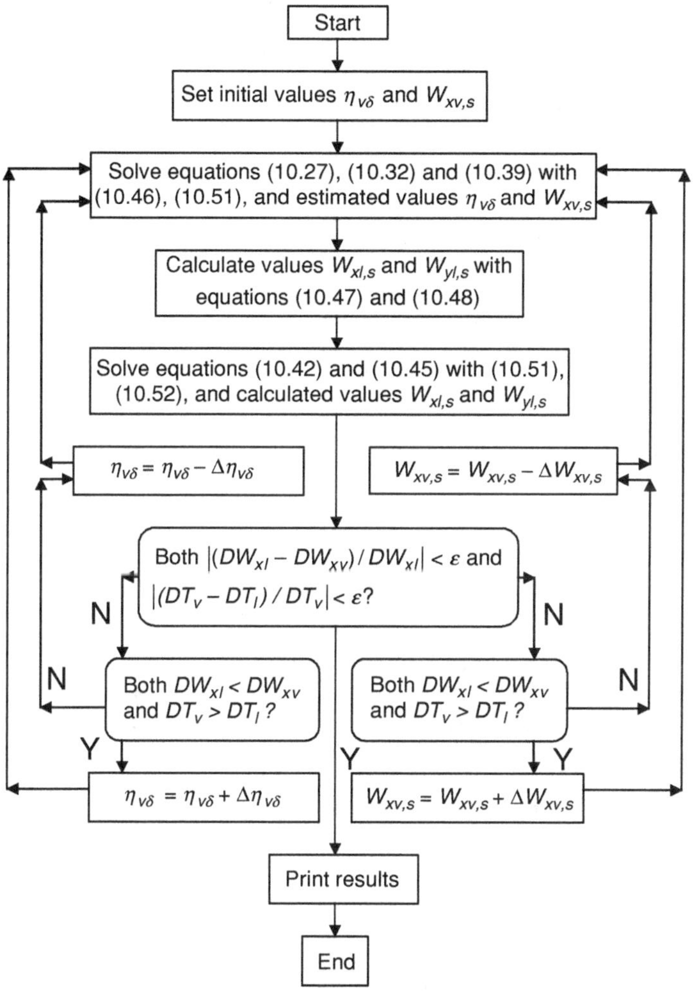

Fig. 10.2. The procedure chart of calculation, cited from Shang, Wang, and Zhong [10]

equations (10.27), (10.32), and (10.39) for vapor film by using the shooting method presented in Chap. 4. From the solutions we can gain two values: $W_{yv,s}$ and $(\mathrm{d}W_{xv}/\mathrm{d}\eta_v)_s$ at the vapor–liquid interface. With the values $\eta_{v\delta}$, $W_{xv,s}$, and $W_{yv,s}$, the values of $W_{xl,s}$ and $W_{yl,s}$ can be calculated from the corresponding boundary condition equations, (10.47) and (10.48). Then, the values $W_{xl,s}$ and $W_{yl,s}$ together with the boundary conditions (10.51) and (10.52) are used to solve the governing equations for liquid film (10.42) and (10.45) by using the shooting method. The solutions will yield the values of $(\mathrm{d}W_{xl}/\mathrm{d}\eta_l)_s$. Equations (10.49) and (10.50) are taken to adjust the convergence of the solutions for the two-phase boundary governing equations. Thus, the calculation is successively iterated by changing the values of $\eta_{v\delta}$ and $W_{xv,s}$.

Table 10.1. Density of water vapor of different temperature at atmospheric pressure

t (°C)	377	477	577	677	827	927
ρ (kg m^{-3})	0.338	0.2931	0.2579	0.2312	0.1996	0.1830

Table 10.2. Physical values of saturated water and water vapor at $t_s = 100$°C

phase	t (°C)	ρ (kg m^{-3})	ν (m^2 s^{-1})	λ (W (K m)$^{-1}$)	Pr
water	100	958.4	0.296×10^{-6}	0.6773	1.76
vapor	100	0.5974	20.55×10^{-6}	0.02478	1

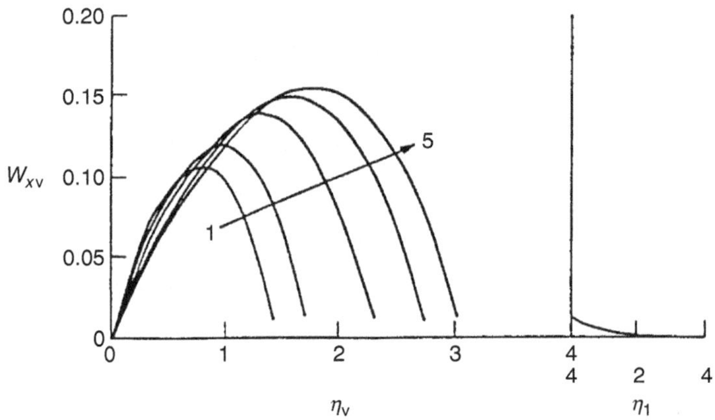

Fig. 10.3. Velocity profiles for laminar film boiling of saturated water, cited from Shang, Wang, and Zhong [10] 1–5: $\Delta t_w = 277, 377, 577, 727$, and 827°C, respectively

10.4.3 Numerical Results

As an example of application for solving the theoretical model, the numerical calculation of laminar film boiling of saturated water on an isothermal vertical plate was carried out. From Chap. 4 we know that the temperature parameters n_μ and n_λ of water vapor are 1.04 and 1.185. By using the earlier procedure, the numerical calculations have been done at wall superheated temperatures $\Delta t_w = 277, 377, 477, 577, 727$, and 827°C. The densities of water vapor at the earlier specified temperatures and the thermophysical properties of saturated water at $t_s = 100$°C needed in the calculations, are taken from [12] and given in Tables 10.1 and 10.2, respectively. Some numerical results of velocity and temperature profiles for the film boiling of saturated water are shown in Figs. 10.3 and 10.4, respectively.

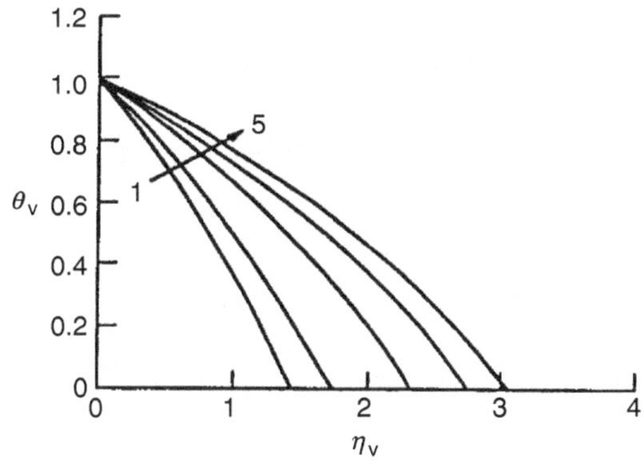

Fig. 10.4. Temperature profiles for laminar film boiling of saturated water, cited from Shang, Wang, and Zhong [10] 1–5: $\Delta t_w = 277, 377, 577, 727$, and 827°C, respectively

It is obviously known from Figs. 10.3 and 10.4 that the vapor film thickness will increase with increasing wall superheated temperature $\Delta t_w (= t_w - t_s)$. With increase of the wall superheated temperature Δt_w, the velocity profile of the vapor film will increase, the maximum velocity will shift to far from the wall, meanwhile, the temperature gradient on the wall, $\left(-\left(\frac{d\theta_v}{d\eta_v}\right)_{\eta_v=0}\right)_{\Delta t_\infty=0}$, will decrease. The effect of the wall superheated degree on the velocity and temperature fields of the vapor film reflects the influences of variable thermophysical properties of the vapor medium.

10.5 Heat Transfer

10.5.1 Heat Transfer Analysis

The local heat transfer rate at position x per unit area from the plate to the vapor film for the pool laminar film boiling of saturated liquid can be expressed by Fourier's law, i.e.,

$$q_x = -\lambda_{v,w}\left(\frac{\partial t}{\partial y}\right)_{y=0}.$$

Combining (10.13)–(10.15), q_x is rewritten as

$$\begin{aligned} q_x &= -\lambda_{v,w}\left(\frac{\partial((T_w - T_s)\theta_v + T_s)}{\partial y}\right)_{y=0} \\ &= -\lambda_{v,w}(T_w - T_s)(\frac{\partial\theta_v)}{\partial y})_{y=0} \\ &= \lambda_{v,w}(T_w - T_s)\left(\left(-\frac{\partial\theta_v)}{\partial\eta_v}\right)_{\eta=0}\right)_{\Delta t_\infty=0}\left(\frac{\partial\eta_v}{\partial y}\right)_{y=0}, \end{aligned}$$

where

$$\left(\frac{\partial \eta_v}{\partial y}\right)_{y=0} = \left(\frac{1}{4}Gr_{x\mathrm{v,s}}\right)^{1/4} x^{-1}.$$

Then,

$$q_x = \lambda_{\mathrm{v,w}}(T_{\mathrm{w}} - T_{\mathrm{s}})\left(\frac{1}{4}Gr_{x\mathrm{v,s}}\right)^{1/4} x^{-1}\left(\left(-\frac{\mathrm{d}\theta_{\mathrm{v}}}{\mathrm{d}\eta_{\mathrm{v}}}\right)_{\eta=0}\right)_{\Delta t_\infty=0}. \tag{10.62}$$

With the Newtonian cooling law defined as $q_x = \alpha_x(T_{\mathrm{w}} - T_{\mathrm{s}})$, the local heat transfer coefficient on the surface will be

$$\alpha_x = \lambda_{\mathrm{v,w}}\left(\frac{1}{4}Gr_{x\mathrm{v,s}}\right)^{1/4} x^{-1}\left(\left(-\frac{\mathrm{d}\theta_{\mathrm{v}}}{\mathrm{d}\eta_{\mathrm{v}}}\right)_{\eta_{\mathrm{v}}=0}\right)_{\Delta t_\infty=0}, \tag{10.63}$$

where the subscript $\Delta t_\infty (= t_s - t_\infty) = 0$ expresses the saturated state of bulk liquid.

Total heat transfer rate for position $x = 0$ to x with width of b on the plate is a integration, $Q_x = \iint\limits_A q_x\,\mathrm{d}A = \int_0^x q_x b\,\mathrm{d}x$, and hence

$$Q_x = \lambda_{\mathrm{v,w}} b(T_{\mathrm{w}} - T_\infty)\left(\left(-\frac{\mathrm{d}\theta_{\mathrm{v}}}{\mathrm{d}\eta_{\mathrm{v}}}\right)_{\eta_{\mathrm{v}}=0}\right)_{\Delta t_\infty=0}\int_0^x \left(\frac{1}{4}Gr_{x\mathrm{v,s}}\right)^{1/4} x^{-1}\,\mathrm{d}x.$$

With the definition of local Grashof number $Gr_{x\mathrm{v,s}}$ we obtain

$$Q_x = \frac{4}{3}b\lambda_{\mathrm{v,w}}(T_{\mathrm{w}} - T_\infty)\left(\frac{1}{4}Gr_{x,\mathrm{v,s}}\right)^{1/4}\left(-\left(\frac{\mathrm{d}\theta_{\mathrm{v}}}{\mathrm{d}\eta_{\mathrm{v}}}\right)_{\eta_{\mathrm{v}}=0}\right)_{\Delta t_\infty=0}. \tag{10.64}$$

The average heat transfer coefficient $\overline{\alpha_x}$ defined as $Q_x = \overline{\alpha_x}(T_{\mathrm{w}} - T_\infty) \times b \times x$ is expressed as

$$\overline{\alpha_x} = \frac{4}{3}\lambda_{\mathrm{v,w}}\left(\frac{1}{4}Gr_{x\mathrm{v,s}}\right)^{1/4} x^{-1}\left(\left(-\frac{\mathrm{d}\theta_{\mathrm{v}}}{\mathrm{d}\eta_{\mathrm{v}}}\right)_{\eta_{\mathrm{v}}=0}\right)_{\Delta t_\infty=0}. \tag{10.65}$$

The local Nusselt number, defined as $Nu_{x\mathrm{v,w}} = \alpha_x x/\lambda_{\mathrm{v,w}}$, is expressed by

$$Nu_{x\mathrm{v,w}} = \left(\frac{1}{4}Gr_{x\mathrm{v,s}}\right)^{1/4}\left(-\left(\frac{\mathrm{d}\theta_{\mathrm{v}}}{\mathrm{d}\eta_{\mathrm{v}}}\right)_{\eta_{\mathrm{v}}=0}\right)_{\Delta t_\infty=0.} \tag{10.66}$$

The average Nusselt number is defined as $\overline{Nu_{x\mathrm{v,w}}} = \overline{\alpha_x}x/\lambda_{\mathrm{v,w}}$ and hence

$$\overline{Nu_{x\mathrm{v,w}}} = \frac{4}{3}\left(\frac{1}{4}Gr_{x\mathrm{v,s}}\right)^{1/4}\left(\left(-\frac{\mathrm{d}\theta_{\mathrm{v}}}{\mathrm{d}\eta_{\mathrm{v}}}\right)_{\eta_{\mathrm{v}}=0}\right)_{\Delta t_\infty=0}.$$

It is seen that, for practical calculation of heat transfer, only $\left(\left(-\frac{d\theta_v}{d\eta_v}\right)_{\eta_v=0}\right)_{\Delta t_\infty=0}$ dependent on numerical solution is no-given variable.

Comparing the relationships between α_x and $\overline{\alpha}_x$ and between $Nu_{xv,w}$ and $\overline{Nu}_{xv,w}$, we have

$$\overline{\alpha_x} = \frac{4}{3}\alpha_x$$

$$\overline{Nu}_{xv,w} = \frac{4}{3}Nu_{xv,w}$$

10.5.2 Curve-fit Equation for Heat Transfer

Equations (10.63)–(10.66) show that the heat transfer is in direct proportion to the dimensionless temperature gradient $\left(-\left(\frac{d\theta_v}{d\eta_v}\right)_{\eta_v=0}\right)_{\Delta t_\infty=0}$ and the fourth power of the defined local Grashof number $Gr_{xv,s}$.

The solutions for the dimensionless temperature gradient $\left(-\left(\frac{d\theta_v}{d\eta_v}\right)_{\eta_v=0}\right)_{\Delta t_\infty=0}$ for the film boiling of saturated water on the vertical surface are obtained by calculating equations (10.27), (10.32), (10.39), (10.42), and (10.45) and their boundary conditions equations (10.46)–(10.52) and are shown in Table 10.3 and plotted in Fig. 10.5. On this basis, heat transfer for the film boiling of saturated water can be evaluated rigorously by using (10.62).

Furthermore, according to the rigorous numerical solutions of $\left(-\left(\frac{d\theta_v}{d\eta_v}\right)_{\eta_v=0}\right)_{\Delta t_\infty=0}$ in Table 10.3 the following curve-fit equation was obtained by Shang, Wang, and Zhong [10]:

$$\left(-\left(\frac{d\theta_v}{d\eta_v}\right)_{\eta_v=0}\right)_{\Delta t_\infty=0} = \frac{\exp\left(4.7356 + 0.1228\frac{\Delta t_w}{t_s} - 0.0086(\frac{\Delta t_w}{t_s})^2\right)}{\frac{\Delta t_w}{t_s}} \times 10^{-2}. \tag{10.67}$$

The results of $\left(-\left(\frac{d\theta_v}{d\eta_v}\right)_{\eta_v=0}\right)_{\Delta t_\infty=0}$ calculated by (10.67) are also described in Table 10.3 and plotted in Fig. 10.5, and agree very well with the corresponding rigorous numerical solutions. It follows that the curve-fit equation (10.67) is reliable for prediction of heat transfer of the film boiling of saturated water.

Table 10.3. Numerical solutions and predicted values of $\eta_{v\delta}$, $\left(\left(-\frac{d\theta_v}{d\eta_v}\right)_{\eta_v=0}\right)_{\Delta t_\infty=0}$, $W_{xv,s}$, $-W_{yv,s}$ and $(\eta_{v\delta}W_{xv,s}-4W_{yv,s})_{\Delta t_\infty=0}$ with $\Delta t_w/t_s$ for film boiling of saturated water on isothermal surface (parts of data are cited from Shang, Wang, and Zhong [10])

t_w(°C)	377	477	577	677	827	927
$\Delta t_w(=t_w-t_s)$(°C)	277	377	477	577	727	827
t_s(°C)	100	100	100	100	100	100
$\frac{\Delta t_w}{t_s}$	2.77	3.77	4.77	5.77	7.27	8.27
$\eta_{v\delta}$ (1)	1.437	1.728	2.019	2.308	2.745	3.038
$\eta_{v\delta}$ (2)	1.4371	1.7281	2.0191	2.3101	2.7466	3.0376
$\left(\left(-\frac{d\theta_v}{d\eta_v}\right)_{\eta_v=0}\right)_{\Delta t_\infty=0}$ (1)	0.5411	0.4285	0.3531	0.2997	0.2433	0.2158
$\left(\left(-\frac{d\theta_v}{d\eta_v}\right)_{\eta_v=0}\right)_{\Delta t_\infty=0}$ (2)	0.5410	0.4249	0.3528	0.3012	0.2429	0.2112
$W_{xv,s}$ (1)	0.01084	0.01080	0.01068	0.01057	0.01035	0.01020
$-W_{yv,s}$ (1)	0.05362	0.06569	0.07634	0.08598	0.09868	0.10625
$(\eta_{v\delta}W_{xv,s}-4W_{yv,s})_{\Delta t_\infty=0}$ (1)	0.23006	0.28142	0.32692	0.36832	0.42313	0.45599
$(\eta_{v\delta}W_{xv,s}-4W_{yv,s})_{\Delta t_\infty=0}$ (2)	0.23105	0.28147	0.32789	0.37031	0.42644	0.45886

(1) Numerical solution, (2) predicted value by the related curve-fit equation

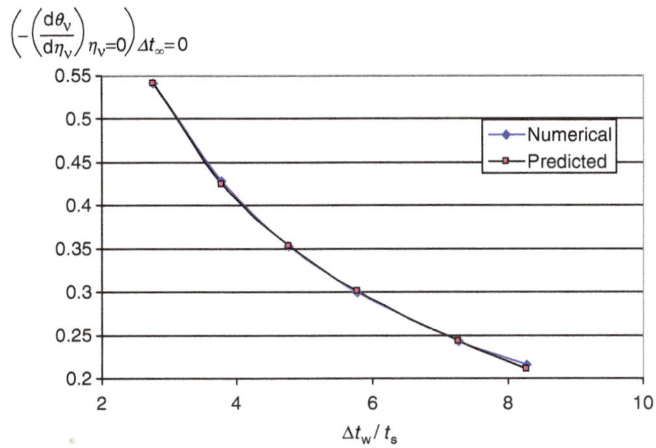

Fig. 10.5. The dimensionless temperature gradients $\left(-\left(\frac{d\theta_v}{d\eta_v}\right)_{\eta_v=0}\right)_{\Delta t_\infty=0}$ obtained from numerical solutions and results predicted by using (10.67), respectively, for the film condensation of saturated water on the isothermal plate

10.6 Mass Transfer

10.6.1 Mass Transfer Analysis

Let us set g_x to be a local mass flow rate entering the vapor film at position x per unit area of the plate. According to the boundary layer theory of fluid mechanics, g_x is expressed as

$$g_x = \rho_{v,s}\left(w_{xv,s}\frac{d\delta_v}{dx} - w_{yv,s}\right)_s .$$

With the corresponding dimensionless variables in (10.16) and (10.17), the earlier is changed into the following one:

$$g_x = \rho_{v,s}\left[2\sqrt{gx}\left(\frac{\rho_{l,s}}{\rho_{v,w}} - 1\right)^{1/2} W_{xv,s}\left(\frac{d\delta_v}{dx}\right)_s \right.$$
$$\left. -2\sqrt{gx}\left(\frac{\rho_{l,s}}{\rho_{v,w}} - 1\right)^{1/2}\left(\frac{1}{4}Gr_{xv,s}\right)^{-1/4} W_{yv,s}\right], \quad (10.68)$$

where the boiled vapor film thickness is expressed as follows according to (10.13):

$$\delta_v = \eta_{v\delta}\left(\frac{1}{4}Gr_{xv,s}\right)^{-1/4} x. \quad (10.69)$$

With the definition of the local Grashof number $Gr_{x\mathrm{v,s}}$, (10.69) is changed into

$$\delta_{\mathrm{v}} = \eta_{\mathrm{v}\delta}\left(\frac{1}{4}\frac{g(\rho_{\mathrm{l,s}}/\rho_{\mathrm{v,w}}-1)x^3}{\nu_{\mathrm{v,s}}^2}\right)^{-1/4} x,$$

or

$$\left(\frac{\mathrm{d}\delta_{\mathrm{v}}}{\mathrm{d}x}\right)_{\mathrm{s}} = \frac{1}{4}\eta_{\mathrm{v}\delta}\left(\frac{1}{4}Gr_{x\mathrm{v,s}}\right)^{-1/4}. \tag{10.70}$$

With (10.70), equation (10.68) will be simplified to

$$\begin{aligned}
g_x &= \rho_{\mathrm{v,s}}\left[2\sqrt{gx}\left(\frac{\rho_{\mathrm{l,s}}}{\rho_{\mathrm{v,w}}}-1\right)^{1/2} W_{x\mathrm{v,s}}\frac{1}{4}\eta_{\mathrm{v}\delta}\left(\frac{1}{4}Gr_{x\mathrm{v,s}}\right)^{-1/4}\right.\\
&\qquad \left. -2\sqrt{gx}\left(\frac{\rho_{\mathrm{l,s}}}{\rho_{\mathrm{v,w}}}-1\right)^{1/2}\left(\frac{1}{4}Gr_{x\mathrm{v,s}}\right)^{-1/4} W_{y\mathrm{v,s}}\right]_{\mathrm{s}}\\
&= \rho_{\mathrm{v,s}}\left[2\sqrt{gx}\left(\frac{\rho_{l,\mathrm{s}}}{\rho_{\mathrm{v,w}}}-1\right)^{1/2} W_{x\mathrm{v,s}}\left(\frac{\mathrm{d}\delta_{\mathrm{v}}}{\mathrm{d}x}\right)_{\mathrm{s}}\right.\\
&\qquad \left. -2\sqrt{gx}\left(\frac{\rho_{l,\mathrm{s}}}{\rho_{\mathrm{v,w}}}-1\right)^{1/2}\left(\frac{1}{4}Gr_{x\mathrm{v,s}}\right)^{-1/4} W_{y\mathrm{v,s}}\right]\\
&= 4\mu_{\mathrm{v,s}}x^{-1}\left(\frac{1}{4}Gr_{x\mathrm{v,s}}\right)^{1/2}\left(\frac{1}{4}Gr_{x\mathrm{v,s}}\right)^{-1/4}\left[\frac{1}{4}\eta_{\mathrm{v},\delta}W_{x\mathrm{v,s}} - W_{x\mathrm{v,s}}\right]_{\mathrm{s}}\\
&= \mu_{\mathrm{v,s}}x^{-1}\left(\frac{1}{4}Gr_{x\mathrm{v,s}}\right)^{1/4}\left(\eta_{\mathrm{v}\delta}W_{x\mathrm{v,s}} - 4W_{y\mathrm{v,s}}\right)_{\Delta t_\infty=0}.
\end{aligned} \tag{10.71}$$

If G_x is taken to express total mass flow rate entering the boundary layer for position $x=0$ to x with width of b of the plate, it should be the following integration:

$$\begin{aligned}
G_x &= \iint\limits_A (g_x)_{\mathrm{i}}\,\mathrm{d}A\\
&= b\int_0^x (g_x)_{\mathrm{i}}\,\mathrm{d}x,
\end{aligned} \tag{10.72}$$

where $A = b\cdot x$ is area of the plate.

With (10.71), G_x is expressed as

$$G_x = b\int_0^x \left[\mu_{\mathrm{v,s}}x^{-1}\left(\frac{1}{4}Gr_{x\mathrm{v,s}}\right)^{1/4}\left(\eta_{\mathrm{v}\delta}W_{x\mathrm{v,s}} - 4W_{y\mathrm{v,s}}\right)_{\Delta t_\infty=0}\right]\mathrm{d}x.$$

With the definition of the local Grashof number $Gr_{x\mathrm{v,s}}$, G_x should be

$$G_x = \frac{4}{3}b\cdot\mu_{\mathrm{v,s}}\left(\frac{1}{4}Gr_{x\mathrm{v,s}}\right)^{1/4}\left(\eta_{\mathrm{v}\delta}W_{x\mathrm{v,s}} - 4W_{y\mathrm{v,s}}\right)_{\Delta t_\infty=0}, \tag{10.73}$$

or in dimensionless form can be rewritten as

$$\frac{G_x}{b \cdot \mu_{\mathrm{v,s}}} = \frac{4}{3}\left(\frac{1}{4}Gr_{x\mathrm{v,s}}\right)^{1/4}(\eta_{\mathrm{v}\delta}W_{x\mathrm{v,s}} - 4W_{y\mathrm{v,s}}), \quad (10.74)$$

where $(\eta_{\mathrm{v}\delta}W_{x\mathrm{v,s}} - 4W_{y\mathrm{v,s}})_{\Delta t_\infty=0}$ is mass flow rate parameter of film boiling of saturated liquid.

10.6.2 Curve-Fit Formulae for Mass Transfer

It is found from (10.73) that mass flow rate G_x is direct proportional to mass flow rate parameter $(\eta_{\mathrm{v}\delta}W_{x\mathrm{v,s}} - 4W_{y\mathrm{v,s}})_{\Delta t_\infty=0}$. The rigorous numerical solutions of $\eta_{\mathrm{v}\delta}, W_{x\mathrm{v,s}}, -W_{y\mathrm{v,s}}$, and $(\eta_{\mathrm{v}\delta}W_{x\mathrm{v,s}} - 4W_{y\mathrm{v,s}})_{\Delta t_\infty=0}$ for film boiling free convection of saturated water on isothermal surface are listed in Table 10.3 and plotted in Figs. 10.6–10.9, respectively. It is seen that with increase of the wall superheated grade $\Delta t_{\mathrm{w}}/t_{\mathrm{s}}$, the values of $\eta_{\mathrm{v}\delta}, W_{y\mathrm{v,s}}$ and $(\eta_{\mathrm{v}\delta}W_{x\mathrm{v,s}} - 4W_{y\mathrm{v,s}})_{\Delta t_\infty=0}$ increase, but the value of $W_{x\mathrm{v,s}}$ decreases. Based on the rigorous numerical solutions, the following curve-fit equations are obtained for water vapor film thickness $\eta_{\mathrm{v}\delta}$ and mass flow rate parameter $(\eta_{\mathrm{v}\delta}W_{x\mathrm{v,s}} - 4W_{y\mathrm{v,s}})_{\Delta t_\infty=0}$, respectively:

$$\eta_{\mathrm{v}\delta} = 0.291\frac{\Delta t_{\mathrm{w}}}{t_{\mathrm{s}}} + 0.631, \quad (10.75)$$

$$(\eta_{\mathrm{v}\delta}W_{x\mathrm{v,s}} - 4W_{y\mathrm{v,s}})_{\Delta t_\infty=0} = -0.002\left(\frac{\Delta t_{\mathrm{w}}}{t_{\mathrm{s}}}\right)^2 + 0.0635\frac{\Delta t_{\mathrm{w}}}{t_{\mathrm{s}}} + 0.0705 \quad (10.76)$$

It is obvious that (10.75) can be used to simply and reliably predict the vapor film thickness for film boiling of saturated water. While, (10.76) with (10.73) can be used to predict the mass flow rate for film boiling of saturated water very accurately.

10.7 Remarks

From the analysis and calculated results, the following points can be concluded:

- The dimensionless velocity components W_x and W_y have been put forward further in this chapter. As W_x and W_y have definite physical meanings, the corresponding solutions of the models can be understood easily. Therefore, velocity component method has its special advantage over the traditional Falkner–Skan transformation for the theoretical and calculative models of the film boiling of saturated liquids.
- The numerical procedure shown in Fig. 10.2 is convenient and reliable for rigorous solutions of the theoretical models of the film boiling of saturated liquids with consideration of a system of physical conditions including variable thermophysical properties.

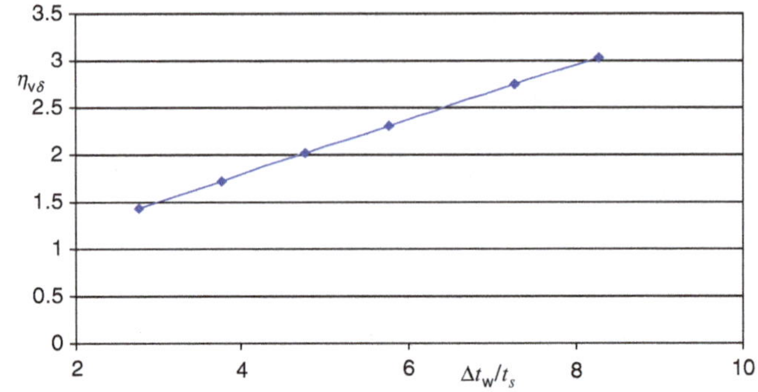

Fig. 10.6. Numerical solutions for $\eta_{v\delta}$ with $\Delta t_w/t_s$ for film boiling of saturated water on isothermal surface

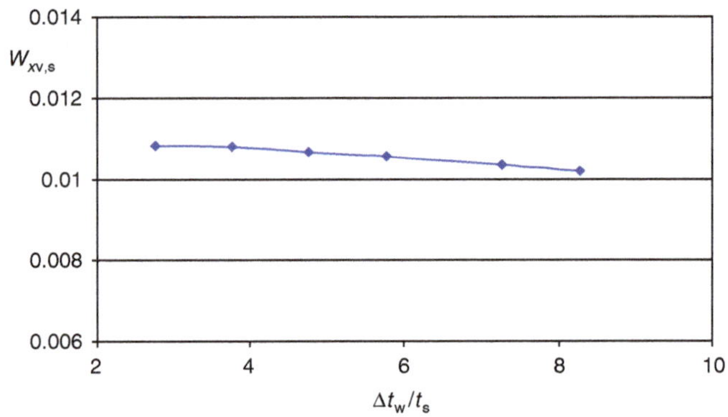

Fig. 10.7. Numerical solutions for $W_{xv,s}$ with $\frac{\Delta t_w}{t_s}$ for film boiling of saturated water on isothermal surface

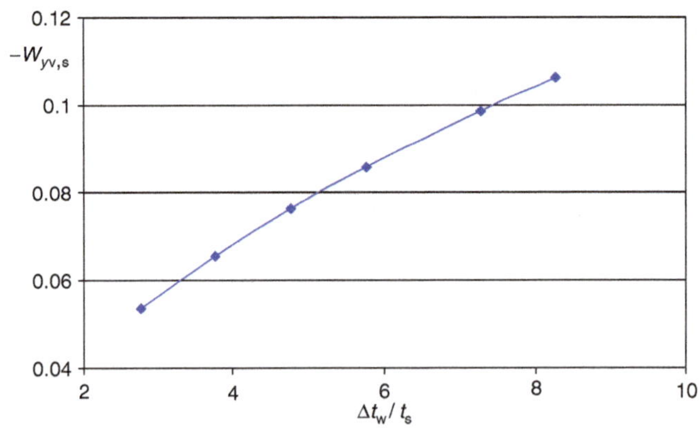

Fig. 10.8. Numerical solutions for $-W_{yv,s}$ with $\frac{\Delta t_w}{t_s}$ for film boiling of saturated water on isothermal surface

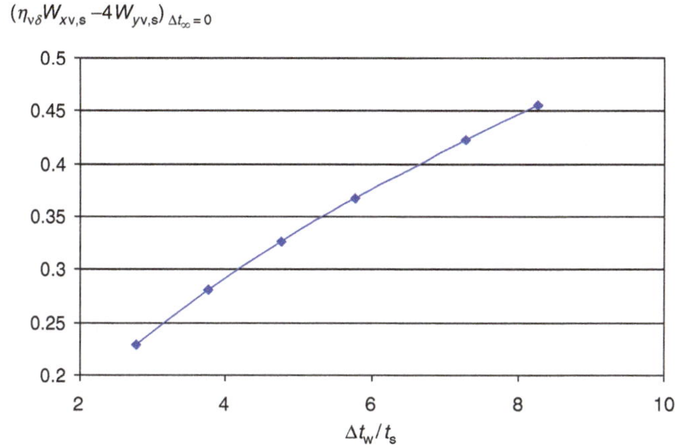

Fig. 10.9. Numerical solutions for $(\eta_{v\delta}W_{xv,s} - 4W_{yv,s})_{\Delta t_\infty=0}$ with with $\frac{\Delta t_w}{t_s}$ for film boiling saturated water on isothermal surface

- For the pool laminar film boiling of saturated water, the vapor film thickness $\eta_{v\delta}$, velocity component at the interface, $W_{yv,s}$, and mass flow rate parameter $(\eta_{v\delta}W_{xv,s} - 4W_{yv,s})_{\Delta t_\infty=0}$ increase with increasing increasing wall superheated degree $\Delta t_w (= t_w - t_s)$. Meanwhile, the temperature gradient $\left(-\left(\frac{d\theta_v}{d\eta_v}\right)_{\eta_v=0}\right)_{\Delta t_\infty=0}$ on the wall at the will decrease with increasing the wall superheated degree, $\Delta t_w (= t_w - t_s)$. Meanwhile, $W_{xv,s}$ will decrease very slightly with increasing $\Delta t_w (= t_w - t_s)$.
- The equations obtained by heat and mass transfer analysis are universally suitable to the film boiling of saturated liquids. The curve-fit equations (10.67), (10.75), and (10.76) are reliable for simple and reliable prediction of temperature gradient, vapor film thickness, and mass flow rate parameter, respectively, for laminar film boiling of saturated water. Of course, such prediction does not include radiative heat transfer across the vapor film, which needs to be further investigated.

10.8 Calculation Example

Example. A flat plate with 0.3 m in width and 0.3 m in length is suspended vertically in water. The plate temperature $t_w = 577°C$, and the water temperature is $t_\infty = t_s = 100°C$. Assume that the boiling is the laminar film boiling, please calculate:

1. Boiling heat and mass transfer of the plate,
2. Vapor film thicknesses at $x = 0, 0.01, 0.05, 0.1, 0.15, 0.2$, and 0.3 m

Solution. The wall superheated grade is

$$\Delta t_{\mathrm{w}}/t_{\mathrm{s}} = (t_{\mathrm{w}} - t_{\mathrm{s}})/t_{\mathrm{s}} = (577 - 100)/100 = 4.77,$$

and the water bulk subcooled grade is

$$\Delta t_{\infty}/t_{\mathrm{s}} = (t_{\mathrm{s}} - t_{\infty})/t_{\mathrm{s}} = 100 - 100/100 = 0,$$

which shows that it is the film boilng of saturated water. For water saturated physical properties at $t_{\mathrm{s}} = 100°\mathrm{C}$ we obtain $\rho_{\mathrm{l,s}} = 958.4\,\mathrm{kg\ m^{-3}}$, and for saturated water vapor at 100°C, we obtain $\nu_{\mathrm{v,s}} = 20.55 \times 10^{-6}\,\mathrm{m^2\ s^{-1}}, \rho_{\mathrm{v,s}} = 0.5974\ \mathrm{kg\ m^{-3}}, \mu_{\mathrm{v,s}} = 12.28 \times 10^{-6}$. In addition, for water vapor at the wall temperature $t_{\mathrm{w}} = 577°\mathrm{C}$ we obtain $\rho_{\mathrm{v,w}} = 0.2579\,\mathrm{kg\ m^{-3}}$ and $\lambda_{\mathrm{v,w}} = 0.00637\,\mathrm{kg\ m^{-3}}$.

(i) Calculate the condensate heat and mass transfer

For heat transfer. With (10.14) the local Grashof number is evaluated as

$$\begin{aligned} Gr_{x\mathrm{v,s}} &= \frac{g(\rho_{\mathrm{l,s}}/\rho_{\mathrm{v,w}} - 1)x^3}{\nu_{\mathrm{v,s}}^2} \\ &= \frac{9.8 \times (958.4/0.2579 - 1) \times 0.3^3}{(20.55 \times 10^{-6})^2} \\ &= 2.3278 \times 10^{12}. \end{aligned}$$

With (10.67), the dimensionless temperature gradient of the film boiling of saturated liquid is evaluated as

$$\begin{aligned} \left(-\left(\frac{\mathrm{d}\theta_{\mathrm{v}}}{\mathrm{d}\eta_{\mathrm{v}}}\right)_{\eta_{\mathrm{v}}=0}\right)_{\Delta t_{\infty}=0} &= \frac{\exp\left(4.7356 + 0.1228\frac{\Delta t_{\mathrm{w}}}{t_{\mathrm{s}}} - 0.0086(\frac{\Delta t_{\mathrm{w}}}{t_{\mathrm{s}}})^2\right)}{\frac{\Delta t_{\mathrm{w}}}{t_{\mathrm{s}}}} \times 10^{-2} \\ &= 10^{-2} \times \exp(4.7356 + 0.1228 \times 4.77 \\ &\qquad -0.0086 \times 4.77^2)/4.77 \\ &= 0.352807. \end{aligned}$$

With (10.64), the local Nusselt number is evaluated as

$$\begin{aligned} Nu_{x\mathrm{v,w}} &= -\left(\frac{1}{4}Gr_{x\mathrm{v,s}}\right)^{1/4}\left(\left(\frac{\mathrm{d}\theta}{\mathrm{d}\eta}\right)_{\eta=0}\right)_{\Delta t_{\infty}=0} \\ &= \left(\frac{1}{4} \times 2.3278 \times 10^{12}\right)^{1/4} \times 0.352807 \\ &= 305.02. \end{aligned}$$

With (10.66), the mean Nusselt number is evaluated as

$$\overline{Nu}_{x\mathrm{v,w}} = \frac{4}{3} Nu_{x\mathrm{v,w}} = \frac{4}{3} \times 305.02 = 406.69.$$

With definition of the mean Nusselt number, the mean heat transfer coefficient is evaluated as

$$\overline{\alpha}_x = \overline{Nu}_{x\mathrm{v,w}} \frac{\lambda_{\mathrm{v,w}}}{x} = 406.69 \times 0.0637/0.3 = 86.35 \ \mathrm{W\ (mK)^{-1}}.$$

With Newtonian cooling law, the total heat transfer rate of plate at the plate temperature $t_{\mathrm{w}} = 577°\mathrm{C}$ is calculated as follows:

$$\begin{aligned} Q_x &= \overline{\alpha_x}(t_{\mathrm{w}} - t_{\mathrm{s}})A \\ &= \overline{\alpha_x}(t_{\mathrm{w}} - t_{\mathrm{s}}) \times b \times x \\ &= 86.35 \times (577 - 100) \times 0.3 \times 0.3 \\ &= 3707 \ \mathrm{W}. \end{aligned}$$

For mass flow rate of the boiling. With (10.76), the mass flow rate parameter $(\eta_{\mathrm{v}\delta} W_{x\mathrm{v,s}} - 4W_{y\mathrm{v,s}})_{\Delta t_\infty = 0}$ of film boiling of saturated water at 100°C can be evaluated as

$$\begin{aligned} (\eta_{\mathrm{v}\delta} W_{x\mathrm{v,s}} - 4W_{y\mathrm{v,s}})_{\Delta t_\infty = 0} &= -0.002 \left(\frac{\Delta t_{\mathrm{w}}}{t_{\mathrm{s}}}\right)^2 + 0.0635 \frac{\Delta t_{\mathrm{w}}}{t_{\mathrm{s}}} + 0.0705 \\ &= -0.002 \times 4.77^2 + 0.0635 \times 4.77 + 0.0705 \\ &= 0.327889. \end{aligned}$$

Then, the total mass flow rate entering the boiled vapor film through the area with width of b and with length from $x = 0$ to x for the film boiling is

$$\begin{aligned} G_x &= \frac{4}{3} b \cdot \mu_{\mathrm{v,s}} \left(\frac{1}{4} Gr_{x\mathrm{v,s}}\right)^{1/4} (\eta_{\mathrm{v}\delta} W_{x\mathrm{v,s}} - 4W_{y\mathrm{v,s}})_{\Delta t_\infty = 0} \\ &= \frac{4}{3} \times 0.3 \times 12.28 \times 10^{-6} \times \left(\frac{1}{4} \times 2.3278 \times 10^{12}\right)^{1/4} \times 0.327889 \\ &= 0.001407 \ \mathrm{kg\ s^{-1}} \\ &= 5.0652 \ \mathrm{kg\ h^{-1}}. \end{aligned}$$

(ii) Calculate the vapor film thicknesses

Equation (10.75) is taken to evaluated $\eta_{l\delta}$ as

$$\begin{aligned} \eta_{\mathrm{v}\delta} &= 0.291 \frac{\Delta t_{\mathrm{w}}}{t_{\mathrm{s}}} + 0.631 \\ &= 0.291 \times 4.77 + 0.631 \\ &= 2.01907. \end{aligned}$$

Table 10.4. The variation of condensate film thickness y with the position x

x (m)	0	0.01	0.05	0.1	0.15	0.2	0.25	0.3
δ_v (m)	0	0.000296	0.000443	0.0005268	0.000583	0.000627	0.000663	0.000694

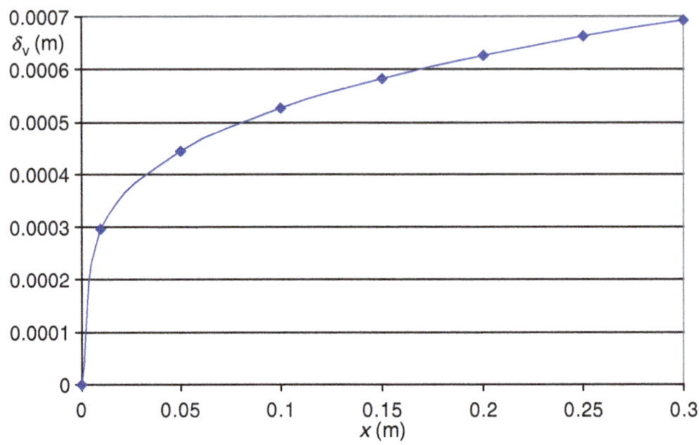

Fig. 10.10. The variation of condensate film thickness δ_v with the position x

From (12.13), the condensate film thickness δ_v is expressed as

$$
\begin{aligned}
\delta_V &= \eta_{\delta_V} x \left(\frac{1}{4} Gr_{xv,s}\right)^{-1/4} \\
&= \eta_{\delta_v} x \left(\frac{1}{4}\frac{g(\rho_{l,s}/\rho_{v,w}-1)x^3}{\nu_{v,s}^2}\right)^{-1/4} \\
&= \eta_{\delta_v} \left(\frac{1}{4}\frac{g(\rho_{l,s}/\rho_{v,w}-1)}{\nu_{v,s}^2}\right)^{-1/4} x^{1/4} \\
&= 2.01907 \times \left(\frac{1}{4} \times \frac{9.81 \times (958.4/0.2579-1)}{(20.55\times 10^{-6})^2}\right)^{-1/4} \times x^{1/4} \\
&= 0.00093707 \times x^{1/4}.
\end{aligned}
$$

For $x = 0$, $\delta_v = 0\,\mathrm{m}$

For $x = 0.01\,\mathrm{m}$, $\delta_v = 0.00093707 \times 0.05^{1/4} = 0.000296\,\mathrm{m}$

For $x = 0.05\,\mathrm{m}$, $\delta_v = 0.00093707 \times 0.01^{1/4} = 0.000443\,\mathrm{m}$

For $x = 0.1\text{m}$, $\delta_v = 0.00093707 \times 0.1^{1/4} = 0.0005268\,\text{m}$

For $x = 0.15\,\text{m}$, $\delta_v = 0.00093707 \times 0.15^{1/4} = 0.000583\,\text{m}$

For $x = 0.2\,\text{m}$, $\delta_v = 0.00093707 \times 0.2^{1/4} = 0.000627\,\text{m}$

For $x = 0.25\,\text{m}$, $\delta_v = 0.00093707 \times 0.25^{1/4} = 0.000663\,\text{m}$

For $x = 0.3\,\text{m}$, $\delta_v = 0.00093707 \times 0.3^{1/4} = 0.000694\,\text{m}$

For clear expression, the variation of condensate film thickness y with the position x is listed in Table 10.4 and plotted in Fig. 10.10.

References

1. L. Bromley, Heat transfer in stable film boiling, Chem. Eng. Prog. 46 (5), pp. 221–227, 1950
2. J.C.H. Koh, Analysis of film boiling on vertical surface, J. Heat Transfer, 84C (1), pp. 55–62, 1962
3. R.D. Cess, Analysis of laminar film boiling from a vertical flat plate, Research report 405 FF 340-R2-x, Westinghouse Research Laboratory, Pittsburgh, PA, March, 1959, surface, J. Heat Transfer
4. E.M. Sparrow and R.D. Cess, The effect of subcooled liquid on laminar film boiling, J. Heat Transfer, 84C (2), pp. 149–156, 1962
5. P.M. McFadden and R.J. Grosh, An analysis of laminar film boiling with variable properties, Int. J. Heat Mass Transfer, 1 (4), pp. 325–335, 1961
6. K. Nishikawa and T. Ito, Two-phase boundary-layer treatment of free convection film boiling, Int. J. Heat Mass Transfer, 9 (2), pp. 103–115, 1966
7. K. Nishikawa, T. Ito and K. Matsumoto, Investigation of variable thermophysical property problem concerning pool film boiling from vertical plate with prescribed uniform temperature, Int. J. Heat Mass Transfer, 19 (10), pp. 1173–1182, 1976
8. D.Y. Shang, L.C. Zhong and B.X. Wang, Study of film boiling of water with variable thermophysical properties along a vertical plate, transfer phenomena, Science and Technology, 1992 (Ed. by B.X. Wang), Higher Education Press, pp. 427–432, 1992
9. L.C. Zhong, D.Y. Shang and B.X. Wang, Effect of variable thermophysical properties on film boiling of liquid. J. Northeastern Univ., 15 (88), pp. 6–11, 1994
10. D.Y. Shang, B.X. Wang and L.C. Zhong, A study on laminar film boiling of liquid along an isothermal vertical plates in a pool with consideration of variable thermophysical properties, Int. J. Heat Mass Transfer 37 (5), pp. 819–828, 1994
11. D.Y. Shang and B.X. Wang, Effect of variable thermophysical properties on laminar free convection of gas, Int. J. Heat Mass Transfer 33 (7), pp. 1387–1395, 1990
12. VDI – Warmeatlas, Berechnungs blätter fur Warmeubertragung, 5, erweiterte Auflage, VDI Verlage GmbH, Dusseldorf, 1988

11

Laminar Film Boiling of Subcooled Liquid

a	thermal diffusive coefficient, $\mathrm{m^2\,s^{-1}}$
b	width of plate, m
c_p	specific heat at constant pressure, $\mathrm{J\,(kg\ K)^{-1}}$
g	gravitation acceleration, $\mathrm{m\,s^{-2}}$
$Gr_{x\mathrm{l},\infty}$	local Grashof number of liquid film for film boiling of subcooled liquid, $\frac{g(\rho_{\mathrm{l},\infty}/\rho_{\mathrm{l,s}}-1)x^3}{\nu_{\mathrm{l},\infty}^2}$
$Gr_{x\mathrm{v,s}}$	local Grashof number of vapor film for film boiling of subcooled liquid, $\frac{g(\rho_{\mathrm{l},\infty}/\rho_{\mathrm{v,w}}-1)x^3}{\nu_{\mathrm{v,s}}^2}$
g_x	local mass flow rate entering the vapor film at position x per unit area of the plate, $\mathrm{kg\,(m^2\,s)^{-1}}$
G_x	total mass flow rate entering the vapor film for position $x=0$ to x with width b of the plate, $\mathrm{kg\,s^{-1}}$
$Nu_{x\mathrm{v,w}}$	local Nusselt number for film boiling when wall temperature t_w is taken as reference temperature, $\alpha_x x/\lambda_\mathrm{w}$
$\overline{Nu}_{x\mathrm{v,w}}$	average Nusselt number, $\overline{\alpha}_x x/\lambda_\mathrm{w}$
n_{c_p}	specific heat parameter of gas
n_λ	thermal conductivity parameter of gas
n_μ	viscosity parameter of gas
Pr	Prandtl number
q_x	local heat transfer rate at position x per unit area on the plate, $\mathrm{W\,m^{-2}}$
Q	total heat transfer rate for position $x=0$ to x with width b on the plate, W
t	temperature, °C
T	absolute temperature, K
w_x, w_y	velocity components in the x- and y-directions, respectively, $\mathrm{m\,s^{-1}}$
W_x, W_y	dimensionless velocity components in the x- and y-directions, respectively

Greek symbols

α_x	local heat transfer coefficient, W (m^2 K)$^{-1}$
$\overline{\alpha}_x$	average heat transfer coefficient, W (m^2 K)$^{-1}$
δ	boundary layer thickness, m
δ_l	thickness of liquid film, m
δ_v	thickness of vapor film, m
η	dimensionless coordinate variable for boundary layer
θ_v	dimensionless temperature of vapor film, $\frac{T_\mathrm{v}-T_\mathrm{s}}{T_\mathrm{w}-T_\mathrm{s}}$
θ_l	dimensionless temperature of liquid film, $\frac{t_\mathrm{l}-t_\infty}{t_\mathrm{s}-t_\infty}$
λ	thermal conductivity, W (m K)$^{-1}$
μ	absolute viscosity, kg (m s)$^{-1}$
ν	kinetic viscosity, m^2 s^{-1}
ρ	density, kg m^{-3}
Δt_w	wall superheated temperature, $t_\mathrm{w} - t_\mathrm{s}$, °C
$\frac{\Delta t_\mathrm{w}}{t_\mathrm{s}}$	wall superheated grade
Δt_∞	subcooled temperature of liquid bulk, $t_\mathrm{s} - t_\infty$, °C
$\frac{\Delta t_\infty}{t_\mathrm{s}}$	subcooled grade of liquid bulk
$-\left(\frac{\mathrm{d}\theta_\mathrm{v}}{\mathrm{d}\eta_\mathrm{v}}\right)_{\eta_\mathrm{v}=0}$	dimensionless temperature gradient on the plate for film boiling of subcooled liquid
$\eta_{\mathrm{v}\delta}$	dimensionless thickness of vapor film
$\eta_{\mathrm{v}\delta} W_{x\mathrm{v,s}} - 4W_{y\mathrm{v,s}}$	mass flow rate parameter for film boiling
$\frac{\rho_\mathrm{v,w}}{\rho_\mathrm{v}} \frac{\rho_{\mathrm{l},\infty} - \rho_\mathrm{v}}{\rho_{\mathrm{l},\infty} - \rho_\mathrm{v,w}}$	buoyancy factor of film boiling
$\frac{1}{\rho_\mathrm{v}} \frac{\mathrm{d}\rho_\mathrm{v}}{\mathrm{d}\eta_\mathrm{v}}$	density factor
$\frac{1}{\mu_\mathrm{v}} \frac{\mathrm{d}\mu_\mathrm{v}}{\mathrm{d}\eta_\mathrm{v}}$	viscosity factor
$\frac{1}{\lambda_\mathrm{v}} \frac{\mathrm{d}\lambda_\mathrm{v}}{\mathrm{d}\eta_\mathrm{v}}$	thermal conductivity factor
$\frac{\nu_\mathrm{v,s}}{\nu_\mathrm{v}}$	kinetic viscosity factor

Subscripts

i	liquid film
s	saturated state, or at the vapor–liquid interface
v	vapor film
w	at wall
∞	far from the wall surface

11.1 Introduction

In this chapter the studies of film boiling free convection of subcooled liquid provided by Shang, Wang and Zhong [1–3] are further presented. Rigorous theoretical models for film boiling of subcooled liquid along an isothermal vertical plate are further presented. Meanwhile, all matching conditions including the variable thermophysical properties of vapor and liquid films are taken into account. The method for the numerical calculations of the two-phase boundary-layer problem is further developed. The theoretical models have been rigorously solved by the successive iterative procedure at different wall superheated degree with different liquid subcooled degree for the film boiling of subcooled water. The distributions of velocity and temperature fields of the film boiling of subcooled water are rigorously evaluated at different conditions. The theoretical equations of heat transfer coefficient and mass flow rate are derived, including the curve-fit equations for prediction of the temperature gradient on the wall.

11.2 Governing Partial Differential Equations

In Chap. 10 we have given the theoretical model and the similarity transformation for film boiling of saturated liquid. In this chapter we further introduce the theoretical model and the similarity transformation for the film boiling of subcooled liquid. The analytical model and coordinating system used in this problem are shown in Fig. 11.1. The heated plate with uniform temperature T_{w} is submerged vertically in stagnant liquid whose temperature is higher than the saturated temperature T_{s}, while the bulk temperature T_{∞} is less than the saturated temperature T_{s}. We assume that the heating surface of the plate is covered with continuous laminar vapor film, which moves upward together with the vapor. Thus a two-phase boundary layer is formed. Heat flux produced from the heating surface of the plate transfers through the two-phase boundary layer to the bulk liquid. Meanwhile, mass transfer is produced at the vapor–liquid interface due to the film boiling of the liquid.

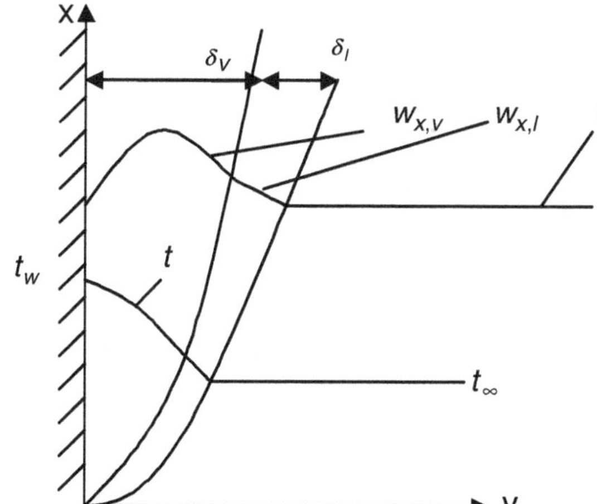

Fig. 11.1. Physical model and coordinate system of film boiling of subcooled liquid

For vapor film. The governing partial differential equations for the vapor film of the film boiling of subcooled liquid are as follows:

$$\frac{\partial}{\partial x}(\rho_{\mathrm{v}} w_{x\mathrm{v}}) + \frac{\partial}{\partial y}(\rho_{\mathrm{v}} w_{y\mathrm{v}}) = 0, \tag{11.1}$$

$$\rho_{\mathrm{v}}\left(w_{x\mathrm{v}}\frac{\partial w_{x\mathrm{v}}}{\partial x} + w_{y\mathrm{v}}\frac{\partial w_{x\mathrm{v}}}{\partial y}\right) = \frac{\partial}{\partial y}\left(\mu_{\mathrm{v}}\frac{\partial w_{x\mathrm{v}}}{\partial y}\right) + g(\rho_{\mathrm{l},\infty} - \rho_{\mathrm{v}}), \tag{11.2}$$

$$\rho_{\mathrm{v}} c_{p_{\mathrm{v}}}\left(w_{x\mathrm{v}}\frac{\partial T_{\mathrm{v}}}{\partial x} + w_{y\mathrm{v}}\frac{\partial T_{\mathrm{v}}}{\partial y}\right) = \frac{\partial}{\partial y}\left(\lambda_{\mathrm{v}}\frac{\partial T_{\mathrm{v}}}{\partial y}\right). \tag{11.3}$$

On comparing (11.1)–(11.3) with (10.1) and (10.2) it is found that their difference is that liquid density $\rho_{\mathrm{l},\infty}$ at the liquid bulk temperature is used in the buoyancy term $g(\rho_{\mathrm{l},\infty} - \rho_{\mathrm{v}})$ of (11.2), instead of liquid density $\rho_{\mathrm{l,s}}$ at the saturated temperature.

For liquid film. For liquid film, because the temperature of the bulk liquid is lower than its saturated one, the thermal boundary layer of liquid appears together besides the velocity boundary layer. Also, the variable thermophysical properties must be considered in the following governing partial differential equations for the liquid film:

$$\frac{\partial}{\partial x}(\rho_{\mathrm{l}} w_{x\mathrm{l}}) + \frac{\partial}{\partial y}(\rho_{\mathrm{l}} w_{y\mathrm{l}}) = 0, \tag{11.4}$$

$$\rho_{\mathrm{l}}\left(w_{x\mathrm{l}}\frac{\partial w_{x\mathrm{l}}}{\partial x} + w_{y\mathrm{l}}\frac{\partial w_{x\mathrm{l}}}{\partial y}\right) = \frac{\partial}{\partial y}\left(\mu_{\mathrm{l}}\frac{\partial w_{x\mathrm{l}}}{\partial y}\right) + g(\rho_{\mathrm{l},\infty} - \rho_{\mathrm{l}}), \tag{11.5}$$

$$\rho_{\mathrm{l}} c_{p_{\mathrm{l}}}\left(w_{x\mathrm{l}}\frac{\partial t_{\mathrm{l}}}{\partial x} + w_{y\mathrm{l}}\frac{\partial t_{\mathrm{l}}}{\partial y}\right) = \frac{\partial}{\partial y}\left(\lambda_{\mathrm{l}}\frac{\partial t_{\mathrm{l}}}{\partial y}\right). \tag{11.6}$$

For boundary conditions. For the boundary conditions, the variable thermophysical properties of both liquid and vapor films will be considered:

$$y = 0: \tag{11.7}$$

$$w_{x\mathrm{v}} = 0, \quad w_{y\mathrm{v}} = 0, \quad T = T_{\mathrm{w}};$$

$$y = \delta_{\mathrm{v}}:$$

$$w_{x\mathrm{v,s}} = w_{x\mathrm{l,s}}, \tag{11.8}$$

$$\rho_{\mathrm{v,s}}\left(w_{x\mathrm{v}}\frac{\partial \delta_{\mathrm{v}}}{\partial x} - w_{y\mathrm{v}}\right)_{\mathrm{s}} = \rho_{\mathrm{l,s}}\left(w_{x\mathrm{l}}\frac{\partial \delta_{\mathrm{l}}}{\partial x} - w_{y\mathrm{l}}\right)_{\mathrm{s}}, \tag{11.9}$$

$$\mu_{\mathrm{v,s}}\left(\frac{\partial w_{x\mathrm{v}}}{\partial y}\right)_{\mathrm{s}} = \mu_{\mathrm{l,s}}\left(\frac{\partial w_{x\mathrm{l}}}{\partial y}\right)_{\mathrm{s}}, \tag{11.10}$$

$$-\lambda_{\mathrm{v,s}}\left(\frac{\partial t_{\mathrm{v}}}{\partial y}\right)_{y=\delta_{\mathrm{v}}} = h_{fg}\rho_{\mathrm{v,s}}\left(w_{x\mathrm{v}}\frac{\partial \delta_{x\mathrm{v}}}{\partial x} - w_{y\mathrm{v}}\right)_{\mathrm{s}} - \lambda_{\mathrm{l,s}}\left(\frac{\partial t_{\mathrm{l}}}{\partial y}\right)_{y=\delta_{\mathrm{v}}}, \tag{11.11}$$

$$T_{\mathrm{v}} = T_{\mathrm{s}}, \qquad t_{\mathrm{l}} = t_{\mathrm{s}}; \tag{11.12}$$

$y \to \infty$:

$$w_{x\mathrm{l}} \to 0, \quad t_{\mathrm{l}} \to t_{\infty}; \tag{11.13}$$

where (11.8)–(11.12) express the physical matching conditions such as velocity, local mass flux, shear force, heat flux, and temperature at the vapor–liquid interface, respectively. It is seen that in the energy balance equation (11.11), the heat conduction on the liquid film is further considered.

11.3 Similarity Transformation

For the similarity transformation of the partial differential equations of the film boiling of subcooled liquid, the velocity component method will be further used on the basis of that for the film boiling of saturated liquid introduced in Chap. 10. The similarity transformation is introduced, respectively, as follows:

11.3.1 Transformation Variables

For vapor film. For vapor film of the film boiling of subcooled liquid, the similarity transformation variables are assumed as follows:

η_{v} is set up at first as the dimensionless coordinate variable

$$\eta_{\mathrm{v}} = \left(\frac{1}{4} Gr_{x\mathrm{v,s}}\right)^{1/4} \frac{y}{x}, \tag{11.14}$$

where the local Grashof number $Gr_{x\mathrm{v,s}}$ is assumed as

$$Gr_{x\mathrm{v,s}} = \frac{g(\rho_{\mathrm{l},\infty}/\rho_{\mathrm{v,w}} - 1)x^3}{\nu_{\mathrm{v,s}}^2}. \tag{11.15}$$

The dimensionless temperature is given as

$$\theta_{\mathrm{v}} = \frac{T_{\mathrm{v}} - T_{\mathrm{s}}}{T_{\mathrm{w}} - T_{\mathrm{s}}}. \tag{11.16}$$

The dimensionless velocity components are assumed as

$$W_{x\mathrm{v}} = (2\sqrt{gx}(\rho_{\mathrm{l},\infty}/\rho_{\mathrm{v,w}} - 1)^{1/2})^{-1} w_{x\mathrm{v}}, \tag{11.17}$$

$$W_{y\mathrm{v}} = \left(2\sqrt{gx}(\rho_{\mathrm{l},\infty}/\rho_{\mathrm{v,w}} - 1)^{1/2}\left(\frac{1}{4} Gr_{xv,s}\right)^{-1/4}\right)^{-1} w_{y\mathrm{v}}. \tag{11.18}$$

It is found that the similarity transformation variables in (11.14)–(11.18) for vapor film of the film boiling of subcooled liquid are almost same as those for that of the film boiling of saturated liquid shown in Chap. 10. However the liquid density here in the buoyancy factor $\rho_{\mathrm{l},\infty}/\rho_{\mathrm{v,w}} - 1$ is $\rho_{\mathrm{l},\infty}$ instead of $\rho_{\mathrm{l,s}}$.

For liquid film. For similarity transformation of the governing equation for subcooled liquid film the following transformation variables are set up:

For liquid film the dimensionless coordinate variable η_l is set up at first as follows:

$$\eta_\mathrm{l} = \left(\frac{1}{4} Gr_{x\mathrm{l},\infty}\right)^{1/4} \frac{y}{x}, \tag{11.19}$$

where the local Grashof number $Gr_{x\mathrm{l},\infty}$ is assumed as

$$Gr_{x\mathrm{l},\infty} = \frac{g(\rho_{\mathrm{l},\infty}/\rho_{\mathrm{l,s}} - 1)x^3}{\nu_{\mathrm{l},\infty}^2} \tag{11.20}$$

The dimensionless temperature is given as

$$\theta_\mathrm{l} = \frac{t_\mathrm{l} - t_\infty}{t_\mathrm{s} - t_\infty}. \tag{11.21}$$

The dimensionless velocity components are assumed as

$$W_{x\mathrm{l}} = (2\sqrt{gx}(\rho_{\mathrm{l},\infty}/\rho_{\mathrm{l,s}} - 1)^{1/2})^{-1} w_{x\mathrm{l}}, \tag{11.22}$$

$$W_{y\mathrm{l}} = \left(2\sqrt{gx}(\rho_{\mathrm{l},\infty}/\rho_{\mathrm{l,s}} - 1)^{1/2} \left(\frac{1}{4} Gr_{x\mathrm{l},\infty}\right)^{-1/4}\right)^{-1} w_{y\mathrm{l}}. \tag{11.23}$$

11.3.2 Similarity Transformation

After the assumptions of the dimensionless variables, the similarity transformations of the governing partial differential equations for vapor and liquid films and the equations for their boundary conditions will be done as follows:

For vapor film. The similarity transformation of (11.1)–(11.3) for vapor film of film boiling of subcooled liquid is similar to that shown in Chap. 10 for vapor film of film boiling of saturated liquid. Similar to the derivation of Chap. 10, (11.1)–(11.3) are, respectively, transformed to following dimensionless ones by using the variables assumed in (11.14)–(11.18),

$$2W_{x\mathrm{v}} - \eta_\mathrm{v} \frac{\mathrm{d}W_{x\mathrm{v}}}{\mathrm{d}\eta_\mathrm{v}} + 4\frac{\mathrm{d}W_{y\mathrm{v}}}{\mathrm{d}\eta_\mathrm{v}} - \frac{1}{\rho_\mathrm{v}} \frac{\mathrm{d}\rho_\mathrm{v}}{\mathrm{d}\eta_\mathrm{v}} (\eta_\mathrm{v} W_{x\mathrm{v}} - 4W_{y\mathrm{v}}) = 0, \tag{11.24}$$

$$\frac{\nu_\mathrm{v,s}}{\nu_\mathrm{v}} \left(W_{x\mathrm{v}} \left(2W_{x\mathrm{v}} - \eta_\mathrm{v} \frac{\mathrm{d}W_{x\mathrm{v}}}{\mathrm{d}\eta_\mathrm{v}}\right) + 4W_{y\mathrm{v}} \frac{\mathrm{d}W_{x\mathrm{v}}}{\mathrm{d}\eta_\mathrm{v}}\right) = \frac{\mathrm{d}^2 W_{x\mathrm{v}}}{\mathrm{d}\eta_\mathrm{v}^2} + \frac{1}{\mu_\mathrm{v}} \frac{\mathrm{d}\mu_\mathrm{v}}{\mathrm{d}\eta_\mathrm{v}} \frac{\mathrm{d}W_{x\mathrm{v}}}{\mathrm{d}\eta_\mathrm{v}} + \frac{\nu_\mathrm{v,s}}{\nu_\mathrm{v}} \frac{\rho_\mathrm{v,w}}{\rho_\mathrm{v}} \frac{\rho_{\mathrm{l},\infty} - \rho_\mathrm{v}}{\rho_{\mathrm{l},\infty} - \rho_\mathrm{v,w}}, \tag{11.25}$$

$$Pr_\mathrm{v} \frac{\nu_\mathrm{v,s}}{\nu_\mathrm{v}} (-\eta_\mathrm{v} W_{x\mathrm{v}} + 4W_{y\mathrm{v}}) \frac{\mathrm{d}\theta_\mathrm{v}}{\mathrm{d}\eta_\mathrm{v}} = \frac{\mathrm{d}^2\theta_\mathrm{v}}{\mathrm{d}\eta_\mathrm{v}^2} + \frac{1}{\lambda_\mathrm{v}} \frac{\mathrm{d}\lambda_\mathrm{v}}{\mathrm{d}\eta_\mathrm{v}} \frac{\mathrm{d}\theta_\mathrm{v}}{\mathrm{d}\eta_\mathrm{v}}. \tag{11.26}$$

For liquid film. For transformation of (11.4):
The similarity transformation of (11.4) is done, initially yielding

$$\rho_{\mathrm{v}}\left(\frac{\partial w_{x\mathrm{l}}}{\partial x}+\frac{\partial w_{y\mathrm{l}}}{\partial y}\right)+w_{x\mathrm{l}}\frac{\partial \rho_{\mathrm{l}}}{\partial x}+w_{y\mathrm{l}}\frac{\partial \rho_{\mathrm{l}}}{\partial y}=0. \tag{11.27}$$

With the similarity variables assumed in (11.19), (11.20), (11.22) and (11.23) we can obtain the following correlations:

$$\frac{\partial w_{x\mathrm{l}}}{\partial x}=\sqrt{\frac{g}{x}}(\rho_{\mathrm{l},\infty}/\rho_{\mathrm{l,s}}-1)^{1/2}\left(W_{x\mathrm{l}}-\frac{1}{2}\eta_{\mathrm{l}}\frac{\mathrm{d}W_{x\mathrm{l}}}{\mathrm{d}\eta_{\mathrm{l}}}\right), \tag{11.28}$$

$$\frac{\partial w_{y\mathrm{l}}}{\partial y}=2\sqrt{\frac{g}{x}}(\rho_{\mathrm{l},\infty}/\rho_{\mathrm{l,s}}-1)^{1/2}\frac{\mathrm{d}W_{y\mathrm{l}}}{\mathrm{d}\eta_{\mathrm{l}}}, \tag{11.29}$$

$$\frac{\partial \rho_{\mathrm{l}}}{\partial x}=-\frac{1}{4}\eta_{\mathrm{l}}x^{-1}\frac{\mathrm{d}\rho_{\mathrm{l}}}{\mathrm{d}\eta_{\mathrm{l}}}, \tag{11.30}$$

$$\frac{\partial \rho_{\mathrm{l}}}{\partial y}=\frac{\mathrm{d}\rho_{\mathrm{l}}}{\mathrm{d}\eta_{\mathrm{l}}}\left(\frac{1}{4}Gr_{x\mathrm{l},\infty}\right)^{1/4}x^{-1}. \tag{11.31}$$

With the earlier equations (11.28)–(11.31), (11.27) can be changed into

$$\rho_{\mathrm{l}}\left(\sqrt{\frac{g}{x}}(\rho_{\mathrm{l},\infty}/\rho_{\mathrm{l,s}}-1)\right)^{1/2}\left(W_{x\mathrm{l}}-\frac{1}{2}\eta_{\mathrm{l}}\frac{\mathrm{d}W_{x\mathrm{l}}}{\mathrm{d}\eta_{\mathrm{l}}}\right)+2\sqrt{\frac{g}{x}}\left(\rho_{\mathrm{l},\infty}/\rho_{\mathrm{l,s}}-1)^{1/2}\frac{\mathrm{d}W_{y\mathrm{l}}}{\mathrm{d}\eta_{\mathrm{l}}}\right)$$

$$+2\sqrt{gx}(\rho_{\mathrm{l},\infty}/\rho_{\mathrm{l,s}}-1)^{1/2}W_{x\mathrm{l}}\left(-\frac{1}{4}\eta_{\mathrm{l}}x^{-1}\frac{\mathrm{d}\rho_{\mathrm{l}}}{\mathrm{d}\eta_{\mathrm{l}}}\right)$$

$$+2\sqrt{gx}(\rho_{\mathrm{l},\infty}/\rho_{\mathrm{l,s}}-1)^{1/2}\left(\frac{1}{4}Gr_{x\mathrm{l},\infty}\right)^{-1/4}W_{y\mathrm{l}}\frac{\mathrm{d}\rho_{\mathrm{l}}}{\mathrm{d}\eta_{\mathrm{l}}}\left(\frac{1}{4}Gr_{x\mathrm{l},\infty}\right)^{1/4}x^{-1}=0.$$

The earlier equation is divided by

$$\sqrt{\frac{g}{x}}(\rho_{\mathrm{l},\infty}/\rho_{\mathrm{l,s}}-1)^{1/2}$$

and is further simplified to

$$2W_{x\mathrm{l}}-\eta_{\mathrm{l}}\frac{\mathrm{d}W_{x}}{\mathrm{d}\eta}+4\frac{\mathrm{d}W_{y\mathrm{l}}}{\mathrm{d}\eta_{\mathrm{l}}}-\frac{1}{\rho_{\mathrm{l}}}\frac{\mathrm{d}\rho_{\mathrm{l}}}{\mathrm{d}\eta_{\mathrm{l}}}(\eta_{\mathrm{l}}W_{x\mathrm{l}}-4W_{y\mathrm{l}})=0. \tag{11.32}$$

For transformation of (11.5). Equation (11.5) is firstly rewritten as

$$\rho_{l}\left(w_{\mathrm{xl}}\frac{\partial w_{x\mathrm{l}}}{\partial x}+w_{y\mathrm{l}}\frac{\partial w_{x\mathrm{l}}}{\partial y}\right)=\mu_{l}\frac{\partial^{2}w_{x\mathrm{l}}}{\partial y^{2}}+\frac{\partial w_{x\mathrm{l}}}{\partial y}\frac{\partial \mu_{\mathrm{l}}}{\partial y}+g\left(\rho_{\mathrm{l},\infty}-\rho_{\mathrm{l}}\right). \tag{11.33}$$

With the similarity variables assumed in (11.19), (11.20), (11.22) and (11.23) we can obtain the following correlations:

$$\frac{\partial w_{xl}}{\partial y} = 2\sqrt{gx}\left(\rho_{l,\infty}/\rho_{l,s}-1\right)^{1/2}\frac{dW_{xl}}{d\eta_l}x^{-1}\left(\frac{1}{4}Gr_{xl,\infty}\right)^{1/4}, \tag{11.34}$$

$$\frac{\partial^2 w_{xl}}{\partial y^2} = 2\sqrt{gx}\left(\rho_{l,\infty}/\rho_{l,s}-1\right)^{1/2}\frac{d^2W_{xl}}{d\eta_l^2}x^{-1}\left(\frac{1}{4}Gr_{xl,\infty}\right)^{1/4}\left(\frac{1}{4}Gr_{xl,\infty}\right)^{1/4}x^{-1}$$

$$= 2\sqrt{gx}\left(\rho_{l,\infty}/\rho_{l,s}-1\right)^{1/2}\frac{d^2W_{xl}}{d\eta_l^2}\left(\frac{1}{4}Gr_{xl,\infty}\right)^{1/2}x^{-2}, \tag{11.35}$$

$$\frac{\partial \mu_l}{\partial y} = \frac{d\mu_l}{d\eta_l}\left(\frac{1}{4}Gr_{xl,\infty}\right)^{1/4}x^{-1}. \tag{11.36}$$

With (11.22), (11.23), (11.28), (11.34), (11.35), and (11.36), (11.33) becomes

$$\rho_l\left[2\sqrt{gx}\left(\rho_{l,\infty}/\rho_{l,s}-1\right)^{1/2}W_{xl}\sqrt{\frac{g}{x}}\left(\rho_{l,\infty}/\rho_{l,s}-1\right)^{1/2}\left(W_{xl}-\frac{1}{2}\eta_l\frac{dW_{xl}}{d\eta_l}\right)\right.$$

$$+2\sqrt{gx}\left(\rho_{l,\infty}/\rho_{l,s}-1\right)^{1/2}\left(\frac{1}{4}Gr_{xl,\infty}\right)^{-1/4}$$

$$\left.\times W_{yl}2\sqrt{gx}\left(\rho_{l,\infty}/\rho_{l,s}-1\right)^{1/2}\frac{dW_{xl}}{d\eta_l}x^{-1}\left(\frac{1}{4}Gr_{xl,\infty}\right)^{1/4}\right]$$

$$= \mu_l 2\sqrt{gx}\left(\rho_{l,\infty}/\rho_{l,s}-1\right)^{1/2}\frac{d^2W_{xl}}{d\eta_l^2}\left(\frac{1}{4}Gr_{xl,\infty}\right)^{1/2}x^{-2}$$

$$+2\sqrt{gx}\left(\rho_{l,\infty}/\rho_{l,s}-1\right)^{1/2}\frac{dW_{xl}}{d\eta_l}x^{-1}\left(\frac{1}{4}Gr_{xl,\infty}\right)^{1/4}$$

$$\times\frac{d\mu_l}{d\eta_l}\left(\frac{1}{4}Gr_{xl,\infty}\right)^{1/4}x^{-1}+g\left(\rho_{l,\infty}-\rho_l\right).$$

With the definition of $Gr_{xl,\infty}$, the earlier equation is simplified to

$$\rho_l\left[2\sqrt{gx}\left(\rho_{l,\infty}/\rho_{l,s}-1\right)^{1/2}W_{xl}\sqrt{\frac{g}{x}}\left(\rho_{l,\infty}/\rho_{l,s}-1\right)^{1/2}\left(W_{xl}-\frac{1}{2}\eta_l\frac{dW_{xl}}{d\eta_l}\right)\right.$$

$$\left.+2\sqrt{gx}\left(\rho_{l,\infty}/\rho_{l,s}-1\right)^{1/2}W_{yl}2\sqrt{gx}\left(\rho_{l,\infty}/\rho_{l,s}-1\right)^{1/2}\frac{dW_{xl}}{d\eta_l}x^{-1}\right]$$

$$=\mu_l 2\sqrt{gx}\left(\rho_{l,\infty}/\rho_{l,s}-1\right)^{1/2}\frac{d^2W_{xl}}{d\eta_l^2}\left(\frac{1}{4}\frac{g\left(\rho_{l,\infty}/\rho_{l,s}-1\right)x^3}{v_{l,\infty}^2}\right)^{1/2}x^{-2}$$

$$+2\sqrt{gx}\left(\rho_{l,\infty}/\rho_{l,s}-1\right)^{1/2}\frac{dW_{xl}}{d\eta_l}x^{-1}\frac{d\mu_l}{d\eta_l}\left(\frac{1}{4}\frac{g\left(\rho_{l,\infty}/\rho_{l,s}-1\right)x^3}{v_{l,\infty}^2}\right)^{1/2}x^{-1}+g\left(\rho_{l,\infty}-\rho_l\right).$$

The earlier equation is divided by $g\left(\rho_{\mathrm{l},\infty}/\rho_{\mathrm{l,s}}-1\right)$ and simplified to

$$\rho_{\mathrm{l}}\left[2W_{x\mathrm{l}}\left(W_{x\mathrm{l}}-\frac{1}{2}\eta_{\mathrm{l}}\frac{\mathrm{d}W_{x\mathrm{l}}}{\mathrm{d}\eta_{\mathrm{l}}}\right)+4W_{y\mathrm{l}}\frac{\mathrm{d}W_{x\mathrm{l}}}{\mathrm{d}\eta_{\mathrm{l}}}\right]$$
$$=\mu_{\mathrm{l}}\frac{\mathrm{d}^2W_{x\mathrm{l}}}{\mathrm{d}\eta_{\mathrm{l}}^2}\frac{1}{\nu_{\mathrm{l},\infty}}+\frac{\mathrm{d}W_{x\mathrm{l}}}{\mathrm{d}\eta_{\mathrm{l}}}\frac{\mathrm{d}\mu_{\mathrm{l}}}{\mathrm{d}\eta_{\mathrm{l}}}\left(\frac{1}{\nu_{\mathrm{l},\infty}}\right)+\rho_{\mathrm{l,s}}\frac{(\rho_{\mathrm{l},\infty}-\rho_{\mathrm{l}})}{(\rho_{\mathrm{l},\infty}-\rho_{\mathrm{l,s}})}.$$

The earlier equation is multipled by $\nu_{\mathrm{l},\infty}/\mu_{\mathrm{l}}$ and simplified to

$$\frac{\nu_{\mathrm{l},\infty}}{\nu_{\mathrm{l}}}\left[W_{x\mathrm{l}}\left(2W_{x\mathrm{l}}-\eta_{\mathrm{l}}\frac{\mathrm{d}W_{x\mathrm{l}}}{\mathrm{d}\eta_{\mathrm{l}}}\right)+4W_{y\mathrm{l}}\frac{\mathrm{d}W_{x\mathrm{l}}}{\mathrm{d}\eta_{\mathrm{l}}}\right]$$
$$=\frac{\mathrm{d}^2W_{x\mathrm{l}}}{\mathrm{d}\eta_{\mathrm{l}}^2}+\frac{1}{\mu_{\mathrm{l}}}\frac{\mathrm{d}W_{x\mathrm{l}}}{\mathrm{d}\eta_{\mathrm{l}}}\frac{\mathrm{d}\mu_{\mathrm{l}}}{\mathrm{d}\eta_{\mathrm{l}}}+\frac{\nu_{\mathrm{l},\infty}}{\nu_{\mathrm{l}}}\frac{\left(\frac{\rho_{\mathrm{l},\infty}}{\rho_{\mathrm{l}}}-1\right)}{\left(\frac{\rho_{\mathrm{l},\infty}}{\rho_{\mathrm{l,s}}}-1\right)}. \tag{11.37}$$

For transformation of (11.6). Equation (11.6) is firstly rewritten as

$$\rho_{\mathrm{l}}c_{p\mathrm{l}}\left(w_{x\mathrm{l}}\frac{\partial t_{\mathrm{l}}}{\partial x}+w_{y\mathrm{l}}\frac{\partial t_{\mathrm{l}}}{\partial y}\right)=\lambda_{\mathrm{l}}\frac{\partial^2 t_{\mathrm{l}}}{\partial y^2}+\frac{\partial\lambda_{\mathrm{l}}}{\partial y}\frac{\partial t_{\mathrm{l}}}{\partial y} \tag{11.38}$$

With the similarity variables assumed in (11.19)–(11.23) the following correlations are produced:

$$t_{\mathrm{l}}=(t_{\mathrm{s}}-t_\infty)\,\theta_{\mathrm{l}}+t_\infty, \tag{11.39}$$

$$\frac{\partial t_{\mathrm{l}}}{\partial x}=-\left(t_{\mathrm{s}}-t_\infty\right)\frac{\mathrm{d}\theta_{\mathrm{l}}}{\mathrm{d}\eta_{\mathrm{l}}}\left(\frac{1}{4}\right)\eta_{\mathrm{l}}x^{-1}, \tag{11.40}$$

$$\frac{\partial t_{\mathrm{l}}}{\partial y}=\left(t_{\mathrm{s}}-t_\infty\right)\frac{\mathrm{d}\theta_{\mathrm{l}}}{\mathrm{d}\eta_{\mathrm{l}}}\left(\frac{1}{4}Gr_{x\mathrm{l},\infty}\right)^{1/4}x^{-1}, \tag{11.41}$$

$$\frac{\partial^2 t_{\mathrm{l}}}{\partial y^2}=\left(t_{\mathrm{s}}-t_\infty\right)\frac{\mathrm{d}^2\theta_{\mathrm{l}}}{\mathrm{d}\eta_{\mathrm{l}}^2}\left(\frac{1}{4}Gr_{x\mathrm{l},\infty}\right)^{1/2}x^{-2}, \tag{11.42}$$

$$\frac{\partial\lambda_{\mathrm{l}}}{\partial y}=\frac{\mathrm{d}\lambda_{\mathrm{l}}}{\mathrm{d}\eta}\left(\frac{1}{4}Gr_{x\mathrm{l},\infty}\right)^{1/4}x^{-1}. \tag{11.43}$$

With (11.22), (11.23), and (11.39)–(11.43), (11.38) is transformed to

$$\rho_{\mathrm{l}}c_{p\mathrm{l}}\left[2\sqrt{gx}\left(p_{\mathrm{l},\infty}/\rho_{\mathrm{l,s}}-1\right)^{1/2}W_{x\mathrm{l}}\left(-\left(t_{\mathrm{s}}-t_\infty\right)\frac{\mathrm{d}\theta_{\mathrm{l}}}{\mathrm{d}\eta_{\mathrm{l}}}\left(\frac{1}{4}\right)\eta_{\mathrm{l}}x^{-1}\right)\right.$$
$$\left.+2\sqrt{gx}\left(\rho_{\mathrm{l},\infty}/\rho_{\mathrm{l,s}}-1\right)^{1/2}\left(\frac{1}{4}Gr_{x\mathrm{l},\infty}\right)^{-1/4}W_{y\mathrm{l}}\left(t_{\mathrm{s}}-t_\infty\right)\frac{\mathrm{d}\theta_{\mathrm{l}}}{\mathrm{d}\eta_{\mathrm{l}}}\left(\frac{1}{4}Gr_{x\mathrm{l},\infty}\right)^{1/4}x^{-1}\right]$$
$$=\lambda_{\mathrm{l}}\left(t_{\mathrm{s}}-t_\infty\right)\frac{\mathrm{d}^2\theta_{\mathrm{l}}}{\mathrm{d}\eta_{\mathrm{l}}^2}\left(\frac{1}{4}Gr_{x\mathrm{l},\infty}\right)^{1/2}x^{-2}$$
$$+\frac{\mathrm{d}\lambda_{\mathrm{l}}}{\mathrm{d}\eta}\left(\frac{1}{4}Gr_{x\mathrm{l},\infty}\right)^{1/4}x^{-1}\left(t_{\mathrm{s}}-t_\infty\right)\frac{\mathrm{d}\theta_{\mathrm{l}}}{\mathrm{d}\eta_{\mathrm{l}}}\left(\frac{1}{4}Gr_{x\mathrm{l},\infty}\right)^{1/4}x^{-1}.$$

With the definition of $Gr_{x\mathrm{l},\infty}$, the earlier equation is simplified to

$$\rho_\mathrm{l} c_{p\mathrm{l}} \left[2\sqrt{gx}\left(\rho_{\mathrm{l},\infty}/\rho_{\mathrm{l,s}}-1\right)^{1/2} W_{x\mathrm{l}}\left(-(t_\mathrm{s}-t_\infty)\frac{\mathrm{d}\theta_\mathrm{l}}{\mathrm{d}\eta_\mathrm{l}}\left(\frac{1}{4}\right)\eta_\mathrm{l} x^{-1}\right)\right.$$
$$\left.+2\sqrt{gx}\left(\rho_{\mathrm{l},\infty}/\rho_{\mathrm{l,s}}-1\right)^{1/2} W_{y\mathrm{l}}(t_\mathrm{s}-t_\infty)\frac{\mathrm{d}\theta_\mathrm{l}}{\mathrm{d}\eta_\mathrm{l}}x^{-1}\right]$$
$$=\lambda_\mathrm{l}(t_\mathrm{s}-t_\infty)\frac{\mathrm{d}^2\theta_\mathrm{l}}{\mathrm{d}\eta_\mathrm{l}^2}\left(\frac{1}{4}\frac{g\left(\rho_{\mathrm{l},\infty}/\rho_{\mathrm{l,s}}-1\right)x^3}{v_{\mathrm{l},\infty}^2}\right)^{1/2}x^{-2}$$
$$+\frac{\mathrm{d}\lambda_\mathrm{l}}{\mathrm{d}\eta}x^{-1}(t_\mathrm{s}-t_\infty)\frac{\mathrm{d}\theta_\mathrm{l}}{\mathrm{d}\eta_\mathrm{l}}\left(\frac{1}{4}\frac{g\left(\rho_{\mathrm{l},\infty}/\rho_{\mathrm{l,s}}-1\right)x^3}{v_{\mathrm{l},\infty}^2}\right)^{1/2}x^{-1}.$$

The earlier equation is divided by

$$\sqrt{\frac{g}{x}}\left(\rho_{\mathrm{l},\infty}/\rho_{\mathrm{l,s}}-1\right)^{1/2}(t_\mathrm{s}-t_\infty),$$

and simplified to

$$\rho_\mathrm{l} c_{p\mathrm{l}}\left[-W_{x\mathrm{l}}\frac{\mathrm{d}\theta_\mathrm{l}}{\mathrm{d}\eta_\mathrm{l}}\eta_\mathrm{l}+4W_{y\mathrm{l}}\frac{\mathrm{d}\theta_\mathrm{l}}{\mathrm{d}\eta_\mathrm{l}}\right]=\lambda_\mathrm{l}\frac{\mathrm{d}^2\theta_\mathrm{l}}{\mathrm{d}\eta_\mathrm{l}^2}\frac{1}{v_{\mathrm{l},\infty}}+\frac{\mathrm{d}\lambda_\mathrm{l}}{\mathrm{d}\eta}\frac{\mathrm{d}\theta_\mathrm{l}}{\mathrm{d}\eta_\mathrm{l}}\frac{1}{v_{\mathrm{l},\infty}}.$$

This equation is multiplied by $\nu_{\mathrm{l},\infty}/\lambda_\mathrm{l}$ and simplified to

$$Pr_\mathrm{l}\frac{v_{\mathrm{l},\infty}}{v_\mathrm{l}}\left(-\eta W_{x\mathrm{l}}+4W_{y\mathrm{l}}\right)\frac{\mathrm{d}\theta_\mathrm{l}}{\mathrm{d}\eta_\mathrm{l}}=\frac{\mathrm{d}^2\theta_\mathrm{l}}{\mathrm{d}\eta_\mathrm{l}^2}+\frac{1}{\lambda_\mathrm{l}}\frac{\mathrm{d}\lambda_\mathrm{l}}{\mathrm{d}\eta_\mathrm{l}}\frac{\mathrm{d}\theta_\mathrm{l}}{\mathrm{d}\eta_\mathrm{l}}. \tag{11.44}$$

Equations (11.32), (11.37), and (11.44) are rewritten as follows:

$$2W_{x\mathrm{l}}-\eta_\mathrm{l}\frac{\mathrm{d}W_x}{\mathrm{d}\eta}+4\frac{\mathrm{d}W_{y\mathrm{l}}}{\mathrm{d}\eta_\mathrm{l}}-\frac{1}{\rho_\mathrm{l}}\frac{\mathrm{d}\rho_\mathrm{l}}{\mathrm{d}\eta_\mathrm{l}}\left(\eta_\mathrm{l}W_{x\mathrm{l}}-4W_{y\mathrm{l}}\right)=0 \tag{11.32}$$

$$\frac{\nu_{\mathrm{l},\infty}}{\nu}\left[W_{x\mathrm{l}}\left(2W_{x\mathrm{l}}-\eta_\mathrm{l}\frac{\mathrm{d}W_{x\mathrm{l}}}{\mathrm{d}\eta_\mathrm{l}}\right)+4W_{y\mathrm{l}}\frac{\mathrm{d}W_{x\mathrm{l}}}{\mathrm{d}\eta_\mathrm{l}}\right]$$
$$=\frac{\mathrm{d}^2W_{x\mathrm{l}}}{\mathrm{d}\eta_\mathrm{l}^2}+\frac{1}{\mu_\mathrm{l}}\frac{\mathrm{d}W_{x\mathrm{l}}}{\mathrm{d}\eta_\mathrm{l}}\frac{\mathrm{d}\mu_\mathrm{l}}{\mathrm{d}\eta_\mathrm{l}}+\frac{\nu_{\mathrm{l},\infty}}{\nu_\mathrm{l}}\frac{\left(\frac{\rho_{\mathrm{l},\infty}}{\rho_\mathrm{l}}-1\right)}{\left(\frac{\rho_{\mathrm{l},\infty}}{\rho_{\mathrm{l,s}}}-1\right)}, \tag{11.37}$$

$$Pr_\mathrm{l}\frac{v_{\mathrm{l},\infty}}{v_\mathrm{l}}\left(-\eta W_{x\mathrm{l}}+4W_{y\mathrm{l}}\right)\frac{\mathrm{d}\theta_\mathrm{l}}{\mathrm{d}\eta_\mathrm{l}}=\frac{\mathrm{d}^2\theta_\mathrm{l}}{\mathrm{d}\eta_\mathrm{l}^2}+\frac{1}{\lambda_\mathrm{l}}\frac{\mathrm{d}\lambda_\mathrm{l}}{\mathrm{d}\eta_\mathrm{l}}\frac{\mathrm{d}\theta_\mathrm{l}}{\mathrm{d}\eta_\mathrm{l}}. \tag{11.44}$$

For boundary conditions. With the corresponding assumed variable equations mentioned earlier the physical boundary conditions (11.7)–(11.13) for the film boiling of subcooled liquid are transformed equivalently to the following ones, respectively:

$\eta_{\mathrm{v}} = 0$:

$$W_{x\mathrm{v}} = 0, W_{y\mathrm{v}} = 0, \theta_{\mathrm{v}} = 1, \tag{11.45}$$

$\eta_{\mathrm{v}} = \eta_{\mathrm{v}\delta}(\eta_{\mathrm{l}} = 0)$:

$$W_{x\mathrm{l,s}} = \left(\frac{\rho_{\mathrm{l},\infty}}{\rho_{\mathrm{v,w}}} - 1\right)^{1/2} \left(\frac{\rho_{\mathrm{l},\infty}}{\rho_{\mathrm{l,s}}} - 1\right)^{-1/2} W_{x\mathrm{v,s}}, \tag{11.46}$$

$$W_{y\mathrm{l,s}} = -0.25\frac{\mu_{\mathrm{v,s}}}{\mu_{\mathrm{l,s}}}\left(\frac{\nu_{\mathrm{l},\infty}}{\nu_{\mathrm{v,s}}}\right)^{1/2}\left(\frac{\rho_{\mathrm{l},\infty}}{\rho_{\mathrm{v,w}}} - 1\right)^{1/4} \times \left(\frac{\rho_{\mathrm{l},\infty}}{\rho_{\mathrm{l,s}}} - 1\right)^{-1/4} \left(W_{x\mathrm{v,s}}\eta_{\mathrm{v}\delta} - 4W_{y\mathrm{v,s}}\right), \tag{11.47}$$

$$\left(\frac{\mathrm{d}W_{x\mathrm{l}}}{\mathrm{d}\eta_{\mathrm{l}}}\right)_{\eta_{\mathrm{l}}=0} = \frac{\mu_{\mathrm{v,s}}}{\mu_{\mathrm{l,s}}}\left(\frac{\rho_{\mathrm{l},\infty}}{\rho_{\mathrm{v,w}}} - 1\right)^{3/4}\left(\frac{\rho_{\mathrm{l},\infty}}{\rho_{\mathrm{l,s}}} - 1\right)^{-3/4} \times \left(\frac{\nu_{\mathrm{l},\infty}}{\nu_{\mathrm{v,s}}}\right)^{1/2}\left(\frac{\mathrm{d}W_{x\mathrm{v}}}{\mathrm{d}\eta_{\mathrm{v}}}\right)_{\eta_{\mathrm{v}}=\eta_{\mathrm{v}\delta}}, \tag{11.48}$$

$$\left(\frac{\mathrm{d}\theta_{\mathrm{l}}}{\mathrm{d}\eta_{\mathrm{l}}}\right)_{\eta_{\mathrm{l}}=0} = \left(\frac{\rho_{\mathrm{l},\infty}}{\rho_{\mathrm{v,w}}} - 1\right)^{1/4}\left(\frac{\rho_{\mathrm{l},\infty}}{\rho_{\mathrm{l,s}}} - 1\right)^{-1/4}\left(\frac{\nu_{\mathrm{l},\infty}}{\nu_{\mathrm{v,s}}}\right)^{1/2} \times \left(\frac{h_{fg}\mu_{\mathrm{v,s}}\left(W_{x\mathrm{v,s}}\eta_{\mathrm{v}\delta} - 4W_{y\mathrm{v,s}}\right) + \lambda_{\mathrm{v,s}}\left(t_{\mathrm{w}} - t_{\mathrm{s}}\right)\left(\frac{\mathrm{d}\theta_{\mathrm{v}}}{\mathrm{d}\eta_{\mathrm{v}}}\right)_{\eta_{\mathrm{v}}=\eta_{\mathrm{v}\delta}}}{\lambda_{\mathrm{l,s}}\left(t_{\mathrm{s}} - t_{\infty}\right)}\right), \tag{11.49}$$

$$\theta_{\mathrm{v}} = 0, \tag{11.50}$$

$$\theta_{\mathrm{l}} = 1, \tag{11.51}$$

$\eta_{\mathrm{l}} \to \infty$:

$$W_{x\mathrm{l}} \to 0, \quad \theta_{\mathrm{l}} \to 0. \tag{11.52}$$

11.4 Numerical Calculation

11.4.1 Treatment of Variable Thermophysical Properties

The treatment of variable thermophysical properties can be carried out for the equations of both liquid and gas films for the film boiling of subcooled liquid. Equations (10.53)–(10.61) can also be applied to treatment of variable thermophysical properties for the vapor film of film boiling of subcooled liquid.

Here we further introduce the treatment of variable thermophysical properties of the medium of the liquid film. For this purpose the treatment method for variable thermophysical properties of liquid presented in Chap. 6 is used.

For example, for water the temperature-dependent expressions of density, thermal conductivity, and absolute viscosity can be expressed as follows:

$$\rho = -4.48 \times 10^{-3} t^2 + 999.9, \tag{11.53}$$

$$\lambda = -8.01 \times 10^{-6} t^2 + 1.94 \times 10^{-3} t + 0.563, \tag{11.54}$$

$$\mu = \exp\left[-1.6 - \frac{1150}{T} + \left(\frac{690}{T}\right)^2\right] \times 10^{-3}. \tag{11.55}$$

Then the thermophysical property factors

$$\frac{1}{\rho_l}\frac{d\rho_l}{d\eta_l}, \frac{1}{\mu_l}\frac{d\mu_l}{d\eta_l}, \text{ and } \frac{1}{\lambda_l}\frac{d\lambda_l}{d\eta_l}$$

in the governing ordinary differential equations of liquid film, (12.32), (11.37), and (12.44), can be transformed, respectively, as later:

At first, the physical factor

$$\frac{1}{\rho_l}\frac{d\rho_l}{d\eta_l}$$

is expressed as

$$\frac{1}{\rho_l}\frac{d\rho_l}{d\eta_l} = \frac{1}{\rho_l}\frac{d\rho_l}{dt}\frac{dt}{d\eta_l},$$

where

$$\frac{d\rho_l}{dt} = -2 \times 4.48 \times 10^{-3}\, t.$$

With (11.21) we obtain

$$\frac{dt}{d\eta_l} = (t_s - t_\infty)\frac{d\theta_l}{d\eta_l}.$$

Therefore,

$$\frac{1}{\rho_l}\frac{d\rho_l}{d\eta_l} = \frac{-2 \times 4.48 \times 10^{-3}\, t(t_s - t_\infty)\frac{d\theta_l}{d\eta_l}}{-4.48 \times 10^{-3}\, t^2 + 999.9}. \tag{11.56}$$

Similarly, with (11.54) and (11.55) we obtain

$$\frac{1}{\lambda_l}\frac{d\lambda_l}{d\eta_l} = \frac{1}{\lambda_l}\frac{d\lambda_l}{dT}\frac{dt}{d\eta_l},$$

$$= \frac{[-2 \times 8.01 \times 10^{-6}\, t + 1.94 \times 10^{-3}](t_s - t_\infty)\frac{d\theta_l}{d\eta_l}}{-8.01 \times 10^{-6}\, t^2 + 1.94 \times 10^{-3}\, t + 0.5623}, \tag{11.57}$$

and

$$\frac{1}{\mu_l}\frac{d\mu_l}{d\eta_l} = \frac{1}{\mu_l}\frac{d\mu_l}{dT}\frac{dT}{d\eta_l}$$

$$= \frac{10^{-3} \times \exp\left[-1.6 - \frac{1150}{T} + \left(\frac{690}{T}\right)^2\right]\left[\frac{1150}{(T)^2} - 2 \times \frac{690^2}{(T)^3}\right](T_s - T_\infty)\frac{d\theta_l}{d\eta_l}}{10^{-3} \times \exp\left[-1.6 - \frac{1150}{T} + \left(\frac{690}{T}\right)^2\right]}$$

$$= \left[\frac{1150}{T^2} - 2 \times \frac{690^2}{T^3}\right](t_s - t_\infty)\frac{d\theta_l}{d\eta_l}, \tag{11.58}$$

where $t = (t_s - t_\infty)\theta_l + t_\infty$.

In addition, from the analysis of Chap. 6, it is known that the physical factor $Pr_{\mathrm{l}}\,(\nu_{\mathrm{l},\infty}/\nu_{\mathrm{l}})$ in (11.44) can be expressed as

$$Pr_{\mathrm{l}}\frac{\nu_{\mathrm{l},\infty}}{\nu_{\mathrm{l}}} = Pr_{\mathrm{l},\infty}\frac{\rho_{\mathrm{l}}}{\rho_{\mathrm{l},\infty}}\frac{\lambda_{\mathrm{l},\infty}}{\lambda_{\mathrm{l}}}$$

for water and a lot of liquids in the special temperature range for engineering application.

11.4.2 Numerical Calculation

Calculation Procedures. The general procedure of the calculation with the theoretical model for the film boiling of the subcooled liquid is similar to that of the saturated liquids, and is described as follows: First the values of $\eta_{\mathrm{v}\delta}$ and $W_{x\mathrm{v,s}}$ of the vapor film at the vapor–liquid interface are guessed. These two values combined with (11.45) and (11.50) allow us to solve the governing equations (11.24)–(11.26) for vapor film, so as to obtain $W_{x\mathrm{v,s}}$, $W_{y\mathrm{v,s}}$, $(\mathrm{d}W_{x\mathrm{v}}/\mathrm{d}\eta_{\mathrm{v}})_{\mathrm{s}}$ and $(\mathrm{d}\theta_{\mathrm{v}}/\mathrm{d}\eta_{\mathrm{v}})_{\mathrm{s}}$ at the interface. With $W_{x\mathrm{v,s}}$ and $W_{y\mathrm{v,s}}$, values, the values of $W_{x\mathrm{l,s}}$ and $W_{y\mathrm{l,s}}$ can be calculated from (11.46) and (11.47). Then, the values $W_{x\mathrm{l,s}}$ and $W_{y\mathrm{l,s}}$ together with the boundary conditions (11.51) and (11.52) are used to solve the liquid film governing (11.32), (11.37), and (11.44), and will yield the values of $(\mathrm{d}W_{x\mathrm{l}}/\mathrm{d}\eta_{\mathrm{l}})_{\mathrm{s}}$ and $(\mathrm{d}\theta_{\mathrm{l}}/\mathrm{d}\eta_{\mathrm{l}})_{\mathrm{s}}$. Equations (11.48) and (11.49) are taken to adjudge convergence of the solutions for the two-phase boundary layer governing equations, and by using shooting method, the calculation is successively iterated by changing the values of $\eta_{\mathrm{v}\delta}$ and $W_{x\mathrm{v,s}}$.

As an example of application for the theoretical model, the solutions of laminar film boiling of subcooled water on an isothermal vertical plate is presented here [1–3].

Velocity and temperature profiles. Let us take the film boiling of subcooled water as an example for the solutions. From Chap. 5, the water vapor temperature parameters n_μ, n_λ, and n_{c_p} are 1.04, 1.185, and 0.003, respectively. Such low value of n_{c_p} makes it possible to actually treat n_{c_p} of water vapor as zero, i.e., c_p is taken as constant.

The numerical method mentioned earlier is applied to obtaining the solutions of (11.24)–(11.26), (11.32), (11.37), and (11.44) with their boundary conditions (11.45)–(11.52) and equations (11.53)–(11.58) for the variable thermophysical property factors both of water vapor and water films. The numerical calculations have been carried out at wall superheated grades of $\Delta t_{\mathrm{w}}/t_{\mathrm{s}}\,(=(t_{\mathrm{w}}-t_{\mathrm{s}})/t_{\mathrm{s}}) = 2.77, 3.77, 4.77, 5.77, 7.27$, and 8.27, and water bulk subcooled grades of $\Delta t_\infty/t_{\mathrm{s}}(=(t_{\mathrm{s}}-t_\infty)/t_{\mathrm{s}}) = 0, 0.1, 0.3, 0.5, 0.7, 0.9$, and 1.0, respectively [3]. The thermophysical properties of subcooled water described in Chap. 6 are shown, respectively, in Table 11.1. Based on the earlier procedures, the numerical calculations were carried out by using the shooting method. Some numerical results of the velocity and temperature profiles for the film boiling of subcooled water are listed in Table 11.2 and shown in Figs. 11.2 and 11.3, respectively.

Figure 11.2 shows the velocity profiles of vapor film with wall superheated grades of $\Delta t_{\mathrm{w}}/t_{\mathrm{s}}(=(t_{\mathrm{w}}-t_{\mathrm{s}})/t_{\mathrm{s}}) = 2.77, 3.77, 4.77, 5.77$, and 7.27 and water

Table 11.1. Physical values of water [4]

$t(°\mathrm{C})$	0	10	20	30	40	50	60	70	80	90	100
ρ ($\mathrm{kg\,m^{-3}}$)	999.8	999.8	998.3	995.8	992.3	988.1	983.2	977.7	971.4	965.1	958.4
$\nu \times 10^{-6}$ ($\mathrm{m^2\,s^{-1}}$)	1.792	1.308	1.004	0.798	0.658	0.554	0.457	0.414	0.365	0.326	0.296
λ ($\mathrm{W\,(K\,m)^{-1}}$)	0.562	0.582	0.5996	0.6151	0.6287	0.6405	0.6507	0.6595	0.6668	0.6728	0.6773
Pr	13.44	9.42	6.99	5.42	4.34	3.57	3.00	2.57	2.23	1.97	1.76

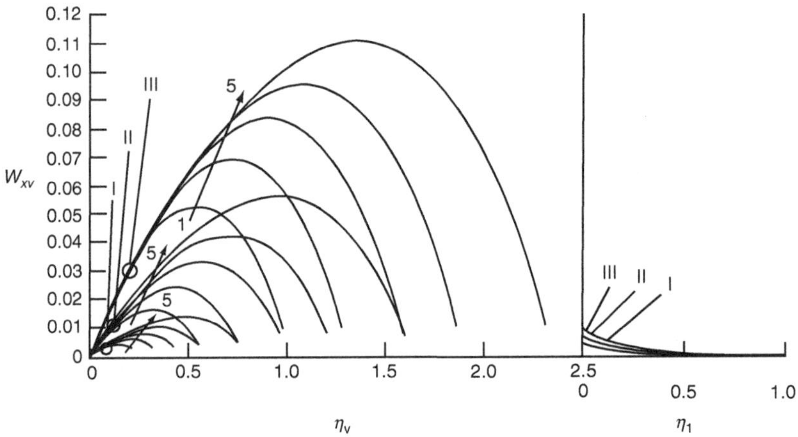

Fig. 11.2. Velocity profiles for laminar film boiling of subcooled water, cited from Shang, Wang, and Zhong [3] I–III: $\Delta t_\infty/t_s (= (t_s - t_\infty)/t_s) = 1, 0.3$ and 0.1; 1–5: $\Delta t_w/t_s (= (t_w - t_s)/t_s) = 2.77, 3.77, 4.77, 5.77$, and 7.27

bulk subcooled grades of $\Delta t_\infty/t_s\ (= (t_s - t_\infty)/t_s) = 1, 0.3, 0.1$, respectively. It is clear that, velocity components of vapor film increase with increasing wall superheated grade, $\Delta t_w/t_s (= (t_w - t_s)/t_s)$, or with decreasing water subcooled grade, $\Delta t_{\infty w}/t_s = (t_s - t_\infty/t_s)$.

Figure 11.3 shows the temperature profiles of vapor film with wall superheated grades of $\Delta t_w/t_s (= (t_w - t_s)/t_s) = 2.77, 3.77, 4.77, 5.77$, and 7.27 and water bulk subcooled grades of $\Delta t_\infty/t_s (= (t_s - t_\infty)/t_s) = 1, 0.3, 0.1$, respectively.

It is clear that, temperature profiles of vapor film increase with increasing wall superheated grade,

$$\Delta t_w/t_s = (t_w - t_s)/t_s,$$

Table 11.2. Numerical results of $\eta_{v\delta}, W_{xv,s}, W_{yv,s}, (d\theta_v/d\eta_v)_{\eta_v=0}$ and $(\eta_{v\delta}W_{xv,s} - 4W_{yv,s})$ for the film boiling of subcooled water (A part of the data are cited from Shang, Wang, and Zhong [3])

t_w (°C)	$\frac{\Delta t_w}{t_s}$	t_∞ (°C)	$\frac{\Delta t_\infty}{t_s}$	$\eta_{v\delta}$	$-W_{yv,s}$	$W_{xv,s}$	$-(\frac{d\theta_v}{d\eta_v})_{\eta_v=0}$ (1)	$-(\frac{d\theta_v}{d\eta_v})_{\eta_v=0}$ (2)	$\Delta(\frac{d\theta_v}{d\eta_v})_{\eta_v=0}$	$\eta_{v\delta}W_{xv,s} - 4W_{yv,s}$
377	2.77	100	0	1.437	0.05362	0.01084	0.5411	0.5410	0.000185	0.23006
477	3.77	100	0	1.728	0.06569	0.01080	0.4285	0.4249	0.008401	0.28142
577	4.77	100	0	2.019	0.07634	0.01068	0.3531	0.3528	0.00085	0.32692
677	5.77	100	0	2.308	0.08598	0.01057	0.2997	0.3012	−0.00501	0.36832
827	7.27	100	0	2.745	0.09868	0.01035	0.2433	0.2429	0.001644	0.42313
927	8.27	100	0	3.038	0.10625	0.01020	0.2158	0.2112	0.021316	0.45599
377	2.77	90	0.10	0.9828	0.01756	0.00855	0.7733	0.7731	0.000259	0.07865
477	3.77	90	0.10	1.2832	0.02746	0.00910	0.5621	0.5611	0.001779	0.12153
577	4.77	90	0.10	1.5800	0.03729	0.00939	0.4387	0.4383	0.000912	0.16399
677	5.77	90	0.10	1.8728	0.04675	0.00957	0.3587	0.3582	0.001394	0.20491
827	7.27	90	0.10	2.3135	0.05999	0.00967	0.2802	0.2797	0.001784	0.26232
927	8.27	90	0.10	2.6084	0.06820	0.00968	0.2439	0.2433	0.00246	0.29803
377	2.77	70	0.30	0.5240	0.00271	0.00543	1.4414	1.4410	0.000278	0.01368
477	3.77	70	0.30	0.7369	0.00524	0.00616	0.9684	0.9673	0.001136	0.02549
577	4.77	70	0.30	0.9675	0.00859	0.00672	0.7054	0.7054	0	0.04085
677	5.77	70	0.30	1.2114	0.01268	0.00717	0.5438	0.5435	0.000552	0.05939
827	7.27	70	0.30	1.5990	0.01984	0.00769	0.3954	0.3955	−0.00025	0.09166
927	8.27	70	0.30	1.8682	0.02511	0.00795	0.3313	0.3322	−0.00272	0.11528
377	2.77	50	0.50	0.3652	0.00093	0.00410	2.0672	2.0551	0.005853	0.00521
477	3.77	50	0.50	0.5191	0.00183	0.00470	1.3734	1.3639	0.006917	0.00975
577	4.77	50	0.50	0.6900	0.00309	0.00518	0.9872	0.9819	0.005369	0.01593
677	5.77	50	0.50	0.8760	0.00474	0.00561	0.7496	0.7456	0.005336	0.02411
827	7.27	50	0.50	1.1827	0.00791	0.00613	0.5317	0.5292	0.004702	0.03891
927	8.27	50	0.50	1.4033	0.01048	0.00644	0.4379	0.4364	0.003425	0.05096

Table 11.2. *Continue*

t_w (°C)	$\frac{\Delta t_w}{t_s}$	t_∞ (°C)	$\frac{\Delta t_\infty}{t_s}$	$\eta_{v\delta}$	$-W_{yv,s}$	$W_{xv,s}$	$-(\frac{d\theta_v}{d\eta_v})_{\eta_v=0}$ (1)	$-(\frac{d\theta_v}{d\eta_v})_{\eta_v=0}$ (2)	$\Delta(\frac{d\theta_v}{d\eta_v})_{\eta_v=0}$	$\eta_{v\delta}W_{xv,s}-4W_{yv,s}$
377	2.77	30	0.70	0.2881	0.00046	0.00335	2.6202	2.6541	−0.01294	0.00281
477	3.77	30	0.70	0.4113	0.00091	0.00385	1.7328	1.7523	−0.01125	0.00522
577	4.77	30	0.70	0.5484	0.00154	0.00426	1.2416	1.2539	−0.00991	0.00849
677	5.77	30	0.70	0.6992	0.00238	0.00463	0.9384	0.9457	−0.00778	0.01275
827	7.27	30	0.70	0.9504	0.00404	0.00509	0.6607	0.6634	−0.00409	0.02101
927	8.27	30	0.70	1.1332	0.00543	0.00537	0.5412	0.5422	−0.00185	0.02779
377	2.77	10	0.90	0.2450	0.00029	0.00284	3.0810	3.1041	−0.0075	0.00184
477	3.77	10	0.90	0.3497	0.00056	0.00327	2.0381	2.0512	−0.00643	0.00337
577	4.77	10	0.90	0.4671	0.00094	0.00362	1.4577	1.4690	−0.00775	0.00547
677	5.77	10	0.90	0.5964	0.00146	0.00394	1.0999	1.1088	−0.00809	0.00820
827	7.27	10	0.90	0.8130	0.00250	0.00434	0.7720	0.7787	−0.00868	0.01352
927	8.27	10	0.90	0.9705	0.00336	0.00459	0.6315	0.6369	−0.00855	0.01790
377	2.77	0	1.00	0.2309	0.00024	0.00262	3.2686	3.2344	0.010463	0.00157
477	3.77	0	1.00	0.3298	0.00047	0.00302	2.1609	2.1431	0.008237	0.00287
577	4.77	0	1.00	0.4400	0.00079	0.00335	1.5472	1.5394	0.005041	0.00462
677	5.77	0	1.00	0.5613	0.00122	0.00365	1.1686	1.1657	0.002482	0.00691
827	7.27	0	1.00	0.7643	0.00207	0.00403	0.8211	0.8231	−0.00244	0.01135
927	8.27	0	1.00	0.9163	0.00282	0.00426	0.6688	0.6759	−0.01062	0.01517

$-(d\theta_v/d\eta_v)_{\eta_v=0}$ (1) Numerical solution; $-(d\theta_v/d\eta_v)_{\eta_v=0}$ (2) Predicted results of $-(d\theta_v/d\eta_v)_{\eta_v=0}$ by using (11.64); and

$$\Delta\left(\frac{d\theta_v}{d\eta_v}\right)_{\eta_v=0} = \frac{[-\left(\frac{d\theta_v}{d\eta_{v=0}}\right)_{\eta_{v0}}(1)]-[-\left(\frac{d\theta_v}{d\eta_v}\right)_{\eta_v=0}(2)]}{-\left(\frac{d\theta_v}{d\eta_v}\right)_{\eta_v=0}(1)}$$

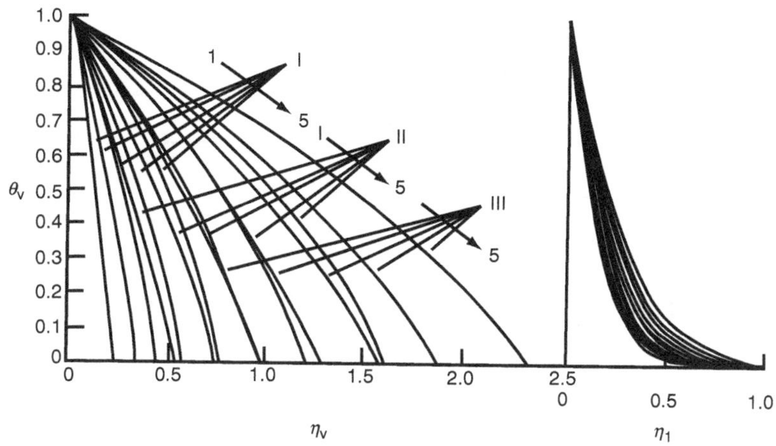

Fig. 11.3. Temperature profiles for laminar film boiling of subcooled water, cited from Shang, Wang, and Zhong [3] I–III: $\Delta t_\infty/t_s (= (t_s - t_\infty)/t_s) = 1, 0.3$ and 0.1; 1–5: $\Delta t_w/t_s (= (t_w - t_s)/t_s) = 2.77, 3.77, 4.77, 5.77$, and 7.27

or with decreasing water subcooled grade,

$$\Delta t_\infty/t_s = (t_s - t_\infty)/t_s.$$

The effects of the wall superheated grade $\Delta t_w/t_s$ and liquid bulk subcooled grade $\Delta t_\infty/t_s$ on the velocity and temperature fields show effects of variable thermophysical properties of both vapor and liquid films on the film boiling of subcooled liquid.

Vapor film thickness. The numerical results for vapor film thickness $\eta_{v\delta}$ of the film boiling of subcooled water are listed in Table 11.2 and plotted in Fig. 11.4 together with wall superheated grade $\Delta t_w/t_s$ and water bulk subcooled grade $\Delta t_\infty/t_s$. It is seen that $\eta_{v\delta}$ will increase with increasing wall superheated grade $\Delta t_w/t_s$. The reason is easy to be understood that with increasing the wall superheated grade $\Delta t_w/t_s$, the vaporization rate will increase, thus, the vapor film thickness $\eta_{v\delta}$ will increase.

Meanwhile from Fig. 11.3 it is seen that with increasing the water bulk subcooled grade $\Delta t_\infty/t_s$, the vapor film thickness $\eta_{v\delta}$ will decrease. The reason is that with increasing the water bulk subcooled grade $\Delta t_\infty/t_s$, the vaporization of the bulk liquid will become more difficult at the liquid–vapor interface.

It should be indicated that in the iterative calculation of the film boiling problem it is a key work to correctly determine suitable value $\eta_{v\delta}$, The solutions of the models are converged in very rigorous values of $\eta_{v\delta}$ as shown in Table 11.2 and Fig. 11.4, otherwise the convergence solutions will not be obtained.

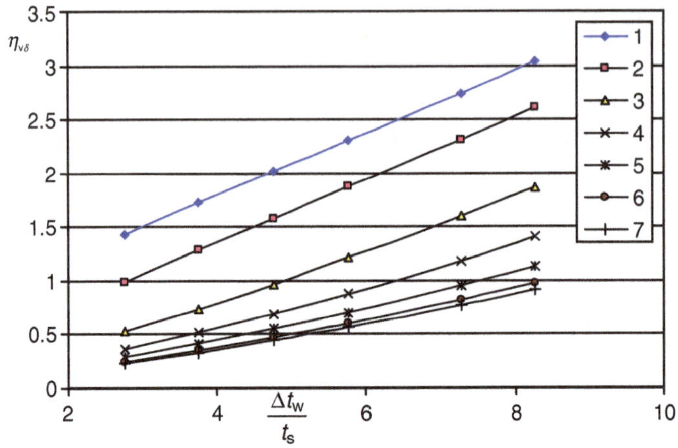

Fig. 11.4. The film thickness $\eta_{\mathrm{v}\delta}$ with $\Delta t_{\mathrm{w}}/t_{\mathrm{s}}$ and $\Delta t_\infty/t_{\mathrm{s}}$ for film boiling of subcooled water. 1–7: $\Delta t_\infty/t_{\mathrm{s}} = 0$, 0.1, 0.3, 0.5, 0.7, 0.9, and 1

11.5 Heat and Mass transfer

11.5.1 Heat Transfer Analysis

Same as the heat transfer analysis in Chap. 10 for film boiling of saturated vapor, the corresponding equations for the heat transfer analysis of the film boiling of subcooled liquid are as follows:

The local heat transfer rate at position x per unit area from the plate to the vapor film for the pool laminar film boiling of subcooled liquid can be expressed by Fourier's law, i.e., $q_x = -\lambda_{\mathrm{v,w}}\left(\frac{\partial t}{\partial y}\right)_{y=0}$, and with (11.14) and (11.16) further described as

$$q_x = -\lambda_{\mathrm{v,w}}(T_{\mathrm{w}} - T_{\mathrm{s}})\left(\frac{1}{4}Gr_{x\mathrm{v,s}}\right)^{1/4} x^{-1}\left(\frac{\mathrm{d}\theta_{\mathrm{v}}}{\mathrm{d}\eta_{\mathrm{v}}}\right)_{\eta_{\mathrm{v}}=0}. \tag{11.59}$$

With the Newtonian cooling law defined as $q_x = \alpha_x(t_{\mathrm{w}} - t_{\mathrm{s}})$ the related local heat transfer coefficient α_x on the surface will be

$$\alpha_x = -\lambda_{\mathrm{v,w}}\left(\frac{1}{4}Gr_{x\mathrm{v,s}}\right)^{1/4} x^{-1}\left(\frac{\mathrm{d}\theta_{\mathrm{v}}}{\mathrm{d}\eta_{\mathrm{v}}}\right)_{\eta_{\mathrm{v}}=0}. \tag{11.60}$$

The local Nusselt number, defined as $Nu_{x\mathrm{v,w}} = \alpha_x x/\lambda_{\mathrm{v,w}}$ is expressed by

$$\frac{Nu_{x\mathrm{v,w}}}{\left(\frac{1}{4}Gr_{x\mathrm{v,s}}\right)^{1/4}} = -\left(\frac{\mathrm{d}\theta_{\mathrm{v}}}{\mathrm{d}\eta_{\mathrm{v}}}\right)_{\eta_{\mathrm{v}}=0}. \tag{11.61}$$

It is seen that, for practical calculation of heat transfer, only $(\mathrm{d}\theta_{\mathrm{v}}/\mathrm{d}\eta_{\mathrm{v}})_{\eta_{\mathrm{v}}=0}$ dependent on numerical solution is no-given variable.

The average heat transfer coefficient $\overline{\alpha}_x$ defined as $Q_x = \overline{\alpha}_x(t_{\mathrm{w}} - t_\infty)A$ is expressed as

$$\overline{\alpha}_x = \frac{4}{3}\alpha_x. \tag{11.62}$$

Here, Q_x is total heat transfer rate for position $x = 0$ to x with width of b on the plate.

The average Nusselt number $\overline{Nu}_{x\mathrm{v,w}}$ defined as $\overline{Nu}_{x\mathrm{v,w}} = \overline{\alpha}_x x/\lambda_{\mathrm{v,w}}$ is expressed as

$$\overline{Nu}_{x\mathrm{v,w}} = \frac{4}{3}Nu_{x,\mathrm{w}}. \tag{11.63}$$

11.5.2 Curve-Fit Equations for Heat Transfer

It is seen from (11.59)–(11.61) that the heat transfer for the film boiling of subcooled water is also in direct proportion to dimensionless temperature gradient $(\mathrm{d}\theta_{\mathrm{v}}/\mathrm{d}\eta_{\mathrm{v}})_{\eta_{\mathrm{v}}=0}$ and to the fourth power of local Grashof number $Gr_{x\mathrm{v,s}}$.

The rigorous solutions on the dimensionless temperature gradients $(\mathrm{d}\theta_{\mathrm{v}}/\mathrm{d}\eta_{\mathrm{v}})_{\eta_{\mathrm{v}}=0}$ for the film boiling of subcooled water are computed and the results are tabulated and plotted respectively in Table 11.2 and Fig. 11.5, respectively.

It is obviously seen from Fig.11.5 that the temperature gradient $(\mathrm{d}\theta_{\mathrm{v}}/d\eta_{\mathrm{v}})_{\eta_{\mathrm{v}}=0}$ will decrease with increasing the wall superheated grade,

$$\Delta t_{\mathrm{w}}/t_{\mathrm{s}}(= (t_{\mathrm{w}} - t_{\mathrm{s}})/t_{\mathrm{s}}),$$

and will increase with increasing the bulk subcooled grade $\Delta t_\infty/t_{\mathrm{s}}$. On the other hand, the temperature profile of the vapor film has the results

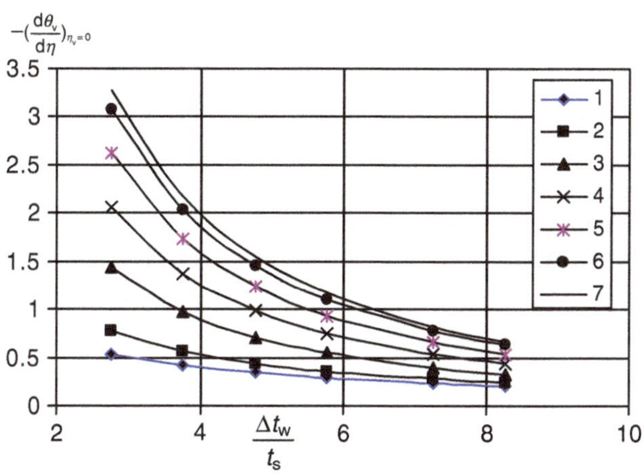

Fig. 11.5. Temperature gradient $(\mathrm{d}\theta_{\mathrm{v}}/\mathrm{d}\eta_{\mathrm{v}})_{\eta_{\mathrm{v}}=0}$ with $\Delta t_{\mathrm{w}}/t_{\mathrm{s}}$ and $\Delta t_\infty/t_{\mathrm{s}}$ for film boiling of subcooled water 1–7: $\Delta t_\infty/t_{\mathrm{s}} = 0$, 0.1, 0.3, 0.5, 0.7, 0.9, and 1

$d^2\theta_v/d\eta_v^2 < 0$, which are quite different from those of other solutions with constant thermophysical properties where the temperature profile in the vapor film has the results $d^2\theta_v/d\eta_v^2 > 0$ [5, 6].

Based on the rigorous numerical solutions of $(d\theta_v/d\eta_v)_{\eta_v=0}$ in Table 11.2, the following correlations were obtained by Shang, Wang, and Zhong [3] by means of a curve-fit method for laminar film boiling of subcooled water:

$$-\left(\frac{d\theta_v}{d\eta_v}\right)_{\eta_v=0} = \frac{\exp\left(A + B\frac{\Delta t_w}{t_s} + C\left(\frac{\Delta t_w}{t_s}\right)^2\right)}{\frac{\Delta t_w}{t_s}} \times 10^{-2}, \qquad (11.64)$$

where for $0 \le \Delta t_\infty/t_s \le 0.3$

$$A = 4.7356 + 7.407\frac{\Delta t_\infty}{t_s} - 7.4\left(\frac{\Delta t_\infty}{t_s}\right)^2$$

$$B = 0.1228 - 1.633\frac{\Delta t_\infty}{t_s} + 2.71\left(\frac{\Delta t_\infty}{t_s}\right)^2$$

$$C = -0.0086 + 1.092 \times 10^{-1}\frac{\Delta t_\infty}{t_s} - 2.132 \times 10^{-1}\left(\frac{\Delta t_\infty}{t_s}\right)^2,$$

for $0.3 < \Delta t_\infty/t_s \le 1$

$$A = 5.515 + 3.01\frac{\Delta t_\infty}{t_s} - 1.4\left(\frac{\Delta t_\infty}{t_s}\right)^2$$

$$B = -0.1073 - 6.6 \times 10^{-2}\frac{\Delta t_\infty}{t_s} + 4.437 \times 10^{-2}\left(\frac{\Delta t_\infty}{t_s}\right)^2$$

$$C = 0.0069 - 7.812 \times 10^{-3}\frac{\Delta t_\infty}{t_s} + 4.82 \times 10^{-3}\left(\frac{\Delta t_\infty}{t_s}\right)^2.$$

The results of $-(d\theta_v/d\eta_v)_{\eta_v=0}$ calculated by (11.64) are also listed in Table 11.2. It is shown that the calculated results by the correlation (11.64) coincide very well with the corresponding rigorous numerical solutions $-(d\theta_v/d\eta_v)_{\eta_v=0}$.

11.5.3 Mass Transfer Analysis

Consulting the related derivation in Chap. 10 for mass transfer entering the vapor film for film boiling of saturated liquid, the mass transfer entering the vapor film for film boiling of subcooled liquid can be easily derived as follows:

Set g_x to be a local mass flow rate entering the vapor film at position x per unit area of the plate. According to the boundary layer theory of fluid mechanics, g_x is expressed as

$$g_x = \rho_{\mathrm{v,s}} \left(w_{x\mathrm{v,s}} \frac{\mathrm{d}\delta_{\mathrm{v}}}{\mathrm{d}x} - w_{y\mathrm{v,s}} \right)_{\mathrm{s}}.$$

With the corresponding dimensionless variables in (10.17) and (10.18), the earlier equation is changed into the following one:

$$\begin{aligned} g_x = \; & \rho_{\mathrm{v,s}} \left[2\sqrt{gx} \left(\frac{\rho_{\mathrm{l},\infty}}{\rho_{\mathrm{v,w}}} - 1 \right)^{1/2} W_{x\mathrm{v,s}} \left(\frac{\mathrm{d}\delta_{\mathrm{v}}}{\mathrm{d}x} \right)_{\mathrm{s}} \right. \\ & \left. -2\sqrt{gx} \left(\frac{\rho_{\mathrm{l},\infty}}{\rho_{\mathrm{v,w}}} - 1 \right)^{1/2} \left(\frac{1}{4} Gr_{x\mathrm{v,s}} \right)^{-1/4} W_{y\mathrm{v,s}} \right], \end{aligned} \quad (11.65)$$

where the boiled vapour film thickness is expressed as

$$\delta_{\mathrm{v}} = \eta_{\mathrm{v}\delta} \left(\frac{1}{4} Gr_{x\mathrm{v,s}} \right)^{1/4} x. \quad (11.66)$$

With the definition of the local Grashof number $Gr_{x\mathrm{v,s}}$, (11.66) is changed into

$$\delta_{\mathrm{v}} = \eta_{\mathrm{v}\delta} \left(\frac{1}{4} \frac{g(\rho_{\mathrm{l},\infty}/\rho_{\mathrm{v,w}} - 1)x^3}{\nu_{\mathrm{v,s}}^2} \right)^{1/4} x.$$

Hence,

$$\left(\frac{\mathrm{d}\delta_{\mathrm{v}}}{\mathrm{d}x} \right)_{\mathrm{s}} = \frac{1}{4} \eta_{\mathrm{v}\delta} \left(\frac{1}{4} Gr_{x\mathrm{v,s}} \right)^{-1/4}. \quad (11.67)$$

Then, (10.65) will be simplified to

$$g_x = \mu_{\mathrm{v,s}} x^{-1} \left(\frac{1}{4} Gr_{x\mathrm{v,s}} \right)^{1/4} (\eta_{\mathrm{v}\delta} W_{x\mathrm{v,s}} - 4W_{y\mathrm{v,s}}), \quad (11.68)$$

where the dimensionless expression $(\eta_{\mathrm{v}\delta} W_{x\mathrm{v,s}} - 4W_{y\mathrm{v,s}})$ is regarded as mass flow rate parameter of the film boiling of subcooled vapor, and depends on the solutions $\eta_{\mathrm{v}\delta}$, $W_{x\mathrm{v,s}}$, and $W_{y\mathrm{v,s}}$. If G_x is taken to express total mass flow rate entering the boundary layer for position $x = 0$ to x with width of b of the plate, it should be the following integration:

$$\begin{aligned} G_x &= \iint\limits_A (g_x)_i \mathrm{d}A \\ &= b \int_0^x (g_x)_i \mathrm{d}x. \end{aligned}$$

where $A = b \cdot x$ is area of the plate.

Then, G_x is expressed as

$$G_x = b\int_0^x \left[\mu_{\mathrm{v,s}} x^{-1}\left(\frac{1}{4}Gr_{x\mathrm{v,s}}\right)^{1/4}(\eta_{\mathrm{v}\delta}W_{x\mathrm{v,s}} - 4W_{y\mathrm{v,s}})_{\Delta t_\infty=0}\right]\mathrm{d}x$$

$$= \frac{4}{3}b\cdot\mu_{\mathrm{v,s}}\left(\frac{1}{4}Gr_{x\mathrm{v,s}}\right)^{1/4}(\eta_{\mathrm{v}\delta}W_{x\mathrm{v,s}} - 4W_{y\mathrm{v,s}}). \qquad (11.69)$$

or in dimensionless form can be rewritten as

$$\frac{G_x}{b\cdot\mu_{\mathrm{v,s}}} = \frac{4}{3}\left(\frac{1}{4}Gr_{x\mathrm{v,s}}\right)^{1/4}(\eta_{\mathrm{v}\delta}W_{x\mathrm{v,s}} - 4W_{y\mathrm{v,s}}). \qquad (11.70)$$

Actually, the film boiling of saturated liquid can be regarded as a special case of the film boiling of subcooled liquid when the liquid subcooled grade $\Delta t_\infty/t_\mathrm{s}$ tends to zero, then, (10.68)–(10.70) are suitable for the film boiling of subcooled or saturated liquid.

The rigorous numerical solutions of interfacial velocity components $W_{x\mathrm{v,s}}$ and $W_{y\mathrm{v,s}}$ and mass flow rate parameter $\eta_{\mathrm{v}\delta}W_{x\mathrm{v,s}} - 4W_{y\mathrm{v,s}}$ for the film boiling of subcooled water are described in Table 11.2, and plotted in Figs. 11.6–11.8, respectively, with variations of wall superheated grade $\Delta t_\mathrm{w}/t_\mathrm{s}$ and the water bulk subcooled grade $\Delta t_\infty/t_\mathrm{s}$. It is found that the variations of interfacial velocity components $W_{x\mathrm{v,s}}$ and $W_{y\mathrm{v,s}}$ and mass flow rate parameter $\eta_{\mathrm{v}\delta}W_{x\mathrm{v,s}} - W_{y\mathrm{v,s}}$ are as follows with wall superheated grade $\Delta t_\mathrm{w}/t_\mathrm{s}$ and the water bulk subcooled degree $\Delta t_\infty/t_\mathrm{s}$:

The interfacial velocity component $W_{x\mathrm{v,s}}$ will increase with increasing the wall superheated grade $\Delta t_\mathrm{w}/t_\mathrm{s}$ except for very low water bulk subcooled grade

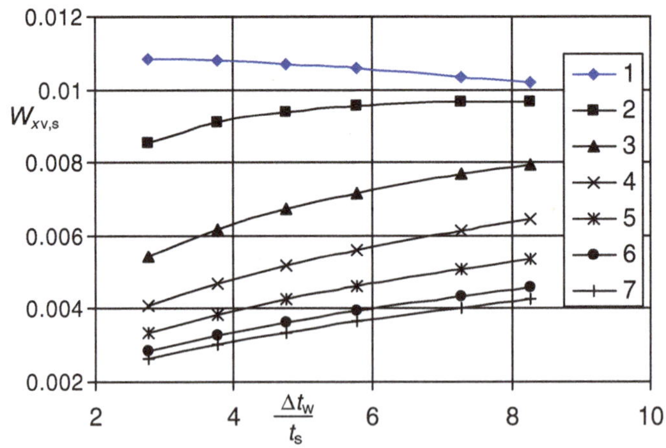

Fig. 11.6. Variation of $W_{x\mathrm{v,s}}$ with $\Delta t_\mathrm{w}/t_\mathrm{s}$ and $\Delta t_\infty/t_\mathrm{s}$ for film boiling of subcooled water 1–7: $\Delta t_\infty/t_\mathrm{s} = 0$, 0.1, 0.3, 0.5, 0.7, 0.9, and 1

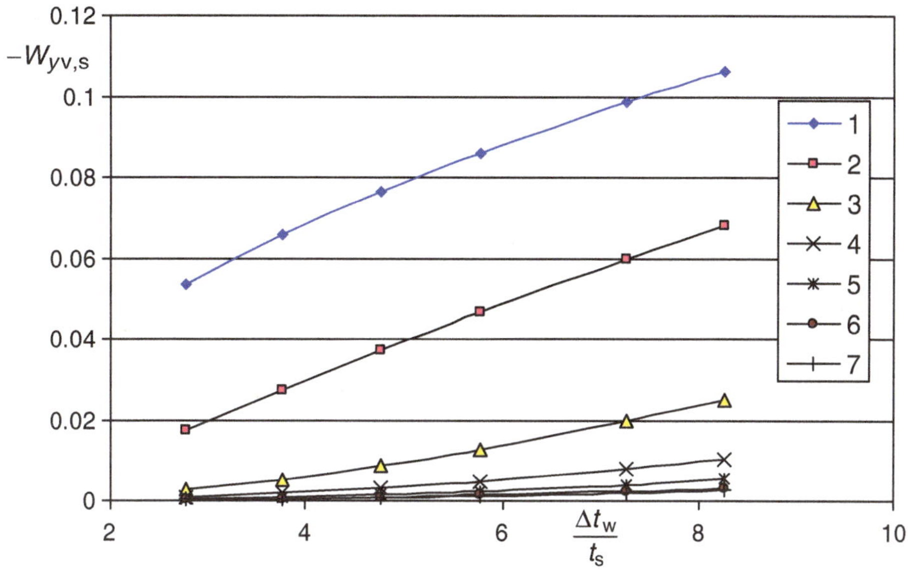

Fig. 11.7. Variation of $W_{yv,s}$ with $\Delta t_w/t_s$ and $\Delta t_\infty/t_s$ for film boiling of subcooled water 1–7: $\Delta t_\infty/t_s = 0$, 0.1, 0.3, 0.5, 0.7, 0.9, and 1

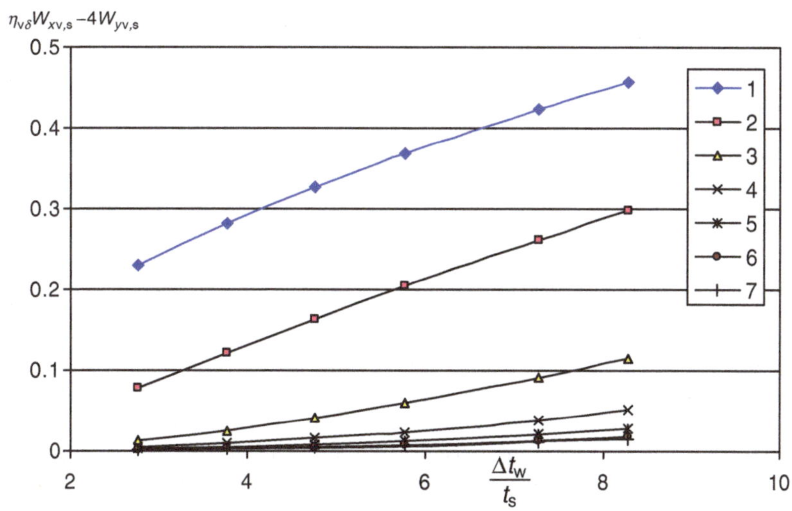

Fig. 11.8. Variation of mass flow rate parameter $\eta_{v\delta}W_{xv,s} - 4W_{yv,s}$ with $\Delta t_w/t_s$ and $\Delta t_\infty/t_s$ for film boiling of subcooled water 1–7: $\Delta t_\infty/t_s = 0$, 0.1, 0.3, 0.5, 0.7, 0.9, and 1

$\Delta t_\infty/t_s$. Meanwhile, the interfacial velocity component $W_{xv,s}$ will decrease with increasing the water bulk subcooled grade $\Delta t_\infty/t_s$.

The interfacial velocity component $-W_{yv,s}$ and mass flow rate parameter $(\eta_{v\delta}W_{xv,s} - 4W_{yv,s})$ will increase with increasing the wall superheated grade $\Delta t_w/t_s$ especially in the range of lower water subcooled grade $\Delta t_\infty/t_s$. Meanwhile, they will decrease with increasing the water bulk subcooled grade $\Delta t_\infty/t_s$. It is seen that the value of the interfacial velocity component $-W_{yv,s}$ is much more than that of the interfacial velocity component $W_{xv,s}$ usually. Then, it follows that the interfacial velocity component $-W_{yv,s}$ will dominate the interfacial mass flow rate in general. However, the effect of the interfacial velocity component $W_{xv,s}$ on the interfacial mass flow rate can never be ignored.

11.6 Summary

So far, governing equations, and the equations for heat and mass transfer of the film boiling of subcooled liquid can be summarized in Tables 11.3 and 11.4. Meanwhile, the film boiling of saturated liquid can be regarded as a special case of the film boiling of subcooled liquid with zero for the liquid bulk subcooled grade $\Delta t_\infty/t_s$.

11.7 Remarks

The film boiling of saturated liquid is a special case of the film boiling of subcooled liquid only with $\Delta t_\infty/t_s = 0$. The velocity component method is successfully applied to in the theoretical model of the film boiling. The dimensionless velocity components W_x and W_y of vapor and liquid films have definite physical meanings, and then the solutions of the models can be understood easily. It follows that the velocity component method is appropriate for the treatment of the two-phase boundary layer problems with the three-point value problem.

In the theoretical models of the film boiling of subcooled liquids, the various physical matching conditions including variable thermophysical properties are taken into account. On this basis, the numerical solutions for film boiling of subcooled water at different wall superheated grades $\Delta t_w/t_s$ and liquid bulk subcooled grades $\Delta t_\infty/t_s$ are theoretically reliable. On this basis, a system of rigorous solutions for momentum, heat, and mass transfer are calculated for taking the film boiling of subcooled water as the example. From these numerical results, the following points can be concluded:

Generally, the vapor film thickness $\eta_{v\delta}$ and the velocity components of vapor film will increase with increasing wall superheated grade $\Delta t_w/t_s$ or with decreasing the water bulk subcooled grade $\Delta t_\infty/t_s$.

The temperature gradients $-(d\theta_v/d\eta_v)_{\eta_v=0}$ on the plate will decrease with increasing the wall superheated grade $\Delta t_w/t_s$, and increase with increasing the

Table 11.3. Summary of governing equations of film boiling of subcooled liquid

term	expression
	partial differential equations for vapor film
mass equation	$\frac{\partial}{\partial x}(\rho_v w_{xv}) + \frac{\partial}{\partial y}(\rho_v w_{yv}) = 0$
momentum equation	$\rho_v \left(w_{xv} \frac{\partial w_{xv}}{\partial x} + w_{yv} \frac{\partial w_{xv}}{\partial y} \right) = \frac{\partial}{\partial y}\left(\mu_v \frac{\partial w_{xv}}{\partial y} \right) + g(\rho_{l,\infty} - \rho_v)$
energy equation	$\rho_v c_{p_v} \left(w_{xv} \frac{\partial T_v}{\partial x} + w_{yv} \frac{\partial T_v}{\partial y} \right) = \frac{\partial}{\partial y}\left(\lambda_v \frac{\partial T_v}{\partial y} \right)$
	partial differential equations for subcooled liquid film
mass equation	$\frac{\partial}{\partial x}(\rho_l w_{xl}) + \frac{\partial}{\partial y}(\rho_l w_{yl}) = 0$
momentum equation	$\rho_l(w_{xl} \frac{\partial w_{xl}}{\partial x} + w_{yl} \frac{\partial w_{xl}}{\partial y}) = \frac{\partial}{\partial y}(\mu_l \frac{\partial w_{xl}}{\partial y}) + g(\rho_{l,\infty} - \rho_l)$
Energy equation	$\rho_l c_{p_l} \left(w_{xl} \frac{\partial t_l}{\partial x} + w_{yl} \frac{\partial t_l}{\partial y} \right) = \frac{\partial}{\partial y}\left(\lambda_l \frac{\partial t_l}{\partial y} \right)$
	boundary conditions
$y = 0$	$w_{xv} = 0, \quad w_{yv} = 0, \quad T = T_w$
$y = \delta_v$	$w_{xv,s} = w_{xl,s}$
	$\rho_{v,s} \left(w_{xv} \frac{\partial \delta_{xv}}{\partial x} - w_{yv} \right)_s = \rho_{l,s}(w_{xl} \frac{\partial \delta_{xl}}{\partial x} - w_{yl})_s$
	$\mu_{v,s} \left(\frac{\partial w_{xv}}{\partial y} \right)_s = \mu_{l,s} \left(\frac{\partial w_{xl}}{\partial y} \right)_s$
	$-\lambda_{v,s}(\frac{\partial t_v}{\partial y})_{y=\delta_v} = h_{fg}\rho_{v,s}(w_{xv} \frac{\partial \delta_{xv}}{\partial x} - w_{yv})_s - \lambda_{l,s}(\frac{\partial T_l}{\partial y})_{y=\delta_v}$
	$T = T_s$
$y \to \infty$	$w_{xl} \to 0, \quad t_l \to t_\infty$

Table 11.3. *Continue...*

term	expression
	assumed similarity variables for vapor film
η_v	$\left(\frac{1}{4}Gr_{x\mathrm{v,s}}\right)^{1/4}\frac{y}{x}$
$Gr_{x\mathrm{v,s}}$	$\frac{g(\rho_{\mathrm{l},\infty}/\rho_{\mathrm{v,w}}-1)x^3}{\nu_{\mathrm{v,s}}^2}$
θ_v	$\frac{T-T_\mathrm{s}}{T_\mathrm{w}-T_\mathrm{s}}$
$W_{x\mathrm{v}}$	$(2\sqrt{gx}(\rho_{\mathrm{l},\infty}/\rho_{\mathrm{v,w}}-1)^{1/2})^{-1}w_{x\mathrm{v}}$
$W_{y\mathrm{v}}$	$\left(2\sqrt{gx}(\rho_{\mathrm{l},\infty}/\rho_{\mathrm{v,w}}-1)^{1/2}\left(\frac{1}{4}Gr_{x\mathrm{v,s}}\right)^{-1/4}\right)^{-1}w_{y\mathrm{v}}$
	ordinary differential equations for vapor film
mass equation	$2W_{x\mathrm{v}}-\eta_\mathrm{v}\frac{\mathrm{d}W_{x\mathrm{v}}}{\mathrm{d}\eta_\mathrm{v}}+4\frac{\mathrm{d}W_{y\mathrm{v}}}{\mathrm{d}\eta_\mathrm{v}}-\frac{1}{\rho_\mathrm{v}}\frac{\mathrm{d}\rho_\mathrm{v}}{\mathrm{d}\eta_\mathrm{v}}(\eta_\mathrm{v}W_{x\mathrm{v}}-4W_{y\mathrm{v}})=0$
momentum equation	$\frac{\nu_{\mathrm{v,s}}}{\nu_\mathrm{v}}\left(W_{x\mathrm{v}}\left(2W_{x\mathrm{v}}-\eta_\mathrm{v}\frac{\mathrm{d}W_{x\mathrm{v}}}{\mathrm{d}\eta_\mathrm{v}}\right)+4W_{y\mathrm{v}}\frac{\mathrm{d}W_{x\mathrm{v}}}{\mathrm{d}\eta_\mathrm{v}}\right)=\frac{\mathrm{d}^2W_{x\mathrm{v}}}{\mathrm{d}\eta_\mathrm{v}^2}+\frac{1}{\mu_\mathrm{v}}\frac{\mathrm{d}\mu_\mathrm{v}}{\mathrm{d}\eta_\mathrm{v}}\frac{\mathrm{d}W_{x\mathrm{v}}}{\mathrm{d}\eta_\mathrm{v}}+\frac{\nu_{\mathrm{v,s}}}{\nu_\mathrm{v}}\frac{\rho_{\mathrm{v,w}}}{\rho_\mathrm{v}}\frac{\rho_{\mathrm{l},\infty}-\rho_\mathrm{v}}{\rho_{\mathrm{l},\infty}-\rho_{\mathrm{v,w}}}$
energy equation	$Pr_\mathrm{v}\frac{\nu_{\mathrm{v,s}}}{\nu_\mathrm{v}}(-\eta_\mathrm{v}W_{x\mathrm{v}}+4W_{y\mathrm{v}})\frac{\mathrm{d}\theta_\mathrm{v}}{\mathrm{d}\eta_\mathrm{v}}=\frac{\mathrm{d}^2\theta_\mathrm{v}}{\mathrm{d}\eta_\mathrm{v}^2}+\frac{1}{\lambda_\mathrm{v}}\frac{\mathrm{d}\lambda_\mathrm{v}}{\mathrm{d}\eta_\mathrm{v}}\frac{\mathrm{d}\theta_\mathrm{v}}{\mathrm{d}\eta_\mathrm{v}}$
	assumed similarity variables for liquid film
η_l	$\left(\frac{1}{4}Gr_{x\mathrm{l},\infty}\right)^{1/4}\frac{y}{x}$
$Gr_{x\mathrm{l},\infty}$	$\frac{g(\rho_{\mathrm{l},\infty}/\rho_{\mathrm{l,s}}-1)x^3}{\nu_{\mathrm{l},\infty}^2}$
θ_l	$\frac{t-t_\infty}{t_\mathrm{s}-t_\infty}$
$W_{x\mathrm{l}}$	$\left(2\sqrt{gx}(\rho_{\mathrm{l},\infty}/\rho_{\mathrm{l,s}}-1)^{1/2}\right)^{-1}w_{x\mathrm{l}}$
$W_{y\mathrm{l}}$	$\left(2\sqrt{gx}(\rho_{\mathrm{l},\infty}/\rho_{\mathrm{l,s}}-1)^{1/2}\left(\frac{1}{4}Gr_{x\mathrm{l},\infty}\right)^{-1/4}\right)^{-1}w_{y\mathrm{l}}$

Table 11.3. *Continue...*

term	expression
	ordinary differential equations for liquid film
mass equations	$2W_{xl} - \eta_l \frac{dW_x}{d\eta} + 4\frac{dW_{yl}}{d\eta_l} - \frac{1}{\rho_l}\frac{d\rho_l}{d\eta_l}(\eta_l W_{xl} - 4W_{yl}) = 0$
momentum equation	$\frac{\nu_{l,\infty}}{\nu_l}\left[W_{xl}\left(2W_{xl} - \eta_l \frac{dW_{xl}}{d\eta_l}\right) + 4W_{yl}\frac{dW_{xl}}{d\eta_l}\right] = \frac{d^2 W_{xl}}{d\eta_l^2} + \frac{1}{\mu_l}\frac{dW_{xl}}{d\eta_l}\frac{d\mu_l}{d\eta_l} + \frac{\nu_{l,\infty}}{\nu_l}\frac{\left(\frac{\rho_{l,\infty}}{\rho_l} - 1\right)}{\left(\frac{\rho_{l,\infty}}{\rho_{l,s}} - 1\right)}$
energy equation	$Pr_l \frac{\nu_{l,\infty}}{\nu_l}(-\eta W_{xl} + 4W_{yl})\frac{d\theta_l}{d\eta_l} = \frac{d^2\theta_l}{d\eta_l} + \frac{1}{\lambda_l}\frac{d\lambda_l}{d\eta_l}\frac{d\theta_l}{d\eta_l}$
	dimensionless boundary conditions
$\eta_v = 0$:	$W_{xv} = 0, \quad W_{yv} = 0, \ \theta_v = 1,$
$\eta_v = \eta_{v\delta}$	$W_{xl,s} = \left(\frac{\rho_{l,\infty}}{\rho_{v,w}} - 1\right)^{1/2}\left(\frac{\rho_{l,\infty}}{\rho_{l,s}} - 1\right)^{-1/2} W_{xv,s}$
$(\eta_l = 0)$:	
	$W_{yl,s} = -0.25\frac{\mu_{v,s}}{\mu_{l,s}}\left(\frac{\nu_{v,s}}{\nu_{l,\infty}}\right)^{-1/2}\left(\frac{\rho_{l,\infty}}{\rho_{v,w}} - 1\right)^{1/4}\left(\frac{\rho_{l,\infty}}{\rho_{l,s}} - 1\right)^{-1/4}(W_{xv,s}\eta_{v\delta} - 4W_{yv,s})$
	$\left(\frac{dW_{xl}}{d\eta_l}\right)_{\eta_l=0} = \frac{\mu_{v,s}}{\mu_{l,s}}\left(\frac{\rho_{l,\infty}}{\rho_{v,w}} - 1\right)^{3/4}\left(\frac{\rho_{l,\infty}}{\rho_{l,s}} - 1\right)^{-3/4}\left(\frac{\nu_{l,\infty}}{\nu_{v,s}}\right)^{1/2}\left(\frac{dW_{xv}}{d\eta_v}\right)_{\eta_v=\eta_{v\delta}}$
	$\left(\frac{d\theta_l}{d\eta_l}\right)_{\eta_l=0} =$
	$\left(\frac{\rho_{l,\infty}}{\rho_{v,w}} - 1\right)^{1/4}\left(\frac{\rho_{l,\infty}}{\rho_{l,s}} - 1\right)^{-1/4}\left(\frac{\nu_{l,\infty}}{\nu_{v,s}}\right)^{1/2}\left(\frac{h_{fg}\mu_{v,s}(W_{xv,s}\eta_{v\delta} - 4W_{yv,s}) + \lambda_{v,s}(t_w - t_s)\left(\frac{d\theta_v}{d\eta_v}\right)_{\eta_v=\eta_{v\delta}}}{\lambda_{l,s}(t_s - t_\infty)}\right)$
	$\theta_v = 0$
	$\theta_l = 1$
$\eta_l \to \infty$:	$W_{xl} = 0, \quad \theta_l = 0$

Table 11.4. Summary of the heat transfer and mass flow rate of the film boiling of subcooled liquid

term	expression
q_x (defined as $-\lambda_{\mathrm{v,w}}\left(\frac{\partial t}{\partial y}\right)_{y=0}$)	$-\lambda_{\mathrm{v,w}}(T_{\mathrm{w}}-T_{\mathrm{s}})(\frac{1}{4}Gr_{x\mathrm{v,s}})^{1/4}x^{-1}\left(\frac{\mathrm{d}\theta_{\mathrm{v}}}{\mathrm{d}\eta_{\mathrm{v}}}\right)_{\eta_{\mathrm{v}}=0}$
α_x (defined as $\alpha_x(t_{\mathrm{w}}-t_{\mathrm{s}})$	$-\lambda_{\mathrm{v,w}}\left(\frac{1}{4}Gr_{x\mathrm{v,s}}\right)^{1/4}x^{-1}\left(\frac{\mathrm{d}\theta_{\mathrm{v}}}{\mathrm{d}\eta_{\mathrm{v}}}\right)_{\eta_{\mathrm{v}}=0}$
$\overline{\alpha_x}$ (defined as $\overline{\alpha_x}(t_{\mathrm{w}}-t_{\mathrm{s}})$)	$-\frac{4}{3}\lambda_{\mathrm{v,w}}\left(\frac{1}{4}Gr_{x\mathrm{v,s}}\right)^{1/4}x^{-1}(\frac{\mathrm{d}\theta_{\mathrm{v}}}{\mathrm{d}\eta_{\mathrm{v}}})_{\eta_{\mathrm{v}}=0}$
$Nu_{x\mathrm{v,w}}$ $\left(\text{defined as } \frac{\alpha_x x}{\lambda_{\mathrm{v,w}}}\right)$	$-\left(\frac{1}{4}Gr_{x\mathrm{v,s}}\right)^{1/4}\left(\frac{\mathrm{d}\theta_{\mathrm{v}}}{\mathrm{d}\eta_{\mathrm{v}}}\right)_{\eta_{\mathrm{v}}=0}$
$\overline{Nu}_{x\mathrm{v,w}}$ $\left(\text{defined as } \frac{\overline{\alpha}_x x}{\lambda_{\mathrm{v,w}}}\right)$	$-\frac{4}{3}\left(\frac{1}{4}Gr_{x\mathrm{v,s}}\right)^{1/4}(\frac{\mathrm{d}\theta_{\mathrm{v}}}{\mathrm{d}\eta_{\mathrm{v}}})_{\eta_{\mathrm{v}}=0}$
$-\left(\frac{\mathrm{d}\theta_{\mathrm{v}}}{\mathrm{d}\eta_{\mathrm{v}}}\right)_{\eta_{\mathrm{v}}=0}$	$\frac{\exp\left(A+B\frac{\Delta t_{\mathrm{w}}}{t_{\mathrm{s}}}+C\left(\frac{\Delta t_{\mathrm{w}}}{t_{\mathrm{s}}}\right)^2\right)}{\frac{\Delta t_{\mathrm{w}}}{t_{\mathrm{s}}}}\times 10^{-2}$ (for film boiling of water)
$A=4.7356+7.407\frac{\Delta t_\infty}{ts}-7.4\left(\frac{\Delta t_\infty}{ts}\right)^2$	$(0\le\frac{\Delta t_\infty}{ts}\le 0.3)$
$A=5.515+3.01\frac{\Delta t_\infty}{ts}-1.4\left(\frac{\Delta t_\infty}{ts}\right)^2$	$(0.3<\frac{\Delta t_\infty}{t_{\mathrm{s}}}\le 1)$
$B=0.1228-1.633\frac{\Delta t_\infty}{ts}+2.71\left(\frac{\Delta t_\infty}{ts}\right)^2$	$(0\le\frac{\Delta t_\infty}{t_{\mathrm{s}}}\le 0.3)$
$B=-0.1073-6.6\times 10^{-2}\frac{\Delta t_\infty}{t_{\mathrm{s}}}+4.437\times 10^{-2}\left(\frac{\Delta t_\infty}{t_{\mathrm{s}}}\right)^2$	$(0.3<\frac{\Delta t_\infty}{t_{\mathrm{s}}}\le 1)$
$C=-0.0086+1.092\times 10^{-1}\frac{\Delta t_\infty}{ts}-2.132\times 10^{-1}\left(\frac{\Delta t_\infty}{ts}\right)^2$	$(0\le\frac{\Delta t_\infty}{t_{\mathrm{s}}}\le 0.3)$
$C=0.0069-7.812\times 10^{-3}\frac{\Delta t_\infty}{ts}+4.82\times 10^{-3}\left(\frac{\Delta t_\infty}{ts}\right)^2$	$(0.3<\frac{\Delta t_\infty}{t_{\mathrm{s}}}\le 1)$
g_x	$\mu_{\mathrm{v,s}}x^{-1}\left(\frac{1}{4}Gr_{x\mathrm{v,s}}\right)^{1/4}(\eta_{\mathrm{v}\delta}W_{x\mathrm{v,s}}-4W_{y\mathrm{v,s}})$
G_x	$\frac{4}{3}b\cdot\mu_{\mathrm{v,s}}\left(\frac{1}{4}Gr_{x\mathrm{v,s}}\right)^{1/4}(\eta_{\mathrm{v}\delta}W_{x\mathrm{v,s}}-4W_{y\mathrm{v,s}})$
$\eta_{\mathrm{v}\delta}$, defined as $\left(\frac{1}{4}Gr_{x\mathrm{v,s}}\right)^{1/4}\frac{\delta_{\mathrm{v}}}{x}$	$0.291\frac{\Delta t_{\mathrm{w}}}{t_{\mathrm{s}}}+0.631$ (for film boiling of saturated water)
$(\eta_{\mathrm{v}\delta}W_{x\mathrm{v,s}}-4W_{y\mathrm{v,s}})$	$-0.002\left(\frac{\Delta t_{\mathrm{w}}}{t_{\mathrm{s}}}\right)^2+0.0635\frac{\Delta t_{\mathrm{w}}}{t_{\mathrm{s}}}+0.0705$ (for film boiling of saturated water)

bulk subcooled grade $\Delta t_\infty/t_s$. In fact, the temperature gradient $(\mathrm{d}\theta_v/\mathrm{d}\eta_v)_{\eta_v=0}$ on the plates is steeper with higher liquid bulk subcooled grade $\Delta t_\infty/t_s$ and with lower wall superheated grade $\Delta t_w/t_s$.

The interfacial velocity component $W_{xv,s}$ will increase with increasing the wall superheated grade $\Delta t_w/t_s$ except the case for very low water bulk subcooled grade $\Delta t_\infty/t_s$. Meanwhile, the interfacial velocity component $W_{xv,s}$ will decrease with increasing the liquid bulk subcooled grade $\Delta t_\infty/t_s$.

The interfacial velocity component $-W_{yv,s}$ and mass flow rate parameter $(\eta_{v\delta}W_{xv,s} - 4W_{yv,s})$ will increase with increasing the wall superheated grade $\Delta t_w/t_s$, and will increase obviously with decreasing the liquid subcooled grade $\Delta t_\infty/t_s$. The value of the interfacial velocity component $-W_{yv,s}$ is much larger than that of the interfacial velocity component $W_{xv,s}$ usually. Then, it follows that $-W_{yv,s}$ will dominate the interfacial mass flow rate in general. However, the effect of the interfacial velocity component $W_{xv,s}$ on the interfacial mass flow rate can never be ignored.

The effects of the wall superheated grade $\Delta t_w/t_s$ and liquid bulk subcooled grade $\Delta t_\infty/t_s$ on the momentum, heat, and mass transfer presented here also reveal effects of variable thermophysical properties of both vapor and liquid film on the film boiling of subcooled liquid.

The curve-fit equations introduced in this chapter agree very well with the related rigorous numerical solutions, and useful for a simple and accurate prediction of heat and mass transfer of the laminar film boiling of subcooled water.

11.8 Calculation Example

Example: A flat plate with 0.3 m width and 0.3 m length is suspended vertically in water. The plate temperature is kept at $t_w = 577°\mathrm{C}$. The water bulk temperature is $t_\infty = 90°\mathrm{C}$. Assume the steady laminar film boiling occurs on the plates. Calculate the heat transfer and mass flow rate of the film boiling.

Solution. The wall superheated grade is $\Delta t_w/t_s = (t_w - t_s)/t_s = (577 - 100)/100 = 4.77$, and the water bulk subcooled grade is $\Delta t_\infty/t_s = (t_s - t_\infty)/t_s = (100 - 90)/100 = 0.1$.

The related physical properties are water saturated density $\rho_{l,s} = 958.4\,\mathrm{kg\,m^{-3}}$ at $t_s = 100°\mathrm{C}$, saturated water vapor kinetic viscosity $\nu_{v,s} = 20.55 \times 10^{-6}\,\mathrm{m^2\,s^{-1}}$, density $\rho_{v,s} = 0.5974\,\mathrm{kg\,m^{-3}}$, and absolute viscosity $\mu_{v,s} = 12.28 \times 10^{-6}$ at $t_s = 100°\mathrm{C}$, water vapor density $\rho_{v,w} = 0.2579\,\mathrm{kg\,m^{-3}}$ and thermal conductivity $\lambda_{v,w} = 0.0637\,\mathrm{W\,(m\,K)^{-1}}$ at $t_w = 577°\mathrm{C}$, and subcooled water density $\rho_{l,\infty} = 965.3\,\mathrm{kg\,m^{-3}}$ at $t_\infty = 90°\mathrm{C}$.

1. For heat transfer. With (11.64) the temperature gradient of the film boiling of subcooled water vapor is evaluated as

$$-\left(\frac{\mathrm{d}\theta_v}{\mathrm{d}\eta_v}\right)_{\eta_v=0} = \frac{\exp\left(A + B\frac{\Delta t_w}{t_s} + C\left(\frac{\Delta t_w}{t_s}\right)^2\right)}{\frac{\Delta t_w}{t_s}} \times 10^{-2}.$$

Since $\Delta t_\infty/t_\mathrm{s} = 0.1$.

The following formulae and calculations for the coefficients A, B, and C are available:

$$\begin{aligned}A &= 4.7356 + 7.407\frac{\Delta t_\infty}{t_\mathrm{s}} - 7.4 \times \left(\frac{\Delta t_\infty}{t_\mathrm{s}}\right)^2 \\ &= 4.7356 + 7.407 \times 0.1 - 7.4 \times 0.1^2 \\ &= 5.4023,\end{aligned}$$

$$\begin{aligned}B &= 0.1228 - 1.633\frac{\Delta t_\infty}{t_\mathrm{s}} + 2.71\left(\frac{\Delta t_\infty}{t_\mathrm{s}}\right)^2 \\ &= 0.1228 - 1.633 \times 0.1 + 2.71 \times (0.1)^2 \\ &= -0.0134,\end{aligned}$$

$$\begin{aligned}C &= -0.0086 + 1.092 \times 10^{-1}\frac{\Delta t_\infty}{t_\mathrm{s}} - 2.132 \times 10^{-1}\left(\frac{\Delta t_\infty}{t_\mathrm{s}}\right)^2 \\ &= 0.0086 + 1.092 \times 10^{-1} \times 0.1 - 2.131 \times 10^{-1}(0.1)^2 \\ &= 0.000188\end{aligned}$$

$$\begin{aligned}-\left(\frac{\mathrm{d}\theta_\mathrm{v}}{\mathrm{d}\eta_\mathrm{v}}\right)_{\eta_\mathrm{v}=0} &= \frac{\exp\left(A + B\frac{\Delta t_\mathrm{w}}{t_\mathrm{s}} + C\left(\frac{\Delta t_\mathrm{w}}{t_\mathrm{s}}\right)^2\right)}{\frac{\Delta t_\mathrm{w}}{t_\mathrm{s}}} \times 10^{-2} \\ &= \frac{\exp(5.4023 - 0.0134 \times 4.77 + 0.000188 \times (4.77^2))}{4.77} \times 10^{-2} \\ &= 0.438298.\end{aligned}$$

With (11.15) local Grashof number $Gr_{x\mathrm{v,s}}$ is evaluated as

$$\begin{aligned}Gr_{x\mathrm{v,s}} &= \frac{g(\rho_{\mathrm{l},\infty}/\rho_\mathrm{v,w} - 1)x^3}{\nu_\mathrm{v,s}^2} \\ &= \frac{9.8(965.3/0.2579 - 1) \times 0.3^3}{(20.55 \times 10^{-6})^2} \\ &= 2.34456 \times 10^{12}.\end{aligned}$$

With (11.61) the local Nusselt number is evaluated as

$$\begin{aligned}Nu_{x\mathrm{v,w}} &= -\left(\frac{1}{4}Gr_{x\mathrm{v,s}}\right)^{1/4}\left(\frac{\mathrm{d}\theta_\mathrm{v}}{\mathrm{d}\eta_\mathrm{v}}\right)_{\eta_\mathrm{v}=0} \\ &= \left(\frac{1}{4} \times 2.34456 \times 10^{12}\right)^{1/4} \times 0.438298 \\ &= 383.50.\end{aligned}$$

According to the definition of the local Nusselt number, $Nu_{x\mathrm{v,w}} = \alpha_x x/\lambda_\mathrm{v,w}$, then

$$\begin{aligned}\alpha_x &= \frac{Nu_{x\mathrm{v,w}}\lambda_{\mathrm{v,w}}}{x}\\ &= \frac{383.50 \times 0.0637}{0.3}\\ &= 81.43\,\mathrm{W\,(m^2\,{}^\circ C)^{-1}}.\end{aligned}$$

At last, average heat transfer coefficient $\overline{\alpha}_x$ and total heat transfer rate Q_x of the film boiling of the plate are calculated

$$\begin{aligned}\overline{\alpha}_x &= \frac{4}{3}\alpha_x\\ &= \frac{4}{3} \times 81.43\\ &= 108.57\,\mathrm{W\,(m^2\,{}^\circ C)^{-1}},\\ Q_x &= \overline{\alpha}_x(t_{\mathrm{w}} - t_{\mathrm{s}})A\\ &= 108.57 \times (577 - 100) \times 0.3 \times 0.3\\ &= 4660.91\,\mathrm{W}.\end{aligned}$$

2. For mass flow rate of the boiling. The total mass flow rate of the film boiling of subcooled water is expressed as

$$G_x = \frac{4}{3}b \cdot \mu_{\mathrm{v,s}}\left(\frac{1}{4}Gr_{x\mathrm{v,s}}\right)^{1/4}(\eta_{\mathrm{v}\delta}W_{x\mathrm{v,s}} - 4W_{y\mathrm{v,s}}).$$

From Table 11.2, the related mass flow rate parameter $(\eta_{\mathrm{v}\delta}W_{x\mathrm{v,s}} - 4W_{y\mathrm{v,s}})$ is obtained as 0.16399 for $\Delta t_{\mathrm{w}}/t_{\mathrm{s}} = 4.77$ and $\Delta t_{\infty}/t_{\mathrm{s}} = 0.1$.

Then,

$$\begin{aligned}G_x &= \frac{4}{3}b \cdot \mu_{\mathrm{v,s}}\left(\frac{1}{4}Gr_{x\mathrm{v,s}}\right)^{1/4}(\eta_{\mathrm{v}\delta}W_{x\mathrm{v,s}} - 4W_{y\mathrm{v,s}})\\ &= \frac{4}{3} \times 0.3 \times 12.28 \times 10^{-6} \times \left(\frac{1}{4} \times 2.34456 \times 10^{12}\right)^{1/4} \times 0.16399\\ &= 0.000705\,\mathrm{kg\,s^{-1}}\\ &= 2.537\,\mathrm{kg\,h^{-1}}.\end{aligned}$$

References

1. D. Y. Shang, L. C. Zhong, and B. X. Wang, Study of film boiling of water with variable thermophysical properties along a vertical plate, transfer phenomena, Science and Technology, 1992 (Ed. by B. X. Wang), Higher Education Press, Beijing, pp. 427–432, 1992
2. L. C. Zhong, D. Y. Shang, and B. X. Wang, Effect of variable thermophysical properties on film boiling of liquid. J. Northeast. Univ., 15, No. 88, pp. 6–11, 1994

3. D. Y. Shang , B. X. Wang, and L. C. Zhong, A study on laminar film boiling of liquid along an isothermal vertical plates in a pool with consideration of variable thermophysical properties, Int. J. Heat Mass Transfer, 37, No. 5, pp. 819–828, 1994
4. VDI – Warmeatlas, Berechnungs blätter fur Warmeubertragung, 5, erweiterte Auflage, VDI Verlage GmbH, Dusseldorf, 1988
5. J. C. H. Koh, Analysis of film boiling on vertical surface, J. Heat Transfer, 84C, No. 1, pp. 55–62, 1962
6. K. Nishikawa and T. Ito, Two-phase boundary-layer treatment of free convection film boiling, Int. J. Heat Mass Transfer, 9, No. 2, pp. 103–115, 1966

12

Laminar Film Condensation of Saturated Vapor

Nomenclature

a	thermal diffusive coefficient, $m^2\ s^{-1}$
b	plate width, m
c_p	specific heat at constant pressure, J $(kg\ K)^{-1}$
g	gravitation acceleration, $m\ s^{-2}$
$(g_x)_i$	local mass flow rate entering the liquid film at position x per unit area of the plate, kg $(m^2\ s)^{-1}$
G_x	total mass flow rate entering the liquid film for position $x = 0$ to x with width of b of the plate, kg s^{-1}
$Gr_{x\mathrm{l,s}}$	local Grashof number of liquid film for film condensation of saturated vapor, $\frac{g(\rho_{\mathrm{l,w}}-\rho_{\mathrm{v}})x^3}{\nu_{\mathrm{l,s}}^2\rho_{\mathrm{l,s}}}$
$Gr_{x\mathrm{v}}$	local Grashof number of vapor film for film condensation of saturated vapor, gx^3/ν_{v}^2
$\overline{Nu_{x,\mathrm{w}}}$	average Nusselt number, $\overline{\alpha}_x x/\lambda_{\mathrm{w}}$
$Nu_{xv,w}$	local Nusselt number, $a_x x/\lambda_{\mathrm{w}}$
n_{c_p}	specific heat parameter of gas
n_λ	thermal conductivity parameter of gas
n_μ	viscosity parameter of gas
Pr	Prandtl number
q_x	local heat transfer rate at position x per unit area on the plate, W m^{-2}
Q_x	total heat transfer rate for position $x = 0$ to x with width of b on the plate, W
t	temperature, °C
T	absolute temperature, K
w_x, w_y	velocity components in the x- and y- directions, respectively
W_x, W_y	dimensionless velocity components in the x- and y- directions, respectively

$W_{x\mathrm{l,s}}, W_{y\mathrm{l,s}}$	dimensionless velocity components of liquid film at liquid–vapor interface
x, y	dimensional coordinate variables

Greek symbols

α_x	local heat transfer coefficient, W (m^2 K)$^{-1}$
$\overline{\alpha}_x$	average heat transfer coefficient, W (m^2 K)$^{-1}$
δ	boundary layer thickness, m
δ_l	thickness of liquid film, m
δ_v	thickness of vapor film, m
η	dimensionless coordinate variable for boundary layer
θ	dimensionless temperature
λ	thermal conductivity, W (m K)$^{-1}$
μ	absolute viscosity, kg (m s)$^{-1}$
ν	kinetic viscosity, m^{-2} s^{-1}
ρ	density, kg m^{-3}
Δt_w	wall subcooled temperature, $t_\mathrm{s} - t_\mathrm{w}$,° C
$\frac{\Delta t_\mathrm{w}}{t_\mathrm{s}}$	wall subcooled grade, $(t_\mathrm{s} - t_\mathrm{w})/t_\mathrm{s}$
Δt_∞	superheated temperature of vapor, $t_\infty - t_\mathrm{s}$,° C
$-\left(\left(\frac{\mathrm{d}\theta_\mathrm{l}}{\mathrm{d}\eta_\mathrm{l}}\right)_{\eta=0}\right)_{\Delta t_\infty=0}$	dimensionless temperature gradient on the plate for film condensation of saturated vapor
$\eta_{\mathrm{l}\delta}$	dimensionless thickness of liquid film
$\eta_{\mathrm{l}\delta} W_{x\mathrm{l,s}} - 4W_{y\mathrm{l,s}}$	mass flow rate parameter for film condensation
$\dfrac{\rho_\mathrm{l} - \rho_\mathrm{v}}{\rho_\mathrm{l,w} - \rho_\mathrm{v}}$	buoyancy factor
$\dfrac{1}{\rho}\dfrac{\mathrm{d}p}{\mathrm{d}x}$	density factor
$\dfrac{1}{\mu}\dfrac{\mathrm{d}\mu}{\mathrm{d}\eta}$	viscosity factor
$\dfrac{1}{\lambda}\dfrac{\mathrm{d}\lambda}{\mathrm{d}\eta}$	thermal conductivity factor

Subscripts

i	liquid
s	saturate state, or at the liquid-vapor interface
v	vapor
w	at wall
∞	far from the wall surface

12.1 Introduction

It was Nusselt [1] who first treated the laminar film condensation of saturated steam on a vertical isothermal flat plate. His theory was based on the assumption that the inertia and thermal convection of condensate film, the vapor drag due to the shear force at the liquid–vapor interface, the dependence of the thermophysical properties of the condensate medium on temperature, and the effect of the liquid–vapor interfacial wave are neglected. Bromley [2] and Rohsenow [3] first investigated the effects of thermal convection. Later on, the study of Sparrow and Gregg [4] included also the effects of thermal convection and inertia forces in the liquid film by using the boundary layer analysis, and Koh et al. [5] further solved numerically a boundary-layer model for both the condensate and vapor films. Chen [6] has considered analytically the effect of thermal convection, the inertia, and the interfacial shear force.

On the basis of foregoing studies of the independent-temperature physical properties Mc Adams [7], Voskresenskiy [8], and Labuntsov [9] made relatively simple modifications for variable thermophysical properties. Then, Poots and Miles [10] studied the effects of variable thermophysical properties on laminar film condensation of saturated steam on a vertical flat plate. They have simplified the governing equations of the liquid and vapor phases by neglecting the effects of surface tension at the liquid–vapor interface, and have got solutions of the ordinary differential equations. Nevertheless the results obtained do not allow heat and mass transfer prediction, probably due to the difficulty of getting a solution.

In this book, I will use three chapters (Chaps. 12–14) to introduce an extended study development of Shang, Wang and Adamek [11–13] for analyzing and calculating the film condensation of vapor and to further clarify the character of its heat and mass transfer.

In this chapter, the extended theory of steady state laminar film condensation process of pure saturated vapor at atmospheric pressure on an isothermal vertical flat plate is established. Its equations provide a complete account of the physical process for consideration of various physical factors including variable thermophysical properties, except for surface tension at the liquid–vapor film interface.

First, similarity considerations are proposed to transform the governing system of partial differential equations and its boundary conditions into the corresponding dimensionless system.

Then, the dimensionless new system is computed numerically in two steps: First, neglecting shear force at the interface, so that the initial values of the boundary conditions $W_{x\mathrm{l,s}}$ and $W_{y\mathrm{l,s}}$ are obtained. Then, the calculations of a problem of the three-point boundary values for coupling the equations of liquid film with those of vapor film are carried out.

Furthermore, the correlations for the heat transfer coefficient and mass flow rate are proposed by analysis of heat and mass transfer and it is found that the heat transfer coefficient is function of dimensionless temperature gradient $(\mathrm{d}\theta_\mathrm{l}/\mathrm{d}\eta_\mathrm{l})_{\eta_\mathrm{l}=0}$, and condensate mass flow rate is a function of the mass flow rate parameter $(\eta_{\mathrm{l}\delta} W_{x\mathrm{l,s}} - W_{y\mathrm{l,s}})$. In addition, the curve-fitting formulae corresponding to heat and mass transfer correlations are developed based on the rigorous numerical solutions.

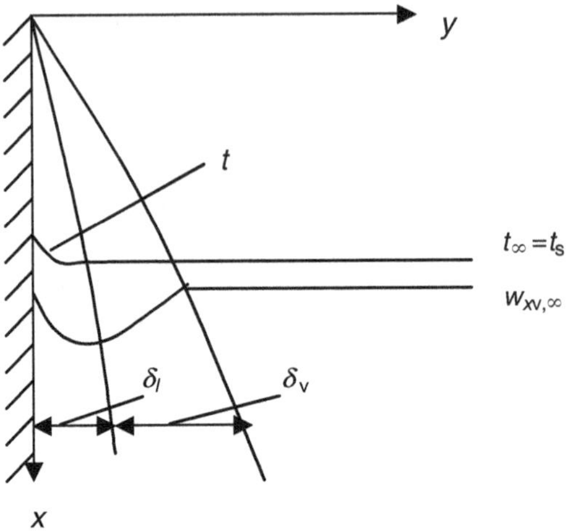

Fig. 12.1. Physical model and coordinate system of film condensation of saturate vapor

12.2 Governing Partial Differential Equations

The analytical model and coordinate system used for the laminar film condensation of saturated vapor on an isothermal vertical flat plate is shown in Fig. 12.1. An isothermal vertical flat plate is suspended in a large volume of quiescent pure saturated vapor at atmospheric pressure. The plate temperature is t_w and the saturation temperature of the fluid is t_s. If provided condition for the model is $t_w < t_s$, a steady film condensation will occur on the plate. We assume that laminar flow within the liquid and vapor phases is induced by gravity, and take into account the shear force at the liquid–vapor interface, the inertia force and thermal convection of the condensate film, and the variable thermophysical properties. Then the conservation governing equations of mass, momentum, and energy for steady laminar condensation in two-phase boundary layer are as follows:

With consideration of variable thermophysical properties, the governing partial differential equations for condensate liquid film are

$$\frac{\partial}{\partial x}(\rho_l w_{xl}) + \frac{\partial}{\partial y}(\rho_l w_{yl}) = 0, \tag{12.1}$$

$$\rho_l \left(w_{xl}\frac{\partial w_{xl}}{\partial x} + w_{yl}\frac{\partial w_{xl}}{\partial y} \right) = \frac{\partial}{\partial y}\left(\mu_l \frac{\partial w_{xl}}{\partial y} \right) + g(\rho_l - \rho_v), \tag{12.2}$$

$$\rho_l c_{p_l} \left(w_{xl}\frac{\partial t_l}{\partial x} + w_{yl}\frac{\partial t_l}{\partial y} \right) = \frac{\partial}{\partial y}\left(\lambda_l \frac{\partial t_l}{\partial y} \right). \tag{12.3}$$

For vapor film the condition of saturation implies that the vapor temperature t_∞ is essentially constant, and equals t_s. Consequently, the variable thermophysical properties and the energy equation need not be considered, and only the continuity and momentum equations remain:

$$\frac{\partial}{\partial y}(w_{y\mathrm{v}}) + \frac{\partial}{\partial y}(w_{y\mathrm{v}}) = 0, \tag{12.4}$$

$$w_{x\mathrm{v}}\frac{\partial w_{x\mathrm{v}}}{\partial x} + w_{y\mathrm{v}}\frac{\partial w_{x\mathrm{v}}}{\partial y} = \nu_{\mathrm{v}}\frac{\partial^2 w_{x\mathrm{v}}}{\partial y^2}. \tag{12.5}$$

The boundary conditions are

$$y = 0: \quad w_{x\mathrm{l}} = 0, \quad w_{y\mathrm{l}} = 0, \quad t_{\mathrm{l}} = t_{\mathrm{w}}, \tag{12.6}$$

$$y = \delta_{\mathrm{l}}$$

$$w_{x\mathrm{l,s}} = w_{x\mathrm{v,s}}, \tag{12.7}$$

$$\rho_{\mathrm{l,s}}\left(w_{x\mathrm{l}}\frac{\partial \delta_{\mathrm{l}}}{\partial x} - w_{y\mathrm{l}}\right)_{\mathrm{s}} = \rho_{\mathrm{v}}\left(w_{x\mathrm{v}}\frac{\partial \delta_{\mathrm{v}}}{\partial x} - w_{y\mathrm{v}}\right)_{\mathrm{s}}, \tag{12.8}$$

$$\mu_{\mathrm{l,s}}\left(\frac{\partial w_{x\mathrm{l}}}{\partial y}\right)_{\mathrm{s}} = \mu_{\mathrm{v}}\left(\frac{\partial w_{x\mathrm{v}}}{\partial y}\right)_{\mathrm{s}}, \tag{12.9}$$

$$\lambda_{\mathrm{l,s}}\left(\frac{\partial t_{\mathrm{l}}}{\partial y}\right)_{y=\delta_{\mathrm{l}}} = h_{fg}\rho_{\mathrm{l,s}}\left(w_{x\mathrm{l}}\frac{\partial \delta_{\mathrm{l}}}{\partial x} - w_{y\mathrm{l}}\right)_{\mathrm{s}}, \tag{12.10}$$

$$t_{\mathrm{l}} = t_{\mathrm{s}}, \tag{12.11}$$

$$y \to \infty:$$

$$w_{x\mathrm{v}} \to 0, \tag{12.12}$$

where (12.7) expresses a continuity of the interfacial velocity components of liquid and vapor phases, (12.8) expresses that the local mass flux crossing the liquid–vapour interface is continuous, (12.9) shows the balance condition for shear force at the liquid–vapor interface which also implies a consideration of the drag of the vapor film, (12.10) describes that the latent heat of condensation is in balance with the heat conduction of liquid at the interface, and (12.11) expresses the temperature continuity at the liquid–vapor interface. Obviously, the subscript v related to the vapor film in the above equations refers to the case of saturated vapor.

12.3 Similarity Variables

Using the velocity component method we assume the following dimensionless variables for the similarity transformation of the governing partial differential equations of the film condensation of saturated vapor. This assumption is subdivided into the following phases:

For liquid film. For liquid film the similarity transformation variables are assumed as follows:

The dimensionless coordinate variable η_{l} of liquid film is set up as

$$\eta_{\mathrm{l}} = \left(\frac{1}{4}Gr_{x\mathrm{l,s}}\right)^{1/4}\frac{y}{x}, \tag{12.13}$$

where the local Grashof number $Gr_{x\mathrm{l,s}}$ is assumed as

$$Gr_{x\mathrm{l,s}} = \frac{g(\rho_{\mathrm{l,w}} - \rho_{\mathrm{v,s}})x^3}{\nu_{\mathrm{l,s}}^2 \rho_{\mathrm{l,s}}}. \tag{12.14}$$

The dimensionless temperature is given as

$$\theta_\mathrm{l} = \frac{t - t_\mathrm{s}}{t_\mathrm{w} - t_\mathrm{s}}. \tag{12.15}$$

The dimensionless velocity components are assumed as

$$W_{x\mathrm{l}} = \left(2\sqrt{gx}\left(\frac{\rho_{\mathrm{l,w}} - \rho_{\mathrm{v,s}}}{\rho_{\mathrm{l,s}}}\right)^{1/2}\right)^{-1} w_{x\mathrm{l}}, \tag{12.16}$$

$$W_{y\mathrm{l}} = \left(2\sqrt{gx}\left(\frac{\rho_{\mathrm{l,w}} - \rho_{\mathrm{v,s}}}{\rho_{\mathrm{l,s}}}\right)^{1/2}\left(\frac{1}{4}Gr_{x\mathrm{l,s}}\right)^{-1/4}\right)^{-1} w_{y\mathrm{l}}. \tag{12.17}$$

For vapor film. For the similarity transformations for vapor film of the governing partial differential equations of the film condensation of saturated vapor we introduce the following similarity variables:

The dimensionless coordinate variable η_v of vapor film and the local Grashof number $Gr_{x\mathrm{v,s}}$ are assumed as, respectively,

$$\eta_\mathrm{v} = \left(\frac{1}{4}Gr_{\mathrm{xv}}\right)^{1/4}\frac{y}{x}, \quad Gr_{x\mathrm{v}} = \frac{gx^3}{\nu_\mathrm{v}^2}. \tag{12.18}$$

The dimensionless velocity components are assumed as

$$W_{x\mathrm{v}} = (2\sqrt{gx})^{-1} w_{x\mathrm{v}}, \tag{12.19}$$

$$W_{y\mathrm{v}} = (2\sqrt{gx}\left(\frac{1}{4}Gr_{x\mathrm{v}}\right)^{-1/4})^{-1} w_{y\mathrm{v}}. \tag{12.20}$$

12.4 Similarity Transformation of Governing Equations

The similarity transformations of the governing partial differential equations and the boundary conditions for the film condensation of saturated vapor are done in Appendix 1, and the transformed results are shown as follows:

For liquid film. From (12.59), (12.64), and (12.71) of Appendix A, the transformed governing equations for the liquid film are shown as follows, respectively:

$$2W_{x\mathrm{l}} - \eta_\mathrm{l}\frac{\mathrm{d}W_{x\mathrm{l}}}{\mathrm{d}\eta_\mathrm{l}} + 4\frac{\mathrm{d}W_{y\mathrm{l}}}{\mathrm{d}\eta_\mathrm{l}} + \frac{1}{\rho_\mathrm{l}}\frac{\mathrm{d}\rho_\mathrm{l}}{\mathrm{d}\eta_\mathrm{l}}(-\eta_\mathrm{l} W_{x\mathrm{l}} + 4W_{y\mathrm{l}}) = 0, \tag{12.21}$$

$$\frac{\nu_{\rm l,s}}{\nu_{\rm l}}\left(W_{x\rm l}\left(2W_{x\rm l}-\eta_{\rm l}\frac{{\rm d}W_{x\rm l}}{{\rm d}\eta_{\rm l}}\right)+4W_{y\rm l}\frac{{\rm d}W_{x\rm l}}{{\rm d}\eta_{\rm l}}\right)$$

$$=\frac{{\rm d}^2W_{x\rm l}}{{\rm d}\eta_{\rm l}^2}+\frac{1}{\mu_{\rm l}}\frac{{\rm d}\mu_{\rm l}}{{\rm d}\eta_{\rm l}}\frac{{\rm d}W_{x\rm l}}{{\rm d}\eta_{\rm l}}+\frac{\mu_{\rm l,s}}{\mu_{\rm l}}\frac{\rho_{\rm l}-\rho_{\rm v}}{\rho_{\rm l,w}-\rho_{\rm v}}, \tag{12.22}$$

$$Pr_{\rm l}\frac{\nu_{\rm l,s}}{\nu_{\rm l}}\left[-W_{x\rm l}\frac{{\rm d}\theta_{\rm l}}{{\rm d}\eta_{\rm l}}\eta_{\rm l}+4W_{y\rm l}\frac{{\rm d}\theta_{\rm l}}{{\rm d}\eta_{\rm l}}\right]=\frac{{\rm d}^2\theta_{\rm l}}{{\rm d}\eta_{\rm l}^2}+\frac{1}{\lambda_{\rm l}}\frac{{\rm d}\lambda_{\rm l}}{{\rm d}\eta_{\rm l}}\frac{{\rm d}\theta_{\rm l}}{{\rm d}\eta_{\rm l}}. \tag{12.23}$$

For vapor film. From Appendix A, the transformed governing equations for the vapor film are shown as follows, respectively:

$$2W_{x\rm v}-\eta_{\rm v}\frac{{\rm d}W_{x\rm v}}{{\rm d}\eta_{\rm v}}+4\frac{{\rm d}W_{y\rm v}}{{\rm d}\eta_{\rm v}}=0, \tag{12.24}$$

$$W_{x\rm v}\left(2W_{x\rm v}-\eta_{\rm v}\frac{{\rm d}W_{x\rm v}}{{\rm d}\eta_{\rm v}}\right)+4W_{y\rm v}\frac{{\rm d}W_{x\rm v}}{{\rm d}\eta_{\rm v}}=\frac{{\rm d}^2W_{x\rm v}}{{\rm d}\eta_{\rm v}^2}. \tag{12.25}$$

For boundary conditions. From (12.26)–(12.33) of Appendix A, the transformed boundary conditions are shown as follows, respectively:

$\eta_{\rm l}=0:$

$$W_{x\rm l}=0,\quad W_{y\rm l}=0,\quad \theta_{\rm l}=1, \tag{12.26}$$

$\eta_{\rm v}=\eta_{\rm l\delta}(\eta_{\rm v}=0):$

$$W_{x\rm v,s}=\left(\frac{\rho_{\rm l,w}-\rho_{\rm v}}{\rho_{\rm l,s}}\right)^{1/2}W_{x\rm l,s}, \tag{12.27}$$

$$W_{y\rm v,s}=-0.25\frac{\mu_{\rm l,s}}{\mu_{\rm v}}\left(\frac{\nu_{\rm v,s}}{\nu_{\rm l,s}}\right)^{1/2}\left(\frac{\rho_{\rm l,w}-\rho_{\rm v}}{\rho_{\rm l,s}}\right)^{1/4}(W_{x\rm l,s}\eta_{\rm l\delta}-4W_{y\rm l,s}), \tag{12.28}$$

$$\left(\frac{{\rm d}W_{x\rm v}}{{\rm d}\eta_{\rm v}}\right)_{\eta_{\rm v}=0}=\frac{\mu_{\rm l,s}}{\mu_{\rm v}}\left(\frac{\rho_{\rm l,w}-\rho_{\rm v}}{\rho_{\rm l,s}}\right)^{3/4}\left(\frac{\nu_{\rm v}}{\nu_{\rm l,s}}\right)^{1/2}\left(\frac{{\rm d}W_{x\rm l}}{{\rm d}\eta_{\rm l}}\right)_{\eta{\rm l}=\eta{\rm l}\delta}, \tag{12.29}$$

$$h_{fg}\mu_{\rm l,s}(W_{x\rm l,s}\eta_{\rm l\delta}-4W_{y\rm l,s})-\lambda_{p_{\rm l,s}}t_{\rm w}-t_{\rm s})\left(\frac{{\rm d}\theta_{\rm l}}{{\rm d}\eta_{\rm l}}\right)_{\eta_{\rm l}=\eta_{\rm l\delta}}=0, \tag{12.30}$$

$$\theta_{\rm l}=0, \tag{12.31}$$

$\eta_{\rm v}\to\infty:$

$$W_{x\rm v}\to 0. \tag{12.32}$$

12.5 Numerical Solutions

12.5.1 Treatment of Variable Thermophysical Properties

The variable thermophysical properties of the governing ordinary differential equations (12.21)–(12.23) and the boundary conditions must be treated.

For example, the corresponding predictive expressions for density ρ_l, thermal conductivity λ_l, and absolute viscosity μ_l, for water are

$$\rho_\mathrm{l} = -4.48\times10^{-3}t^2 + 999.9, \tag{12.33}$$

$$\lambda_\mathrm{l} = -8.01\times10^{-6}t^2 + 1.94\times10^{-3}t + 0.563, \tag{12.34}$$

$$\mu_\mathrm{l} = \exp\left[-1.6 - \frac{1150}{T} + \left(\frac{690}{T}\right)^2\right]\times10^{-3}. \tag{12.35}$$

Similar to the related derivations in Chap. 11, the thermophysical property factors $(1/\rho_\mathrm{l})(\mathrm{d}\rho_\mathrm{l}/\mathrm{d}\eta_\mathrm{l})$, $(1/\mu_\mathrm{l})(\mathrm{d}\mu_\mathrm{l}/\mathrm{d}\eta_\mathrm{l})$, $(1/\lambda_\mathrm{l})(\mathrm{d}\lambda_\mathrm{l}/\mathrm{d}\eta_\mathrm{l})$, $\mu_\mathrm{l,s}/\mu_\mathrm{l}$, $\nu_\mathrm{l,s}/\nu_\mathrm{l}$, and $Pr_\mathrm{l}(\mu_\mathrm{l,s}/\mu_\mathrm{l})(\rho_\mathrm{l}/\rho_\mathrm{l,s})Pr_\mathrm{l,s}(\rho_\mathrm{l}/\rho_\mathrm{l,s})(\lambda_\mathrm{l,s}/\lambda_\mathrm{l})$ in the governing ordinary differential equations of liquid film, (12.21)–(12.23) are shown as later:

$$\frac{1}{\rho_\mathrm{l}}\frac{\mathrm{d}\rho_\mathrm{l}}{\mathrm{d}\eta_\mathrm{l}} = \frac{[-2\times4.48\times10^{-3}t](t_\mathrm{w}-t_\infty)\frac{\mathrm{d}\theta_\mathrm{l}}{\mathrm{d}\eta_\mathrm{l}}}{-4.48\times10^{-3}t^2+999.9}, \tag{12.36}$$

$$\frac{1}{\lambda_\mathrm{l}}\frac{\mathrm{d}\lambda_\mathrm{l}}{\mathrm{d}\eta_\mathrm{l}} = \frac{[-2\times8.01\times10^{-6}t+1.94\times10^{-3}](t_\mathrm{w}-t_\infty)\frac{\mathrm{d}\theta_\mathrm{l}}{\mathrm{d}\eta_\mathrm{l}}}{-8.01\times10^{-6}t^2+1.94\times10^{-3}t+0.563}, \tag{12.37}$$

$$\frac{1}{\mu_\mathrm{l}}\frac{\mathrm{d}\mu_\mathrm{l}}{\mathrm{d}\eta_\mathrm{l}} = \left[\frac{1150}{T^2} - 2\times\frac{690^2}{T^3}\right](t_\mathrm{w}-t_\infty)\frac{\mathrm{d}\theta_\mathrm{l}}{\mathrm{d}\eta_\mathrm{l}}. \tag{12.38}$$

Additionally, with (12.33) and (12.35) we obtain

$$\frac{\mu_\mathrm{l,s}}{\mu_\mathrm{l}} = \exp\left[-1150\left(\frac{1}{T_\mathrm{l,s}}-\frac{1}{T_\mathrm{l}}\right)+690^2\left(\frac{1}{T_\mathrm{l,s}}-\frac{1}{T_\mathrm{l}}\right)\right] \tag{12.39}$$

and

$$\frac{\nu_\mathrm{l,s}}{\nu_\mathrm{l}} = \frac{\mu_\mathrm{l,s}}{\mu_\mathrm{l}}\frac{\rho_\mathrm{l}}{\rho_\mathrm{l,s}} = \exp\left[-1150\left(\frac{1}{T_\mathrm{l,s}}-\frac{1}{T_\mathrm{l}}\right)+690^2\left(\frac{1}{T_\mathrm{l,s}}-\frac{1}{T_\mathrm{l}}\right)\right]\frac{4.48\times10^{-3}t_\mathrm{l}^2+999.9}{4.48\times10^{-3}t_\mathrm{l,s}^2+999.9}. \tag{12.40}$$

In addition, from the analysis of Chap. 6, it is known that the physical factor $Pr_\mathrm{l}(\nu_\mathrm{l,\infty}/\nu_\mathrm{l})$ in (12.23) can be expressed as

$$Pr_\mathrm{l}\frac{\nu_\mathrm{l,s}}{\nu_\mathrm{l}} = Pr_\mathrm{l,s}\frac{\rho_\mathrm{l}}{\rho_\mathrm{l,s}}\frac{\lambda_\mathrm{l,s}}{\lambda_\mathrm{l}}$$

for water and a lot of liquids in a special temperature range for engineering application.
where

$$Pr_\mathrm{l,s}\frac{\rho_\mathrm{l}}{\rho_\mathrm{l,s}}\frac{\lambda_\mathrm{l,s}}{\lambda_\mathrm{l}} = Pr_\mathrm{l,s}\frac{-4.48\times10^{-3}t_\mathrm{l}^2+999.9}{-4.48\times10^{-3}t_\mathrm{l,s}^2+999.9} \times\frac{-8.01\times10^{-6}t_\mathrm{l,s}^2+1.94\times10^{-3}t_\mathrm{l,s}+0.563}{-8.01\times10^{-6}t_\mathrm{l}^2+1.94\times10^{-3}t_\mathrm{l}+0.563} \tag{12.41}$$

for water. Here, $t = (t_w - t_s)\theta_l + t_s$ and $T = t + 273$, while, the Prandtl number at the saturated temperature, $Pr_{l,s}$, can be taken from the experimental data.

12.5.2 Calculation Procedure

It is a three-point boundary value problem for solving (12.21)–(12.25) with the boundary condition equations (12.26)–(12.32). The calculations are carried out numerically by two processes, and in each process, the shooting method is applied for the solutions. In the first process the solutions of (12.21)–(12.23) are assumed to be without shear force of vapor at the liquid–vapor interface, and in this case the boundary condition (12.29) must be changed to

$$\left(\frac{\mathrm{d}W_{xl}}{\mathrm{d}\eta}\right)_{\eta_l=\eta_{l\delta}} = 0. \tag{12.42}$$

In this case, (12.26) and (12.31), are regarded as the boundary conditions for getting the solutions of (12.21)–(12.23). Then, (12.30) is taken to adjudge convergence of the solution, and in virtue of changing the value $\eta_{l\delta}$ the successively iterative calculation is continued. After this convergence, the second process for carrying out calculation of three-point boundary problem for coupling equations of liquid film with equations of vapor film is started. In this process, first the boundary values $W_{xv,s}$ and $W_{yv,s}$ are found out by (12.27) and (12.28), respectively. Then (12.24) and (12.25) of vapor film are calculated with the boundary condition (12.32) and the earlier values of $W_{xv,s}$ and $W_{yv,s}$. Equations (12.29) and (12.30) are used to adjudge convergence of the solutions. By means of the adjudgement equations the calculation is iterated by changing the values $W_{xl,s}$ and $\eta_{l\delta}$. In each iteration, calculations of (12.21)–(12.23) for liquid film and (12.24) and (12.25) for vapor film are made successively by the shooting method.

12.5.3 Solution

It will be expected from (12.21)–(12.23) that for consideration of variable thermophysical properties the dimensionless velocity and temperature fields in the condensate film will depend on the temperature-dependent properties ρ_l, ν_l, λ_l, and Pr_l and hence will depend on temperature t, and furthermore, depend on θ, t_w and t_s.

For convenience some special values of the thermophysical properties of water [14] used in the calculation are listed in Table 12.1, and water densities at different temperatures are listed in Table 12.2.

For laminar film condensation of saturated water vapor the numerical calculations have been carried out for wall temperatures $t_w = 0, 20, 40, 60, 80, 90, 95, 97.5$, and 99.9°C (i.e., wall subcooled grade $\Delta t_w/t_s = (t_s - t_w)/t_s = 1$,

Table 12.1. Thermophysical property values for water and water vapor at saturated temperature

term	value	
	for water	for water vapor
t_s(°C)	100	100
c_p (J (kg K)$^{-1}$)	4,216	
h_{fg} (kJ kg^{-1})		2,257.3
Pr	1.76	1
ρ (kg m^{-3})	958.1	0.5974
μ (kg (m s)$^{-1}$)	282.2×10^{-6}	12.28×10^{-6}
ν (m^2 s^{-1})	0.294×10^{-6}	20.55×10^{-6}
λ (W (m K)$^{-1}$)	0.677	0.02478

Table 12.2. Water densities at different temperatures

t(°C)	0	20	40	60	80	90	95	97.5	99.9
ρ (kg m^{-3})	999.8	998.3	992.3	983.2	971.4	965.1	961.7	960.05	958.1

$0.8, 0.6, 0.4, 0.2, 0.1, 0.05, 0.025$, and 0.001), respectively. The calculated results of velocity and temperature fields are plotted in Figs. 12.2 and 12.3, respectively.

It is seen from Figs. 12.2 and 12.3 that with increasing wall subcooled grade $\Delta t_{\mathrm{w}}/t_{\mathrm{s}} = (t_{\mathrm{s}} - t_{\mathrm{w}})/t_{\mathrm{s}}$, the thickness of condensate film will increase, the velocity component $w_{x\mathrm{l,s}}$ at the liquid–vapor interface will increase, but the temperature gradient will decrease.

12.6 Heat and Mass Transfer

12.6.1 Analysis for Heat and Mass Transfer

Heat transfer analysis. The local heat transfer rate q_x of the film condensation at position x per unit area from the surface of the plate to the gas can be calculated by Pourier's law as

$$q_x = -\lambda_{\mathrm{l},w} \left(\frac{\partial t}{\partial y} \right)_{y=0} .$$

With (12.13) and (12.15) we have

$$\left(\frac{\partial t}{\partial y} \right)_{y=0} = \left(\frac{\mathrm{d}t}{\mathrm{d}\eta} \right)_{\eta=0} \frac{\partial \eta}{\partial y}$$
$$= (t_{\mathrm{w}} - t_{\infty}) \left(\left(\frac{\mathrm{d}\theta_{\mathrm{l}}}{\mathrm{d}\eta_{\mathrm{l}}} \right)_{\eta_{\mathrm{l}}=0} \right)_{\Delta t_{\infty}=0} \left(\frac{1}{4} Gr_{x\mathrm{l,s}} \right)^{1/4} x^{-1}.$$

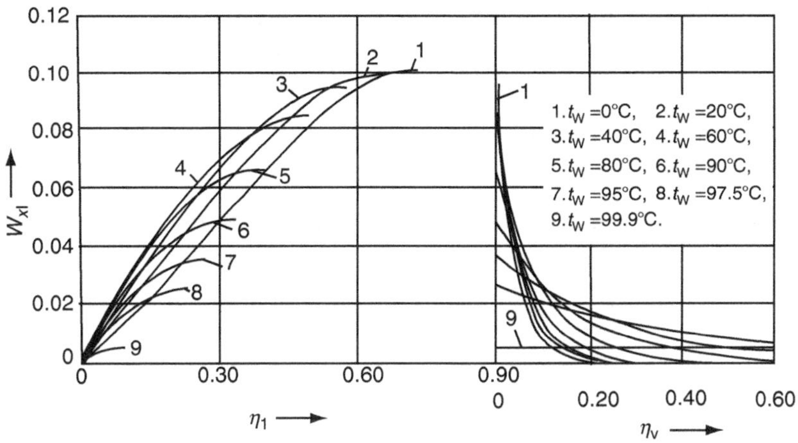

Fig. 12.2. Velocity profiles of liquid and vapor films for laminar film condensation of saturated water vapor, cited from Shang and Adamek [12]
Lines 1–9: $\Delta t_\mathrm{w}/t_\mathrm{s} = (t_\mathrm{s} - t_\mathrm{w})/t_\mathrm{s} = 1, 0.8, 0.6, 0.4, 0.2, 0.1, 0.05, 0.025$, and 0.001

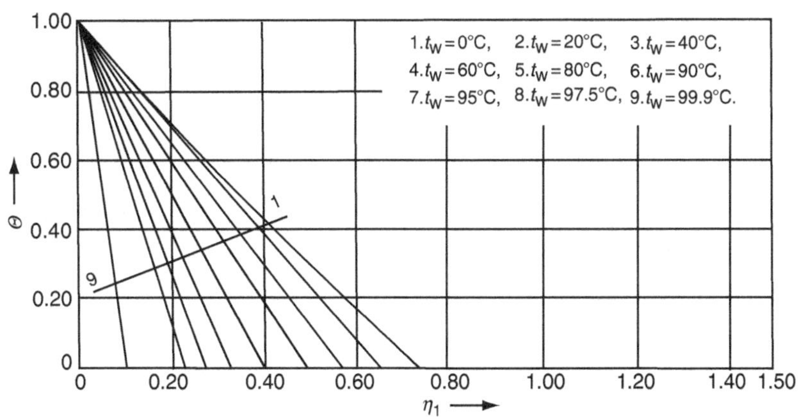

Fig. 12.3. Temperature profiles of liquid and vapor films for laminar film condensation of saturated water vapor, cited from Shang and Adamek [12]
Lines 1–9: $\Delta t_\mathrm{w}/t_\mathrm{s}(= (t_\mathrm{s} - t_\mathrm{w})/t_\mathrm{s}) = 1, 0.8, 0.6, 0.4, 0.2, 0.1, 0.05, 0.025$, and 0.001

Then,

$$q_x = -\lambda_{\mathrm{l,w}}(T_\mathrm{w} - T_\mathrm{s}) \left(\frac{1}{4} Gr_{x\mathrm{l,s}}\right)^{1/4} x^{-1} \left(\left(\frac{\mathrm{d}\theta_\mathrm{l}}{\mathrm{d}\eta_\mathrm{l}}\right)_{\eta_\mathrm{l}=0}\right)_{\Delta t_\infty=0}, \qquad (12.43)$$

where $((-\mathrm{d}\theta_\mathrm{l}/\mathrm{d}\eta_\mathrm{l})_{\eta_\mathrm{l}=0})_{\Delta t_\infty=0}$ is defined as dimensionless temperature gradient on the wall for film condensation of saturated vapor, while the subscript $\Delta t_\infty = 0$ denotes the case with saturated bulk vapor.

Total heat transfer rate for position $x = 0$ to x with width of b on the plate is a integration $Q_x = \iint_A q_x \, \mathrm{d}A = \int_0^x q_x b \, \mathrm{d}x$, and hence

$$Q_x = -\lambda_{\mathrm{l,w}} b(t_{\mathrm{w}} - t_{\infty}) \left(\left(\frac{\mathrm{d}\theta_{\mathrm{l}}}{\mathrm{d}\eta_{\mathrm{l}}} \right)_{\eta_{\mathrm{l}}=0} \right)_{\Delta t_{\infty}=0} \int_0^x \left(\frac{1}{4} Gr_{x\mathrm{l,s}} \right)^{1/4} x^{-1} \, \mathrm{d}x.$$

With definition of local Grashof number $Gr_{x\mathrm{l,s}}$ we obtain

$$Q_x = -\frac{4}{3} b \lambda_{\mathrm{l,w}} (t_{\mathrm{w}} - t_{\infty}) \left(\frac{1}{4} Gr_{x,\infty} \right)^{1/4} \left(\left(\frac{\mathrm{d}\theta_{\mathrm{l}}}{\mathrm{d}\eta_{\mathrm{l}}} \right)_{\eta=0} \right)_{\Delta t_{\infty}=0}. \quad (12.44)$$

The local heat transfer coefficient α_x, defined as $q_x = \alpha_x(t_{\mathrm{w}} - t_{\infty})$, will be given by

$$\alpha_x = -\lambda_{\mathrm{l,w}} \left(\frac{1}{4} Gr_{x\mathrm{l,s}} \right)^{1/4} x^{-1} \left(\left(\frac{\mathrm{d}\theta_{\mathrm{l}}}{\mathrm{d}\eta_{\mathrm{l}}} \right)_{\eta_{\mathrm{l}}=0} \right)_{\Delta t_{\infty}=0}. \quad (12.45)$$

The average heat transfer coefficient $\overline{\alpha_x}$ defined as $Q_x = \overline{\alpha_x}(t_{\mathrm{w}} - t_{\infty}) \times b \times x$ is expressed as

$$\overline{\alpha_x} = -\frac{4}{3} \lambda_{\mathrm{l,w}} \left(\frac{1}{4} Gr_{x\mathrm{l,s}} \right)^{1/4} x^{-1} \left(\left(\frac{\mathrm{d}\theta_{\mathrm{l}}}{\mathrm{d}\eta_{\mathrm{l}}} \right)_{\eta_{\mathrm{l}}=0} \right)_{\Delta t_{\infty}=0}. \quad (12.46)$$

The local Nusselt number defined as $Nu_{x\mathrm{l,w}} = \alpha_x x / \lambda_{\mathrm{l,w}}$ is expressed by

$$Nu_{x\mathrm{l,w}} = -\left(\frac{1}{4} Gr_{x\mathrm{l,s}} \right)^{1/4} \left(\left(\frac{\mathrm{d}\theta_{\mathrm{l}}}{\mathrm{d}\eta_{\mathrm{l}}} \right)_{\eta_{\mathrm{l}}=0} \right)_{\Delta t_{\infty}=0}. \quad (12.47)$$

The average Nusselt number is defined as $\overline{Nu_{x\mathrm{l,w}}} = \overline{\alpha_x} x / \lambda_{\mathrm{l,w}}$, and hence

$$\overline{Nu_{x,\mathrm{w}}} = -\frac{4}{3} \left(\frac{1}{4} Gr_{x\mathrm{l,s}} \right)^{1/4} \left(\left(\frac{\mathrm{d}\theta_{\mathrm{l}}}{\mathrm{d}\eta_{\mathrm{l}}} \right)_{\eta_{\mathrm{l}}=0} \right)_{\Delta t_{\infty}=0}. \quad (12.48)$$

It is seen that, for practical calculation of heat transfer, only $\left(\left(\frac{\mathrm{d}\theta_{\mathrm{l}}}{\mathrm{d}\eta_{\mathrm{l}}} \right)_{\eta_{\mathrm{l}}=0} \right)_{\Delta t_{\infty}=0}$ dependent on numerical solution is no-given variable.

The average heat transfer coefficient $\overline{\alpha_x}$ defined as $Q_x = \overline{\alpha}_x(t_{\mathrm{w}} - t_{\infty})A$ and average Nusselt number $\overline{Nu}_{x\mathrm{l,w}}$ defined as $\overline{Nu}_{x\mathrm{l,w}} = \overline{\alpha_x} x / \lambda_{\mathrm{l,w}}$ will be, respectively,

$$\overline{\alpha_x} = \frac{4}{3} \alpha_x \quad \overline{Nu}_{x\mathrm{l,w}} = \frac{4}{3} Nu_{x\mathrm{l,w}}.$$

Mass transfer analysis. Consulting the mass transfer analysis in Chap. 10 for film boiling of saturated liquid, the related analytic expressions of mass transfer for film condensation of saturated vapor are obtained later:

Set g_x to be a local mass flow rate entering the liquid film at position x per unit area of the plate. According to the boundary layer theory of fluid mechanics, g_x is expressed as

$$g_x = \rho_{\mathrm{l,s}}\left(w_{x\mathrm{l,s}}\frac{\mathrm{d}\delta_{\mathrm{l}}}{\mathrm{d}x} - w_{y\mathrm{l,s}}\right)_{\mathrm{s}}.$$

With the corresponding dimensionless variables in (12.16) and (12.17), the earlier equation is changed into the following one:

$$g_x = \rho_{\mathrm{l,s}}\left[2\sqrt{gx}\left(\frac{\rho_{\mathrm{l,w}}-\rho_{\mathrm{v,s}}}{\rho_{\mathrm{l,s}}}\right)^{1/2} W_{x\mathrm{l,s}}\frac{\mathrm{d}\delta_{\mathrm{l}}}{\mathrm{d}x} - 2\sqrt{gx}\left(\frac{\rho_{\mathrm{l,w}}-\rho_{\mathrm{v,s}}}{\rho_{\mathrm{l,s}}}\right)^{1/2} \times \left(\frac{1}{4}Gr_{x\mathrm{l,s}}\right)^{-1/4} W_{y\mathrm{l,s}}\right]_{\mathrm{s}},$$

where the condensate liquid film thickness is expressed as follows according to (12.13):

$$\delta_{\mathrm{l}} = \eta_{\mathrm{l}\delta}\left(\frac{1}{4}Gr_{x\mathrm{l,s}}\right)^{-1/4} x.$$

With the definition of the local Grashof number $Gr_{x\mathrm{l,s}}$, the earlier equation is changed into

$$\delta_{\mathrm{l}} = \eta_{\mathrm{l}\delta}\left(\frac{1}{4}\frac{g(\rho_{\mathrm{l,w}}-\rho_{\mathrm{v,s}})x^3}{\nu_{\mathrm{l,s}}^2\rho_{\mathrm{l,s}}}\right)^{-1/4} x,$$

or

$$\frac{\mathrm{d}\delta_{\mathrm{l}}}{\mathrm{d}x} = \frac{1}{4}\eta_{\mathrm{l}\delta}\left(\frac{1}{4}Gr_{x\mathrm{l,s}}\right)^{-1/4}.$$

Therefore

$$\begin{aligned}
g_x &= \rho_{\mathrm{l,s}}\left\{2\sqrt{gx}\left(\frac{\rho_{\mathrm{l,w}}-\rho_{\mathrm{v,s}}}{\rho_{\mathrm{l,s}}}\right)^{1/2} W_{x\mathrm{l,s}}\left[\frac{1}{4}\eta_{\mathrm{l}\delta}\left(\frac{1}{4}Gr_{x\mathrm{l,s}}\right)^{-1/4}\right]\right. \\
&\quad \left. - 2\sqrt{gx}\left(\frac{\rho_{\mathrm{l,w}}-\rho_{\mathrm{v,s}}}{\rho_{\mathrm{l,s}}}\right)^{1/2} \left(\tfrac{1}{4}Gr_{x\mathrm{l,s}}\right)^{-1/4} W_{y\mathrm{l,s}}\right\}_{\mathrm{s}} \\
&= 2\rho_{\mathrm{l,s}}\sqrt{gx}\left(\frac{\rho_{\mathrm{l,w}}-\rho_{\mathrm{v,s}}}{\rho_{\mathrm{l,s}}}\right)^{1/2}\left(\frac{1}{4}Gr_{x\mathrm{l,s}}\right)^{-1/4}\left\{\frac{1}{4}\eta_{\mathrm{l}\delta}W_{x\mathrm{l,s}} - W_{y\mathrm{l,s}}\right\}_{-\Delta t_\infty=0} \\
&= 4\mu_{\mathrm{l,s}}x^{-1}\left(\frac{1}{4}Gr_{x\mathrm{l,s}}\right)^{1/2}\left(\frac{1}{4}Gr_{x\mathrm{l,s}}\right)^{-1/4}\left\{\frac{1}{4}\eta_{\mathrm{l}\delta}W_{x\mathrm{l,s}} - W_{y\mathrm{l,s}}\right\}_{\Delta t_\infty=0} \\
&= \mu_{\mathrm{l,s}}x^{-1}\left(\frac{1}{4}Gr_{x\mathrm{l,s}}\right)^{1/4}\left\{\eta_{\mathrm{l}\delta}W_{x\mathrm{l,s}} - 4W_{y\mathrm{l,s}}\right\}_{\Delta t_\infty=0}.
\end{aligned}$$

If G_x is taken to express total mass flow rate entering the boundary layer for position $x = 0$ to x with width of b of the plate, it should be the following integration:

$$G_x = \iint_A g_x \, \mathrm{d}A = b \int_0^x g_x \, \mathrm{d}x,$$

where $A = b \cdot x$ is area of the plate.

Then, we obtain

$$G_x = \frac{4}{3} b \cdot \mu_{\mathrm{l,s}} \left(\frac{1}{4} Gr_{x\mathrm{l,s}}\right)^{1/4} (\eta_{\mathrm{l}\delta} W_{x\mathrm{l,s}} - 4 W_{y\mathrm{l,s}})_{\Delta t_\infty = 0} \tag{12.49}$$

or as dimensionless form it can be rewritten as

$$\frac{G_x}{b \cdot \mu_{\mathrm{l,s}}} = \frac{4}{3} \left(\frac{1}{4} Gr_{x\mathrm{l,s}}\right)^{1/4} (\eta_{\mathrm{l}\delta} W_{x\mathrm{l,s}} - 4 W_{y\mathrm{l,s}})_{\Delta t_\infty = 0},$$

where $(\eta_{\mathrm{l}\delta} W_{x\mathrm{l,s}} - W_{y\mathrm{l,s}})_{\Delta t_\infty = 0}$ is defined as mass flow rate parameter through the liquid–vapor interface for the film condensation of saturated vapor, while the subscript $\Delta t_\infty = 0$ denotes the case with saturated vapor.

12.6.2 Curve-Fit Equations for Heat and Mass Transfer

The corresponding numerical solutions of condensate liquid film thickness $\eta_{\mathrm{l}\delta}$, dimensionless temperature gradient $((\mathrm{d}\theta_{\mathrm{l}}/\mathrm{d}\eta_{\mathrm{l}})_{\eta_{\mathrm{l}}=0})_{\Delta t_{\mathrm{w}}=0}$, velocity components at liquid–vapor interface, $W_{x\mathrm{l,s}}$ and $W_{y\mathrm{l,s}}$ and mass flow rate parameter $(\eta_{\mathrm{l}\delta} W_{x\mathrm{l,s}} - 4 W_{y\mathrm{l,s}})_{\Delta t_\infty = 0}$ of the film condensation of saturated water vapor are listed in Tables 12.3 with the corresponding wall subcooled grade $\Delta t_{\mathrm{w}}/t_{\mathrm{s}} = (t_{\mathrm{s}} - t_{\mathrm{w}})/t_{\mathrm{s}}$, and are plotted in Figs. 12.4–12.8, respectively.

It is seen from Fig. 12.4 that the condensate liquid film thickness $\eta_{\mathrm{l}\delta}$will increase with increasing the wall subcooled grade $\Delta t_{\mathrm{w}}/t_{\mathrm{s}}$. In the range with smaller wall subcooled grade $\Delta t_{\mathrm{w}}/t_{\mathrm{s}}$, the liquid film thickness will increase rapidly, and then will increase slowly with increasing the wall subcooled grade $\Delta t_{\mathrm{w}}/t_{\mathrm{s}}$. The maximum value of liquid film thickness is 0.73561 at the maximum wall subcooled grade of $\Delta t_{\mathrm{w}}/t_{\mathrm{s}} = 1$, for the film condensation of saturated water vapor. The rigorous numerical solutions of the condensate liquid film thickness $\eta_{\mathrm{l}\delta}$ of the film condensation of saturated water vapor are formulated as following by using a curve-fitting method:

$$\eta_{\mathrm{l}\delta} = 0.5934 \left(\frac{\Delta t_{\mathrm{w}}}{t_{\mathrm{s}}}\right)^{0.2562} \quad (0.001 \le \Delta t_{\mathrm{w}}/t_{\mathrm{s}} < 0.2), \tag{12.50}$$

$$\eta_{\mathrm{l}\delta} = 0.417 \frac{\Delta t_{\mathrm{w}}}{t_{\mathrm{s}}} + 0.3223 \quad (0.2 \le \Delta t_{\mathrm{w}}/t_{\mathrm{s}} \le 1). \tag{12.51}$$

Table 12.3. Numerical and predicted solutions of $((-\mathrm{d}\theta_l/\mathrm{d}\eta_l)_{\eta_l=0})_{\Delta t_\infty=0}, \eta_{l\delta}, W_{xl,s}, W_{yl,s}$, and $(\eta_{l\delta} W_{xl,s} - 4W_{yl,s})_{\Delta t_\infty=0}$ for film condensation of saturated water vapor, cited from Shang and Adamek [12]

t_w (°C)	99.9	97.5	95	90	80	60	40	20	0
$\Delta t_w (= t_s - t_w)$ (°C)	0.1	2.5	5	10	20	40	60	80	100
$\frac{\Delta t_w}{t_s}$	0.001	0.025	0.05	0.10	0.20	0.40	0.60	0.80	1.00
$\eta_{l\delta}$ (1)	0.10170	0.22850	0.27307	0.32804	0.39344	0.49580	0.57700	0.65545	0.73561
$\eta_{l\delta}$ (2)	0.1011	0.23062	0.27544	0.32896	0.4057	0.4891	0.5725	0.6559	0.7393
$\left(\left(-\frac{\mathrm{d}\theta_l}{\mathrm{d}\eta_l}\right)_{\eta_l=0}\right)_{\Delta t_w=0}$ (1)	9.8338	4.3814	3.6751	3.0641	2.5398	2.0789	1.8324	1.6679	1.5511
$-\left(\left(\frac{\mathrm{d}\theta_l}{\mathrm{d}\eta_l}\right)_{\eta_l=0}\right)_{\Delta t_w=0}$ (2)	9.7795	4.3621	3.6580	3.0591	2.5440	2.0916	1.8468	1.6785	1.5495
$W_{xl,s}$	0.00513	0.02537	0.03546	0.04901	0.06619	0.08510	0.09463	0.09925	0.10111
$W_{yl,s}$	0.00013	0.00145	0.00242	0.00400	0.00650	0.01009	0.01237	0.01355	0.01371
$(\eta_{l\delta} \cdot W_{xl,\delta} - 4W_{yl,\delta})_{\Delta t_\infty=0}$ (1)	0.00104	0.01159	0.01934	0.03208	0.05237	0.08255	0.10408	0.11925	0.12922
$(\eta_{l\delta} \cdot W_{xl,\delta} - 4W_{yl,\delta})_{\Delta t_\infty=0}$ (1)	0.00105	0.01161	0.01938	0.03209	0.05226	0.08217	0.10361	0.11893	0.12921

(1) numerical solution, (2) results predicted by using the related curve-fitting formulae

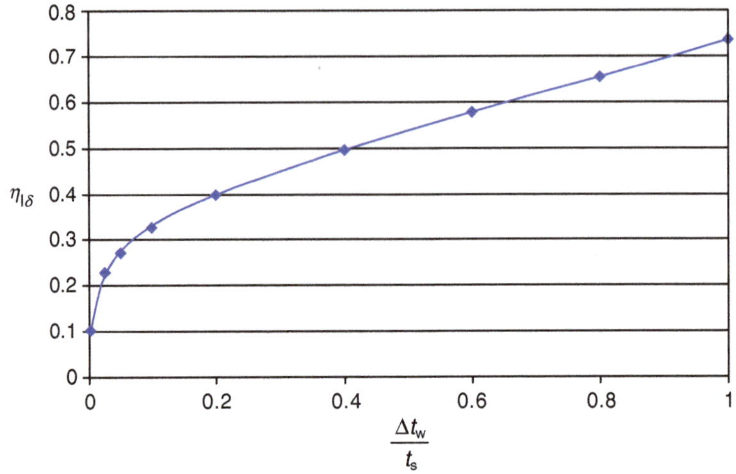

Fig. 12.4. The distribution of $\eta_{l\delta}$ with $\Delta t_w/t_s$ for film condensed of saturated water vapor.

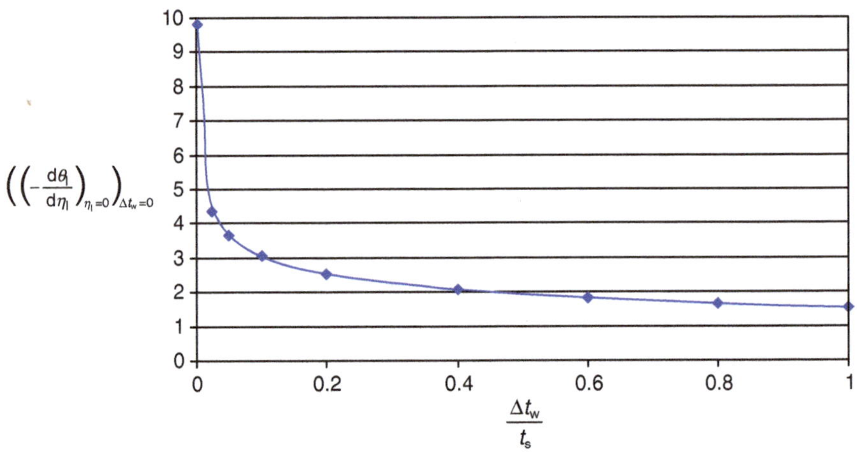

Fig. 12.5. The distribution of $((-\mathrm{d}\theta_l/\mathrm{d}\eta_l)_{\eta_l=0})_{\Delta t_\infty=0}$ with $\Delta t_w/t_s$ for film condensation of saturated water vapor

The predicted results by using (12.50) and (12.51) are listed in Table 12.3 with the wall subcooled grade $\Delta t_w/t_s$, and it is found that they agree very well with the numerical solutions.

It is found from Fig. 12.5 that the temperature gradient $((-\mathrm{d}\theta_l/\mathrm{d}\eta_l)_{\eta_l=0})_{\Delta t_\infty=0}$ of film condensation of saturated vapor will decrease with increasing the wall subcooled grade $\Delta t_w/t_s$. In the range with smaller value of $\Delta t_w/t_s$, the temperature gradient will decrease rapidly, and then decrease slower with increasing the wall subcooled temperature. The minimum value of the temperature gradient $((-\mathrm{d}\theta_l/\mathrm{d}\eta_l)_{\eta_l=0})_{\Delta t_\infty=0}$ is 1.5511, which oc-

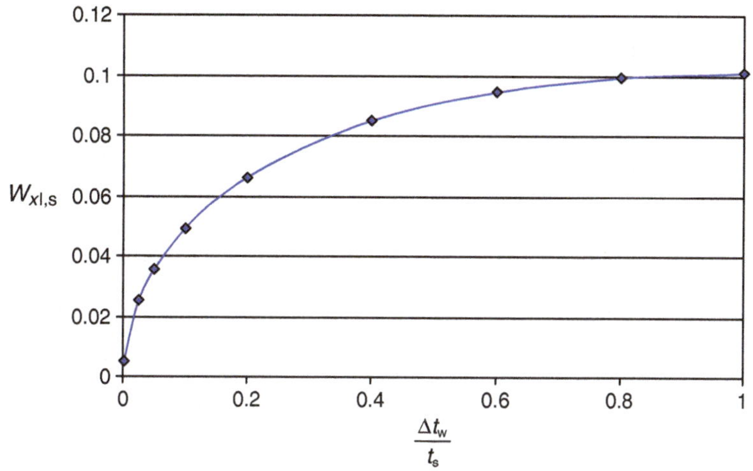

Fig. 12.6. The distribution of $W_{x\mathrm{l,s}}$ with $\Delta t_{\mathrm{w}}/t_{\mathrm{s}}$ for film condensation of saturated water vapor

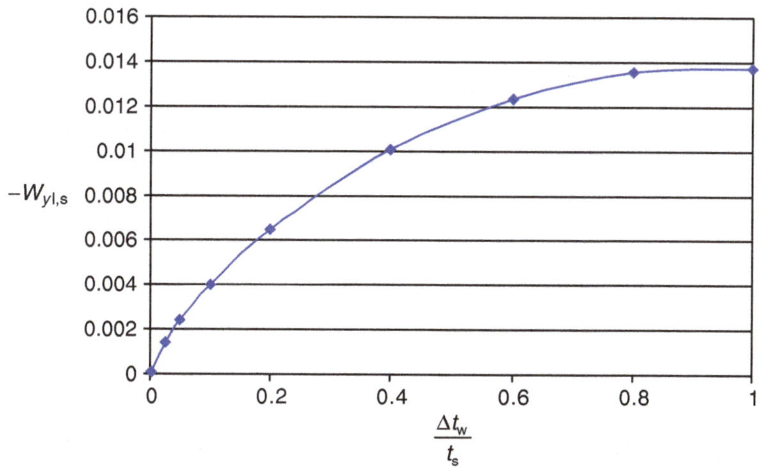

Fig. 12.7. The distribution of $-W_{y\mathrm{l,s}}$ with $\Delta t_{\mathrm{w}}/t_{\mathrm{s}}$ for film condensation of saturated water vapor

curs at the maximum wall subcooled grade $\Delta t_{\mathrm{w}}/t_{\mathrm{s}} = 1$, for the film condensation of saturated water vapor.

Based on the earlier related numerical solutions, the formula for $((-\mathrm{d}\theta_{\mathrm{l}}/\mathrm{d}\eta_{\mathrm{l}})_{\eta_{\mathrm{l}}=0})_{\Delta t_\infty=0}$ is obtained as follows by using a curve-matching method for film condensation of saturated water vapor:

$$\left(\left(-\frac{\mathrm{d}\theta_{\mathrm{l}}}{\mathrm{d}\eta_{\mathrm{l}}}\right)_{\eta_{\mathrm{l}}=0}\right)_{\Delta t_\infty=0} = \frac{1.74 - 0.19\frac{\Delta t_{\mathrm{w}}}{t_{\mathrm{s}}}}{\left(\frac{\Delta t_{\mathrm{w}}}{t_{\mathrm{s}}}\right)^{1/4}}. \tag{12.52}$$

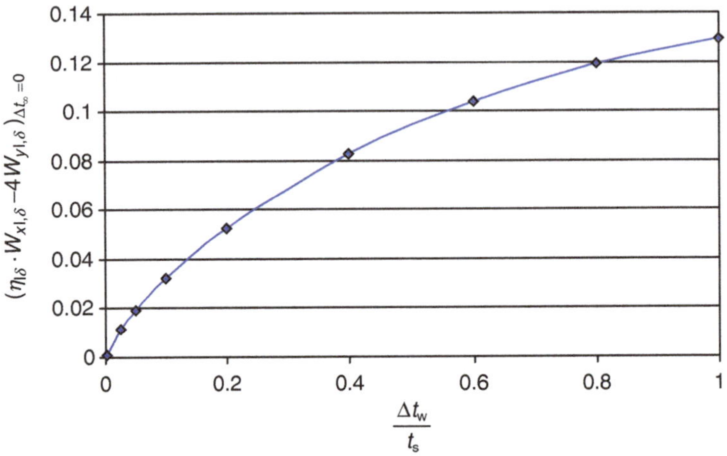

Fig. 12.8. Distribution of $(\eta_{l\delta} \cdot W_{xl,\delta} - 4W_{yl,\delta})_{\Delta t_\infty=0}$, with $\Delta t_w/t_s$ for film condensation of saturated vapor

The values $((-d\theta_l/d\eta_l)_{\eta_l=0})_{\Delta t_\infty=0}$ predicted by (12.52) are listed in Table 12.3 for the film condensation of saturated water vapor, and it is found that they agree very well with the numerical solutions.

It is found from Figs. 12.6 and 12.7 that the liquid–vapor interfacial velocity components $W_{xl,s}$ and $-W_{yl,s}$ will increase with increasing the wall subcooled grade $\Delta t_w/t_s$.

It is found from Fig. 12.8 that the mass flow rate parameter $(\eta_{l\delta}W_{xl,s} - 4W_{yl,s})_{\Delta t_\infty=0}$ will increase monotonously with increasing the subcooled grade $\Delta t_w/t_s$. The maximum value of the mass flow rate parameter is 0.12922, which occurs at the maximum wall subcooled grade of $\Delta t_w/t_s = 1$, for the film condensation of saturated water vapor.

Based on the numerical solutions, the formula for condensate mass flow rate parameter $(\eta_{l\delta}W_{xl,s} - 4W_{yl,s})_{\Delta t_\infty=0}$ of film condensation of saturated water vapor is obtained as follows by using a curve-matching method:

$$(\eta_{l\delta} \cdot W_{xl,\delta} - 4W_{yl,\delta})_{\Delta t_\infty=0} = \left(0.186 - 0.057\frac{\Delta t_w}{t_s}\right)\left(\frac{\Delta t_w}{t_s}\right)^{3/4}. \quad (12.53)$$

The values $(\eta_{l\delta}W_{xl,s} - 4W_{yl,s})_{\Delta t_\infty=0}$ predicted by (12.53) are listed in Table 12.3 for mass flow rate parameter $(\eta_{l\delta}W_{xl,s} - 4W_{yl,s})_{\Delta t_\infty=0}$ of film condensation of saturated water vapor. The comparison shows that the good agreements have been obtained between the numerical solutions and the results evaluated by the corresponding curve-fit equation (12.53).

12.7 Remarks

This chapter deals with the theory of laminar film condensation of saturated vapor on a vertical flat plate, with consideration of the various factors including the variable thermophysical properties. Meanwhile, the corresponding expressions for prediction of heat and mass transfer characteristics are obtained. The analysis presented here is an extension of former studies.

In the first part a similarity transformation with the velocity component method is used to transform the system of partial differential equations associated with the two-phase boundary problem into a system of dimensionless ordinary differential equations.

In the second part the system of ordinary differential equations and its related boundary conditions is computed by a successively iterative procedure with the shooting method adopted for the numerical solution of the three-point boundary value problem.

On the basis of rigorous numerical solutions the following important conclusions are remarked for the film condensation of saturated water vapor. The condensate liquid film thickness $\eta_{l\delta}$ will increase with increasing the wall subcooled grade $\Delta t_w/t_s$. In the range with smaller wall subcooled grade $\Delta t_w/t_s$, the liquid film thickness will increase rapidly, and increase slower with increasing the wall subcooled grade $\Delta t_w/t_s$.

The temperature gradient $((d\theta_l/d\eta_l)_{\eta_l=0})_{\Delta t_\infty=0}$ will decrease with increasing the wall subcooled grade $\Delta t_w/t_s$. In the range with smaller value of $\Delta t_w/t_s$, the temperature gradient will decrease rapidly, and decrease slower with increasing the wall subcooled grade $\Delta t_w/t_s$.

The liquid–vapor interfacial velocity components $W_{xl,s}$, $-W_{yl,s}$, and the mass flow rate parameter $(\eta_{l\delta}W_{xl,s} - 4W_{yl,s})$ will increasing with the wall subcooled grade $\Delta t_w/t_s$.

The curve-fitting formulae of the condensate liquid film thickness $\eta_{l\delta}$, temperature gradient $((d\theta_l/d\eta_l)_{\eta_l=0})_{\Delta t_\infty=0}$ and mass flow rate parameter $(\eta_{l\delta}W_{xl,s} - 4W_{yl,s})$ are developed based on the related rigorous numerical solutions and simple and available for their predictions.

12.8 Calculation Example

Example 1: A flat plate, 0.3 m in width and 0.3 m in length, is suspended vertically in the superheated water vapor. The wall temperature of the plate is $t_w = 98°C$, and the vapor bulk temperature is $t_\infty = t_s = 100°C$. Suppose the condensate film is laminar, please calculate the free convection condensation heat and mass transfer on the plate.

Calculation. The vapor superheated grade is $\Delta t_\infty/t_s = (t_\infty - t_s)/t_s = 0$, then, the water vapor bulk is at the saturated state with $\rho_{v,s} = 0.5974\,\mathrm{kg\,m^{-3}}$.

The wall subcooled grade is $\Delta t_w/t_s = (t_s - t_w)/t_s = (100-98)/100 = 0.02$, and $\rho_{l,w} = 961.6\,\mathrm{kg\,m^{-3}}$, and $\lambda_{l,w} = 0.6824\mathrm{W}\,(\mathrm{m}^\circ\mathrm{C})^{-1}$ at $t_w = 98^\circ\mathrm{C}$.

Additionally, for saturated condition of water at 100°C, there should be the following property data, i.e., $\rho_{l,s} = 958.1\,\mathrm{kg\,m^{-3}}$, $\nu_{l,s} = 0.294 \times 10^{-6}\,\mathrm{m^2\,s^{-1}}$, $\mu_{l,s} = 281.7 \times 10^{-6}\,\mathrm{kg\,(ms)^{-1}}$.

1. For heat transfer. From (12.44) and (12.46) the average heat transfer coefficient of film condensation of saturated vapor is evaluated as

$$\overline{\alpha_x} = -\frac{4}{3}\lambda_{l,w}\left(\frac{1}{4}Gr_{xl,s}\right)^{1/4} x^{-1}\left(-\left(\frac{d\theta_l}{d\eta_l}\right)_{\eta_l=0}\right)_{\Delta t_\infty=0}.$$

From (12.14), the local Grashof number of the film condensation is evaluated as

$$\begin{aligned} Gr_{xl,s} &= \frac{g(\rho_{l,w} - \rho_{v,s})x^3}{\nu_{l,s}^2\rho_{l,s}} \\ &= \frac{9.8 \times (961.6 - 0.5974) \times 0.3^3}{(0.294 \times 10^{-6})^2 \times 958.1} \\ &= 3.0705 \times 10^{12}. \end{aligned}$$

From (12.52), the temperature gradient $((d\theta_l/d\eta_l)_{\eta_l=0})_{\Delta t_\infty=0}$ for the film condensation of saturated water vapor is calculated as

$$\begin{aligned} \left(-\left(\frac{d\theta_l}{d\eta_l}\right)_{\eta_l=0}\right)_{\Delta t_\infty=0} &= \frac{1.74 - 0.19\frac{\Delta t_w}{t_s}}{\left(\frac{\Delta t_w}{t_s}\right)^{1/4}} \\ &= \frac{1.74 - 0.19 \times 0.02}{0.02^{1/4}} \\ &= 4.6168. \end{aligned}$$

Then, the local heat transfer coefficient is evaluated as

$$\begin{aligned} \overline{\alpha_x} &= \frac{4}{3}\lambda_{l,w}\left(\frac{1}{4}Gr_{xl,s}\right)^{1/4} x^{-1}\left(-\left(\frac{d\theta_l}{d\eta_l}\right)_{\eta_l=0}\right)_{\Delta t_\infty=0} \\ &= \frac{4}{3} \times 0.6824 \times \left(\frac{1}{4} \times 3.0705 \times 10^{12}\right)^{1/4} \times 0.3^{-1} \times 4.6168 \\ &= 13106.44\,\mathrm{W}\,(\mathrm{m}^{-2\circ}\mathrm{C})^{-1}. \end{aligned}$$

The total heat transfer of laminar film condensation of the superheated water vapor on the vertical plate is calculated

$$
\begin{aligned}
Q_x &= \overline{\alpha_x}(t_\mathrm{w} - t_\mathrm{s})A \\
&= 13106.44 \times (98 - 100) \times 0.3 \times 0.3 \\
&= -2359.16\,\mathrm{W}.
\end{aligned}
$$

The negative means that the heat transfer direction is to the plate from the condensate film.

2. For Mass Flow Rate of the Condensation. From (12.48), the total mass flow rate G_x of the laminar film condensate of saturated vapour for position $x = 0$ to x with width of b of the plate is evaluated as

$$
G_x = \frac{4}{3} b \cdot \mu_{\mathrm{l,s}} \left(\frac{1}{4} Gr_{x\mathrm{l,s}}\right)^{1/4} (\eta_{\mathrm{l}\delta} W_{x\mathrm{l},\delta} - 4W_{y\mathrm{l},\delta})_{\Delta t_\infty = 0}.
$$

The mass flow rate parameter $(\eta_{\mathrm{l}\delta} \cdot W_{x\mathrm{l},\delta} - 4W_{y\mathrm{l},\delta})_{\Delta t_\infty=0}$ is calculated as

$$
\begin{aligned}
(\eta_{\mathrm{l}\delta} \cdot W_{x\mathrm{l},\delta} - 4W_{y\mathrm{l},\delta})_{\Delta t_\infty=0} &= \left(0.186 - 0.057 \frac{\Delta t_\mathrm{w}}{t_\mathrm{s}}\right) \left(\frac{\Delta t_\mathrm{w}}{t_\mathrm{s}}\right)^{3/4} \\
&= (0.186 - 0.057 \times 0.02)(0.02)^{3/4} \\
&= 0.0098314.
\end{aligned}
$$

Then, the total mass flow rate G_x of the laminar film condensation of superheated water vapor is calculated as

$$
\begin{aligned}
G_x &= \frac{4}{3} b \cdot \mu \mathrm{l,s} \left(\frac{1}{4} Gr_{x_{\mathrm{l,s}}}\right)^{1/4} (\eta_{\mathrm{l}\delta} W_{x\mathrm{l},\delta} - 4W_{y\mathrm{l},\delta})_{\Delta t_\infty=0} \\
&= \frac{4}{3} \times 0.3 \times 281.7 \times 10^{-6} \times \left(\frac{1}{4} \times 3.0705 \times 10^{12}\right)^{1/4} \times 0.0098314 \\
&= 0.001037\,\mathrm{kg\,s^{-1}} \\
&= 3.7329\,\mathrm{kg\,h^{-1}}.
\end{aligned}
$$

Example 2: A flat plate with 0.3 m in width and 0.3 m in height is suspended vertically in the saturated water vapor, i.e., $\Delta t_\infty = t_\infty - t_\mathrm{s} = 0°\mathrm{C}$. The wall temperature is $t_\mathrm{w} = 90°\mathrm{C}$ and then the wall subcooled grade is $\Delta t_\mathrm{w}/t_\mathrm{s} = (t_\mathrm{s} - t_\mathrm{w})/t_\mathrm{s} = 100 - 90/100 = 0.1$. Assume laminar film condensation occurs on the plate, please calculate

(i) heat transfer and mass flow rate of the film boiling of saturated water vapor on the plate
(ii) condensate film thickness at $x = 0$, 0.01, 0.05, 0.1, 0.15, 0.2, 0.25, and 0.3 m from the top ($x = 0$) of the plate

Calculation: At first, the related physical properties are given as follows: $\rho_{\mathrm{v,s}} = 0.5974\,\mathrm{g\,m^{-3}}$ for saturated water vapor at $t_{\mathrm{s}} = 100°\mathrm{C}$, $\rho_{\mathrm{l,s}} = 958.1\,\mathrm{kg\,m^{-3}}$, $\nu_{\mathrm{l,s}} = 0.294 \times 10^{-6}\mathrm{m^2\,s^{-1}}$, and $\mu_{\mathrm{l,s}} = 281.7 \times 10^{-6}\,\mathrm{kg\,(m\ s)^{-1}}$ for saturated water at $t_{\mathrm{s}} = 100°\mathrm{C}$, and $\rho_{\mathrm{l,w}} = 965.3\,\mathrm{kg\,m^{-3}}$ and $\lambda_{\mathrm{l,w}} = 0.68$ for water at $t_{\mathrm{w}} = 90°\mathrm{C}$.

(i) Heat transfer and mass flow rate of the film boiling of saturated water steam on two sides of the plate:

For heat transfer. With (12.13), the local Grashof number $Gr_{x\mathrm{l,s}}$ is evaluated as

$$\begin{aligned} Gr_{x\mathrm{l,s}} &= \frac{g(\rho_{\mathrm{l,w}} - \rho_{\mathrm{v,s}})x^3}{\nu_{\mathrm{l,s}}^2 \rho_{\mathrm{l,s}}} \\ &= \frac{9.8 \times (965.3 - 0.5974) \times 0.3^3}{(0.294 \times 10^{-6})^2 \times 958.1} \\ &= 3.08232 \times 10^{12}. \end{aligned}$$

Additionally, with (12.52) the temperature gradient of the laminar film condensation of the saturated water vapor is calculated as

$$\begin{aligned} \left(-\left(\frac{\mathrm{d}\theta_{\mathrm{l}}}{\mathrm{d}\eta_{\mathrm{l}}}\right)_{\eta_{\mathrm{l}}=0}\right)_{\Delta t_{\infty}=0} &= \frac{1.74 - 0.19\frac{\Delta t_{\mathrm{w}}}{t_{\mathrm{s}}}}{\left(\frac{\Delta t_{\mathrm{w}}}{t_{\mathrm{s}}}\right)^{1/4}} \\ &= \frac{1.74 - 0.19 \times \frac{10}{100}}{\left(\frac{10}{100}\right)^{1/4}} \\ &= 3.060. \end{aligned}$$

With (12.44) and (12.46) the average heat transfer coefficient is evaluated as

$$\begin{aligned} \overline{\alpha_x} &= \frac{4}{3}\lambda_{\mathrm{l,w}}\left(\frac{1}{4}Gr_{x\mathrm{l,s}}\right)^{1/4} x^{-1}\left(\left(-\frac{\mathrm{d}\theta_{\mathrm{l}}}{\mathrm{d}\eta_{\mathrm{l}}}\right)_{\eta_{\mathrm{l}}=0}\right)_{\Delta t_{\infty}=0} \\ &= \frac{4}{3} \times 0.68 \times \left(\frac{1}{4} \times 3.08232 \times 10^{12}\right)^{1/4} \times (0.3)^{-1} \times 3.060 \\ &= 8664.67\,\mathrm{W\,(m^{2°}C)^{-1}}. \end{aligned}$$

The heat transfer of laminar film condensation of the saturated water vapor on the vertical plate is calculated

$$\begin{aligned} Q_x &= \overline{\alpha_x}(t_{\mathrm{w}} - t_{\mathrm{s}})A \\ &= 8664.67 \times (90 - 100) \times 0.3 \times 0.3 \\ &= -7798.2\,\mathrm{W}. \end{aligned}$$

The negative sign means that the heat flux is to the plate from the condensate film.

For mass flow rate of the condensation. The mass flow rate parameter is evaluated as

$$(\eta_{l\delta} \cdot W_{xl,\delta} - 4W_{yl,\delta})_{\Delta t_\infty=0} = \left(0.186 - 0.057\frac{\Delta t_w}{t_s}\right)\left(\frac{\Delta t_w}{t_s}\right)^{3/4}$$

$$= \left(0.186 - 0.057 \times \frac{10}{100}\right)\left(\frac{10}{100}\right)^{3/4}$$

$$= 0.032062.$$

The total mass flow rate of the film condensation of saturated water vapor is

$$G_x = \frac{4}{3}b \cdot \mu_{l,s}\left(\frac{1}{4}Gr_{xl,s}\right)^{1/4}(\eta_{l\delta} \cdot W_{xl,\delta} - 4W_{yl,\delta})_{\Delta t_\infty=0}$$

$$= \frac{4}{3} \times 0.3 \times 281.7 \times 10^{-6} \times \left(\frac{1}{4} \times 3.08232 \times 10^{12}\right)^{1/4} \times 0.032062$$

$$= 0.0033849\,\mathrm{kg\,s^{-1}}$$

$$= 12.186\,\mathrm{kg\,h^{-1}}$$

(ii) Condensate film thickness:
The wall subcooled grade is 0.1, then, (12.51) is taken to evaluate $\eta_{l\delta}$ as

$$\eta_{l\delta} = 0.5934\left(\frac{\Delta t_w}{t_s}\right)^{0.2562}$$

$$= 0.5934 \times (0.1)^{0.2562}$$

$$= 0.32896.$$

From (12.13), the condensate film thickness δ_l is expressed as

$$\delta_l = \eta_l x\left(\frac{1}{4}Gr_{xl,s}\right)^{-1/4}$$

$$= \eta_l x\left(\frac{1}{4}\frac{g(\rho_{l,w} - \rho_{v,s})x^3}{\nu_{l,s}^2\rho_{l,s}}\right)^{-1/4}$$

$$= \eta_l\left(\frac{1}{4}\frac{g(\rho_{l,w} - \rho_{v,s})}{\nu_{l,s}^2\rho_{l,s}}\right)^{-1/4} x^{1/4}$$

$$= 0.32896 \times \left(\frac{1}{4} \times \frac{9.8 \times (965.3 - 0.5974)}{(0.294 \times 10^{-6})^2 \times 958.1}\right)^{-1/4} \times x^{1/4}$$

$$= 0.000142324 \times \mathrm{x}^{1/4}.$$

Table 12.4. The condensate film thickness y with the position x

x (m)	0	0.01	0.05	0.1	0.15	0.2	0.25	0.3
δ_l (m)	0	4.5×10^{-5}	6.73×10^{-5}	8×10^{-5}	8.86×10^{-5}	9.52×10^{-5}	0.000101	0.000105

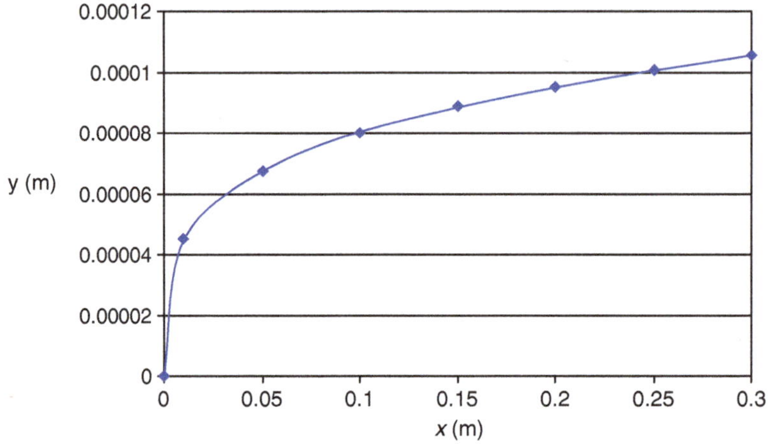

Fig. 12.9. The condensate film thickness δ_l with the position x

$$\begin{aligned}
&\text{For } x = 0, && \delta_l = 0 \\
&\text{For } x = 0.01\,\text{m}, && \delta_l = 0.000142324 \times 0.01^{1/4} = 4.5 \times 10^{-5}\,\text{m} \\
&\text{For } x = 0.05\,\text{m}, && \delta_l = 0.000142324 \times 0.05^{1/4} = 6.73 \times 10^{-5}\,\text{m} \\
&\text{For } x = 0.1\,\text{m}, && \delta_l = 0.000142324 \times 0.1^{1/4} = 8 \times 10^{-5}\,\text{m} \\
&\text{For } x = 0.15\,\text{m}, && \delta_l = 0.000142324 \times 0.15^{1/4} = 8.86 \times 10^{-5}\,\text{m} \\
&\text{For } x = 0.2\,\text{m}, && \delta_l = 0.000142324 \times 0.2^{1/4} = 9.52 \times 10^{-}\,\text{m} \\
&\text{For } x = 0.25\,\text{m}, && \delta_l = 0.000142324 \times 0.25^{1/4} = 0.000101\,\text{m} \\
&\text{For } x = 0.3\,\text{m}, && \delta_l = 0.000142324 \times 0.3^{1/4} = 0.000105\,\text{m},
\end{aligned}$$

For clear expression, the condensate film thickness δ_l with the position x is listed and plotted as Table 12.4 and Fig. 12.9

Appendix A. Derivation of Similarity Transformation of Governing Equations (12.1)–(12.5)

The derivation of similarity transformation of governing equations for film condensation of saturated vapor is similar to that for film boiling of saturated liquid. So the derivation can be presented briefly.

1. *Transformation of (12.1).* Equation (12.1) is rewritten as

$$\rho_l \left(\frac{\partial w_{xl}}{\partial x} + \frac{\partial w_{yl}}{\partial y} \right) + w_{xl} \frac{\partial \rho_l}{\partial x} + w_{yl} \frac{\partial \rho_l}{\partial y} = 0. \qquad (12.54)$$

With (10.13)–(10.16) we can obtain the following correlations:

$$\frac{\partial w_{x\mathrm{l}}}{\partial x} = \sqrt{\frac{g}{x}} \left(\frac{\rho_{\mathrm{l,w}} - \rho_{\mathrm{v}}}{\rho_{\mathrm{l,s}}} \right)^{1/2} \left(W_{x\mathrm{l}} - \frac{1}{2} \eta_{\mathrm{l}} \frac{\mathrm{d}W_{x\mathrm{l}}}{\mathrm{d}\eta_{\mathrm{l}}} \right), \tag{12.55}$$

$$\frac{\partial w_{y\mathrm{l}}}{\partial y} = 2\sqrt{\frac{g}{x}} \left(\frac{\rho_{\mathrm{l,w}} - \rho_{\mathrm{v}}}{\rho_{\mathrm{l,s}}} \right)^{1/2} \frac{\mathrm{d}W_{y\mathrm{l}}}{\mathrm{d}\eta_{\mathrm{l}}}, \tag{12.56}$$

$$\frac{\partial \rho_{\mathrm{l}}}{\partial x} = -\frac{1}{4} \eta_{\mathrm{l}} x^{-1} \frac{\mathrm{d}\rho_{\mathrm{l}}}{\mathrm{d}\eta_{\mathrm{l}}}, \tag{12.57}$$

$$\frac{\partial \rho_{\mathrm{l}}}{\partial y} = \frac{\mathrm{d}\rho_{\mathrm{l}}}{\mathrm{d}\eta_{\mathrm{l}}} \left(\frac{1}{4} Gr_{x\mathrm{l,s}} \right)^{1/4} x^{-1}. \tag{12.58}$$

With (12.15), (12.16), and (12.55)–(12.58), (12.54) can be changed into

$$\rho_{\mathrm{l}} \left[\sqrt{\frac{g}{x}} \left(\frac{\rho_{\mathrm{l,w}} - \rho_{\mathrm{v}}}{\rho_{\mathrm{l,s}}} \right)^{1/2} \left(W_{x\mathrm{l}} - \frac{1}{2} \eta_{\mathrm{l}} \frac{\mathrm{d}W_{x\mathrm{l}}}{\mathrm{d}\eta_{\mathrm{l}}} \right) + 2\sqrt{\frac{g}{x}} \left(\frac{\rho_{\mathrm{l,w}} - \rho_{\mathrm{v}}}{\rho_{\mathrm{l,s}}} \right)^{1/2} \frac{\mathrm{d}W_{y\mathrm{l}}}{\mathrm{d}\eta_{\mathrm{l}}} \right]$$

$$+2\sqrt{gx} \left(\frac{\rho_{\mathrm{l,w}} - \rho_{\mathrm{v}}}{\rho_{\mathrm{l,s}}} \right)^{1/2} W_{x\mathrm{l}} \left(-\frac{1}{4} \eta_{\mathrm{l}} x^{-1} \frac{\mathrm{d}\rho_{\mathrm{l}}}{\mathrm{d}\eta_{\mathrm{l}}} \right)$$

$$+2\sqrt{gx} \left(\frac{\rho_{\mathrm{l,w}} - \rho_{\mathrm{v}}}{\rho_{\mathrm{l,s}}} \right)^{1/2} \left(\tfrac{1}{4} Gr_{x\mathrm{l,s}} \right)^{-1/4} W_{y\mathrm{l}} \frac{\mathrm{d}\rho_{\mathrm{l}}}{\mathrm{d}\eta_{\mathrm{l}}} \left(\frac{1}{4} Gr'_{x\mathrm{l,s}} \right)^{1/4} x^{-1} = 0.$$

The earlier equation is divided by

$$\sqrt{\frac{g}{x}} \left(\frac{\rho_{\mathrm{l,w}} - \rho_{\mathrm{v}}}{\rho_{\mathrm{l,s}}} \right)^{1/2}$$

and simplified to

$$2W_{\mathrm{xl}} - \eta_{\mathrm{l}} \frac{\mathrm{d}W_{x\mathrm{l}}}{\mathrm{d}\eta_{\mathrm{l}}} + 4\frac{\mathrm{d}W_{y\mathrm{l}}}{\mathrm{d}\eta_{\mathrm{l}}} - \frac{1}{\rho_{\mathrm{l}}} \frac{\mathrm{d}\rho_{\mathrm{l}}}{\mathrm{d}\eta_{\mathrm{l}}} (\eta_{\mathrm{l}} W_{x\mathrm{l}} - 4W_{y\mathrm{l}}) = 0. \tag{12.59}$$

2. *Transformation of* (12.2). Equation (12.2) is rewritten as

$$\rho_{\mathrm{l}} \left(w_{x\mathrm{l}} \frac{\partial w_{x\mathrm{l}}}{\partial x} + w_{y\mathrm{l}} \frac{\partial w_{x\mathrm{l}}}{\partial y} \right) = \frac{\partial \mu_{\mathrm{l}}}{\partial y} \frac{\partial w_{x\mathrm{l}}}{\partial y} + \mu_{\mathrm{l}} \frac{\partial^2 w_{x\mathrm{l}}}{\partial y^2}) + g(\rho_{\mathrm{l}} - \rho_{\mathrm{v}}). \tag{12.60}$$

With the dimensionless transformation variables assumed in (10.13), (10.15), (10.16) we get

$$\frac{\partial w_{x\mathrm{l}}}{\partial y} = 2\sqrt{gx} (\frac{\rho_{\mathrm{l,w}} - \rho_{\mathrm{v}}}{\rho_{\mathrm{l,s}}})^{1/2} \frac{\mathrm{d}W_{x\mathrm{l}}}{\mathrm{d}\eta_{\mathrm{l}}} x^{-1} \left(\tfrac{1}{4} Gr_{x\mathrm{l,s}} \right)^{1/4}, \tag{12.61}$$

$$\frac{\partial^2 w_{xl}}{\partial y^2} = 2\sqrt{gx}\left(\frac{\rho_{l,w}-\rho_v}{\rho_{l,s}}\right)^{1/2}\frac{d^2W_{xl}}{d\eta_l^2}x^{-1}\left(\tfrac{1}{4}Gr_{xl,s}\right)^{1/4}\left(\tfrac{1}{4}Gr_{xl,s}\right)^{1/4}x^{-1},$$

$$= 2\sqrt{gx}\left(\frac{\rho_{l,w}-\rho_v}{\rho_{l,s}}\right)^{1/2}\frac{d^2W_{xl}}{d\eta_l^2}\left(\tfrac{1}{4}Gr_{xl,s}\right)^{1/2}x^{-2}, \quad (12.62)$$

$$\frac{\partial\mu_v}{\partial y} = \frac{d\mu_v}{d\eta_v}\left(\tfrac{1}{4}Gr_{xv,s}\right)^{1/4}x^{-1}. \quad (12.63)$$

With (12.15), (12.16), (12.55), and (12.61)–(12.63), (12.60) is changed into

$$\rho_l\left[2\sqrt{gx}\left(\frac{\rho_{l,w}-\rho_v}{\rho_{l,s}}\right)^{1/2}W_{xl}\sqrt{\frac{g}{x}}\left(\frac{\rho_{l,w}-\rho_v}{\rho_{l,s}}\right)^{1/2}\left(W_{xl}-\frac{1}{2}\eta_l\frac{dW_{xl}}{d\eta_l}\right)\right.$$

$$+2\sqrt{gx}\left(\frac{\rho_{l,w}-\rho_v}{\rho_{l,s}}\right)^{1/2}\left(\tfrac{1}{4}Gr_{xl,s}\right)^{-1/4}W_{yl}2\sqrt{gx}\left(\frac{\rho_{l,w}-\rho_v}{\rho_{l,s}}\right)^{1/2}$$

$$\left.\times\frac{dW_{xl}}{d\eta_l}x^{-1}\left(\tfrac{1}{4}Gr_{xl,s}\right)^{1/4}\right]$$

$$= 2\frac{d\mu_v}{d\eta_v}\left(\tfrac{1}{4}Gr_{xv,s}\right)^{1/4}x^{-1}\sqrt{gx}\left(\frac{\rho_{l,w}-\rho_v}{\rho_{l,s}}\right)^{1/2}\frac{dW_{xl}}{d\eta_l}x^{-1}\left(\tfrac{1}{4}Gr_{xl,s}\right)^{1/4}$$

$$+2\mu_l\sqrt{gx}\left(\frac{\rho_{l,w}-\rho_v}{\rho_{l,s}}\right)^{1/2}\frac{d^2W_{xl}}{d\eta_l^2}\left(\tfrac{1}{4}Gr_{xl,s}\right)^{1/2}x^{-2}) + g(\rho_l-\rho_v).$$

The earlier equation is divided by $g((\rho_{l,w}-\rho_v)/\rho_{l,s})$, and simplified to the following equation with consideration of $Gr_{xl,s}$ definition:

$$\rho_l\left[2W_{xl}\left(W_{xl}-\frac{1}{2}\eta_l\frac{dW_{xl}}{d\eta_l}\right)+2W_{yl}2\frac{dW_{xl}}{d\eta_l}\right]$$

$$= 2\frac{d\mu_v}{d\eta_v}\frac{dW_{xl}}{d\eta_l}\left(\frac{1}{4}\frac{g}{\nu_{l,s}^2}\right)^{1/2}+2\mu_l\frac{d^2W_{xl}}{d\eta_l^2}\left(\frac{1}{4}\frac{1}{\nu_{l,s}^2}\right)^{1/2}) + \rho_{l,s}\frac{\rho_l-\rho_v}{\rho_{l,w}-\rho_v}.$$

The earlier equation is multiplied by

$$\frac{1}{\rho_l}\frac{\nu_{l,s}}{\nu_l}$$

and further simplified to

$$\frac{\nu_{l,s}}{\nu_l}\left[W_{xl}\left(2W_{xl}-\eta_l\frac{dW_{xl}}{d\eta_l}\right)+4W_{yl}\frac{dW_{xl}}{d\eta_l}\right] = \frac{1}{\mu}\frac{d\mu_v}{d\eta_v}\frac{dW_{xl}}{d\eta_l}+\frac{d^2W_{xl}}{d\eta_l^2}$$

$$+\frac{\mu_{l,s}}{\mu_l}\frac{\rho_l-\rho_v}{\rho_{l,w}-\rho_v}. \quad (12.64)$$

3. *Transformation of* (12.3). Equation (12.3) is rewritten as

$$\rho_l c_{p_l}\left(w_{xl}\frac{\partial t_l}{\partial x}+w_{yl}\frac{\partial t_l}{\partial y}\right)=\frac{\partial}{\partial y}\left(\lambda_l\frac{\partial t_l}{\partial y}\right), \tag{12.65}$$

where

$$t_l=(t_w-t_s)\theta_l+t_s, \tag{12.66}$$

$$\frac{\partial t_l}{\partial x}=-(t_w-t_s)\frac{d\theta_l}{d\eta_l}\left(\frac{1}{4}\right)\eta_l x^{-1}, \tag{12.67}$$

$$\frac{\partial t_l}{\partial y}=(t_w-t_s)\frac{d\theta_l}{d\eta_l}\left(\frac{1}{4}Gr_{xl,s}\right)^{1/4}x^{-1}, \tag{12.68}$$

$$\frac{\partial^2 t_l}{\partial y^2}=(t_w-t_s)\frac{d^2\theta_l}{d\eta_l^2}\left(\frac{1}{4}Gr_{xl,s}\right)^{1/2}x^{-2}, \tag{12.69}$$

$$\frac{\partial \lambda_l}{\partial y}=\frac{d\lambda_l}{d\eta_l}\left(\frac{1}{4}Gr_{xl,s}\right)^{1/4}x^{-1}. \tag{12.70}$$

With (12.15), (12.16), and (12.66)–(12.69), (12.65) will become

$$-\rho_l c_{p_l}[2\sqrt{gx}\left(\frac{\rho_{l,w-\rho_v}}{\rho_{l,s}}\right)^{1/2}W_{xl}(t_w-t_s)\frac{d\theta_l}{d\eta_l}\left(\frac{1}{4}\right)\eta_l x^{-1}$$
$$2\sqrt{gx}\left(\frac{\rho_{l,w-\rho_v}}{\rho_{l,s}}\right)^{1/2}\left(\frac{1}{4}Gr_{xl,s}\right)^{-1/4}W_{yl}(t_W-t_s)\frac{d\theta_l}{d\eta_l}\left(\frac{1}{4}Gr_{xl,s}\right)^{1/4}x^{-1}]$$
$$=\lambda_l(t_W-t_s)\frac{d^2\theta_l}{d\eta_l^2}\left(\frac{1}{4}Gr_{xl,s}\right)^{1/2}x^{-2}$$
$$\frac{d\lambda_l}{d\eta_l}\left(\frac{1}{4}Gr_{xl,s}\right)^{1/4}x^{-1}(t_w-t_s)\frac{d\theta_l}{d\eta_l}\left(\frac{1}{4}Gr_{xl,s}\right)^{1/4}x^{-1}.$$

The earlier equation is divided by

$$(t_w-t_s)\sqrt{\frac{g}{x}}(\rho_{l,w}-\rho_v/\rho_{l,s})^{1/2}$$

and simplified to the following equation with consideration of Grashof number $Gr_{xl,s}$:

$$\rho_l c_{p_l}\left[W_{xl}\frac{d\theta_l}{d\eta_l}\eta_l+4W_{yl}\frac{d\theta_l}{d\eta_l}\right]=\lambda_l\frac{d^2\theta_l}{d\eta_l^2}\frac{1}{\nu_{l,s}}+\frac{d\lambda_l}{d\eta_l}\frac{d\theta_l}{d\eta_l}\frac{1}{\nu_{l,s}}.$$

The earlier equation is multiplied by $\nu_{l,s}/\lambda_l$ and simplified into

$$\rho_l c_{p_l}\frac{\nu_{l,s}}{\lambda_l}\left[-W_{xl}\frac{d\theta_l}{d\eta_l}\eta_l+4W_{yl}\frac{d\theta_l}{d\eta_l}\right]=\frac{d^2\theta_l}{d\eta_l^2}+\frac{1}{\lambda_l}\frac{d\lambda_l}{d\eta_l}\frac{d\theta_l}{d\eta_l}.$$

The earlier equation is simplified to

$$Pr_{\mathrm{l}}\frac{\nu_{\mathrm{l,s}}}{\nu_{\mathrm{l}}}\left(-W_{x\mathrm{l}}\eta_{\mathrm{l}}+4W_{y\mathrm{l}}\right)\frac{\mathrm{d}\theta_{\mathrm{l}}}{\mathrm{d}\eta_{\mathrm{l}}}=\frac{\mathrm{d}^2\theta_{\mathrm{l}}}{\mathrm{d}\eta_{\mathrm{l}}^2}+\frac{1}{\lambda_{\mathrm{l}}}\frac{\mathrm{d}\lambda_{\mathrm{l}}}{\mathrm{d}\eta_{\mathrm{l}}}\frac{\mathrm{d}\theta_{\mathrm{l}}}{\mathrm{d}\eta_{\mathrm{l}}}. \tag{12.71}$$

4. *Transformation of* (12.4). With (12.17)–(12.19) we get

$$\frac{\partial w_{x\mathrm{v}}}{\partial x}=\sqrt{\frac{g}{x}}\left(W_{x\mathrm{v}}-\frac{1}{2}\eta_{\mathrm{l}}\frac{\mathrm{d}W_{x\mathrm{v}}}{\mathrm{d}\eta_{\mathrm{v}}}\right), \tag{12.72}$$

$$\frac{\partial w_{y\mathrm{v}}}{\partial y}=2\sqrt{\frac{g}{x}}\frac{\mathrm{d}W_{y\mathrm{v}}}{\mathrm{d}\eta_{\mathrm{v}}}. \tag{12.73}$$

With (12.72) and (12.73), (12.4) is changed into the following one:

$$\sqrt{\frac{g}{x}}\left(W_{x\mathrm{v}}-\frac{1}{2}\eta_{\mathrm{v}}\frac{\mathrm{d}W_{x\mathrm{v}}}{\mathrm{d}\eta_{\mathrm{v}}}\right)+2\sqrt{\frac{g}{x}}\frac{\mathrm{d}W_{y\mathrm{v}}}{\mathrm{d}\eta_{\mathrm{v}}}=0.$$

The earlier equation is simplified to

$$2W_{x\mathrm{v}}-\eta_{\mathrm{v}}\frac{\mathrm{d}W_{x\mathrm{v}}}{\mathrm{d}\eta_{\mathrm{v}}}+4\frac{\mathrm{d}W_{y\mathrm{v}}}{\mathrm{d}\eta_{\mathrm{v}}}=0. \tag{12.74}$$

5. *Transformation of* (12.5). With (10.18)–(10.21) we obtain

$$\frac{\partial w_{x\mathrm{v}}}{\partial y}=2\sqrt{gx}\frac{\mathrm{d}W_{x\mathrm{v}}}{\mathrm{d}\eta_{\mathrm{v}}}x^{-1}\left(\tfrac{1}{4}Gr_{x\mathrm{v,s}}\right)^{1/4}, \tag{12.75}$$

$$\begin{aligned}\frac{\partial^2 w_{x\mathrm{v}}}{\partial y^2}&=2\sqrt{gx}\frac{\mathrm{d}^2W_{x\mathrm{v}}}{\mathrm{d}\eta_{\mathrm{v}}^2}x^{-1}\left(\tfrac{1}{4}Gr_{x\mathrm{v,s}}\right)^{1/4}\left(\tfrac{1}{4}Gr_{x\mathrm{v,s}}\right)^{1/4}x^{-1},\\&=2\sqrt{gx}\frac{\mathrm{d}^2W_{x\mathrm{v}}}{\mathrm{d}\eta_{\mathrm{v}}^2}\left(\tfrac{1}{4}Gr_{x\mathrm{v,s}}\right)^{1/2}x^{-2}.\end{aligned} \tag{12.76}$$

With (12.18), (12.19), (12.72), (12.74), and (12.76), the earlier equation becomes

$$\begin{aligned}&2\sqrt{gx}W_{x\mathrm{v}}\sqrt{\frac{g}{x}}\left(W_{x\mathrm{v}}-\frac{1}{2}\eta_{\mathrm{l}}\frac{\mathrm{d}W_{x\mathrm{v}}}{\mathrm{d}\eta_{\mathrm{v}}}\right)+2\sqrt{gx}\left(\frac{1}{4}Gr_{x\mathrm{v}}\right)^{-1/4}\\&\quad\times W_{y\mathrm{v}}2\sqrt{gx}\frac{\mathrm{d}W_{x\mathrm{v}}}{\mathrm{d}\eta_{\mathrm{v}}}x^{-1}\left(\frac{1}{4}Gr_{x\mathrm{v,s}}\right)^{1/4}\\&=2\nu_{\mathrm{v}}\sqrt{gx}\frac{\mathrm{d}^2W_{x\mathrm{v}}}{\mathrm{d}\eta_{\mathrm{v}}^2}\left(\tfrac{1}{4}Gr_{x\mathrm{v,s}}\right)^{1/2}x^{-2}.\end{aligned}$$

The earlier equation is divided by g and simplified to the following one according to the definition of Grashof number $Gr_{x\mathrm{v,s}}$

$$W_{xv}\left(2W_{xv}-\eta_l\frac{dW_{xv}}{d\eta_v}\right)+4W_{yv}\frac{dW_{xv}}{d\eta_v}=\frac{d^2W_{xv}}{d\eta_v^2}. \tag{12.77}$$

Summarize the earlier derivation results for the similarity transformation, the dimensionless equations for the two-film saturated condensation are as follows:

For liquid film. The dimensionless equations are (12.59), (12.64), and (12.71) for mass, momentum, and energy conservations:

$$2W_{xl}-\eta_l\frac{dW_{xl}}{d\eta_l}+4\frac{dW_{yl}}{d\eta_l}-\frac{1}{\rho_l}\frac{d\rho_l}{d\eta_l}(-\eta_l W_{xl}+4W_{yl})=0, \tag{12.59}$$

$$\frac{\nu_{l,s}}{\nu_l}\left(W_{xl}\left(2W_{xl}-\eta_l\frac{dW_{xl}}{d\eta_l}\right)+4W_{yl}\frac{dW_{xl}}{d\eta_l}\right)=\frac{d^2W_{xl}}{d\eta_l^2}+\frac{1}{\mu_l}\frac{d\mu_l}{d\eta_l}\frac{dW_{xl}}{d\eta_l}+\frac{\mu_{l,s}}{\mu_l}\frac{\rho_l-\rho_v}{\rho_{l,w}-\rho_v}, \tag{12.64}$$

$$Pr_l\frac{\nu_{l,s}}{\nu_l}[-\eta_l W_{xl}+4W_{yl}]\frac{d\theta_l}{d\eta_l}=\frac{d^2\theta_l}{d\eta_l^2}+\frac{1}{\lambda_l}\frac{d\lambda_l}{d\eta_l}\frac{d\theta_l}{d\eta_l}. \tag{12.71}$$

For vapor film. The dimensionless equations are (12.74) and (12.77) for mass and momentum conservations:

$$2W_{xv}-\eta_v\frac{dW_{xv}}{d\eta_v}+4\frac{dW_{yv}}{d\eta_v}=0, \tag{12.74}$$

$$W_{xv}\left(2W_{xv}-\eta_l\frac{dW_{xv}}{d\eta_v}\right)+4W_{yv}\frac{dW_{xv}}{d\eta_v}=\frac{d^2W_{xv}}{d\eta_v^2}. \tag{12.77}$$

For boundary conditions. With the variables assumed in equations (12.3)–(12.19) for liquid and vapor films, the following boundary conditions can be transformed, respectively:

$$\eta_l=0: W_{xl}=0,\ W_{yl}=0,\ \theta_l=1, \tag{12.78}$$

$\eta_v=\eta_{l\delta}(\eta_v=0)$:

$$W_{xv,s}=\left(\frac{\rho_{l,w}-\rho_v}{\rho_{l,s}}\right)^{1/2}W_{xl,s}, \tag{12.79}$$

$$W_{yv,s}=-0.25\frac{\mu_{l,s}}{\mu_v}\left(\frac{\nu_{v,s}}{\nu_{l,s}}\right)^{1/2}\left(\frac{\rho_{l,w}-\rho_v}{\rho_{l,s}}\right)^{1/4}(W_{xl,s}\eta_{l\delta}-4W_{yl,s}), \tag{12.80}$$

$$\left(\frac{dW_{xv}}{d\eta_v}\right)_{\eta_v=0}=\frac{\mu_{l,s}}{\mu_v}\left(\frac{\rho_{l,w}-\rho_v}{\rho_{l,s}}\right)^{3/4}\left(\frac{\nu_v}{\nu_{l,s}}\right)^{1/2}\left(\frac{dW_{xl}}{d\eta_l}\right)_{\eta l=\eta l\delta}, \tag{12.81}$$

$$h_{fg}\mu_{\mathrm{l,s}}(W_{x\mathrm{l,s}}\eta_{\mathrm{l}\delta} - 4W_{y\mathrm{l,s}}) - \lambda_{\mathrm{l,s}}(t_{\mathrm{w}} - t_{\mathrm{s}})\left(\frac{\mathrm{d}\theta_{\mathrm{l}}}{\mathrm{d}\eta_{\mathrm{l}}}\right)_{\eta_{\mathrm{l}}=\eta_{\mathrm{l}\delta}} = 0, \quad (12.82)$$

$$\theta_{\mathrm{l}} = 0. \quad (12.83)$$

$$\eta_{\mathrm{v}} \to \infty : W_{x\mathrm{v}} \to 0. \quad (12.84)$$

References

1. W. Nusselt, Die Oberflächenkondensation des Wasserdampfes, Z. Ver. D Ing. 60, pp. 541–569,1916
2. L.A. Bromley, Effect of heat capacity of condensate in condensing, Ind. Eng. Chem. 44, pp. 2966–2969, 1952
3. W.M. Rohsenow, Heat transfer and temperature distribution in laminar film condensation, Trans. Am. Soc. Mech. Engrs. 78, pp. 1645–1648, 1956
4. E.M. Sparrow and J.L. Gregg, A boundary-layer treatment of laminar film condensation, J. Heat Transfer 81, pp. 13–18, 1959
5. J.C. Koh, E.M. Sparrow, and J.P. Hartnett, The two phase boundary layer in laminar film condensation, Int. J. Heat Mass Transfer 2, pp. 69–88, 1961
6. M.M. Chen, An analytical study of laminar film condensation: part 1 – Flat plate, J. Heat Transfer 81, pp. 48–54, 1961
7. W.H. McAdams, Heat Transmission, 3rd edn. McGraw-Hill, New York, pp. 331–332, 1954
8. Voskresenskiy, Calculation of heat transfer in film condensation allowing for the temperature dependence of the physical properties of the codensate, U.S.S.A. Acad. Sci., 1948
9. D.A. Labuntsov, Effect of temperature dependence of physical parameters of condensate on heat transfer in film condensation of steam, Teploenergetika 4 (2), pp. 49–51, 1957
10. G. Poots and R.G. Miles, Effects of variable physical properties on laminar film condensation of saturated steam on a vertical flat plate, Int. J. Heat Mass Transfer 10, pp. 1677–1692, 1967
11. D.Y. Shang and T. Adamek, Study of laminar film condensation of saturated steam for consideration of variable thermophysical properties, Transport phenomena, Science and Technology, 1992 (Ed. by B.X. Wang), Higher Education Press, Beijing, pp. 470–475, 1992
12. D.Y. Shang and T. Adamek, Study on laminar film condensation of saturated steam on a vertical flat plate for consideration of various physical factors including variable thermophysical properties, Wärme- und Stoffübertragung 30, pp. 89–100, 1994
13. D.Y. Shang and B.X. Wang, An extended study on steady-state laminar film condensation of a superheated vapor on an isothermal vertical plate, Int. J. Heat Mass Transfer 40, No. 4, pp. 931–941, 1997
14. VDI-Wärmeatlas, Berechnungsblätter für die Wärmeübertragung, 5, erweiterte Auflage, VDI Verlage GmbH, Düsseldorf, 1988

13

Effects of Various Physical Conditions on Film Condensations

Nomenclature

a	thermal diffusive coefficient, $m^2\,s^{-1}$; assumption for ignoring the variable thermophysical properties (or for Boussinesq approximation)
b	assumption for ignoring the shear force at the liquid–vapor interface; plate width, m
c	assumption for ignoring inertia force of condensate film
d	assumption for ignoring the thermal convection of condensate film
m	more complete condition
c_p	specific heat at constant pressure, $J\,(kg\,K)^{-1}$
g	gravitation acceleration, $m\,s^2$
$Gr_{x\mathrm{l,s}}$	local Grashof number of liquid film for film condensation of saturated vapor, $\frac{g(\rho_{\mathrm{l,w}}-\rho_{\mathrm{v,s}})x^3}{\nu_{\mathrm{l,s}}^2\rho_{\mathrm{l,s}}}$
G_x	total mass flow rate entering the liquid film for position $x=0$ to x with width of b of the plate, $kg\,s^{-1}$
$Nu_{x,\mathrm{w}}$	local Nusselt number defined as $\overline{\alpha}_x x/\lambda_{\mathrm{w}}$
$\overline{Nu}_{x,\mathrm{w}}$	average Nusselt number, $\overline{\alpha}_x x/\lambda_{\mathrm{w}}$
n_{c_p}	thermal conductivity parameter of gas
n_μ	viscosity parameter of gas
Pr	Prandtl number
q_x	local heat transfer rate per unit area on the plate, W m^{-2}
Q_x	total heat transfer rate for position $x=0$ to x with width of b on the plate, W
t	temperature, °C
T	absolute temperature, K

w_x, w_y	velocity components in the x- and y- directions, respectively
W_x, W_y	dimensionless velocity component in the x- and y- directions, respectively
$W_{x\mathrm{l,s}}, W_{y\mathrm{l,s}}$	dimensionless velocity component at interface for film condensation
x, y	dimensional coordinate variables

Greek symbols

α_x	local heat transfer coefficient, W $(\mathrm{m}^2\ \mathrm{K})^{-1}$
$\overline{\alpha}_x$	average heat transfer coefficient, W $(\mathrm{m}^2\,\mathrm{K})^1$
δ	boundary layer thickness, m
η	dimensionless coordinate variable for boundary layer
θ	dimensionless temperature
λ	thermal conductivity, W $(\mathrm{m\ K})^{-1}$
μ	absolute viscosity, kg $(\mathrm{m\ s})^{-1}$
ν	kinetic viscosity, $\mathrm{m}^2\,s^{-1}$
ρ	density, $\mathrm{kg\,m}^{-3}$
Δt_w	wall subcooled temperature (for film condensation), $t_\mathrm{s} - t_\mathrm{w}$, °C
$\frac{\Delta t_\mathrm{w}}{t_\mathrm{s}}$	wall subcooled grade, $(t_\mathrm{s} - t_\mathrm{w})/t_\mathrm{s}$
$\left(\left(\frac{\mathrm{d}\theta_l}{\mathrm{d}\eta_l}\right)_{\eta_l=0}\right)_{\Delta t_\infty=0}$	dimensionless temperature gradient on the plate for film condensation of saturated vapor
$\eta_{\mathrm{l}\delta}$	dimensionless liquid film thickness
Φ_s	mass flow rate parameter for film condensation, $\eta_{\mathrm{l}\delta} W_{x\mathrm{l,s}} - 4W_{\mathrm{yl,s}}$
$\frac{\rho_\mathrm{l}-\rho_\mathrm{v}}{\rho_\mathrm{l,w}-\rho_\mathrm{v}}$	buoyancy factor
$\frac{1}{\rho}\frac{\mathrm{d}p}{\mathrm{d}x}$	density factor
$\frac{1}{\mu}\frac{\mathrm{d}\mu}{\mathrm{d}\eta}$	viscosity factor
$\frac{1}{\lambda}\frac{\mathrm{d}\lambda}{\mathrm{d}\eta}$	thermal conductivity factor

Subscripts

a	assumption a
b	assumption b
c	assumption c
d	assumption d
f	film
i	liquid
m	more complete condition
s	saturate state

v	vertical case or vapor
w	at wall
∞	far from the wall surface

13.1 Introduction

In Chap. 12, the extended theory of steady state laminar film condensation process of pure saturated vapor at atmospheric pressure on an isothermal vertical flat plate is established. Its equations provide a complete account of the analyses and calculation of its physical process for consideration of various physical factors including variable thermophysical properties.

In this chapter, effects of various physical conditions on heat and mass transfer of the film condensation of saturated vapor will be further presented [1, 2]. To this end, the film condensation of saturated water vapor is taken as an example, and, four different assumptions, such as Boussinesq approximation of condensate film, ignoring shear force at the liquid–vapor interface, ignoring condensate film inertia force, and ignoring condensate film thermal convection are considered for investigation of their effects on the condensate heat transfer coefficient, condensate film thickness, and mass flow rate of the film condensation. Quantitative comparisons from these results indicate the following points:

Effect of the physical conditions on heat transfer coefficient. The Boussinesq approximation of the condensate film will greatly decrease the heat transfer coefficient of the condensation and cause the largest effect on the heat transfer coefficient compared with those caused by other physical conditions. The thermal convection of condensate film will increase the heat transfer coefficient of the condensation, and its effect on the heat transfer coefficient is larger than those caused by the liquid–vapor interfacial shear force and the inertia force of condensate film. The liquid–vapor interfacial shear force and the inertia force of the condensate film will decrease the heat transfer coefficient very slightly.

Effect of the physical conditions on condensate film thickness. The Boussinesq approximation of the condensate film will greatly decrease the condensate film thickness and cause the largest effect on the condensate film thickness compared with those caused by other physical conditions. The thermal convection of condensate film will decrease the condensate film thickness, and its effect on the condensate film thickness is larger than those caused by the liquid–vapor interfacial shear force and the inertia force of condensate film. The liquid–vapor interfacial shear force and the inertia force of the condensate film will increase the condensate film thickness very slightly.

Effect of the physical conditions on condensate mass flow rate. The condensate film thermal convection will greatly decrease the mass flow rate of the film condensation, and will cause the largest effect on the condensate mass flow rate compared with those caused by other physical conditions. The Boussinesq approximation will increase the condensate mass flow rate very slightly, and the liquid–vapor interfacial shear force, and the condensate film inertia force will decrease the condensate mass flow rate also slightly.

13.2 Review of Governing Equations for Film Condensation of Saturated Vapor

In Chap. 12, the governing equations for film condensation of saturated vapor were presented. In this chapter, it is necessary to have a brief review of those equations for a further analysis.

13.2.1 Partial Differential Equations

The analytical model and coordinate system used for the laminar film condensation of the saturated vapor on an isothermal vertical flat plate is shown in Fig. 12.1. The conservation partial differential equations of mass, momentum, and energy for steady laminar saturated condensation in two-phase boundary layer are as follows:

For condensate liquid film:

$$\frac{\partial}{\partial x}\left(\rho_{\mathrm{l}} w_{x\mathrm{l}}\right)+\frac{\partial}{\partial y}\left(\rho_{\mathrm{l}} w_{y\mathrm{l}}\right)=0, \tag{13.1}$$

$$\rho_{\mathrm{l}}\left(w_{x\mathrm{l}}\frac{\partial w_{x\mathrm{l}}}{\partial x}+w_{y\mathrm{l}}\frac{\partial w_{x\mathrm{l}}}{\partial y}\right)=\frac{\partial}{\partial y}\left(\mu_{\mathrm{l}}\frac{\partial w_{x\mathrm{l}}}{\partial y}\right)+g(\rho_{\mathrm{l}}-\rho_{\mathrm{v}}), \tag{13.2}$$

$$\rho_{\mathrm{l}} c_{p_{\mathrm{l}}}\left(w_{x\mathrm{l}}\frac{\partial t}{\partial x}+w_{y\mathrm{l}}\frac{\partial t}{\partial y}\right)=\frac{\partial}{\partial y}\left(\lambda_{\mathrm{l}}\frac{\partial t}{\partial y}\right). \tag{13.3}$$

For vapor film:

$$\frac{\partial}{\partial x}(w_{x\mathrm{v}})+\frac{\partial}{\partial y}(w_{y\mathrm{v}})=0, \tag{13.4}$$

$$w_{x\mathrm{v}}\frac{\partial w_{x\mathrm{v}}}{\partial x}+w_{y\mathrm{v}}\frac{\partial w_{x\mathrm{v}}}{\partial y}=\nu_{\mathrm{v}}\frac{\partial^2 w_{x\mathrm{v}}}{\partial y^2}. \tag{13.5}$$

For boundary conditions:

$$y=0: w_{x\mathrm{l}}=0,\quad w_{y\mathrm{l}}=0,\quad t_{\mathrm{l}}=t_{\mathrm{w}}, \tag{13.6}$$

$$y=\delta_{\mathrm{v}}$$

$$w_{x\mathrm{l,s}}=w_{x\mathrm{v,s}}, \tag{13.7}$$

$$\rho_{\mathrm{l,s}}\left(w_{x\mathrm{l}}\frac{\partial\delta_{\mathrm{l}}}{\partial x}-w_{y\mathrm{l}}\right)_{\mathrm{s}}=\rho_{\mathrm{v}}\left(w_{x\mathrm{v}}\frac{\partial\delta_{\mathrm{v}}}{\partial x}-w_{y\mathrm{v}}\right)_{\mathrm{s}}, \tag{13.8}$$

$$\mu_{\mathrm{l,s}}\left(\frac{\partial w_{x\mathrm{l}}}{\partial y}\right)_{\mathrm{s}}=\mu_{\mathrm{v}}\left(\frac{\partial w_{x\mathrm{v}}}{\partial y}\right)_{\mathrm{s}}, \tag{13.9}$$

$$\lambda_{\mathrm{l,s}}\left(\frac{\partial t_{\mathrm{l}}}{\partial y}\right)_{y=\delta_{\mathrm{l}}}=h_{fg}\rho_{\mathrm{l,s}}\left(w_{x\mathrm{l}}\frac{\partial\delta_{\mathrm{l}}}{\partial x}-w_{y\mathrm{l}}\right)_{\mathrm{s}}, \tag{13.10}$$

$$t_{\mathrm{l}}=t_{\mathrm{s}}, \tag{13.11}$$

$$y\to\infty: w_{x\mathrm{v}}\to 0, \tag{13.12}$$

13.2.2 Similarity Variables

Same as those in Chap. 12, the following dimensionless variables are assumed for the similarity transformation of the governing partial differential equations of the film condensation of saturated vapor:

For liquid film. For liquid film the similarity transformation variables are assumed as follows:

For liquid film the dimensionless coordinate variable η_{l} is set up at first as follows:

$$\eta_{\mathrm{l}}=\left(\frac{1}{4}\mathrm{Gr}_{\mathrm{x}l,s}\right)^{1/4}\frac{y}{x}, \tag{13.13}$$

where the local Grashof number $Gr_{x\mathrm{l,s}}$ is assumed as

$$Gr_{x\mathrm{l,s}}=\frac{g(\rho_{\mathrm{l,w}}-\rho_{\mathrm{v}})x^3}{\nu_{\mathrm{l,s}}^2\rho_{\mathrm{l,s}}}. \tag{13.14}$$

The dimensionless temperature is given as

$$\theta_{\mathrm{l}}=\frac{t_{\mathrm{l}}-t_{\mathrm{s}}}{t_{\mathrm{w}}-t_{\mathrm{s}}}. \tag{13.15}$$

The dimensionless velocity components are assumed as

$$W_{x\mathrm{l}}=\left(2\sqrt{gx}\left(\frac{\rho_{\mathrm{l,w}}-\rho_{\mathrm{v,s}}}{\rho_{\mathrm{l,s}}}\right)^{1/2}\right)^{-1}w_{x\mathrm{l}}, \tag{13.16}$$

$$W_{y\mathrm{l}}=\left(2\sqrt{gx}\left(\frac{\rho_{\mathrm{l,w}}-\rho_{\mathrm{v,s}}}{\rho_{\mathrm{l,s}}}\right)^{1/2}\left(\frac{1}{4}Gr_{x\mathrm{l,s}}\right)^{-1/4}\right)^{-1}w_{y\mathrm{l}}. \tag{13.17}$$

For vapor film. The vapor film dimensionless coordinate variable $\eta_{\rm v}$ and the local Grashof number $Gr_{x{\rm v,s}}$ are assumed as, respectively,

$$\eta_{\rm v} = \left(\frac{1}{4}Gr_{x{\rm v}}\right)^{1/4}\frac{y}{x}, \quad Gr_{x{\rm v}} = \frac{gx^3}{\nu_{{\rm v,s}^2}}. \tag{13.18}$$

The dimensionless velocity components are assumed as

$$W_{x{\rm v}} = (2\sqrt{gx})^{-1}w_{x{\rm v}}, \tag{13.19}$$

$$W_{y{\rm v}} = \left(2\sqrt{gx}\left(\frac{1}{4}Gr_{x{\rm v}}\right)^{-1/4}\right)^{-1} w_{y{\rm v}}. \tag{13.20}$$

13.2.3 Transformed Dimensionless Differential Equations

The governing partial differential equations and the boundary conditions for the film condensation of saturated vapor are transformed to the following forms:

For liquid film:

$$2W_{x{\rm l}} - \eta_{\rm l}\frac{{\rm d}W_{xl}}{{\rm d}\eta_{\rm l}} + 4\frac{{\rm d}W_{y{\rm l}}}{{\rm d}\eta_{\rm l}} - \frac{1}{\rho_{\rm l}}\frac{{\rm d}\rho_{\rm l}}{{\rm d}\eta_{\rm l}}(\eta_{\rm l}W_{x{\rm l}} - 4W_{y{\rm l}}) = 0, \tag{13.21}$$

$$\frac{\nu_{\rm l,s}}{\nu_{\rm l}}\left(W_{x{\rm l}}\left(2W_{x{\rm l}} - \eta_{\rm l}\frac{{\rm d}W_{x{\rm l}}}{{\rm d}\eta_{\rm l}}\right) + 4W_{y{\rm l}}\frac{{\rm d}W_{x{\rm l}}}{{\rm d}\eta_{\rm l}}\right) = \frac{{\rm d}^2W_{x{\rm l}}}{{\rm d}\eta_{\rm l}^2} + \frac{1}{\mu_{\rm l}}\frac{{\rm d}\mu_{\rm l}}{{\rm d}\eta_{\rm l}}\frac{{\rm d}W_{x{\rm l}}}{{\rm d}\eta_{\rm l}} + \frac{\mu_{\rm l,s}}{\mu_{\rm l}}\frac{\rho_{\rm l} - \rho_{\rm v,s}}{\rho_{\rm l,w} - \rho_{\rm v,s}}, \tag{13.22}$$

$$Pr_{\rm l}\frac{\nu_{\rm l,s}}{\nu_{\rm l}}\left[-\eta_{\rm l}W_{x{\rm l}} + 4W_{y{\rm l}}\right]\frac{{\rm d}\theta_{\rm l}}{{\rm d}\eta_{\rm l}} = \frac{{\rm d}^2\theta_{\rm l}}{{\rm d}\eta_{\rm l}^2} + \frac{1}{\lambda_{\rm l}}\frac{{\rm d}\lambda_{\rm l}}{{\rm d}\eta_{\rm l}}\frac{{\rm d}\theta_{\rm l}}{{\rm d}\eta_{\rm l}}. \tag{13.23}$$

For vapor film:

$$2W_{x{\rm v}} - \eta_{\rm v}\frac{{\rm d}W_{x{\rm v}}}{{\rm d}\eta_{\rm v}} + 4\frac{{\rm d}W_{y{\rm v}}}{{\rm d}\eta_{\rm v}} = 0, \tag{13.24}$$

$$\left(W_{x{\rm v}}\left(2W_{x{\rm v}} - \eta_{\rm v}\frac{{\rm d}W_{x{\rm v}}}{{\rm d}\eta_{\rm v}}\right) + 4W_{y{\rm v}}\frac{{\rm d}W_{x{\rm v}}}{{\rm d}\eta_{\rm v}}\right) = \frac{{\rm d}^2W_{x{\rm v}}}{{\rm d}\eta_{\rm v}^2}. \tag{13.25}$$

For boundary conditions

$$\eta_{\rm l} = 0: W_{x{\rm l}} = 0, \quad W_{y{\rm l}} = 0, \quad \theta_{\rm l} = 1, \tag{13.26}$$

$\eta_{\rm v} = \eta_{{\rm l}\delta}(\eta_{\rm v} = 0):$

$$W_{x{\rm v,s}} = \left(\frac{\rho_{\rm l,w} - \rho_{\rm v,s}}{\rho_{\rm l,s}}\right)^{1/2} W_{x{\rm l,s}}, \tag{13.27}$$

$$W_{yv,s} = -0.25 \frac{\mu_{l,s}}{\mu_v} \left(\frac{\nu_{v,s}}{\nu_v} \right)^{1/2} \left(\frac{\rho_{l,w} - \rho_{v,s}}{\rho_{l,s}} \right)^{1/4} (W_{xl,s}\eta_{l\delta} - 4W_{yl,s}), \tag{13.28}$$

$$\left(\frac{dW_{xv}}{d\eta_v} \right)_{\eta_v=0} = \frac{\mu_{l,s}}{\mu_v} \left(\frac{\rho_{l,w} - \rho_{v,s}}{\rho_{l,s}} \right)^{3/4} \left(\frac{\nu_{v,s}}{\nu_{l,s}} \right)^{1/2} \left(\frac{dW_{xl}}{d\eta_l} \right)_{\eta l=\eta l\delta}, \tag{13.29}$$

$$h_{fg}\mu_{l,s}(W_{xl,s}\eta_{l\delta} - 4W_{yl,s}) - c_{p_{l,s}}(t_w - t_s) \left(\frac{d\theta_l}{d\eta_l} \right)_{\eta_l=\eta_{l\delta}} = 0, \tag{13.30}$$

$$\theta_l = 0, \tag{13.31}$$

$$\eta_v \to \infty : W_{xv} \to 0. \tag{13.32}$$

In the governing equations here various physical conditions, such variable thermophysical properties, shear force at liquid–vapor interface, condensate film inertia force, and condensate film thermal convection are considered. These physical conditions are overall named more complete condition for further investigations.

13.3 Different Physical Assumptions

13.3.1 Assumption a (with Boussinesq Approximation of Condensate Film)

The assumption a is defined that on the basis of the more complete condition the Boussinesq approximation is further considered in the governing differential equations of the condensate film. With assumption a, associated partial differential equations of the condensate film become

$$\frac{\partial w_{xl}}{\partial x} + \frac{\partial w_{yl}}{\partial y} = 0, \tag{13.33}$$

$$w_{xl}\frac{\partial w_{xl}}{\partial x} + w_{yl}\frac{\partial w_{xl}}{\partial y} = \nu_l^* \frac{\partial^2 w_{xl}}{\partial y_l^2} + \frac{g(\rho_l - \rho_v)}{\rho_l^*}, \tag{13.34}$$

$$\frac{\rho_l^* c_{pl}^*}{\lambda_l^*} \left(w_{xl}\frac{\partial t}{\partial x} + w_{yl}\frac{\partial t}{\partial y} \right) = \frac{\partial^2 t}{\partial y^2}, \tag{13.35}$$

where the superscript * implies the value at reference temperature t^* that is described by mean temperature $(\mathrm{t_w} + \mathrm{t_s})/2$.

Of course, the governing partial differential equations for vapor film are also (13.4) and (13.5). Strictly speaking, for the boundary conditions under the Boussinesq approximation, the variable thermophysical properties need not be considered. However, for examining the effect of the variable thermophysical properties we still take the boundary conditions (13.6)–(13.12) with the temperature-dependent thermophysical properties as the associated boundary conditions.

With the expressions (13.13)–(13.17) for the defined variables of the condensate film, the following governing ordinary differential equations can be derived from (13.33)–(13.35) as:

$$2W_{xl} - \eta_l \frac{dW_{xl}}{d\eta_l} + 4\frac{dW_{yl}}{d\eta_l} = 0, \tag{13.36}$$

$$\frac{\nu_{l,s}}{\nu_l^*}\left(W_{xl}\left(2W_{xl} - \eta_l\frac{dW_{xl}}{d\eta_l}\right) + 4W_{yl}\frac{dW_{xl}}{d\eta_l}\right) = \frac{d^2W_{xl}}{d\eta_l^2} + \frac{\mu_{l,s}}{\mu_l}\frac{\rho_l - \rho_v}{\rho_{l,w} - \rho_v}, \tag{13.37}$$

$$Pr_l^* \frac{\rho_l^*}{\rho_{l,s}}\frac{\mu_{l,s}}{\mu_l^*}(-\eta_l W_{lv} + 4W_{yl})\frac{d\theta_l}{d\eta_l} = \frac{d^2\theta_l}{d\eta_l^2}. \tag{13.38}$$

From the analysis of Chap. 6, it is possible to regard the specific heat c_{pl}^* in Pr_l^* as constant for water and a lot of liquids in the special temperature range for engineering application. In this case, the property factors $(Pr_l^*\rho_l^*/\rho_{l,s})(\mu_{l,s}/\mu_l^*)$ in (13.38) can be substituted by $Pr_{l,s}(\rho_l^*/\rho_{l,s})(\lambda_{l,s}/\lambda_l^*)$.

The governing ordinary differential equations of the vapor film are also (13.24) and (13.25). Of course, the transformed dimensionless boundary conditions are also (13.26)–(13.32).

13.3.2 Assumption b (Ignoring Shear Force at Liquid–Vapor Interface)

In assumption *b*, the shear force at the liquid–vapor interface is neglected on the basis of the more complete conditions. The governing partial differential equations for this assumption are only (13.1)–(13.3), since the governing equations of vapor film should be omitted. Consequently, the boundary conditions (13.7), (13.8), and (13.12) are omitted, and (13.6), (13.10), and (13.11) remain. Since the shear force at the liquid–vapor interface is neglected, the boundary condition (13.9) is simplified to

$$y = \delta_l : \left(\frac{\partial w_{xl}}{\partial y}\right)_s = 0. \tag{13.39}$$

With (13.13) and (13.16) for the defined similarity variables, (13.39) is changed into

$$\eta_l = \eta_{l\delta} : \left(\frac{dW_{xl}}{d\eta_l}\right)_{\eta_l=\eta_{l\delta}} = 0. \tag{13.40}$$

Thus, with the similarity transformation the governing partial equations (13.1)–(13.3) are transformed to (13.21)–(13.23), respectively, and their boundary conditions are (13.26), (13.30), (13.31), and (13.40).

13.3.3 Assumption c (Ignoring Inertia Force of the Condensate Film)

The assumption *c* is defined that the inertia force of the condensate film is further omitted on the basis of the assumption *b*. The governing partial differential equations in this assumption are (13.1) and (13.3), as well as the following momentum equation:

$$\frac{\partial}{\partial y}\left(\mu_{\mathrm{l}}\frac{\partial w_{x\mathrm{l}}}{\partial y}\right)+g(\rho_{\mathrm{l}}-\rho_{\mathrm{v,s}})=0. \tag{13.41}$$

Then, by virtue of the expressions for defined similarity variables, (13.13)–(13.17), the associated governing ordinary differential equations will be (13.21) and (13.23), as well as the following equation:

$$\frac{\mathrm{d}^2 W_{x\mathrm{l}}}{\mathrm{d}\eta_{\mathrm{l}}^2}+\frac{1}{\mu_{\mathrm{l}}}\frac{\mathrm{d}\mu_{\mathrm{l}}}{\mathrm{d}\eta_{\mathrm{l}}}\frac{\mathrm{d}W_{x\mathrm{l}}}{\mathrm{d}\eta_{\mathrm{l}}}+\frac{\mu_{\mathrm{l,s}}}{\mu_{\mathrm{l}}}\frac{\rho_{\mathrm{l}}-\rho_{\mathrm{v,s}}}{\rho_{\mathrm{l,w}}-\rho_{\mathrm{v,s}}}=0. \tag{13.42}$$

The boundary conditions for the assumption *c* are same as those for assumption *b*.

13.3.4 Assumption d (Ignoring Thermal Convection of the Condensate Film)

Assumption *d* is that the thermal convection of the condensate film is further omitted on the basis of the assumption *c*. For this further assumption the energy (13.3) is simplified to

$$\frac{\partial}{\partial y}\left(\lambda_{\mathrm{l}}\frac{\partial t}{\partial y}\right)=0. \tag{13.43}$$

Therefore, the governing partial differential equations of this assumption should then be (13.1), (13.41), (13.43) and the associated ordinary differential equations are (13.21) and (13.42), as well as the following energy equation:

$$\frac{\mathrm{d}^2\theta_{\mathrm{l}}}{\mathrm{d}\eta_{\mathrm{l}}^2}=-\frac{1}{\lambda_{\mathrm{l}}}\frac{\mathrm{d}\lambda_{\mathrm{l}}}{\mathrm{d}\eta_{\mathrm{l}}}\frac{\mathrm{d}\theta_{\mathrm{l}}}{\mathrm{d}\eta_{\mathrm{l}}}. \tag{13.44}$$

The associated boundary conditions for the assumption *d* are also the same as those for the assumption *b*.

13.4 Effects of Various Physical Conditions on Velocity and Temperature Fields

The numerical calculations in each assumed physical condition are carried out for different wall subcooled grade, such as $\Delta t_{\mathrm{w}}/t_{\mathrm{s}}(=(t_{\mathrm{s}}-t_{\mathrm{w}})/t_{\mathrm{s}})=$ 0.001, 0.025, 0.05, 0.1, 0.2, 0.4, 0.6, 0.8, 1 for film condensation of saturated

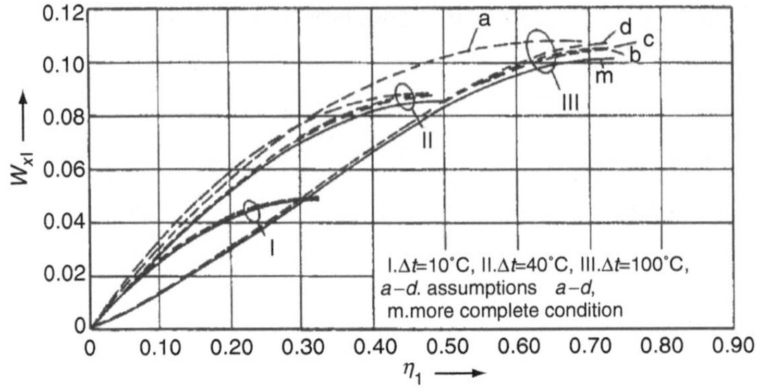

Fig. 13.1. Velocity profiles of film condensation of saturated water vapor in different conditions, cited from Shang and Adamek [1], I. $\Delta t_{\mathrm{w}}/t_{\mathrm{s}} = 0.1$, II. $\Delta t_{\mathrm{w}}/t_{\mathrm{s}} = 0.4$, III. $\Delta t_{\mathrm{w}}/t_{\mathrm{s}} = 1$. Line m: for more complete condition. Lines a–d: for assumptions a–d, respectively

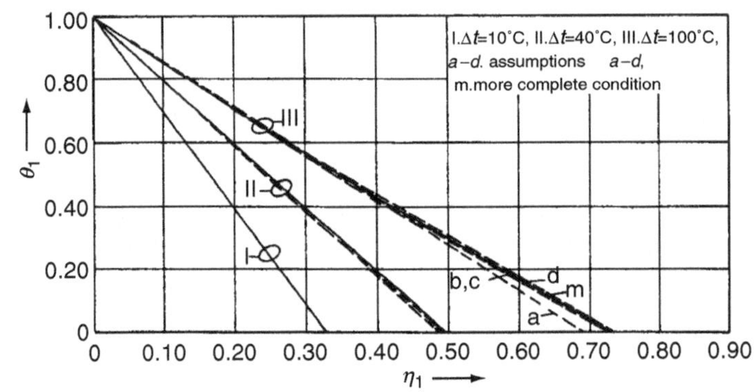

Fig. 13.2. Temperature profiles of film condensation of saturated water vapor in different conditions, cited from Shang and Adamek [1]. I. $\Delta t_{\mathrm{w}}/t_{\mathrm{s}} = 0.1$, II. $\Delta t_{\mathrm{w}}/t_{\mathrm{s}} = 0.4$, III. $\Delta t_{\mathrm{w}}/t_{\mathrm{s}} = 1$. Line m: for more complete condition. Lines a–d: for assumptions a–d respectively

water vapor. Some calculated results of the dimensionless velocity component W_{x1} and temperature fields are plotted in Figs. 13.1 and 13.2, respectively. It is seen that the physical conditions have corresponding influences both on the condensate film velocity and temperature fields. Additionally, the effects of Boussinesq approximation on the condensate film velocity and temperature fields are much larger than those of other physical conditions. While, the effects of the condensate film thermal convection on the condensate film velocity and temperature fields are larger than those of liquid–vapor interfacial shear force and the condensate film inertia force.

13.5 Effects of Various Physical Conditions on Heat Transfer

According to (12.44), the local heat transfer coefficient on the surface for the film condensation of saturated vapor is

$$\alpha_x = -\lambda_{\mathrm{l,w}} \left(\frac{1}{4} Gr_{x\mathrm{l,s}}\right)^{1/4} x^{-1} \left(\left(\frac{\mathrm{d}\theta_\mathrm{l}}{\mathrm{d}\eta_\mathrm{l}}\right)_{\eta_\mathrm{l}=0}\right)_{\Delta t_\infty=0} . \tag{13.45}$$

If we define the same Grashof number $Gr_{x\mathrm{l,s}}$ as (13.14) for that with the different assumed conditions, the deviations of the heat transfer coefficient caused by the related assumed conditions can be expressed as follows, respectively.

Effect of Boussinesq approximation for the condensate film of saturated vapor on heat transfer coefficient can be expressed as

$$\Delta(\alpha_x)_\mathrm{a} = \frac{(\alpha_x)_\mathrm{a} - (\alpha_x)_\mathrm{m}}{(\alpha_x)_\mathrm{m}} = \frac{\left(\left(-\left(\frac{\mathrm{d}\theta_\mathrm{l}}{\mathrm{d}\eta_\mathrm{l}}\right)_{\eta_\mathrm{l}=0}\right)_{\Delta t_\infty=0}\right)_\mathrm{a} - \left(\left(-\left(\frac{\mathrm{d}\theta_\mathrm{l}}{\mathrm{d}\eta_\mathrm{l}}\right)_{\eta_\mathrm{l}=0}\right)_{\Delta t_\infty=0}\right)_\mathrm{m}}{\left(\left(\left(-\frac{\mathrm{d}\theta_\mathrm{l}}{\mathrm{d}\eta_\mathrm{l}}\right)_{\eta_\mathrm{l}=0}\right)_{\Delta t_\infty=0}\right)_\mathrm{m}} . \tag{13.46}$$

Effect of ignoring shear force at the liquid–vapor interface on heat transfer coefficient can be expressed as

$$\Delta(\alpha_x)_\mathrm{b} = \frac{(\alpha_x)_\mathrm{b} - (\alpha_x)_\mathrm{m}}{(\alpha_x)_\mathrm{m}} = \frac{\left(\left(-\left(\frac{\mathrm{d}\theta_\mathrm{l}}{\mathrm{d}\eta_\mathrm{l}}\right)_{\eta_\mathrm{l}=0}\right)_{\Delta t_\infty=0}\right)_\mathrm{b} - \left(\left(\left(-\frac{\mathrm{d}\theta_\mathrm{l}}{\mathrm{d}\eta_\mathrm{l}}\right)_{\eta_\mathrm{l}=0}\right)_{\Delta t_\infty=0}\right)_\mathrm{m}}{\left(\left(\left(-\frac{\mathrm{d}\theta_\mathrm{l}}{\mathrm{d}\eta_\mathrm{l}}\right)_{\eta_\mathrm{l}=0}\right)_{\Delta t_\infty=0}\right)_\mathrm{m}} . \tag{13.47}$$

Effect of ignoring inertia force of the condensate film on heat transfer coefficient can be expressed as

$$\Delta(\alpha_x)_\mathrm{c} = \frac{(\alpha_x)_\mathrm{c} - (\alpha_x)_\mathrm{b}}{(\alpha_x)_\mathrm{b}} = \frac{\left(\left(-\left(\frac{\mathrm{d}\theta_\mathrm{l}}{\mathrm{d}\eta_\mathrm{l}}\right)_{\eta_\mathrm{l}=0}\right)_{\Delta t_\infty=0}\right)_\mathrm{c} - \left(\left(\left(-\frac{\mathrm{d}\theta_\mathrm{l}}{\mathrm{d}\eta_\mathrm{l}}\right)_{\eta_\mathrm{l}=0}\right)_{\Delta t_\infty=0}\right)_\mathrm{b}}{\left(\left(\left(-\frac{\mathrm{d}\theta_\mathrm{l}}{\mathrm{d}\eta_\mathrm{l}}\right)_{\eta_\mathrm{l}=0}\right)_{\Delta t_\infty=0}\right)_\mathrm{b}} . \tag{13.48}$$

Effect of ignoring thermal convection of the condensate film on heat transfer coefficient can be expressed as

$$\Delta(\alpha_x)_\mathrm{d}=\frac{(\alpha_x)_\mathrm{d}-(\alpha_x)_\mathrm{c}}{(\alpha_x)_\mathrm{c}}=\frac{\left(\left(\left(-\frac{\mathrm{d}\theta_\mathrm{l}}{\mathrm{d}\eta_\mathrm{l}}\right)_{\eta_\mathrm{l}=0}\right)_{\Delta t_\infty=0}\right)_\mathrm{d}-\left(\left(\left(-\frac{\mathrm{d}\theta_\mathrm{l}}{\mathrm{d}\eta_\mathrm{l}}\right)_{\eta_\mathrm{l}=0}\right)_{\Delta t_\infty=0}\right)_\mathrm{c}}{\left(\left(\left(-\frac{\mathrm{d}\theta_\mathrm{l}}{\mathrm{d}\eta_\mathrm{l}}\right)_{\eta_\mathrm{l}=0}\right)_{\Delta t_\infty=0}\right)_\mathrm{c}}. \quad (13.49)$$

The numerical solutions of temperature gradient

$$\left(\left(-\frac{\mathrm{d}\theta_\mathrm{l}}{\mathrm{d}\eta_\mathrm{l}}\right)_{\eta_\mathrm{l}=0}\right)_{\Delta t_\infty=0}$$

in the various assumptions are obtained numerically for the film boiling of saturated water vapor, and shown in Table 13.1, and Fig. 13.3 respectively. According to (13.46)–(13.49), the deviation of the heat transfer coefficient related to different assumed conditions are evaluated, shown in Table 13.1, and plotted in Fig. 13.4, respectively.

It is found that condensate film Boussinesq approximation will decrease the condensate heat transfer coefficient, and will cause a largest deviation of heat transfer coefficient compared with those caused by other conditions. The condensate heat transfer coefficient caused by the condensate film Boussinesq approximation will decrease with increasing the wall subcooled grade $\Delta t_\mathrm{w}/t_\mathrm{s}(=t_\mathrm{s}-t_\mathrm{w}/t_\mathrm{s})$, and will decrease 6.8% at $\Delta t_\mathrm{w}/t_\mathrm{s}=1$ for the film condensation of saturated water vapor.

The condensate film thermal convection will increase the condensate heat transfer coefficient with increasing the wall subcooled grade $\Delta t_\mathrm{w}/t_\mathrm{s}(=t_\mathrm{s}-t_\mathrm{w}/t_\mathrm{s})$ for the film condensation of saturated water vapor. The effect of the condensate film thermal convection on the condensate heat transfer coefficient is much smaller than that caused by the condensate film Boussinesq approximation generally, but much larger than those caused by the liquid–vapor interfacial shear force and the condensate film inertial force.

The liquid–vapor interfacial shear force and the condensate film inertia force will very slightly decrease the condensate heat transfer coefficient. Ignoring the liquid–vapor interfacial shear force will cause the deviation of the condensate heat transfer coefficient below 0.826% for the film condensation of saturated water vapor. However, ignoring the condensate film inertia force will cause the deviation of the condensate heat transfer coefficient even below 0.18% for the film condensation of saturated water vapor.

13.6 Effects of Various Physical Conditions on Condensate Film Thickness

The following deviations are defined for expression of the effect of the related physical factors on the condensate film thickness $\eta_{\mathrm{l}\delta}$.

Table 13.1. Temperature gradient $\left(\left(-\frac{d\theta_l}{d\eta_l}\right)_{\eta_l=0}\right)_{\Delta t_\infty=0}$ of film condensation of saturated water vapor for different assumptions and the deviations of heat transfer coefficient caused by the related assumed conditions, cited from Shang and Adamek [1]

$\Delta t_w (= t_s - t_w)$ °C	0.1	2.5	5	10	20	40	60	80	100
$\Delta t_w / t_s$	0.001	0.025	0.05	0.1	0.2	0.4	0.6	0.8	1
$\left(\left(-\frac{d\theta_l}{d\eta_l}\right)_{\eta_l=0}\right)_{\Delta t_\infty=0}$ at different assumed conditions									
$\left(\left(\left(-\frac{d\theta_l}{d\eta_l}\right)_{\eta_l=0}\right)_{\Delta t_\infty=0}\right)_m$	9.8338	4.38135	3.67507	3.06412	2.53975	2.07889	1.83238	1.6679	1.55107
$\left(\left(\left(-\frac{d\theta_l}{d\eta_l}\right)_{\eta_l=0}\right)_{\Delta t_\infty=0}\right)_a$	9.83484	4.3838	3.67408	3.06138	2.52947	2.05258	1.78498	1.59648	1.44567
$\left(\left(\left(-\frac{d\theta_l}{d\eta_l}\right)_{\eta_l=0}\right)_{\Delta t_\infty=0}\right)_b$	9.83424	4.38346	3.677	3.06945	2.5483	2.09063	1.8454	1.68121	1.56388
$\left(\left(\left(-\frac{d\theta_l}{d\eta_l}\right)_{\eta_l=0}\right)_{\Delta t_\infty=0}\right)_c$	9.83424	4.38442	3.67808	3.07104	2.55086	2.09381	1.84874	1.68455	1.56665
$\left(\left(\left(-\frac{d\theta_l}{d\eta_l}\right)_{\eta_l=0}\right)_{\Delta t_\infty=0}\right)_d$	9.83259	4.38085	3.6739	3.062	2.53525	2.06877	1.81672	1.64711	1.52566
deviations of heat transfer coefficient caused by the related assumed conditions									
$\Delta(\alpha_x)_a$	0.000106	0.000559	−0.000269	−0.00089	−0.00405	−0.01266	−0.02587	−0.04282	−0.06795
$\Delta(\alpha_x)_b$	0.000045	0.000482	0.000525	0.001739	0.003366	0.005647	0.007106	0.00798	0.008259
$\Delta(\alpha_x)_c$	0	0.000219	0.000294	0.000518	0.001005	0.001521	0.00181	0.001987	0.001771
$\Delta(\alpha_x)_d$	−0.000168	−0.000814	−0.001136	−0.00294	−0.00612	−0.01196	−0.01732	−0.02223	−0.02616

Line m: for more complete condition. Lines a–d: for assumptions a–d respectively

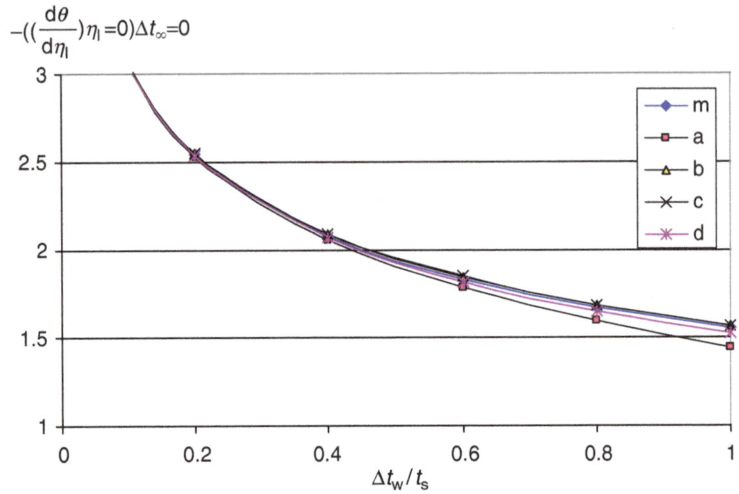

Fig. 13.3. The distributions of $\left(\left(\frac{d\theta}{d\eta_l}\right)_{\eta_l=0}\right)_{\Delta t_\infty=0}$ of film condensation of saturated water vapor with different assumptions. Line m: for more complete condition. Lines a–d: for assumptions a–d, respectively

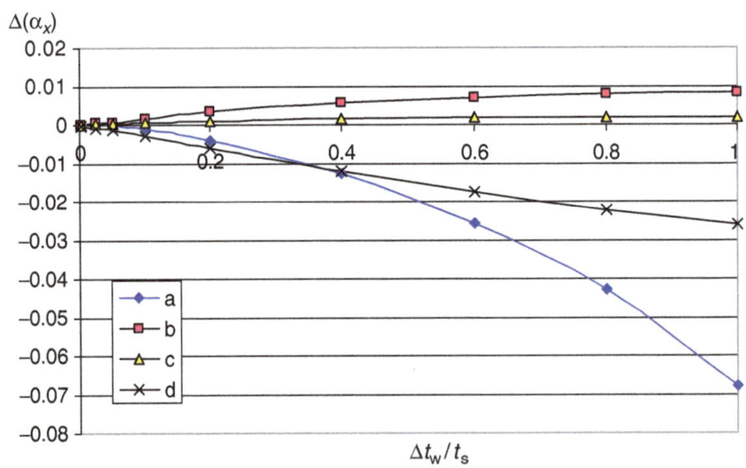

Fig. 13.4. The deviation of α_x for film condensation of saturated water vapor for different assumed conditions. Lines a–d: for $\Delta(\alpha_x)_a$, $\Delta(\alpha_x)_b$, $\Delta(\alpha_x)_c$, and $\Delta(\alpha_x)_d$, respectively

Effect of the condensate film Boussinesq approximation on the condensate film thickness $\eta_{l\delta}$ can be expressed as

$$\Delta(\eta_{l\delta})_{\mathrm{a}} = \frac{(\eta_{l\delta})_{\mathrm{a}} - (\eta_{l\delta})_{\mathrm{m}}}{(\eta_{l\delta})_{\mathrm{m}}}. \tag{13.50}$$

Effect of ignoring shear force at the liquid–vapor interface on the condensate film thickness $\eta_{l\delta}$ can be expressed as

$$\Delta(\eta_{l\delta})_{\mathrm{b}} = \frac{(\eta_{l\delta})_{\mathrm{b}} - (\eta_{l\delta})_{\mathrm{m}}}{(\eta_{l\delta})_{\mathrm{m}}}. \tag{13.51}$$

Effect of ignoring inertia force of the condensate film on the condensate film thickness $\eta_{l\delta}$ can be expressed as

$$\Delta(\eta_{l\delta})_{\mathrm{c}} = \frac{(\eta_{l\delta})_{\mathrm{c}} - (\eta_{l\delta})_{\mathrm{b}}}{(\eta_{l\delta})_{\mathrm{b}}}. \tag{13.52}$$

Effect of ignoring thermal convection of the condensate film on the condensate film thickness $\eta_{l\delta}$ can be expressed as

$$\Delta(\eta_{l\delta})_{\mathrm{d}} = \frac{(\eta_{l\delta})_{\mathrm{d}} - (\eta_{l\delta})_{\mathrm{c}}}{(\eta_{l\delta})_{\mathrm{c}}}. \tag{13.53}$$

The numerical solutions $\eta_{l\delta}$ in the various conditions are shown in Table 13.2 and plotted in Fig. 13.5 for the film condensation of saturated water vapor. Then, with (13.50)–(13.53), the related deviations of liquid film thickness $\eta_{l\delta}$ are evaluated and listed in Table 13.2 and plotted in Fig. 13.6, respectively.

It is found that Boussinesq approximation of condensate film will decrease the predicted condensate film thickness $\eta_{l\delta}$, and will cause the largest deviation of the predicted condensate film thickness $\eta_{l\delta}$ compared with those caused by other conditions. With increasing the wall subcooled grade $\Delta t_{\mathrm{w}}/t_{\mathrm{s}}(= t_{\mathrm{s}} - t_{\mathrm{w}}/t_{\mathrm{s}})$, the predicted deviation of the condensate film thickness $\eta_{l\delta}$ will come to -6% at $\Delta t_{\mathrm{w}} ts = 1$ for the film condensation of saturated water steam.

The condensate film thermal convection will also decrease the condensate film thickness $\eta_{l\delta}$, and its effect is obviously smaller than that of the Boussinesq approximation of condensate film, but obviously larger than those of the liquid–vapor interfacial shear force and the condensate film inertial force.

The liquid–vapor interfacial shear force and the condensate film inertia force will very slightly increase the condensate film thickness $\eta_{l\delta}$. Ignoring the liquid–vapor interfacial shear force will cause the deviation of the condensate film thickness $\eta_{l\delta}$ below 0.83% for the film condensation of saturated water vapor. However, ignoring the condensate film inertia force will cause the deviation of the condensate film thickness $\eta_{l\delta}$ even below 0.18% for the film condensation of saturated water vapor.

Table 13.2. Condensate film thickness $\eta_{l\delta}$ of film condensation of saturated water vapor in different assumed conditions and their deviations caused by the related assumptions, cited from Shang and Adamek [1]

$\Delta t_w (= t_s - t_w)$ °C	0.1	2.5	5	10	20	40	60	80	100
$\Delta t_w / t_s$	0.001	0.025	0.05	0.1	0.2	0.4	0.6	0.8	1
	condensate film thicknesses $\eta_{l\delta}$ for different assumed conditions								
$(\eta_{l\delta})_m$	0.1017	0.2285	0.27307	0.32804	0.39844	0.4958	0.577	0.65545	0.73561
$(\eta_{l\delta})_a$	0.10158	0.228	0.2723	0.32665	0.39534	0.4873	0.56023	0.62638	0.69172
$(\eta_{l\delta})_b$	0.10169	0.22839	0.27283	0.32747	0.3971	0.493	0.5729	0.6502	0.7295
$(\eta_{l\delta})_c$	0.10169	0.22834	0.27275	0.3273	0.3967	0.49225	0.57186	0.6489	0.7282
$(\eta_{l\delta})_d$	0.1017	0.22842	0.2729	0.32774	0.3977	0.49475	0.57619	0.6555	0.7373
	deviations of film thicknesses $\eta_{l\delta}$ caused by the related assumed conditions								
$\Delta(\eta_{l\delta})_a$	−0.0011799	−0.00219	−0.00282	−0.00424	−0.00778	−0.01714	−0.02906	−0.04435	−0.05966
$\Delta(\eta_{l\delta})_b$	−0.0000983	−0.00048	−0.00088	−0.00174	−0.00336	−0.00565	−0.00711	−0.00801	−0.00831
$\Delta(\eta_{l\delta})_c$	0	−0.00022	−0.00029	−0.00052	−0.00101	−0.00152	−0.00182	−0.002	−0.00178
$\Delta(\eta_{l\delta})_d$	0.0000983	0.00035	0.00055	0.001344	0.002521	0.005079	0.007572	0.010171	0.012497

Note: lines m and a–d denote more complete condition, assumed conditions a–d, respectively.

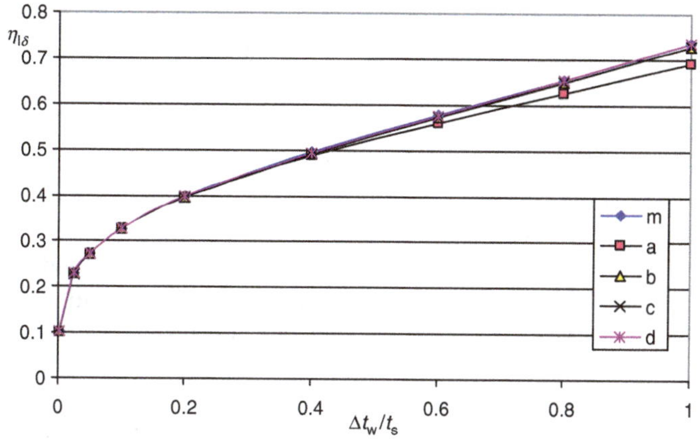

Fig. 13.5. The condensate film thickness of film condensation of saturated water vapor with different assumed conditions. (a) Lines m denotes more complete condition; (b) Lines a–b denote assumed conditions a–d respectively

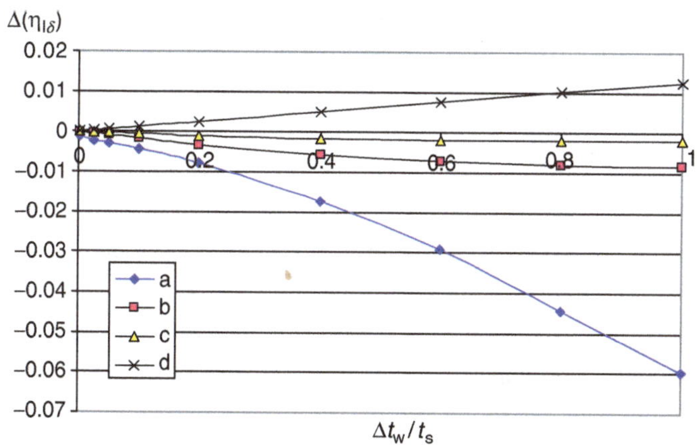

Fig. 13.6. The predicted deviation of condensate film thickness $\eta_{l\delta}$ with different conditions. Note: lines a–d denote $\Delta(\eta_{l\delta})_a$,$\Delta(\eta_{l\delta})_b$, $\Delta(\eta_{l\delta})_c$, and $\Delta(\eta_{l\delta})_d$

13.7 Effect of Various Physical Conditions on Mass Flow Rate of the Condensation

According to (12.48), total mass flow rate G_x entering the liquid film for position $x = 0$ to x with width of b of the plate is described as

$$G_x = \frac{4}{3} b \cdot \mu_{l,s} \left(\frac{1}{4} Gr_{xl,s} \right)^{1/4} \Phi_s \tag{13.54}$$

for film boiling of saturated vapor.

Here,

$$\Phi_{\mathrm{s}} = (\eta_{l\delta} W_{xl,\mathrm{s}} - 4W_{yl,\mathrm{s}})_{\Delta t_\infty = 0} \tag{13.55}$$

is mass flow rate parameter through the liquid–vapor interface for film boiling of saturated vapor.

It is seen from (13.54) that mass flow rate parameter depends on the condensate film thickness $\eta_{l\delta}$ as well as the liquid–vapor interfacial velocity components $W_{xl,\mathrm{s}}$ and $W_{yl,\mathrm{s}}$.

The numerical solutions for liquid–vapor interfacial velocity component $W_{xl,\mathrm{s}}$ with the various physical conditions are obtained, and listed in Table 13.3 and plotted in Fig. 13.7, respectively, for the film condensation of saturated water steam. Meanwhile, the numerical solutions for liquid–vapor interfacial velocity component $W_{yl,\mathrm{s}}$ with the various physical conditions are shown in Table 13.3, and plotted in Fig. 13.8, respectively, for the film condensation of saturated water vapor. It is found that the Boussinesq approximation will increase the liquid–vapor interfacial velocity components $W_{xl,\mathrm{s}}$ and $-W_{yl,\mathrm{s}}$. Other conditions will decrease the values of $W_{xl,\mathrm{s}}$ and $-W_{yl,\mathrm{s}}$.

If the Grashof number $Gr_{xl,\mathrm{s}}$ in (13.13) is defined as that of the film condensation for these assumed different physical conditions, the deviations of the mass flow rate of the film condensation caused by the related assumed conditions can be expressed as follows, respectively, by using (13.54) with (13.55).

Effect of the condensate film Boussinesq approximation on mass flow rate G_x of the film condensation can be expressed as

$$\Delta(G_x)_{\mathrm{a}} = \frac{(G_x)_{\mathrm{a}} - (G_x)_{\mathrm{m}}}{(G_x)_{\mathrm{m}}} = \frac{(\Phi_{\mathrm{s}})_{\mathrm{a}} - (\Phi_{\mathrm{s}})_{\mathrm{m}}}{(\Phi_{\mathrm{s}})_{\mathrm{m}}}. \tag{13.56}$$

Effect of ignoring shear force at the liquid–vapor interface on mass flow rate G_x of the film condensation can be expressed as

$$\Delta(G_x)_{\mathrm{b}} = \frac{(G_x)_{\mathrm{b}} - (G_x)_{\mathrm{m}}}{(G_x)_{\mathrm{m}}} = \frac{(\Phi_{\mathrm{s}})_{\mathrm{b}} - (\Phi_{\mathrm{s}})_{\mathrm{m}}}{(\Phi_{\mathrm{s}})_{\mathrm{m}}}. \tag{13.57}$$

Effect of ignoring inertia force of the condensate film on mass flow rate G_x of the film condensation can be expressed as

$$\Delta(G_x)_{\mathrm{c}} = \frac{(G_x)_{\mathrm{c}} - (G_x)_{\mathrm{b}}}{(G_x)_{\mathrm{b}}} = \frac{(\Phi_{\mathrm{s}})_{\mathrm{c}} - (\Phi_{\mathrm{s}})_{\mathrm{b}}}{(\Phi_{\mathrm{s}})_{\mathrm{b}}}. \tag{13.58}$$

Effect of ignoring thermal convection of the condensate film on mass flow rate G_x of the film condensation can be expressed as

$$\Delta(G_x)_{\mathrm{d}} = \frac{(G_x)_{\mathrm{d}} - (G_x)_{\mathrm{c}}}{(G_x)_{\mathrm{c}}} = \frac{(\Phi_{\mathrm{s}})_{\mathrm{d}} - (\Phi_{\mathrm{s}})_{\mathrm{c}}}{(\Phi_{\mathrm{s}})_{\mathrm{c}}}. \tag{13.59}$$

According to (13.55) with the numerical solutions in Tables 13.2 for $\eta_{l\delta}$ and Table 13.3 for $W_{xl,\mathrm{s}}$ and $W_{yl,\mathrm{s}}$, the mass flow rates parameters Φ_{s} are evaluated

Table 13.3. Interfacial velocity components $W_{xl,s}$ and $W_{yl,s}$, as well as mass flow rate parameter Φ_s for film condensation of saturated water vapor in different physical conditions, cited from Shang and Adamek [1]

$\Delta t_w (= t_s - t_w)$ °C	0.1	2.5	5	10	20	40	60	80	100
$\Delta t_w / t_s$	0.001	0.025	0.05	0.1	0.2	0.4	0.6	0.8	1
			$W_{xl,s}$ for different assumed conditions						
$(W_{xl,s})_m$	0.00513	0.02537	0.03546	0.04901	0.06619	0.0851	0.09463	0.09925	0.10111
$(W_{xl,s})_a$	0.005134	0.02546	0.03567	0.04953	0.06748	0.08824	0.09954	0.1054	0.1077
$(W_{xl,s})_b$	0.00513	0.02541	0.03559	0.04933	0.06698	0.0869	0.09727	0.10245	0.10466
$(W_{xl,s})_c$	0.00513	0.02542	0.03561	0.0494	0.06714	0.08727	0.09776	0.10298	0.10524
$(W_{xl,s})_d$	0.00513	0.02543	0.03565	0.04953	0.06747	0.08807	0.09902	0.10462	0.10708
			$W_{yl,s}$ for different assumed conditions						
$-(W_{yl,s})$	0.00013	0.00145	0.00241	0.004	0.0065	0.01009	0.01237	0.01355	0.01371
$-(W_{yl,s})$	0.00013	0.00145	0.00241	0.004	0.0065	0.01009	0.01237	0.01355	0.01372
$-(W_{yl,s})$	0.00013	0.00145	0.00241	0.004	0.00649	0.01005	0.01228	0.0134	0.0135
$-(W_{yl,s})$	0.00013	0.00145	0.00241	0.004	0.00649	0.01005	0.012287	0.01341	0.0135
$-(W_{yl,s})$	0.00013	0.001452	0.002415	0.004016	0.00654	0.010196	0.01255	0.01377	0.01392
			Φ_s for different assumed conditions						
$(\Phi_s)_m$	0.00104172	0.011597	0.019323	0.032077	0.052373	0.082553	0.104082	0.119253	0.129258
$(\Phi_s)_a$	0.00104151	0.011605	0.019353	0.032179	0.052678	0.083359	0.105245	0.12022	0.129338
$(\Phi_s)_b$	0.00104167	0.011603	0.01935	0.032154	0.052558	0.083042	0.104846	0.120213	0.130349
$(\Phi_s)_c$	0.00104167	0.011604	0.019353	0.032169	0.052594	0.083159	0.105053	0.120464	0.130636
$(\Phi_s)_d$	0.00104172	0.011616	0.01939	0.032297	0.052993	0.084357	0.107254	0.123658	0.13463
		deviations of mass flow rate G_x caused by the related assumed conditions							
$\Delta(G_x)_a$	0	0.000676	0.001546	0.003172	0.00582	0.009773	0.011181	0.008109	0.000624
$\Delta(G_x)_b$	0	0.000547	0.001395	0.002396	0.003533	0.005925	0.007345	0.008047	0.008448
$\Delta(G_x)_c$	0	8.73E-05	0.000135	0.000452	0.000698	0.001408	0.002089	0.001753	0.002196
$\Delta(G_x)_d$	0.000049	0.000993	0.001956	0.00399	0.007575	0.014406	0.020838	0.026861	0.030576

Lines m and a–d denote more complete condition. Assumed conditions a–d, respectively

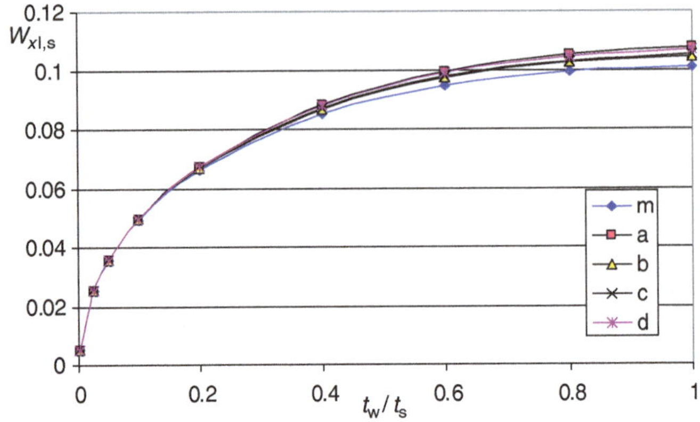

Fig. 13.7. The liquid–vapor interfacial velocity component $W_{xl,s}$ of film condensation of saturated water vapor with different assumed conditions. Lines m denotes more complete condition, and lines a–d denote assumed conditions a–d respectively

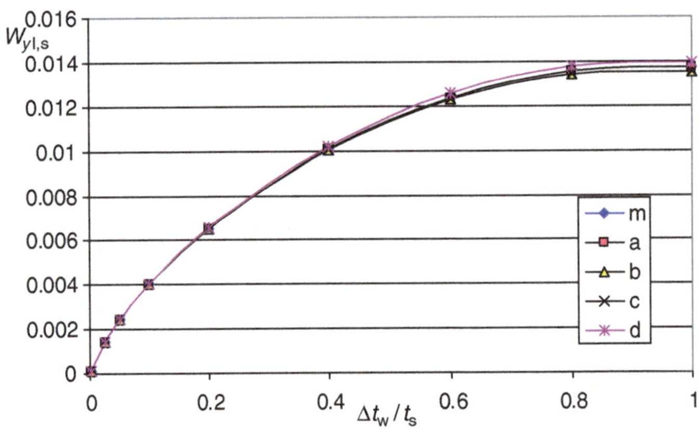

Fig. 13.8. The liquid–vapor interfacial velocity component $W_{xl,s}$ of condensate film thickness of film condensation of saturated water vapor with different assumed conditions. Lines m denotes more complete condition, and lines a–d denote assumed conditions a–d respectively

for the related physical conditions, listed in Table 13.3 and plotted in Fig. 13.9. Furthermore, the deviations of G_x are evaluated by using the (13.56)–(13.59), and listed and plotted in Table 13.3 and in Fig. 13.10, respectively, for the film condensation of saturated water vapor.

It is found that the condensate film Boussinesq approximation will increase the condensate mass flow rate G_x, but the shear force at the liquid–vapor interface, condensate film inertial force, and condensate film thermal convection will decrease the condensate mass flow rate G_x.

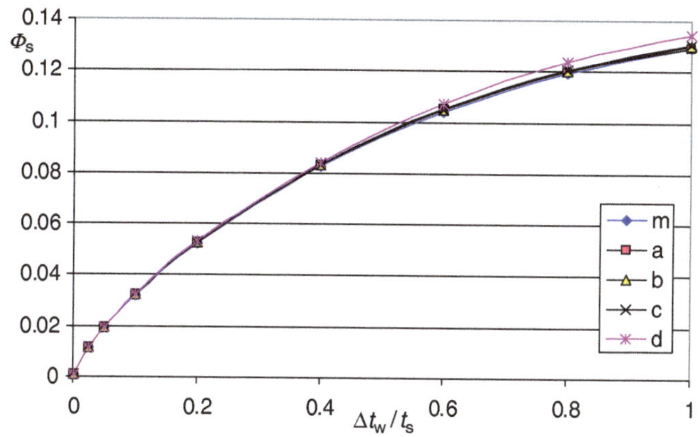

Fig. 13.9. Velocity component Φ_s for film condensation of saturated water vapor in different conditions. Note: lines m denotes more complete condition, and lines a–d denote assumed conditions a–d respectively

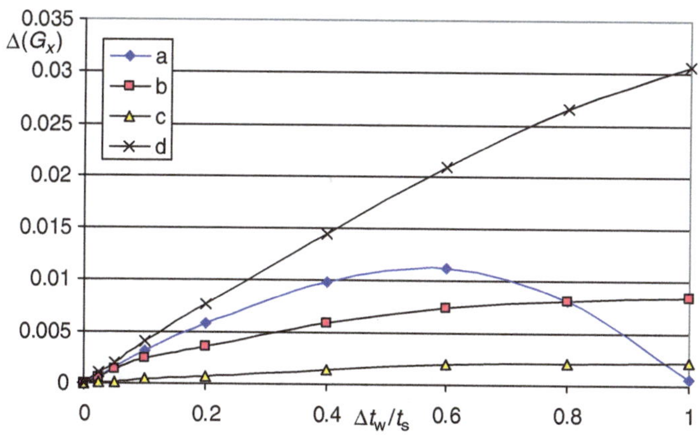

Fig. 13.10. The deviations of mass flow rate $\Delta(G_x)$ for different conditions for film condensation of saturated water vapor. (a) $\Delta(G_x)_\mathrm{a}$, (b) $\Delta(G_x)_\mathrm{b}$, (c) $\Delta(G_x)_\mathrm{c}$, and (d) $\Delta(G_x)_\mathrm{d}$

The thermal convection of the condensate film will cause the largest deviation on the predicted mass flow rates G_x compared with those caused by other physical conditions. With increasing the wall subcooled grade $\Delta t_\mathrm{w}/t_\mathrm{s}$, the deviation of the mass flow rate caused by ignoring the condensate film thermal convection will increase up to 3.06% at the wall subcooled grade $\Delta t_\mathrm{w}/t_\mathrm{s} = 1$ for the film condensation of saturated water vapor.

While, the inertia force of the condensate film will cause the smallest deviation on the mass flow rates G_x compared with those caused by other assumed conditions. Ignoring the inertia force of the condensate film will only cause the predicted deviation of the condensate mass flow rate up to 0.22% at the wall subcooled grade $\Delta t_{\mathrm{w}}/t_{\mathrm{s}} = 1$ for the film condensation of saturated water vapor.

13.8 Remarks

In this chapter, the film condensation of saturated water vapor is taken as an example for analyzing the effects of various physical conditions on the heat transfer coefficient, condensate film thickness, and mass flow rate of the film condensation. In these analyses, the effects of four physical conditions are considered including Boussinesq approximation, shear force at the liquid–vapor interface, inertia force of the condensate film, and the thermal convection of the condensate film. It is found that the effects of these physical conditions on these physical phenomena will increase with the increasing the wall subcooled grade $\Delta t_{\mathrm{w}}/t_{\mathrm{s}}$ generally. From these analyses, the following points can be concluded:

13.8.1 Effects of Boussinesq Approximation

The Boussinesq approximation will decrease predicted heat transfer coefficient and condensate film thickness of the film condensation. The Boussinesq approximation will cause the maximum deviations on heat transfer coefficient and condensate film thickness of the film condensate compared with those caused by other physical conditions. The Boussinesq approximation will cause the predicted deviation of the heat transfer coefficient up to −6.8% and cause the predicted deviation of the condensate film thickness up to −6% with increasing the wall subcooled grade $\Delta t_{\mathrm{w}}/t_{\mathrm{s}}$ for the film condensation of saturated water vapor.

13.8.2 Effects of Shear Force at the Liquid–Vapor Interface

The shear force of the condensate film will decrease the heat transfer coefficient of the film condensation very slightly. Ignoring the shear force of the condensate film will only cause a maximum deviation of the 0.826% at $\Delta t_{\mathrm{w}}/t_{\mathrm{s}} = 1$ on the heat transfer coefficient for the film condensation of saturated water vapor. The effect of the shear force of the condensate film on the heat transfer coefficient of the film condensation is far smaller than those caused by Boussinesq approximation and the condensate thermal convection.

The shear force of the condensate film will increase very slight condensate film thickness, and is far smaller than those caused by Boussinesq approximation. Ignoring the condensate shear force will only cause the predicted deviation of −0.83% at $\Delta t_w/t_s = 1$ on the condensate film thickness for the film condensation of saturated water vapor.

The effect of the shear force of the condensate film on mass flow rate of the film condensation is much smaller than that caused by the thermal convection of the condensate film, but larger than that caused by the inertia force of the condensate film. Ignoring the shear force of the condensate film will cause a maximum deviation with 0.84% at $\Delta t_w/t_s = 1$ on the mass flow rate for the film condensation of saturated water vapor.

13.8.3 Effect of Inertial Force of the Condensate Film

The inertial force of the condensate film will cause the smallest deviation on heat transfer coefficient, condensate film thickness, and mass flow rate of the film condensation compared with those of other conditions. Ignoring the inertial force of the condensate film will cause the maximum predicted deviations of only 0.177% on heat transfer coefficient, −0.178% on the condensate film thickness, and 0.22% on mass flow rate at $\Delta t_w/t_s = 1$ for the film condensation of saturated water vapor.

13.8.4 Effects of Thermal Convection of the Condensate Film

The condensate thermal convection will increase heat transfer coefficient and decrease the condensate film thickness of the film condensation. Ignoring the thermal convection of the condensate film will cause the predicted deviations of the heat transfer coefficient up to −2.6%, and up to 1.25% of the condensate film thickness for the film condensation of saturated water vapor. The effects of condensate thermal convection on heat transfer coefficient and the condensate film thickness are much smaller than those of Boussinasq approximation.

The thermal convection of the condensate film will cause the largest effects on mass flow rate of the condensation compared with those that caused by other conditions. Ignoring the condensate thermal convection will cause the predicted deviation of mass flow rate up to 3.06% for the film condensation of saturated water vapor.

The maximum predicted deviations of heat transfer coefficient of the film condensation, condensate film thickness, and mass flow rate of the film condensation caused by the different assumed physical conditions for the film condensation of saturated water vapor are summarized in Table 13.4.

Table 13.4. Summary for the maximum predicted deviations of the condensate heat transfer coefficient, condensate film thickness, and condensate mass flow rate caused by the different assumed physical conditions for the film condensation of saturated water vapor

assumed physical conditions	maximum deviations caused		
	on heat transfer coefficient (%)	on condensate film thickness (%)	on mass flow rate (%)
boussinesq approximation	−6.8	−6	1.12
ignoring liquid–vapor interfacial force	0.826	−0.83	0.85
ignoring condensate film inertia force	0.177	−0.18	0.22
ignoring condensate film thermal convection	−2.62	1.25	3.06

References

1. D.Y. Shang and T. Adamek, Study on laminar film condensation of saturated steam on a vertical flat plate for consideration of various physical factors including variable thermophysical properties, Wärme- und Stoffübertragung 30, pp. 89–100, 1994
2. D.Y. Shang and T. Adamek, Study of laminar film condensation of saturated steam for consideration of variable thermophysical properties, Transport phenomena, Science and Technology, 1992 (Ed. by B. X. Wang), Higher Education Press, Beijing, pp. 470–475, 1992

14

Laminar Film Condensation of Superheated Vapor

Nomenclature

a	thermal diffusive coefficient, $m^2\ s^{-1}$
b	plate width, m
c_p	specific heat at constant pressure, $J\,(kg\,K)^{-1}$
g	gravitation acceleration, $m\,s^{-2}$
g_x	local mass flow rate entering the liquid film at position x per unit area of the plate, $kg\,(m^2\,s)^{-1}$
G_x	total mass flow rate entering the liquid film for position $x = 0$ to x with width of b of the plate, $kg\,s^{-1}$
$Gr_{x\mathrm{l,s}}$	local Grashof number of condensate film , $\dfrac{g(\rho_{\mathrm{l,w}} - \rho_{\mathrm{v},\infty})x^3}{\nu_{\mathrm{l,s}}^2\ \rho_{\mathrm{l,s}}}$
$Gr_{x\mathrm{v},\infty}$	local Grashof number of vapor film, $\dfrac{g(\rho_{\mathrm{v,s}}/\rho_{\mathrm{v},\infty} - 1)x^3}{\nu_{\mathrm{v},\infty}^2}$
$Nu_{x,\mathrm{w}}$	local Nusselt number, $\overline{\alpha}_x x/\lambda_{\mathrm{w}}$
$\overline{Nu}_{x,\mathrm{w}}$	average Nusselt number, $\overline{\alpha}_x x/\lambda_{\mathrm{w}}$
n_{c_p}	specific heat parameter of gas
n_λ	thermal conductivity parameter of gas
n_μ	viscosity parameter of gas
Pr	Prandtl number
q_x	local heat transfer rate at position x per unit area on the plate, $W\ m^{-2}$
Q_x	total heat transfer rate for position $x = 0$ to x with width of b on the plate, W
t	temperature, °C
T	absolute temperature, K
w_x, w_y	dimensional velocity components in the x- and y-directions, respectively

W_x, W_y	dimensionless velocity components in the x- and y- directions, respectively
$W_{x\mathrm{l,s}}, W_{y\mathrm{l,s}}$	dimensionless velocity components of liquid film at liquid–vapor interface
$W_{x\mathrm{v,s}}, W_{y\mathrm{v,s}}$	dimensionless velocity components of vapor film at liquid–vapor interface

Greek symbols

α_x	local heat transfer coefficient, W $(\mathrm{m}^2\,\mathrm{K})^{-1}$
$\overline{\alpha}_x$	average heat transfer coefficient, W $(\mathrm{m}^2\,\mathrm{K})^{-1}$
δ	boundary layer thickness, m
δ_l	thickness of liquid film, m
δ_v	thickness of vapor film, m
η	dimensionless coordinate variable for boundary layer
θ_l	dimensionless temperature for liquid film, $\dfrac{t_\mathrm{l}-t_\mathrm{s}}{t_\mathrm{w}-t_\mathrm{s}}$
θ_v	dimensionless temperature for vapor film, $\dfrac{T_\mathrm{v}-T_\infty}{T_\mathrm{s}-T_\infty}$
λ	thermal conductivity, W $(\mathrm{m\,K})^{-1}$
μ	absolute viscosity, kg $(\mathrm{m\,s})^{-1}$
ν	kinetic viscosity, $\mathrm{m}^2\ \mathrm{s}^{-1}$
ρ	density, kg m^{-3}
Δt_w	wall subcooled temperature (for film condensation), $t_\mathrm{s}-t_\mathrm{w}$,°C
$\frac{\Delta t_\mathrm{w}}{t_\mathrm{s}}$	wall subcooled grade
Δt_∞	superheated temperature of vapor, $t_\infty - t_\mathrm{s}$,°C
$\frac{\Delta t_\infty}{t_\mathrm{s}}$	superheated grade of vapor
$\left(\left(\frac{\mathrm{d}\theta}{\mathrm{d}\eta}\right)_{\eta=0}\right)_{\Delta t_\infty=0}$	dimensionless temperature gradient on the plate for film condensation of saturated vapor
$\eta_{\mathrm{l}\delta}$	dimensionless thickness of liquid film
$\eta_{\mathrm{l}\delta} W_{x\mathrm{l,s}} - 4W_{y\mathrm{l,s}}$	mass flow rate parameter for film condensation
$\dfrac{\rho_\mathrm{l}-\rho_\mathrm{v}}{\rho_{l,w}-\rho_\mathrm{v}}$	buoyancy factor
$\dfrac{1}{\rho}\dfrac{\mathrm{d}p}{\mathrm{d}x}$	density factor
$\dfrac{1}{\mu}\dfrac{\mathrm{d}\mu}{\mathrm{d}\eta}$	absolute viscosity factor
$\dfrac{1}{\lambda}\dfrac{\mathrm{d}\lambda}{\mathrm{d}\eta}$	thermal conductivity factor

Subscripts

i	liquid
v	vapor
s	saturate state, or at liquid–vapor interface
w	at wall
∞	far from the wall surface

14.1 Introduction

Since the pioneering work of Nusselt [1] in 1916 in treating laminar film condensation of saturated vapor on a vertical isothermal flat plate, a number of studies have been done for successive investigations ignoring variable thermophysical properties [2–6] and with consideration of variable thermophysical properties [7–11]. On these bases, in Chaps. 12 and 13 our recent developments of study on film condensation of saturated vapor were introduced in which a novel approach, the velocity component method, was applied for similarity transformation of the governing partial differential equations for the two-phase boundary layer. Meanwhile, effects of a series of physical factors including variable thermophysical properties on heat transfer and mass flow rate of the film condensation were presented [12, 13].

Actually, a lot of related phenomena are film condensation of superheated vapor. Then, the study on heat and mass transfer of this problem has a strong practical background.

Minkowcs and Sparrow reported their study results for film condensation heat transfer with consideration of superheated vapor [14]. Their work showed that superheated temperature brings about only a slight increase in the heat transfer during the condensation of a pure vapor. They also indicated that for a given degree of superheating, q/q_{Nu} is almost independent on Δt_{w}. Anyway, study of the condensation of superheated vapor is scarcely found in the literature. Then, there is lack of a theoretical development for prediction of heat transfer of the film condensation, and especially, the theoretical study of the effect of the vapor superheated temperature on the condensate mass transfer did not appear in common literature. The reason is that it is difficult to study the two-phase boundary layer problem, because the traditional theoretical methods, such as Falkner–Skan transformation for the similarity transformation of the governing partial differential equations and for treatment of variable thermophysical properties are not suitable for the successive studies.

In this chapter, the extensive study results of Shang and Wang [15] are presented for film condensation free convection of superheated vapor with consideration of various physical factors including variable thermophysical properties. Meanwhile, following the previous chapters, the velocity component method is further applied for a novel similarity transformation of the governing partial equations of the two-phase boundary layers, and the advanced approach for treatment of variable thermophysical properties of the medium in condensate and vapor films is used. Consequently, the effect of superheated temperature on heat and mass transfer of laminar film condensation of superheated vapor is further clarified. On this basis, theoretically rigorous and practically simple formulae are obtained for prediction of heat transfer and mass flow rate of the film condensation of superheated vapor.

14.2 Governing Partial Differential Equations with Two-Phase Film

The analytical model and coordinating system used for the laminar film condensation of superheated vapor on a vertical flat plate is shown in Fig. 14.1. An isothermal vertical flat plate is suspended in a large volume of quiescent pure superheated vapor at atmospheric pressure. The plate temperature is t_w, the saturation temperature of the vapor is t_s, and the ambient temperature is t_∞. If provided condition for the model is $t_w < t_s$ and $t_s < t_\infty$, a steady twodimensional film condensation will occur on the plate. We assume that laminar flow within the liquid and vapor phases is induced by gravity, and take into account the various physical factors including variable thermophysical properties of the medium in the condensate and vapor films. Then the conservation governing partial differential equations of mass, momentum, and energy for steady laminar condensation in two-phase boundary layer are as follows:

For liquid film:

$$\frac{\partial}{\partial x}(\rho_l w_{xl}) + \frac{\partial}{\partial y}(\rho_l w_{yl}) = 0, \tag{14.1}$$

$$\rho_l \left(w_{xl}\frac{\partial w_{xl}}{\partial x} + w_{yl}\frac{\partial w_{xl}}{\partial y} \right) = \frac{\partial}{\partial y}\left(\mu_l \frac{\partial w_{xv}}{\partial y} \right) + g(\rho_l - \rho_{v,\infty}), \tag{14.2}$$

$$\rho_l c_{p_l} \left(w_{xl}\frac{\partial t_l}{\partial x} + w_{yl}\frac{\partial t_l}{\partial y} \right) = \frac{\partial}{\partial y}\left(\lambda_l \frac{\partial t_l}{\partial y} \right). \tag{14.3}$$

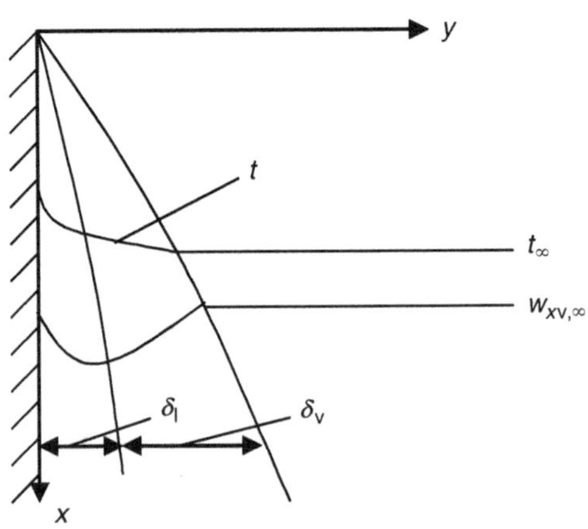

Fig. 14.1. Physical model and coordinate system of film condensation of superheated vapor

For vapor film:

$$\frac{\partial}{\partial x}(\rho_{\mathrm{v}} w_{x\mathrm{v}}) + \frac{\partial}{\partial y}(\rho_{\mathrm{v}} w_{y\mathrm{v}}) = 0, \tag{14.4}$$

$$\rho_{\mathrm{v}}\left(w_{x\mathrm{v}}\frac{\partial w_{x\mathrm{v}}}{\partial x} + w_{y\mathrm{v}}\frac{\partial w_{x\mathrm{v}}}{\partial y}\right) = \frac{\partial}{\partial y}\left(\mu_{\mathrm{v}}\frac{\partial w_{x\mathrm{v}}}{\partial y}\right) + g(\rho_{\mathrm{v}} - \rho_{\mathrm{v},\infty}), \tag{14.5}$$

$$\rho_{\mathrm{v}} c_{p\mathrm{v}}\left(w_{x\mathrm{v}}\frac{\partial T_{\mathrm{v}}}{\partial x} + w_{y\mathrm{v}}\frac{\partial T_{\mathrm{v}}}{\partial y}\right) = \frac{\partial}{\partial y}\left(\lambda_{\mathrm{v}}\frac{\partial T_{\mathrm{v}}}{\partial y}\right). \tag{14.6}$$

For boundary conditions:

$$y = 0: \quad w_{x\mathrm{l}} = 0, \quad w_{y\mathrm{l}} = 0, \quad t_{\mathrm{l}} = t_{\mathrm{w}}, \tag{14.7}$$

$$y = \delta_{\mathrm{l}}:$$

$$w_{x\mathrm{l,s}} = w_{x\mathrm{v,s}}, \tag{14.8}$$

$$\rho_{\mathrm{l,s}}\left(w_{x\mathrm{l}}\frac{\partial \delta_{x\mathrm{l}}}{\partial x} - w_{y\mathrm{l}}\right)_{\mathrm{s}} = \rho_{\mathrm{v,s}}\left(w_{x\mathrm{v}}\frac{\partial \delta_{x\mathrm{v}}}{\partial x} - w_{y\mathrm{v}}\right)_{\mathrm{s}}, \tag{14.9}$$

$$\mu_{\mathrm{l,s}}\left(\frac{\partial w_{x\mathrm{l}}}{\partial y}\right)_{\mathrm{s}} = \mu_{\mathrm{v,s}}\left(\frac{\partial w_{x\mathrm{v}}}{\partial y}\right)_{\mathrm{s}}, \tag{14.10}$$

$$\lambda_{\mathrm{l,s}}\left(\frac{\partial t_{\mathrm{l}}}{\partial y}\right)_{y=\delta_{\mathrm{l}}} = h_{fg}\rho_{\mathrm{l,s}}\left(w_{x\mathrm{l}}\frac{\partial \delta_{\mathrm{l}}}{\partial x} - w_{y\mathrm{l}}\right)_{\mathrm{s}} + \lambda_{\mathrm{v,s}}\left(\frac{\partial t_{\mathrm{v}}}{\partial y}\right)_{y=\delta_{\mathrm{l}}}, \tag{14.11}$$

$$T = T_{\mathrm{s}}, \tag{14.12}$$

$$y \to \infty: \quad w_{x\mathrm{v}} = 0, \quad T_{\mathrm{v}} \to T_{\infty}, \tag{14.13}$$

where equations (14.8)–(14.12) express the physical matching conditions such as velocity, local mass flux, shear force, heat flux, and temperature balances at the vapor–liquid interface, respectively.

14.3 Similarity Transformation

14.3.1 Transformation Variables

Consulting the velocity component method presented in Chap. 12, we assume the following dimensionless variables for the similarity transformation of the governing partial differential equations of the film condensation of superheated vapor:

For liquid film: For liquid film the dimensionless coordinate variable η_{l} and the local Grashof number $Gr_{x\mathrm{l,s}}$ are set up at first as follows:

$$\eta_{\mathrm{l}} = \left(\frac{1}{4}Gr_{x\mathrm{l,s}}\right)^{1/4}\frac{y}{x}, \quad Gr_{x\mathrm{l,s}} = \frac{g(\rho_{\mathrm{l,w}} - \rho_{\mathrm{v},\infty})x^3}{\nu_{\mathrm{l,s}}^2 \rho_{\mathrm{l,s}}}. \tag{14.14}$$

Dimensionless temperature is assumed as

$$\theta_{\rm l} = \frac{t_{\rm l} - t_{\rm s}}{t_{\rm w} - t_{\rm s}}. \tag{14.15}$$

The dimensionless velocity components are given as

$$W_{x{\rm l}} = \left(2\sqrt{gx}\left(\frac{\rho_{\rm l,w} - \rho_{\rm v,\infty}}{\rho_{\rm l,s}}\right)^{1/2}\right)^{-1} w_{x{\rm l}}, \tag{14.16}$$

$$W_{y{\rm l}} = \left(2\sqrt{gx}\left(\frac{\rho_{\rm l,w} - \rho_{\rm v,\infty}}{\rho_{\rm l,s}}\right)^{1/2} \left(\tfrac{1}{4}Gr_{x{\rm l,s}}\right)^{-4}\right)^{-1} w_{y{\rm l}}. \tag{14.17}$$

For vapor film: For vapor film, the dimensionless coordinate variable $\eta_{\rm v}$ and the local Grashof number $Gr_{x{\rm v},\infty}$ are assumed as, respectively,

$$\eta_{\rm v} = \left(\frac{1}{4}Gr_{x{\rm v},\infty}\right)^{1/4} \frac{y}{x}, Gr_{x{\rm v},\infty} = \frac{g(\rho_{\rm v,s}/\rho_{\rm v,\infty} - 1)x^3}{\nu_{\rm v,\infty}^2}. \tag{14.18}$$

The dimensionless temperature is defined as

$$\theta_{\rm v} = \frac{T_{\rm v} - T_\infty}{T_{\rm s} - T_\infty}. \tag{14.19}$$

The dimensionless velocity components are assumed as

$$W_{x{\rm v}} = (2\sqrt{gx}(\rho_{\rm v,s}/\rho_{\rm v,\infty} - 1)^{1/2})^{-1} w_{x{\rm v}}, \tag{14.20}$$

$$W_{y{\rm v}} = \left(2\sqrt{gx}(\rho_{\rm v,s}/\rho_{\rm v,\infty} - 1)^{1/2}\left(\frac{1}{4}Gr_{x{\rm v},\infty}\right)^{-1/4}\right)^{-1} w_{y{\rm v}}. \tag{14.21}$$

14.3.2 Ordinary Differential Equations

Consulting the derivations in Chap. 11 for laminar film boiling of subcooled liquid and by means of the earlier corresponding assumed variables, the following dimensionless equations are obtained from (14.1)–(14.6) for two-phase boundary layer and their boundary conditions of laminar film condensation of superheated vapor:

For liquid film: With the assumed dimensionless similarity transformation variables for the liquid film shown in (14.14)–(14.17), (14.1)–(14.3) are transformed to dimensionless ordinary ones as follows, respectively:

$$2W_{xl} - \eta_l \frac{dW_{xl}}{d\eta_l} + 4\frac{dW_{yl}}{d\eta_l} + \frac{1}{\rho_l}\frac{d\rho_l}{d\eta_l}(-\eta_l W_{xl} + 4W_{yl}) = 0, \qquad (14.22)$$

$$\frac{\nu_{l,s}}{\nu_l}\left(W_{xl}\left(2W_{xl} - \eta_l\frac{dW_{xl}}{d\eta_l}\right) + 4W_{yl}\frac{dW_{xl}}{d\eta_l}\right) = \frac{d^2W_{xl}}{d\eta_l^2}$$
$$+\frac{1}{\mu_l}\frac{d\mu_l}{d\eta_l}\frac{dW_{xl}}{d\eta_l} + \frac{\mu_{l,s}}{\mu_l}\frac{\rho_l - \rho_{v,\infty}}{\rho_{l,w} - \rho_{v,\infty}}, \qquad (14.23)$$

$$Pr_l\frac{\nu_{l,s}}{\nu_l}\left[-W_{xl}\eta_l + 4W_{yl}\right]\frac{d\theta_l}{d\eta_l} = \frac{d^2\theta_l}{d\eta_l^2} + \frac{1}{\lambda_l}\frac{d\lambda_l}{d\eta_l}\frac{d\theta_l}{d\eta_l}. \qquad (14.24)$$

In addition, from the analysis of Chap. 6, it is known that the physical factor $Pr_l(\nu_{l,\infty}/\nu_l)$ in (14.24) can be expressed as

$$Pr_l\frac{\nu_{l,\infty}}{\nu_l} = Pr_{l,\infty}\frac{\rho_l}{\rho_{l,\infty}}\frac{\lambda_{l,\infty}}{\lambda_l}$$

for water and a lot of liquids in the special temperature range for engineering application.

For vapor film: With the assumed similarity transformation variables for vapor film, as shown in (14.18)–(14.21), the governing equations (14.4)–(14.6) for vapor film are transformed into the following dimensionless ordinary ones, respectively:

$$2W_{xv} - \eta_v\frac{dW_{xv}}{d\eta_v} + 4\frac{dW_{yv}}{d\eta_v} + \frac{1}{\rho_v}\frac{d\rho_v}{d\eta_v}(-\eta_v W_{xv} + 4W_{yv}) = 0, \qquad (14.25)$$

$$\frac{\nu_{v,\infty}}{\nu_v}\left(W_{xv}\left(2W_{xv} - \eta_v\frac{dW_{xv}}{d\eta_v}\right) + 4W_{yv}\left(\frac{dW_{xv}}{d\eta_v}\right)\right)$$
$$= \frac{d^2W_{xv}}{d\eta_v^2} + \frac{1}{\mu_v}\frac{d\mu_v}{d\eta_v}\frac{dW_{xv}}{d\eta_v} + \frac{\mu_{v,\infty}}{\mu_v}\frac{\rho_v - \rho_{v,\infty}}{\rho_{v,s} - \rho_{v,\infty}}, \qquad (14.26)$$

$$Pr_v\frac{\nu_{v,\infty}}{\nu_v}(-\eta_v W_{xv} + 4W_{yv})\frac{d\theta_v}{d\eta_v} = \frac{d^2\theta_v}{d\eta_v^2} + \frac{1}{\lambda_v}\frac{d\lambda_v}{d\eta_v}\frac{d\theta_v}{d\eta_v}. \qquad (14.27)$$

For boundary conditions: With the corresponding transformation variables the physical boundary conditions (14.7)–(14.13) are transformed equivalently to the following ones, respectively:

$$\eta_l = 0: \quad W_{xl} = 0, \quad W_{yl} = 0, \quad \theta_l = 1, \qquad (14.28)$$
$$\eta_l = \eta_{l\delta}(\eta_v = 0):$$

$$W_{x\mathrm{v,s}} = \left(\frac{\rho_{\mathrm{l,w}} - \rho_{\mathrm{v},\infty}}{\rho_{\mathrm{l,s}}}\right)^{1/2} \left(\frac{\rho_{\mathrm{v,s}} - \rho_{\mathrm{v},\infty}}{\rho_{\mathrm{v},\infty}}\right)^{-1/2} W_{x\mathrm{l,s}}, \tag{14.29}$$

$$W_{y\mathrm{v,s}} = -0.25 \frac{\mu_{\mathrm{l,s}}}{\mu_{\mathrm{v,s}}} \left(\frac{\nu_{\mathrm{l,s}}}{\nu_{\mathrm{v},\infty}}\right)^{-1/2} \left(\frac{\rho_{\mathrm{l,w}} - \rho_{\mathrm{v},\infty}}{\rho_{\mathrm{l,s}}}\right)^{1/4} \left(\frac{\rho_{\mathrm{v,s}} - \rho_{\mathrm{v},\infty}}{\rho_{\mathrm{v}\infty}}\right)^{-1/4} \times \left(\eta_{\mathrm{l}\delta} W_{x\mathrm{l,s}} - 4W_{y\mathrm{l,s}}\right), \tag{14.30}$$

$$\left(\frac{\mathrm{d}W_{x\mathrm{v}}}{\mathrm{d}\eta_{\mathrm{v}}}\right)_{\eta_{\mathrm{v}}=0} = \frac{\mu_{\mathrm{l,s}}}{\mu_{\mathrm{v,s}}} \left(\frac{\nu_{\mathrm{v,s}}}{\nu_{\mathrm{l,s}}}\right)^{1/2} \left(\frac{\rho_{\mathrm{l,w}} - \rho_{\mathrm{v},\infty}}{\rho_{\mathrm{l,s}}}\right)^{3/4} \left(\frac{\rho_{\mathrm{v,s}}}{\rho_{\mathrm{v},\infty}} - 1\right)^{-3/4} \left(\frac{\mathrm{d}W_{x\mathrm{l}}}{\mathrm{d}\eta_{\mathrm{l}}}\right)_{\eta_{\mathrm{l}}=\eta_{\mathrm{l},\delta}}, \tag{14.31}$$

$$\left(\frac{\mathrm{d}\theta_{\mathrm{v}}}{\mathrm{d}\eta_{\mathrm{v}}}\right)_{\eta_{\mathrm{v}}=0} = \frac{\left(\frac{\rho_{\mathrm{l,w}}-\rho_{\mathrm{v},\infty}}{\rho_{\mathrm{l,s}}}\right)^{1/4} \left(-h_{fg}\mu_{\mathrm{l,s}}(\eta_{\mathrm{v}\delta} W_{x\mathrm{v,s}} - 4W_{y\mathrm{v,s}}) + \lambda_{\mathrm{l,s}}(t_{\mathrm{w}} - t_{\mathrm{s}}) \left(\frac{\mathrm{d}\theta_{\mathrm{l}}}{\mathrm{d}\eta_{\mathrm{l}}}\right)_{\eta_{\mathrm{l}}=\eta_{\mathrm{l}\delta}}\right)}{\left(\frac{\rho_{\mathrm{v,s}}}{\rho_{\mathrm{v},\infty}} - 1\right)^{1/4} \left(\frac{\nu_{\mathrm{l,s}}}{\nu_{\mathrm{v},\infty}}\right)^{1/2} \lambda_{\mathrm{v,s}}(T_{\mathrm{s}} - T_{\infty})}, \tag{14.32}$$

$$\theta_{\mathrm{l}} = 0, \tag{14.33}$$

$$\theta_{\mathrm{v}} = 1, \tag{14.34}$$

$$\eta_{\mathrm{v}} \to \infty: \quad W_{x\mathrm{v}} \to 0, \quad \theta_{\mathrm{v}} \to 0. \tag{14.35}$$

14.4 Treatment of Variable Thermophysical Properties

The treatment of variable thermophysical properties for the medium of the liquid and vapor films must be done for solving the ordinary differential equations with the boundary condition equations. The approaches for the treatment of variable thermophysical properties are presented as follows:

For liquid film: For treatment of the variable thermophysical properties of the liquid film medium, a polynomial method presented Chap. 6 will be used. This method was applied successfully in the corresponding treatment for film condensation of saturated vapor in Chap. 12. With this method the corresponding predictive expressions for density ρ_{l}, thermal conductivity λ_{l} of water are, respectively,

$$\rho_{\mathrm{l}} = -4.48 \times 10^{-3} t_{\mathrm{l}}^2 + 999.9, \tag{14.36}$$

$$\lambda_{\mathrm{l}} = -8.01 \times 10^{-6} t_{\mathrm{l}}^2 + 1.94 \times 10^{-3} t_{\mathrm{l}} + 0.563. \tag{14.37}$$

According to the study in [16], the corresponding predictive expression for absolute viscosity μ_l of water is

$$\mu_l = \exp\left[-1.6 - \frac{1150}{T_l} + \left(\frac{690}{T_l}\right)^2\right] \times 10^{-3}. \qquad (14.38)$$

With (13.36)–(13.38) and the similarity variables assumed in (14.14)–(14.17), the thermophysical property factors $\left(\frac{1}{\rho_l}\right)/\left(\frac{d\rho_l}{d\eta_l}\right)$, $\left(\frac{1}{\mu_l}\right)/\left(\frac{d\mu_l}{d\eta_l}\right)$, $\left(\frac{1}{\lambda_l}\right)/\left(\frac{d\lambda_l}{d\eta_l}\right)$, $\left(\frac{\mu_{l,s}}{\mu_l}\right)$, $\left(\frac{\nu_{l,s}}{\nu_l}\right)$ and $Pr_{l,s}\frac{\rho_l}{\rho_{l,s}}\frac{\lambda_{l,s}}{\lambda_l}$ in the governing ordinary differential equations for condensate water film can be transformed into the following ones, respectively:

$$\frac{1}{\rho_l}\frac{d\rho_l}{d\eta_l} = \frac{\left[-2 \times 4.48 \times 10^{-3}\, t_l(t_w - t_s)\frac{d\theta_l}{d\eta_l}\right]}{(-4.48 \times 10^{-3} t_l^2 + 999.9)}, \qquad (14.39)$$

$$\frac{1}{\mu_l}\frac{d\mu_l}{d\eta_l} = \left(\frac{1150}{T_l^2} - 2 \times \frac{690^2}{T_l^3}\right)(t_w - t_s)\frac{d\theta_l}{d\eta_l}, \qquad (14.40)$$

$$\frac{1}{\lambda_l}\frac{d\lambda_l}{d\eta_l} = \frac{\left[-2 \times 8.01 \times 10^{-6} t_l + 1.94 \times 10^{-3}(t_w - t_s)\frac{d\theta_l}{d\eta_l}\right]}{-8.01 \times 10^{-6} t_l^2 + 1.94 \times 10^{-3} t_l + 0.563}, \qquad (14.41)$$

$$\frac{\mu_{l,s}}{\mu_l} = \exp\left(1150\left(\frac{1}{T_l} - \frac{1}{T_s}\right) + 690^2\left(\frac{1}{T_s} - \frac{1}{T_l}\right)\right), \qquad (14.42)$$

$$\frac{\nu_{l,s}}{\nu_l} = \frac{\mu_{l,s}}{\mu_l}\frac{\rho_l}{\rho_{l,s}} = \exp\left(1150\left(\frac{1}{T_l} - \frac{1}{T_s}\right) + 690^2\left(\frac{1}{T_s} - \frac{1}{T_l}\right)\right)\frac{-4.48 \times 10^{-3} t_l^2 + 999.9}{-4.48 \times 10^{-3} t_s^2 + 999.9}, \qquad (14.43)$$

$$Pr_{l,s}\frac{\rho_l}{\rho_{l,s}}\frac{\lambda_{l,s}}{\lambda_l} = Pr_{l,s}\left(\frac{-4.48 \times 10^{-3} t_l^2 + 999.9}{-4.48 \times 10^{-3} t_s^2 + 999.9}\right)\left(\frac{-8.01 \times 10^{-6} t_s^2 + 1.94 \times 10^{-3} t_s + 0.563}{-8.01 \times 10^{-6} t_l^2 + 1.94 \times 10^{-3} t_l + 0.563}\right), \qquad (14.44)$$

where

$$t_l = (t_w - t_s)\theta_l + t_s, T_v = (T_s - T_\infty)\theta_v + T_\infty \text{ and } T = t + 273. \qquad (14.45)$$

For vapor film: For treatment of the variable thermophysical properties of the vapor film, the temperature parameter method presented in Chaps. 4 and 5 will be used. If the temperature of the bulk vapor T_∞ is taken as the reference temperature, the earlier thermophysical properties of vapor film will be expressed as

$$\frac{\mu_{\mathrm{v}}}{\mu_{\mathrm{v},\infty}} = \left(\frac{T_{\mathrm{v}}}{T_\infty}\right)^{n_\mu}, \tag{14.46}$$

$$\frac{\lambda_{\mathrm{v}}}{\lambda_{\mathrm{v},\infty}} = \left(\frac{T_{\mathrm{v}}}{T_\infty}\right)^{n_\lambda}, \tag{14.47}$$

$$\frac{\rho_{\mathrm{v}}}{\rho_{\mathrm{v},\infty}} = \left(\frac{T_{\mathrm{v}}}{T_\infty}\right)^{-1}, \tag{14.48}$$

while the change of kinematic viscosity at the constant pressure can be expressed as

$$\frac{\nu_{\mathrm{v}}}{\nu_{\mathrm{v},\infty}} = \left(\frac{T_{\mathrm{v}}}{T_\infty}\right)^{n_\mu+1}. \tag{14.49}$$

Then, the thermophysical property factors $(1/\rho_{\mathrm{v}})/(\mathrm{d}\rho_{\mathrm{v}}/\mathrm{d}\eta_{\mathrm{v}})$, $(1/\mu_{\mathrm{v}})/(\mathrm{d}\mu_{\mathrm{v}}/\mathrm{d}\eta_{\mathrm{v}})$, $(1/\lambda_{\mathrm{v}})/(\mathrm{d}\lambda_{\mathrm{v}}/\mathrm{d}\eta_{\mathrm{v}})$, $(\mu_{\mathrm{v},\infty}/\mu_{\mathrm{v}})$ and $(\nu_{\mathrm{v},\infty}/\nu_{\mathrm{v}})$ in the governing ordinary differential equations of vapor film can be transformed, respectively, as later:

$$\frac{1}{\rho_{\mathrm{v}}}\frac{\mathrm{d}\rho_{\mathrm{v}}}{\mathrm{d}\eta_{\mathrm{v}}} = \frac{(T_{\mathrm{s}}/T_\infty - 1)\mathrm{d}\theta_{\mathrm{v}}/\mathrm{d}\eta_{\mathrm{v}}}{(T_{\mathrm{s}}/T_\infty - 1)\theta_{\mathrm{v}} + 1}, \tag{14.50}$$

$$\frac{1}{\mu_{\mathrm{v}}}\frac{\mathrm{d}\mu_{\mathrm{v}}}{\mathrm{d}\eta_{\mathrm{v}}} = \frac{n_\mu(T_{\mathrm{s}}/T_\infty - 1)\mathrm{d}\theta_{\mathrm{v}}/\mathrm{d}\eta_{\mathrm{v}}}{(T_{\mathrm{s}}/T_\infty - 1)\theta_{\mathrm{v}} + 1}, \tag{14.51}$$

$$\frac{1}{\lambda_{\mathrm{v}}}\frac{\mathrm{d}\lambda_{\mathrm{v}}}{\mathrm{d}\eta_{\mathrm{v}}} = \frac{n_\lambda(T_{\mathrm{s}}/T_\infty - 1)\mathrm{d}\theta_{\mathrm{v}}/\mathrm{d}\eta_{\mathrm{v}}}{(T_{\mathrm{s}}/T_\infty - 1)\theta_{\mathrm{v}} + 1}, \tag{14.52}$$

$$\frac{\mu_{\mathrm{v},\infty}}{\mu_{\mathrm{v}}} = [(T_{\mathrm{s}}/T_\infty - 1)\theta_{\mathrm{v}} + 1]^{-n_\mu}, \tag{14.53}$$

$$\frac{\nu_{\mathrm{v},\infty}}{\nu_{\mathrm{v}}} = [(T_{\mathrm{s}}/T_\infty - 1)\theta_{\mathrm{v}} + 1]^{-(n_\mu+1)}. \tag{14.54}$$

14.5 Numerical Solutions

14.5.1 Calculation Procedure

The calculation procedure of the equations of the two-phase boundary layer of the film condensation of superheated vapor belongs to three-point boundary value problem, and is carried out numerically by two processes on the basis of the corresponding approach presented in Chap. 12 for the film condensation of saturated vapor. In the first step, the solutions of (14.22)–(14.24) for the liquid film are assumed to be without shear force of vapor at the liquid–vapor interface. For this case, the boundary condition (14.31) must be changed into

$$\left(\frac{\mathrm{d}W_{x\mathrm{l}}}{\mathrm{d}\eta_{\mathrm{l}}}\right)_{\eta_{\mathrm{l}}=\eta_{\mathrm{l},\delta}} = 0. \tag{14.55}$$

In this case, (14.28), (14.33), and (14.55) are taken as the boundary conditions of the two-point boundary value problem of (14.22)–(14.24) for liquid film, and are solved by the shooting method. Furthermore, the second step for carrying out calculation of three-point boundary value problem for coupling equations of liquid film with equations for vapor films is started. In this step, first the boundary values $W_{x\mathrm{v,s}}$ and $W_{y\mathrm{v,s}}$ are found out by (14.29) and (14.30), respectively. Then (14.25)–(14.27) for the vapor film are calculated with the boundary conditions (14.34) and (14.35) and the earlier values of $W_{x\mathrm{v,s}}$ and $W_{y\mathrm{v,s}}$. On this basis, adjudgement (14.31) and (14.32) are used for checking convergence of the solutions. By means of the adjudgement equations the calculation is iterated with appropriate change of the values $W_{x\mathrm{l,s}}$ and $\eta_{\mathrm{l}\delta}$. In each iteration the calculations of (14.22)–(14.24) for liquid film and (14.25)–(14.27) for vapor film are made successively by the shooting method.

14.5.2 Numerical Solution

From the governing ordinary equations (14.22)–(14.27) and their boundary conditions, (14.28)–(14.35), it will be expected that for consideration of variable thermophysical properties of the liquid and vapor medium, the dimensionless velocity and temperature fields for the film condensation of superheated vapor will depend on the temperature-dependent properties of the liquid and vapor medium, finally on the temperature conditions $t_{\mathrm{w}}, t_{\mathrm{s}}$, and t_{∞}.

All thermophysical properties for water and water vapor at saturated temperature used in the calculation come from [17]. For convenience some special values of the thermophysical properties are listed in Tables 14.1 and 14.2.

Table 14.1. The thermophysical property values for water and water vapor at saturated temperature

term	value	
	for water	for water vapor
t_{s}(°C)	100	100
c_p(J (kg K)$^{-1}$)	4,216	
h_{fg}(kJ kg^{-1})		2,257.3
Pr	1.76	1
ρ(kg m^{-3})	958.1	0.5974
μ(kg (m s)$^{-1}$)	282.2×10^{-6}	12.28×10^{-6}
ν(m^2 s^{-1})	0.294×10^{-6}	20.55×10^{-6}
λ(W (m K)$^{-1}$)	0.677	0.02478

Table 14.2. The values of water density at different temperatures

t (°C)	0	20	40	60	80	95	99.9
ρ (kg m^{-3})	999.8	998.3	992.3	983.2	971.4	961.7	958.1

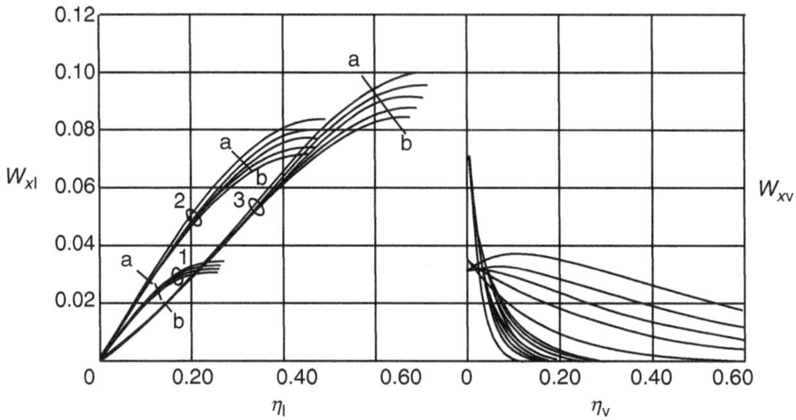

Fig. 14.2. Velocity profiles for film condensation of superheated water vapor, cited from Shang and Wang. [15]. 1. $\frac{\Delta t_w}{t_s}(=\frac{t_s-t_w}{t_s})=0.05$, 2. $\frac{\Delta t_w}{t_s}(=\frac{t_s-t_w}{t_s})=0.40$, and 3. $\frac{\Delta t_w}{t_s}(=\frac{t_s-t_w}{t_s})=1$ $a\to b: \frac{\Delta t_\infty}{t_s}(=\frac{t_\infty-t_s}{t_s})=0.27, 1.27, 2.27, 3.27, 4.27$

For the laminar film condensation of water vapor, the numerical calculations have been carried out for wall subcooled temperatures $\Delta t_w(=t_s-t_w)=0.1, 5, 20, 40, 60, 80, 100°C$ or wall subcooled grades of $\Delta t_w/t_s\left(=\frac{t_s-t_w}{t_s}\right)=0.001, 0.05, 0.2, 0.4, 0.6, 0.8, 1$ and for vapor superheated temperatures of $\Delta t_\infty(=t_\infty-t_s)=0, 27, 127, 227, 327, 427°C$ or vapor superheated grades of $\frac{\Delta t_\infty}{t_s}(=\frac{t_\infty-t_s}{t_s})=0, 0.27, 1.27, 2.27, 3.27, 4.27$. Some of the calculated results for the velocity and temperature fields on the medium in the liquid and vapor films are plotted in Figs. 14.2 and 14.3, respectively.

It is seen that with increasing the wall subcooled grade

$$\frac{\Delta t_w}{t_s}\left(=\frac{t_s-t_w}{t_s}\right),$$

the maximum velocity W_{xl} will increase obviously, which is coincident with the case of the film condensation of saturated vapor. Meanwhile, with increasing the vapor superheated grade

$$\frac{\Delta t_\infty}{t_s}\left(=\frac{t_\infty-t_s}{t_s}\right),$$

the maximum velocity W_{xl} will decrease slightly.

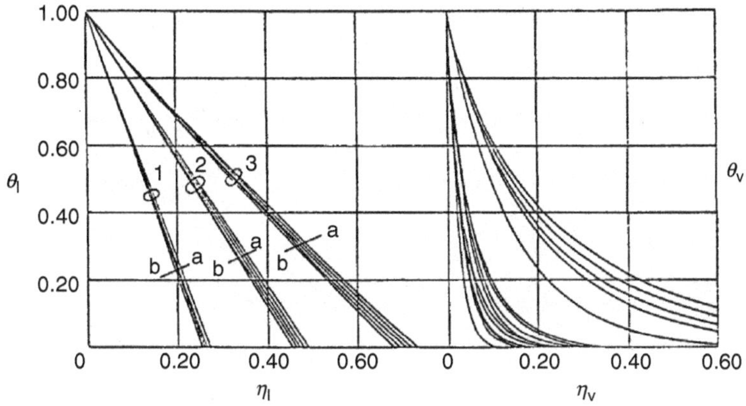

Fig. 14.3. Temperature profiles for film condensation of superheated water steam, cited from Shang and Wang [15]. 1. $\frac{\Delta t_w}{t_s}(=\frac{t_s-t_w}{t_s}) = 0.05$, 2. $\frac{\Delta t_w}{t_s}(=\frac{t_s-t_w}{t_s}) = 0.40$, and 3. $\frac{\Delta t_w}{t_s}(=\frac{t_s-t_w}{t_s}) = 1$ $a \to b$: $\frac{\Delta t_\infty}{t_s}(=\frac{t_\infty-t_s}{t_s}) = 0.27, 1.27, 2.27, 3.27, 4.27$

Additionally, with increasing the wall subcooled grade

$$\frac{\Delta t_w}{t_s}\left(=\frac{t_s - t_w}{t_s}\right),$$

the dimensionless temperature gradient will decrease, while, with increasing the vapor superheated grade

$$\frac{\Delta t_\infty}{t_s}\left(=\frac{t_\infty - t_s}{t_s}\right),$$

the dimensionless temperature gradient will increase slightly.

14.6 Heat Transfer

14.6.1 Heat Transfer

Consulting the analysis in Chap. 12 for film condensation of saturated vapor, the heat transfer expressions for film condensation of superheated vapor can be given later:

Local heat transfer rate at position x per unit area on the plate, defined as $q_x = -\lambda_{l,w}\left(\frac{\partial t}{\partial y}\right)_{y=0}$, is described as

$$q_x = -\lambda_{l,w}(T_w - T_s)\left(\frac{1}{4}Gr_{xl,s}\right)^{1/4} x^{-1}\left(\frac{d\theta_l}{d\eta_l}\right)_{\eta_l=0}. \qquad (14.56)$$

Then, the local heat transfer coefficient, defined as $q_x = \alpha_x(t_\mathrm{w} - t_\mathrm{s})$, will be

$$\alpha_x = -\lambda_{\mathrm{l,w}} \left(\frac{1}{4} Gr_{x\mathrm{l,s}}\right)^{1/4} x^{-1} \left(\frac{\mathrm{d}\theta_\mathrm{l}}{\mathrm{d}\eta_\mathrm{l}}\right)_{\eta_\mathrm{l}=0}. \tag{14.57}$$

The local Nusselt number, defined as $Nu_{x\mathrm{l,w}} = \frac{\alpha_x x}{\lambda_{\mathrm{l,w}}}$, is expressed by

$$\frac{Nu_{x\mathrm{l,w}}}{\left(\frac{1}{4} Gr_{x\mathrm{l,s}}\right)^{1/4}} = -\left(\frac{\mathrm{d}\theta_\mathrm{l}}{\mathrm{d}\eta_\mathrm{l}}\right)_{\eta_\mathrm{l}=0}. \tag{14.58}$$

The average heat transfer coefficient $\overline{\alpha}_x$ defined as $Q_x = \overline{\alpha_x}(t_\mathrm{w} = t_\mathrm{s})A$ and average Nusselt number defined as $\overline{Nu}_{x\mathrm{l,w}} = \frac{\overline{\alpha_x} x}{\lambda_{\mathrm{l,w}}}$ will be, respectively,

$$\overline{\alpha}_x = \frac{4}{3}\alpha_x, \tag{14.59}$$

$$\overline{Nu}_{x\mathrm{l,w}} = \frac{4}{3} Nu_{x\mathrm{l,w}} \tag{14.60}$$

Here, Q_x is total heat transfer rate for position $x = 0$ to x with width of b on the plate. It is seen that, for practical calculation of heat transfer, only $(\mathrm{d}\theta_\mathrm{l}/\mathrm{d}\eta_\mathrm{l})_{\eta_\mathrm{l}=0}$ dependent on numerical solutions is no-given variable.

The numerical solutions for temperature gradient $(\mathrm{d}\theta_\mathrm{l}/\mathrm{d}\eta_\mathrm{l})_{\eta_\mathrm{l}=0}$ for the film condensation of superheated water vapor at the corresponding temperature conditions, $\Delta t_\mathrm{w}/t_\mathrm{s} = (t_\mathrm{s} - t_\mathrm{w})/t_\mathrm{s}$ and $\Delta t_\infty/t_\mathrm{s}(= (t_\infty - t_\mathrm{s})/t_\mathrm{s})$, are described in Table 14.3, and plotted in Fig. 14.4.

It is seen that with increasing the wall subcooled grade $\Delta t_\mathrm{w}/t_\mathrm{s} = (t_\mathrm{s} - t_\mathrm{w})/t_\mathrm{s}$, the temperature gradient $(\mathrm{d}\theta_\mathrm{l}/\mathrm{d}\eta_\mathrm{l})_{\eta_\mathrm{l}=0}$ will decrease rapidly, while, with increasing the vapor bulk superheated grade $\Delta t_\infty/t_\mathrm{s}$ $(= (t_\infty - t_\mathrm{s})/t_\mathrm{s})$, the temperature gradient $(\mathrm{d}\theta_\mathrm{l}/\mathrm{d}\eta_\mathrm{l})_{\eta_\mathrm{l}=0}$ will increase slowly.

Based on these numerical solutions the following formulae are obtained by using a curve matching method for simple and reliable prediction of temperature gradient $(\mathrm{d}\theta_\mathrm{l}/\mathrm{d}\eta_\mathrm{l})_{\eta_\mathrm{l}=0}$ for the laminar film condensation of superheated water vapor:

For $0.05 \le \frac{\Delta t_\mathrm{w}}{t_\mathrm{s}} \le 1$:

$$-\left(\frac{\mathrm{d}\theta_\mathrm{l}}{\mathrm{d}\eta_\mathrm{l}}\right)_{\eta_\mathrm{l}=0} = \left(-\left(\frac{\mathrm{d}\theta_\mathrm{l}}{\mathrm{d}\eta_\mathrm{l}}\right)_{\eta_\mathrm{l}=0}\right)_{\Delta t_\infty = 0_{\Delta t_\infty=0}} + a \cdot \frac{\Delta t_\infty}{t_\mathrm{s}}, \tag{14.61}$$

where $(-\mathrm{d}\theta_\mathrm{l}/\mathrm{d}\eta_\mathrm{l})_{\eta_\mathrm{l}=0}$ is the dimensionless temperature gradient on the wall for film condensation of superheated water vapor, and $\left(-\left(\frac{\mathrm{d}\theta_\mathrm{l}}{\mathrm{d}\eta_\mathrm{l}}\right)_{\eta_\mathrm{l}=0}\right)_{\Delta t_\infty=0}$ is the dimensionless temperature gradient of the film condensation of saturated water vapor. The correlation of $\left(-\left(\frac{\mathrm{d}\theta_\mathrm{l}}{\mathrm{d}\eta_\mathrm{l}}\right)_{\eta_\mathrm{l}=0}\right)_{\Delta t_\infty=0}$ was proposed in

Table 14.3. Numerical solutions $-(\mathrm{d}\theta_l/\mathrm{d}\eta_l)_{\eta_l=0}$ for film condensation of superheated water vapor, cited from Shang and Wang [15]

$\frac{\Delta t_\infty}{t_s} = \left(\frac{t_\infty - t_s}{t_s}\right)$		$\frac{\Delta t_w}{t_s} = \frac{t_s - t_w}{t_s}$					
		0.05	0.20	0.40	0.60	0.80	1.00
0	(1)	3.6707	2.5397	2.0789	1.8324	1.6679	1.5511
	(2)	3.6596	2.5451	2.0924	1.8475	1.6791	1.55
0.27	(1)	3.6913	2.5532	2.0891	1.8411	1.6759	1.5581
	(2)	3.6770	2.5592	2.1032	1.8561	1.6774	1.5576
1.27	(1)	3.7712	2.6054	2.1299	1.8758	1.7066	1.5861
	(2)	3.7416	2.6117	2.1431	1.8878	1.7145	1.5858
2.27	(1)	3.8478	2.6551	2.1688	1.9091	1.7362	1.6131
	(2)	3.8062	2.6641	2.1831	1.9196	1.7423	1.6140
3.27	(1)	3.9245	2.7046	2.2079	1.9425	1.7659	1.6402
	(2)	3.8708	2.7166	2.22303	1.9513	1.7701	1.6422
4.27	(1)	3.9555	2.7512	2.2449	1.9738	1.7938	1.6656
	(2)	3.9354	2.7690	2.2630	2.1086	1.7980	1.6704

(1) Numerical solution and (2) predicted by (14.61)–(14.63)

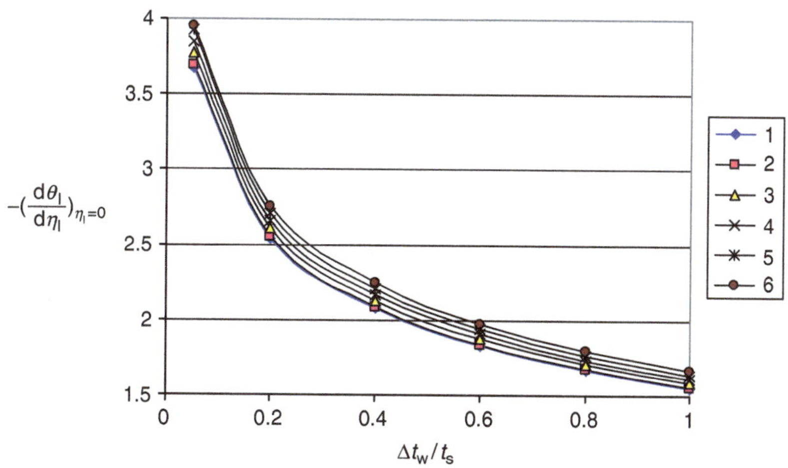

Fig. 14.4. Numerical solutions of $-\left(\frac{\mathrm{d}\theta_l}{\mathrm{d}\eta_l}\right)_{\eta_l=0}$ with $\frac{\Delta t_w}{t_s} = \frac{t_s - t_w}{t_s}$ and $\frac{\Delta t_\infty}{t_s}\left(= \frac{t_\infty - t_s}{t_s}\right)$ for laminar film condensation of superheated water vapor. 1. $\frac{\Delta t_\infty}{t_s}\left(= \frac{t_\infty - t_s}{t_s}\right) = 0$, 2. $\frac{\Delta t_\infty}{t_s}\left(= \frac{t_\infty - t_s}{t_s}\right) = 0.27$, 3. $\frac{\Delta t_\infty}{t_s}\left(= \frac{t_\infty - t_s}{t_s}\right) = 1.27$, 4. $\frac{\Delta t_\infty}{t_s}\left(= \frac{t_\infty - t_s}{t_s}\right) = 2.27$, 5. $\frac{\Delta t_\infty}{t_s}\left(= \frac{t_\infty - t_s}{t_s}\right) = 3.27$, 6. $\frac{\Delta t_\infty}{t_s}\left(= \frac{t_\infty - t_s}{t_s}\right) = 4.27$

Chap. 12 and is rewritten here as

$$\left(-\left(\frac{\mathrm{d}\theta_\mathrm{l}}{\mathrm{d}\eta_\mathrm{l}}\right)_{\eta_\mathrm{l}=0}\right)_{\Delta t_\infty=0}=\frac{1.74-0.19\frac{\Delta t_\mathrm{w}}{t_\mathrm{s}}}{\left(\frac{\Delta t_\mathrm{w}}{t_\mathrm{s}}\right)^{1/4}}. \tag{14.62}$$

Additionally, coefficient a is expressed as follows for the film condensation of superheated water vapor:

$$a=10^{-2}\times\left(6.92-9.45\frac{\Delta t_\mathrm{w}}{t_\mathrm{s}}+5.35\left(\frac{\Delta t_\mathrm{w}}{t_\mathrm{s}}\right)^2\right). \tag{14.63}$$

The results $-(\mathrm{d}\theta_\mathrm{l}/\mathrm{d}\eta_\mathrm{l})_{\eta_\mathrm{l}=0}$, predicted by means of (14.61)–(14.63) for the laminar film condensation of superheated water vapor are also described in Table 14.3. Through the comparison of the predicted values of temperature gradient $-(\mathrm{d}\theta_\mathrm{l}/\mathrm{d}\eta_\mathrm{l})_{\eta_\mathrm{l}=0}$ with the corresponding numerical results, it is seen that the agreements are very good.

14.7 Condensate Mass Flow Rate

Consulting the analysis in Chap. 12 for film condensation of saturated vapor, the condensate mass transfer expressions for film condensation of superheated vapor can be given later:

Set g_x to be a local mass flow rate entering the liquid film at position x per unit area of the plate. According to the boundary layer theory of fluid mechanics, g_x is expressed as

$$g_x=\rho_{\mathrm{l,s}}\left(w_{x\mathrm{l,s}}\frac{\mathrm{d}\delta_\mathrm{l}}{\mathrm{d}x}-w_{y\mathrm{l,s}}\right)_\mathrm{s}.$$

With (14.14), (14.16), and (14.17), the earlier equation is finally changed as

$$g_x=\mu_{\mathrm{l,s}}x^{-1}\left(\frac{1}{4}Gr_{x\mathrm{l,s}}\right)^{1/4}\{\eta_{\mathrm{l}\delta}W_{x\mathrm{l,s}}-4W_{y\mathrm{l,s}}\}_\mathrm{s}.$$

Set G_x is taken to express total mass flow rate entering the boundary layer for position $x=0$ to x with width of b of the plate, it should be the following integration:

$$\begin{aligned}G_x&=\iint\limits_A g_x\,\mathrm{d}A\\&=b\int_0^x g_x\,\mathrm{d}x,\end{aligned}$$

where $A=b\cdot x$ is area of the plate.

Then, we obtain

$$\frac{G_x}{b \cdot \mu_{l,s}} = \frac{4}{3}\left(\frac{1}{4}Gr_{xl,s}\right)^{1/4}(\eta_{l\delta}W_{xl,s} - 4W_{yl,s}). \tag{14.64}$$

From (14.64) it follows that the mass flow rate of the condensate, G_x, depends on the defined local Grashof number $Gr_{xl,s}$, absolute viscosity $\mu_{l,s}$, and mass flow rate parameter $(\eta_{l\delta}W_{xl,s} - 4W_{yl,s})$ of the film condensation of superheated vapor. The numerical solutions for the film condensation of superheated water vapor, such as $\eta_{l\delta}$, $W_{xl,s}$, $W_{yl,s}$ and $(\eta_{l\delta}W_{xl,s} - 4W_{yl,s})$ with the wall subcooled grade $\Delta t_w/t_s = (t_s - t_w)/t_s$ and vapor bulk superheated grade $\Delta t_\infty/t_s(= (t_\infty - t_s)/t_s)$ are listed in Tables 14.4–14.7, and plotted in Figs. 14.5–14.8, respectively. From these figures, it is seen the effects of wall subcooled grade $\Delta t_w/t_s = (t_s - t_w)/t_s$ and vapor bulk superheated temperature $\frac{\Delta t_\infty}{t_s}(= (t_\infty - t_s)/t_s)$ on $\eta_{l\delta}$, $W_{xl,s}$, $W_{yl,s}$, and $(\eta_{l\delta}W_{xl,s} - 4W_{yl,s})$.

From Fig. 14.5 it is seen that with increasing the wall subcooled grade $\Delta t_w/t_s = (t_s - t_w)/t_s$, the condensate film thickness η_{δ_l} will increase rapidly, while with increasing vapor bulk superheated grade $\Delta t_\infty/t_s(= (t_\infty - t_s)/t_s)$, the condensate film thickness η_{δ_l} will decrease slowly.

Table 14.4. Numerical solution of condensate film thickness η_{δ_l} with $\Delta t_w/t_s = (t_s - t_w)/t_s$ and $\Delta t_\infty/t_s = (t_s - t_\infty)/t_s$ for laminar film condensation of superheated water vapor, cited from Shang and Wang [15]

$\frac{\Delta t_\infty}{t_s}$ $(= \frac{t_\infty - t_s}{t_s})$	$\frac{\Delta t_w}{t_s} = \frac{t_s - t_w}{t_s}$							
	0	0.001	0.05	0.20	0.40	0.60	0.80	1.00
0	0	0.1017	0.27307	0.39844	0.4958	0.577	0.65545	0.73561
0.27	0	0.10032	0.27155	0.39631	0.49331	0.57416	0.65217	0.73207
1.27	0	0.09368	0.26577	0.38826	0.4836	0.5631	0.63981	0.71838
2.27	0	0.08623	0.26047	0.3809	0.4747	0.55291	0.6284	0.70572
3.27	0	0.07848	0.25536	0.37384	0.46611	0.5431	0.61739	0.6935
4.27	0	0.07078	0.25335	0.36745	0.45825	0.5342	0.6074	0.68243

Table 14.5. Numerical solutions of $W_{xl,\delta}$ with $\Delta t_w/t_s = \frac{(t_s - t_w)}{t_s}$ and $\Delta t_\infty/t_s = \frac{(t_\infty - t_s)}{t_s}$ for laminar film condensation of superheated water vapor, cited from Shang and Wang [15]

$\frac{\Delta t_\infty}{t_s}$ $(= \frac{t_\infty - t_s}{t_s})$	$\frac{\Delta t_w}{t_s} = \frac{t_s - t_w}{t_s}$							
	0	0.001	0.05	0.20	0.40	0.60	0.80	1.00
0	0	0.00513	0.03546	0.06619	0.0851	0.09463	0.09925	0.10111
0.27	0	0.00504	0.03509	0.06544	0.08413	0.09353	0.09805	0.0999
1.27	0	0.00450	0.03369	0.06270	0.08053	0.08945	0.09373	0.09547
2.27	0	0.00391	0.03245	0.06027	0.07733	0.08585	0.08992	0.09156
3.27	0	0.00333	0.03130	0.05801	0.07435	0.0825	0.08637	0.08794
4.27	0	0.00280	0.03095	0.05603	0.07171	0.07957	0.08328	0.08478

Table 14.6. Numerical solutions of $-W_{yl,\delta}$ with $\Delta t_{\mathrm{w}}/t_{\mathrm{s}} = \frac{(t_{\mathrm{s}}-t_{\mathrm{w}})}{t_{\mathrm{s}}}$ and $\Delta t_{\infty}/t_{\mathrm{s}} = \frac{(t_{\infty}-t_{\mathrm{s}})}{t_{\mathrm{s}}}$ for laminar film condensation of superheated water vapor, cited from Shang and Wang [15]

$\frac{\Delta t_{\infty}}{t_{\mathrm{s}}}$	$\frac{\Delta t_{\mathrm{w}}}{t_{\mathrm{s}}} = \frac{t_{\mathrm{s}}-t_{\mathrm{w}}}{t_{\mathrm{s}}}$						
$\left(= \frac{t_{\infty}-t_{\mathrm{s}}}{t_{\mathrm{s}}}\right)$	0	0.001	0.05	0.20	0.40	0.60	0.80	1.00
0	0	0.00013	0.002416	0.0065	0.010085	0.01237	0.01355	0.01371
0.27	0	0.000126	0.002377	0.006398	0.009928	0.012176	0.01334	0.013509
1.27	0	0.000104	0.002231	0.006012	0.00934	0.011466	0.01257	0.012747
2.27	0	0.000082	0.002103	0.005674	0.008823	0.01084	0.0119	0.01207
3.27	0	0.000063	0.001985	0.005363	0.008344	0.01026	0.011268	0.011443
4.27	0	0.000047	0.001943	0.005092	0.007923	0.009754	0.01072	0.0109

Table 14.7. Numerical solutions of $\eta_{l\delta} \cdot W_{xl,\delta} - 4W_{yl,\delta}$ with $\Delta t_{\mathrm{w}}/t_{\mathrm{s}} = \frac{(t_{\mathrm{s}}-t_{\mathrm{w}})}{t_{\mathrm{s}}}$ and $\Delta t_{\infty}/t_{\mathrm{s}} = ((t_{\infty} - t_{\mathrm{s}})/t_{\mathrm{s}})$ for laminar film condensation of superheated water vapor, cited from Shang and Wang [15]

$\frac{\Delta t_{\infty}}{t_{\mathrm{s}}}$		$\frac{\Delta t_{\mathrm{w}}}{t_{\mathrm{s}}} = \frac{t_{\mathrm{s}}-t_{\mathrm{w}}}{t_{\mathrm{s}}}$						
$\left(= \frac{t_{\infty}-t_{\mathrm{s}}}{t_{\mathrm{s}}}\right)$		0	0.05	0.20	0.40	0.60	0.80	1.00
0	(1)	0	0.019347	0.052373	0.082533	0.104082	0.119253	0.129218
	(2)		0.019367	0.0522186	0.082085	0.103487	0.118764	0.129
0.27	(1)	0	0.019037	0.051526	0.081214	0.102402	0.117305	0.127170
	(2)		0.019132	0.051552	0.080959	0.102029	0.117104	0.12726
1.27	(1)	0	0.017878	0.048391	0.076302	0.096233	0.110249	0.119572
	(2)		0.018264	0.049089	0.076787	0.096629	0.110957	0.12085
2.27	(1)	0	0.016865	0.045654	0.072001	0.090827	0.104103	0.112896
	(2)		0.017396	0.046625	0.072616	0.0912308	0.104809	0.10802
3.27	(1)	0	0.015932	0.043139	0.068030	0.085846	0.098396	0.106758
	(2)		0.016529	0.044162	0.068444	0.085830	0.098661	0.10802
4.27	(1)	0	0.015612	0.040956	0.064551	0.081520	0.093464	0.101456
	(2)		0.015661	0.041698	0.064273	0.080431	0.092514	0.101605

(1) Numerical solution and (2) predicted by (14.64)–(14.66)

It is seen from Figs. 14.6 and 14.7 that with increasing the wall subcooled grade $\Delta t_{\mathrm{w}}/t_{\mathrm{s}} = (t_{\mathrm{s}} - t_{\mathrm{w}})/t_{\mathrm{s}}$, the velocity components $W_{xl,\mathrm{s}}$ and $-W_{yl,\mathrm{s}}$ will increase rapidly, especially for smaller wall subcooled grade $\Delta t_{\mathrm{w}}/t_{\mathrm{s}} = (t_{\mathrm{s}} - t_{\mathrm{w}}/t_{\mathrm{s}})$. While with increasing the vapor bulk superheated grade $\Delta t_{\infty}/t_{\mathrm{s}} = (t_{\infty} - t_{\mathrm{s}})/t_{\mathrm{s}}$, the velocity components $W_{xl,\mathrm{s}}$ and $-W_{yl,\mathrm{s}}$ will decrease.

It is seen from Fig. 14.8 that with increasing the wall subcooled grade $\Delta t_{\mathrm{w}}/t_{\mathrm{s}} = (t_{\mathrm{s}} - t_{\mathrm{w}})/t_{\mathrm{s}}$, the mass flow rate parameter $(\eta_{l\delta} W_{xl,\mathrm{s}} - 4W_{yl,\mathrm{s}})$ will increase rapidly, especially for the smaller wall subcooled grade $\Delta t_{\mathrm{w}}/t_{\mathrm{s}} = (t_{\mathrm{s}} - t_{\mathrm{w}})/t_{\mathrm{s}}$. While with increasing the vapor bulk superheated grade $\Delta t_{\infty}/t_{\mathrm{s}} = (t_{\infty} - t_{\mathrm{s}})/t_{\mathrm{s}}$, the mass flow rate parameter $(\eta_{l\delta} W_{xl,\mathrm{s}} - 4W_{yl,\mathrm{s}})$ will decrease.

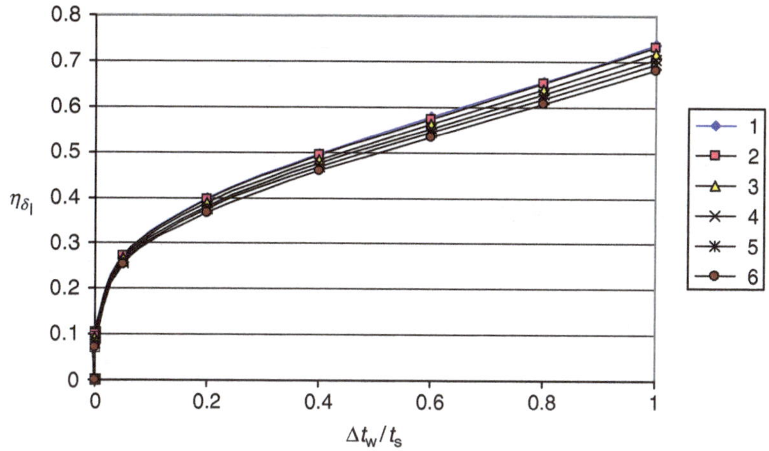

Fig. 14.5. Numerical solution of η_{δ_l} with $\frac{\Delta t_w}{t_s} = \frac{t_s - t_w}{t_s}$ and $\frac{\Delta t_\infty}{t_s} \left(= \frac{t_\infty - t_s}{t_s}\right)$ for laminar film condensation of superheated water vapor. 1. $\frac{\Delta t_\infty}{t_s} \left(= \frac{t_\infty - t_s}{t_s}\right) = 0$, 2. $\frac{\Delta t_\infty}{t_s} \left(= \frac{t_\infty - t_s}{t_s}\right) = 0.27$, 3. $\frac{\Delta t_\infty}{t_s} \left(= \frac{t_\infty - t_s}{t_s}\right) = 1.27$, 4. $\frac{\Delta t_\infty}{t_s} \left(= \frac{t_\infty - t_s}{t_s}\right) = 2.27$, 5. $\frac{\Delta t_\infty}{t_s} \left(= \frac{t_\infty - t_s}{t_s}\right) = 3.27$, 6. $\frac{\Delta t_\infty}{t_s} \left(= \frac{t_\infty - t_s}{t_s}\right) = 4.27$

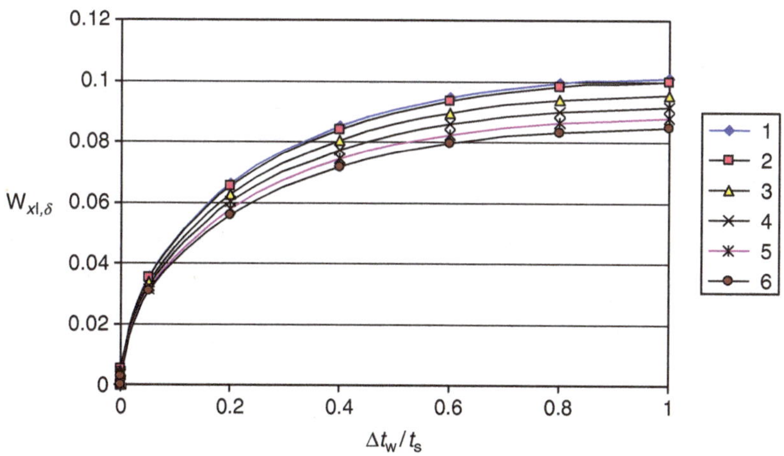

Fig. 14.6. Numerical solution of $W_{xl\delta}$ with $\frac{\Delta t_w}{t_s} = \frac{t_s - t_w}{t_s}$ and $\frac{\Delta t_\infty}{t_s} \left(= \frac{t_\infty - t_s}{t_s}\right)$ for aminar film condensation of superheated water vapor. 1. $\frac{\Delta t_\infty}{t_s} \left(= \frac{t_\infty - t_s}{t_s}\right) = 0$, 2. $\frac{\Delta t_\infty}{t_s} \left(= \frac{t_\infty - t_s}{t_s}\right) = 0.27$, 3. $\frac{\Delta t_\infty}{t_s} \left(= \frac{t_\infty - t_s}{t_s}\right) = 1.27$, 4. $\frac{\Delta t_\infty}{t_s} \left(= \frac{t_\infty - t_s}{t_s}\right) = 2.27$, 5. $\frac{\Delta t_\infty}{t_s} \left(= \frac{t_\infty - t_s}{t_s}\right) = 3.27$, 6. $\frac{\Delta t_\infty}{t_s} \left(= \frac{t_\infty - t_s}{t_s}\right) = 4.27$

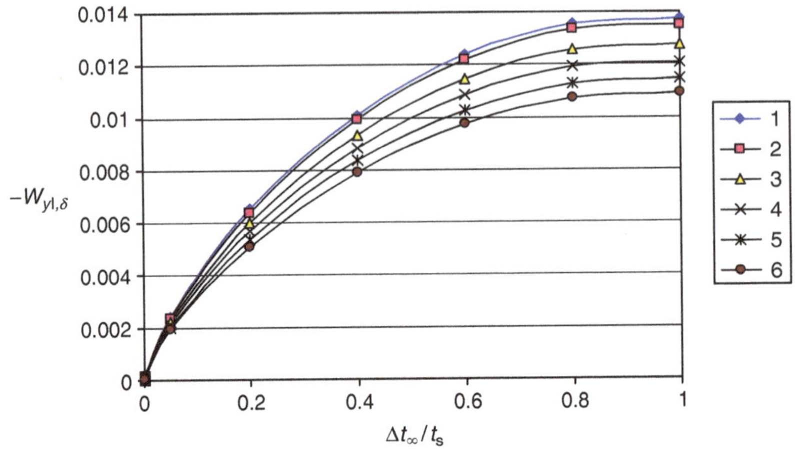

Fig. 14.7. Numerical solution of $-W_{y l\delta}$ with $\frac{\Delta t_w}{t_s} = \frac{t_s - t_w}{t_s}$ and $\frac{\Delta t_\infty}{t_s}\left(=\frac{t_\infty - t_s}{t_s}\right)$ for aminar film condensation of superheated water vapor. 1. $\frac{\Delta t_\infty}{t_s}\left(=\frac{t_\infty - t_s}{t_s}\right) = 0$, 2. $\frac{\Delta t_\infty}{t_s}\left(=\frac{t_\infty - t_s}{t_s}\right) = 0.27$, 3. $\frac{\Delta t_\infty}{t_s}\left(=\frac{t_\infty - t_s}{t_s}\right) = 1.27$, 4. $\frac{\Delta t_\infty}{t_s}\left(=\frac{t_\infty - t_s}{t_s}\right) = 2.27$, 5. $\frac{\Delta t_\infty}{t_s}\left(=\frac{t_\infty - t_s}{t_s}\right) = 3.27$, 6. $\frac{\Delta t_\infty}{t_s}\left(=\frac{t_\infty - t_s}{t_s}\right) = 4.27$

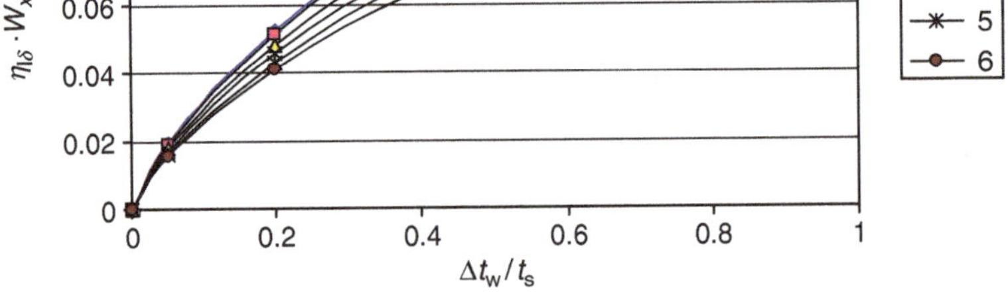

Fig. 14.8. Numerical results of $\eta_{l\delta} W_{xl,\delta} - 4W_{yl,\delta}$ with $\frac{\Delta t_w}{t_s} = \frac{t_s - t_w}{t_s}$ and $\frac{\Delta t_\infty}{t_s}\left(=\frac{t_\infty - t_s}{t_s}\right)$ for aminar film condensation of superheated water vapor. 1. $\frac{\Delta t_\infty}{t_s}\left(=\frac{t_\infty - t_s}{t_s}\right) = 0$, 2. $\frac{\Delta t_\infty}{t_s}\left(=\frac{t_\infty - t_s}{t_s}\right) = 0.27$, 3. $\frac{\Delta t_\infty}{t_s}\left(=\frac{t_\infty - t_s}{t_s}\right) = 1.27$, 4. $\frac{\Delta t_\infty}{t_s}\left(=\frac{t_\infty - t_s}{t_s}\right) = 2.27$, 5. $\frac{\Delta t_\infty}{t_s}\left(=\frac{t_\infty - t_s}{t_s}\right) = 3.27$, 6. $\frac{\Delta t_\infty}{t_s}\left(=\frac{t_\infty - t_s}{t_s}\right) = 4.27$

According to the corresponding numerical solutions and by curve-matching method, the expressions for the mass flow rate parameter $(\eta_{l\delta} W_{x\mathrm{l,s}} - 4W_{y\mathrm{l,s}})$ of the film condensation of superheated water vapor is obtained as follows:

For $0.05 \le \Delta t_{\mathrm{w}}/t_{\mathrm{s}} \le 1$:

$$\eta_{l\delta} \cdot W_{x\mathrm{l},\delta} - 4W_{y\mathrm{l},\delta} = (\eta_{l\delta} \cdot W_{x\mathrm{l},\delta} - 4W_{y\mathrm{l},\delta})_{\Delta t_\infty=0} - b\frac{\Delta t_\infty}{t_{\mathrm{s}}}, \tag{14.65}$$

where $(\eta_{l\delta} \cdot W_{x\mathrm{l},\delta} - 4W_{y\mathrm{l},\delta})_{\Delta t_\infty=0}$ is the mass flow rate parameter for the film condensation of saturated water vapor. Here $(\eta_{l\delta} \cdot W_{x\mathrm{l},\delta} - 4W_{y\mathrm{l},\delta})_{\Delta t_\infty=0}$ and b are expressed by

$$(\eta_{l\delta} \cdot W_{x\mathrm{l},\delta} - 4W_{y\mathrm{l},\delta})_{\Delta t_\infty=0} = \left(0.186 - 0.057\tfrac{\Delta t_{\mathrm{w}}}{t_{\mathrm{s}}}\right)\left(\tfrac{\Delta t_{\mathrm{w}}}{t_{\mathrm{s}}}\right)^{3/4}, \tag{14.66}$$

$$b = 10^{-4} \times \left[2.756 + 121.4\tfrac{\Delta t_{\mathrm{w}}}{t_{\mathrm{s}}} - 60\left(\tfrac{\Delta t_{\mathrm{w}}}{t_{\mathrm{s}}}\right)^2\right]. \tag{14.67}$$

The results of $(\eta_{l\delta} W_{x\mathrm{l,s}} - 4W_{y\mathrm{l,s}})$ predicted by means of (14.65)–(14.67) for the laminar film condensation of superheated water vapor are also described in Table 14.7. Through the comparison of the predicted values of $(\eta_{l\delta} W_{x\mathrm{l,s}} - 4W_{y\mathrm{l,s}})$ with the corresponding numerical results, it is seen that the agreements are very good.

14.8 Summary

So far, we have presented our recent developments on film condensation of saturated superheated and superheated vapors. In principle, the film condensation of saturated vapor is a special case of that of the superheated vapor, when the vapor superheated grade $\frac{\Delta t_\infty}{t_{\mathrm{s}}} = (\frac{t_\infty - t_{\mathrm{s}}}{t_{\mathrm{s}}})$ equals zero. On this basis, governing equations, heat and mass transfer with the related equations of the laminar film condensation of superheated vapor can be summarized in Tables 14.8 and 14.9, respectively.

14.9 Remarks

In this chapter we deal with the theory of film condensation of superheated vapour on a vertical flat plate at atmospheric pressure with consideration of various factors including variable thermophysical properties. The film condensation of saturated vapor presented in Chap. 12 is only a special case of the film condensation of superheated vapor. The present presentation here is an extension of former studies and the following points are concluded:

Table 14.8. Summary of governing equations of the laminar film condensation of superheated vapor

term	equations
	partial differential equations for condensate liquid film
mass equation	$\frac{\partial}{\partial x}(\rho_l w_{xl}) + \frac{\partial}{\partial y}(\rho_l w_{yl}) = 0$
momentum equation	$\rho_l \left(w_{xl}\frac{\partial w_{xl}}{\partial x} + w_{yl}\frac{\partial w_{xl}}{\partial y} \right) = \frac{\partial}{\partial y}\left(\mu_l \frac{\partial w_{xv}}{\partial y} \right) + g(\rho_l - \rho_{v,\infty})$
energy equation	$\rho_l c_{p_l} \left(w_{xl}\frac{\partial t_l}{\partial x} + w_{yl}\frac{\partial t_l}{\partial y} \right) = \frac{\partial}{\partial y}\left(\lambda_l \frac{\partial t_l}{\partial y} \right)$
	partial differential equations for vapor film
mass equation	$\frac{\partial}{\partial x}(\rho_v w_{xv}) + \frac{\partial}{\partial y}(\rho_v w_{yv}) = 0$
momentum equation	$\rho_v \left(w_{xv}\frac{\partial w_{xv}}{\partial x} + w_{yv}\frac{\partial w_{xv}}{\partial y} \right) = \frac{\partial}{\partial y}\left(\mu_v \frac{\partial w_{xv}}{\partial y} \right) + g(\rho_v - \rho_{v,\infty})$
energy equation	$\rho_v c_{pv} \left(w_{xv}\frac{\partial T_v}{\partial x} + w_{yv}\frac{\partial T_v}{\partial y} \right) = \frac{\partial}{\partial y}\left(\lambda_v \frac{\partial T_v}{\partial y} \right)$
	boundary conditions
$y = 0$	$w_{xl} = 0, \quad w_{yl} = 0, \quad t_l = t_w$
$y = \delta_l$	$w_{xl,s} = w_{xv,s}$
	$\rho_{l,s}\left(w_{xl}\frac{\partial \delta_{xl}}{\partial x} - w_{yl} \right)_s = \rho_{v,s}\left(w_{xv}\frac{\partial \delta_{xv}}{\partial x} - w_{yv} \right)_s$
	$\mu_{l,s}\left(\frac{\partial w_{xl}}{\partial y} \right)_s = \mu_{v,s}\left(\frac{\partial w_{xv}}{\partial y} \right)_s$
	$\lambda_{l,s}\left(\frac{\partial t_l}{\partial y} \right)_{y=\delta_l} = h_{fg}\rho_{l,s}\left(w_{xl}\frac{\partial \delta l}{\partial x} - w_{yl} \right)_s + \lambda_{v,s}\left(\frac{\partial t_v}{\partial y} \right)_{y=\delta_l}$
	$T = T_s$
$y \to \infty$	$w_{xv} = 0, \quad T_v = T_\infty$
	assumed similarity variables for liquid film
η_l	$\left(\frac{1}{4} Gr_{xl,s} \right)^{1/4} \frac{y}{x}$
$Gr_{xl,s}$	$\frac{g(\rho_{l,w} - \rho_{v,\infty})x^3}{\nu_{l,s}^2 \rho_{l,s}}$

θ_{l}	$\frac{t_{\mathrm{l}}-t_{\mathrm{s}}}{t_{\mathrm{w}}-t_{\mathrm{s}}}$
$W_{x\mathrm{l}}$	$\left(2\sqrt{gx}\left(\frac{\rho_{\mathrm{l,w}}-\rho_{\mathrm{v},\infty}}{\rho_{\mathrm{l,s}}}\right)^{1/2}\right)^{-1} w_{x\mathrm{l}}$
$W_{y\mathrm{l}}$	$(2\sqrt{gx}\left(\frac{\rho_{\mathrm{l,w}}-\rho_{\mathrm{v},\infty}}{\rho_{\mathrm{l,s}}}\right)^{1/2}\left(\frac{1}{4}Gr_{x\mathrm{l,s}}\right)^{-4})^{-1} w_{y\mathrm{l}}$
	ordinary differential equations for liquid film
mass equation	$2W_{x\mathrm{l}} - \eta_{\mathrm{l}}\frac{\mathrm{d}W_{x\mathrm{l}}}{\mathrm{d}\eta_{\mathrm{l}}} + 4\frac{\mathrm{d}W_{y\mathrm{l}}}{\mathrm{d}\eta_{\mathrm{l}}} + \frac{1}{\rho_{\mathrm{l}}}\frac{\mathrm{d}\rho_{\mathrm{l}}}{\mathrm{d}\eta_{\mathrm{l}}}(-\eta_{\mathrm{l}}W_{x\mathrm{l}} + 4W_{y\mathrm{l}}) = 0$
momentum equation	$\frac{\nu_{\mathrm{l,s}}}{\nu_{\mathrm{l}}}\left(W_{x\mathrm{l}}\left(2W_{x\mathrm{l}} - \eta_{\mathrm{l}}\frac{\mathrm{d}W_{x\mathrm{l}}}{\mathrm{d}\eta_{\mathrm{l}}}\right) + 4W_{y\mathrm{l}}\frac{\mathrm{d}W_{x\mathrm{l}}}{\mathrm{d}\eta_{\mathrm{l}}}\right) = \frac{\mathrm{d}^2W_{x\mathrm{l}}}{\mathrm{d}\eta_{\mathrm{l}}^2} + \frac{1}{\mu_{\mathrm{l}}}\frac{\mathrm{d}\mu_{\mathrm{l}}}{\mathrm{d}\eta_{\mathrm{l}}}\frac{\mathrm{d}W_{x\mathrm{l}}}{\mathrm{d}\eta_{\mathrm{l}}} + \frac{\mu_{\mathrm{l,s}}}{\mu_{\mathrm{l}}}\frac{\rho_{\mathrm{l}}-\rho_{\mathrm{v},\infty}}{\rho_{\mathrm{l,w}}-\rho_{\mathrm{v},\infty}}$
energy equation	$Pr_{\mathrm{l}}\frac{\rho_{\mathrm{l}}}{\rho_{\mathrm{l,s}}}\frac{\mu_{\mathrm{l,s}}}{\mu_{\mathrm{l}}}(-\eta_{\mathrm{l}}W_{x\mathrm{l}} + 4W_{y\mathrm{l}})\frac{\mathrm{d}\theta_{\mathrm{l}}}{\mathrm{d}\eta_{\mathrm{l}}} = \frac{\mathrm{d}^2\theta_{\mathrm{l}}}{\mathrm{d}\eta_{\mathrm{l}}^2} + \frac{1}{\lambda_{\mathrm{l}}}\frac{\mathrm{d}\lambda_{\mathrm{l}}}{\mathrm{d}\eta_{\mathrm{l}}}\frac{\mathrm{d}\theta_{\mathrm{l}}}{\mathrm{d}\eta_{\mathrm{l}}}$
	assumed similarity variables for vapour film
η_{v}	$\eta_{\mathrm{v}} = \left(\frac{1}{4}Gr_{x\mathrm{v},\infty}\right)^{1/4}\frac{y}{x}$
$Gr_{x\mathrm{v},\infty}$	$\frac{g(\rho_{\mathrm{v,s}}/\rho_{\mathrm{v},\infty}-1)x^3}{\nu_{\mathrm{v},\infty}^2}$
θ_{v}	$\frac{T_{\mathrm{v}}-T_{\infty}}{T_{\mathrm{s}}-T_{\infty}}$
$W_{x\mathrm{v}}$	$(2\sqrt{gx}(\rho_{\mathrm{v,s}}/\rho_{\mathrm{v},\infty} - 1)^{1/2})^{-1} w_{x\mathrm{v}}$
$W_{y\mathrm{v}}$	$\left(2\sqrt{gx}(\rho_{\mathrm{v,s}}/\rho_{\mathrm{v},\infty} - 1)^{1/2}\left(\frac{1}{4}Gr_{x\mathrm{v},\infty}\right)^{-1/4}\right)^{-1} w_{y\mathrm{v}}$
	dimensionless equations for vapour film
mass equations	$2W_{x\mathrm{v}} - \eta_{\mathrm{v}}\frac{\mathrm{d}W_{x\mathrm{v}}}{\mathrm{d}\eta_{\mathrm{v}}} + 4\frac{\mathrm{d}W_{y\mathrm{v}}}{\mathrm{d}\eta_{\mathrm{v}}} + \frac{1}{\rho_{\mathrm{v}}}\frac{\mathrm{d}\rho_{\mathrm{v}}}{\mathrm{d}\eta_{\mathrm{v}}}(-\eta_{\mathrm{v}}W_{x\mathrm{v}} + 4W_{y\mathrm{v}}) = 0$
momentum equation	$\frac{\nu_{\mathrm{v},\infty}}{\nu_{\mathrm{v}}}\left(W_{x\mathrm{v}}\left(2W_{x\mathrm{v}} - \eta_{\mathrm{v}}\frac{\mathrm{d}W_{x\mathrm{v}}}{\mathrm{d}\eta_{\mathrm{v}}}\right) + 4W_{y\mathrm{v}}\left(\frac{\mathrm{d}W_{x\mathrm{v}}}{\mathrm{d}\eta_{\mathrm{v}}}\right)\right) = \frac{\mathrm{d}^2W_{x\mathrm{v}}}{\mathrm{d}\eta_{\mathrm{v}}^2} + \frac{1}{\mu_{\mathrm{v}}}\frac{\mathrm{d}\mu_{\mathrm{v}}}{\mathrm{d}\eta_{\mathrm{v}}}\frac{\mathrm{d}W_{x\mathrm{v}}}{\mathrm{d}\eta_{\mathrm{v}}} + \frac{\mu_{\mathrm{v},\infty}}{\mu_{\mathrm{v}}}\frac{\rho_{\mathrm{v}}-\rho_{\mathrm{v},\infty}}{\rho_{\mathrm{v,s}}-\rho_{\mathrm{v},\infty}}$
energy equation	$Pr_{\mathrm{v}}\frac{\nu_{\mathrm{v},\infty}}{\nu_{\mathrm{v}}}(-\eta W_{x\mathrm{v}} + 4W_{y\mathrm{v}})\frac{\mathrm{d}\theta_{\mathrm{v}}}{\mathrm{d}\eta_{\mathrm{v}}} = \frac{\mathrm{d}^2\theta_{\mathrm{v}}}{\mathrm{d}\eta_{\mathrm{v}}^2} + \frac{1}{\lambda_{\mathrm{v}}}\frac{\mathrm{d}\lambda_{\mathrm{v}}}{\mathrm{d}\eta_{\mathrm{v}}}\frac{\mathrm{d}\theta_{\mathrm{v}}}{\mathrm{d}\eta_{\mathrm{v}}}$

Table 14.8. *Continued*

term	equations
	dimensionless boundary conditions
$\eta_l = 0$	$W_{xl} = 0, \quad W_{yl} = 0, \quad \theta_l = 1$
$\eta_l = \eta_{l\delta}$ $(\eta_v = 0)$	
	$W_{xv,s} = \left(\frac{\rho_{l,w}-\rho_{v,\infty}}{\rho_{l,s}}\right)^{1/2} \left(\frac{\rho_{v,s}-\rho_{v,\infty}}{\rho_{v,\infty}}\right)^{-1/2} W_{xl,s}$
	$W_{yv,s} = -0.25 \frac{\mu_{l,s}}{\mu_{v,s}} \left(\frac{\nu_{l,s}}{\nu_{v,\infty}}\right)^{-1/2} \left(\frac{\rho_{l,w}-\rho_{v,\infty}}{\rho_{l,s}}\right)^{1/4} \left(\frac{\rho_{v,s}-\rho_{v,\infty}}{\rho_{v,\infty}}\right)^{-1/4} (\eta_{l\delta} W_{xl,s} - 4W_{yl,s})$
	$\left(\frac{dW_{xv}}{d\eta_v}\right)_{\eta_v=0} = \frac{\mu_{l,s}}{\mu_{v,s}} \left(\frac{\nu_{v,s}}{\nu_{l,s}}\right)^{1/2} \left(\frac{\rho_{l,w}-\rho_{v,\infty}}{\rho_{l,s}}\right)^{3/4} \left(\frac{\rho_{v,s}}{\rho_{v,\infty}} - 1\right)^{-3/4} \left(\frac{dW_{xl}}{d\eta_l}\right)_{\eta_l=\eta_{l,\delta}}$
	$\left(\frac{d\theta_v}{d\eta_v}\right)_{\eta_v=0} = \frac{\left(\frac{\rho_{l,w}-\rho_{v,\infty}}{\rho_{l,s}}\right)^{1/4} (h_{fg}\mu_{l,s}(\eta_{v\delta} W_{xv,s} - 4W_{yv,s}) + \lambda_{l,s}(t_w - t_s)\left(\frac{d\theta_l}{d\eta_l}\right)_{\eta_l=\eta_{l\delta}})}{\left(\frac{\rho_{v,s}}{\rho_{v,\infty}} - 1\right)^{1/4} \left(\frac{\nu_{l,s}}{\nu_{v,\infty}}\right)^{1/2} \lambda_{v,s}(T_s - T_\infty)}$
	$\theta_l = 0, \quad \theta_v = 1$
$\eta_v \to \infty$	$W_{xv} \to 0, \quad \theta_v = 0$

Table 14.9. Summary of the heat transfer and mass flow rate of the laminar film condensation of superheated vapor

term	equations
	for heat transfer
q_x $\left(\text{defined as } -\lambda_{\mathrm{l,w}}\left(\frac{\partial t}{\partial y}\right)_{y=0}\right)$	$-\lambda_{\mathrm{l,w}}(T_{\mathrm{w}}-T_{\mathrm{s}})\left(\frac{1}{4}Gr_{x\mathrm{l,s}}\right)^{1/4}x^{-1}\left(\frac{\mathrm{d}\theta_{\mathrm{l}}}{\mathrm{d}\eta_{\mathrm{l}}}\right)_{\eta_{\mathrm{l}}=0}$
α_x $\left(\text{defined as } \frac{q_x}{T_{\mathrm{w}}-T_x}\right)$	$-\lambda_{\mathrm{l,w}}\left(\frac{1}{4}Gr_{x\mathrm{l,s}}\right)^{1/4}x^{-1}\left(\frac{\mathrm{d}\theta_{\mathrm{l}}}{\mathrm{d}\eta_{\mathrm{l}}}\right)_{\eta_{\mathrm{l}}=0}$
$\overline{\alpha}_x$ (defined as $Q_x=\overline{\alpha_x}(t_{\mathrm{w}}=t_{\mathrm{s}})A$)	$\overline{\alpha}_x=-\frac{4}{3}\lambda_{\mathrm{l,w}}\left(\frac{1}{4}Gr_{x\mathrm{l,s}}\right)^{1/4}x^{-1}\left(\frac{\mathrm{d}\theta_{\mathrm{l}}}{\mathrm{d}\eta_{\mathrm{l}}}\right)_{\eta_{\mathrm{l}}=0}$
$Nu_{x\mathrm{l,w}}$ $\left(\text{defined as} \frac{\alpha_x x}{\lambda_{\mathrm{l,w}}}\right)$	$Nu_{x\mathrm{l,w}}=-\left(\frac{1}{4}Gr_{x\mathrm{l,s}}\right)^{1/4}\left(\frac{\mathrm{d}\theta_{\mathrm{l}}}{\mathrm{d}\eta_{\mathrm{l}}}\right)_{\eta_{\mathrm{l}}=0}$
$-\left(\frac{\mathrm{d}\theta_{\mathrm{l}}}{\mathrm{d}\eta_{\mathrm{l}}}\right)_{\eta_{\mathrm{l}}=0}$	$\left(-\left(\frac{\mathrm{d}\theta_{\mathrm{l}}}{\mathrm{d}\eta_{\mathrm{l}}}\right)_{\eta_{\mathrm{l}}=0}\right)_{\Delta t_\infty=0}+a\cdot\frac{\Delta t_\infty}{t_{\mathrm{s}}}$ (for film condensation of water steam)
$\left(-\left(\frac{\mathrm{d}\theta_{\mathrm{l}}}{\mathrm{d}\eta_{\mathrm{l}}}\right)_{\eta_{\mathrm{l}}=0}\right)_{\Delta t_\infty=0}$ (for $\Delta t_\infty=0$)	$\frac{1.74-0.19\frac{\Delta t_{\mathrm{w}}}{t_{\mathrm{s}}}}{\left(\frac{\Delta t_{\mathrm{w}}}{t_{\mathrm{s}}}\right)^{1/4}}$ $\left(0.005\le\frac{\Delta t_{\mathrm{w}}}{t_{\mathrm{s}}}\le 1\right)$
Coefficient a	$10^{-2}\times\left(6.92-9.45\frac{\Delta t_{\mathrm{w}}}{t_{\mathrm{s}}}+5.35\left(\frac{\Delta t_{\mathrm{w}}}{t_{\mathrm{s}}}\right)^2\right)$ $\left(0.05\le\frac{\Delta t_{\mathrm{w}}}{t_{\mathrm{s}}}\le 1\right)$
	for condensate mass transfer
δ_{l}	$\eta_{\mathrm{l}\delta}\left(\frac{1}{4}Gr_{x\mathrm{l,s}}\right)^{-1/4}x$
$Gr_{x\mathrm{l,s}}$	$\frac{g(\rho_{\mathrm{l,w}}-\rho_{\mathrm{v},\infty})x^3}{\nu_{\mathrm{l,s}}^2\rho_{\mathrm{l,s}}}$
g_x	$\mu_{\mathrm{l,s}}x^{-1}\left(\frac{1}{4}Gr_{x\mathrm{l,s}}\right)^{1/4}(\eta_{\mathrm{l}\delta}W_{x\mathrm{l,s}}-4W_{y\mathrm{l,s}})$
G_x	$\frac{4}{3}b\cdot\mu_{\mathrm{l,s}}\left(\frac{1}{4}Gr_{x\mathrm{l,s}}\right)^{1/4}(\eta_{\mathrm{l}\delta}W_{x\mathrm{l,s}}-4W_{y\mathrm{l,s}})$
$(\eta_{\mathrm{l}\delta}W_{x\mathrm{l,s}}-4W_{y\mathrm{l,s}})$	$(\eta_{\mathrm{l}\delta}\cdot W_{x\mathrm{l},\delta}-4W_{y\mathrm{l},\delta})_{\Delta t_\infty=0}-b\frac{\Delta t_\infty}{t_{\mathrm{s}}}$ (for film condensation of water vapor)
	$b=10^{-4}\times\left[2.756+121.4\frac{\Delta t_{\mathrm{w}}}{t_{\mathrm{s}}}-60\left(\frac{\Delta t_{\mathrm{w}}}{t_{\mathrm{s}}}\right)^2\right]$ $\left(0.05\le\frac{\Delta t_{\mathrm{w}}}{t_{\mathrm{s}}}\le 1\right)$
$(\eta_{\mathrm{l}\delta}\cdot W_{x\mathrm{l},\delta}-4W_{y\mathrm{l},\delta})_{\Delta t_\infty=0}$	$\left(0.186-0.057\frac{\Delta t_{\mathrm{w}}}{t_{\mathrm{s}}}\right)\left(\frac{\Delta t_{\mathrm{w}}}{t_{\mathrm{s}}}\right)^{3/4}$ $\left(0.05\le\frac{\Delta t_{\mathrm{w}}}{t_{\mathrm{s}}}\le 1\right)$

An advanced similarity transformation approach, velocity component method, is used to transform the system of partial differential equations associated with the two-phase boundary problem into a system of dimensionless ordinary equations. This method has an obvious advantages over the corresponding Falkner–Skan transformation [8] for treatment of various physical factors including variable thermophysical properties.

The system of ordinary differential equations and its related boundary conditions is computed by a successively iterative procedure and an iterative method is adopted for the numerical solutions of the three-point boundary value problem. Meanwhile, with heat and mass transfer analysis, the theoretical equations for Nusselt number and mass flow rate are derived for the laminar film condensation of superheated vapor.

Both the Nusselt number and mass flow rate of the film condensation of superheated vapor are proportional to the local Grashof number. In addition, Nusselt number is also proportional to the temperature gradient on the wall and mass flow rate is proportional to the mass flow rate parameter.

Furthermore, on the basis of the rigorous numerical solutions, the temperature gradient $(\mathrm{d}\theta_{\mathrm{l}}/\mathrm{d}\eta_{\mathrm{l}})_{\eta_{\mathrm{l}}=0}$ and then mass flow rate parameter $(\eta_{\mathrm{l}\delta}W_{x\mathrm{l,s}} - 4W_{y\mathrm{l,s}})$ are formulated by using a curve-fitting method for simple and reliable predictions of heat and mass transfer of the film condensation of superheated water vapor.

With increasing the wall subcooled grade $\Delta t_{\mathrm{w}}/t_{\mathrm{s}} = (t_{\mathrm{s}} - t_{\mathrm{w}})/t_{\mathrm{s}}$, the temperature gradient $(\mathrm{d}\theta_{\mathrm{l}}/\mathrm{d}\eta_{\mathrm{l}})_{\eta_{\mathrm{l}}=0}$ will decrease rapidly especially for lower wall subcooled grade $\Delta t_{\mathrm{w}}/t_{\mathrm{s}} = (t_{\mathrm{s}} - t_{\mathrm{w}})/t_{\mathrm{s}}$. While, with increasing the vapor bulk superheated grade $\Delta t_{\infty}/t_{\mathrm{s}} = (t_{\infty} - t_{\mathrm{s}})/t_{\mathrm{s}}$, the temperature gradient $(\mathrm{d}\theta_{\mathrm{l}}/\mathrm{d}\eta_{\mathrm{l}})_{\eta_{\mathrm{l}}=0}$ will increase slowly.

With increasing the wall subcooled grade $\Delta t_{\mathrm{w}}/t_{\mathrm{s}} = (t_{\mathrm{s}} - t_{\mathrm{w}})/t_{\mathrm{s}}$, the condensate film thickness $\eta_{\delta_{\mathrm{l}}}$ will increase rapidly especially for lower wall subcooled grade $\Delta t_{\mathrm{w}}/t_{\mathrm{s}} = (t_{\mathrm{s}} - t_{\mathrm{w}})/t_{\mathrm{s}}$, while with increasing the vapor bulk superheated grade $\Delta t_{\infty}/t_{\mathrm{s}} = (t_{\infty} - t_{\mathrm{s}})/t_{\mathrm{s}}$, the condensate film thickness $\eta_{\delta_{\mathrm{l}}}$ will decrease slowly.

With increasing the wall subcooled grade $\Delta t_{\mathrm{w}}/t_{\mathrm{s}} = (t_{\mathrm{s}} - t_{\mathrm{w}})/t_{\mathrm{s}}$, the velocity components $W_{x\mathrm{l,s}}$ and $-W_{y\mathrm{l,s}}$ will increase rapidly, except for a larger value of $\Delta t_{\mathrm{w}}/t_{\mathrm{s}} = (t_{\mathrm{s}} - t_{\mathrm{w}})/t_{\mathrm{s}}$. While with increasing the vapor bulk superheated grade $\Delta t_{\infty}/t_{\mathrm{s}} = (t_{\infty} - t_{\mathrm{s}})/t_{\mathrm{s}}$, the velocity components $W_{x\mathrm{l,s}}$ and $-W_{y\mathrm{l,s}}$ will decrease slowly.

As the results, with increasing the wall subcooled grade $\Delta t_{\mathrm{w}}/t_{\mathrm{s}} = (t_{\mathrm{s}} - t_{\mathrm{w}})/t_{\mathrm{s}}$, the mass flow rate parameter $(\eta_{\mathrm{l}\delta}W_{x\mathrm{l,s}} - 4W_{y\mathrm{l,s}})$ will increase rapidly, while with increasing the vapor bulk superheated grade $\Delta t_{\infty}/t_{\mathrm{s}} = (t_{\infty} - t_{\mathrm{s}})/t_{\mathrm{s}}$, the mass flow rate parameter $(\eta_{\mathrm{l}\delta}W_{x\mathrm{l,s}} - 4W_{y\mathrm{l,s}})$ will decrease slowly.

14.10 Calculation Example

Example. A flat plate with 0.3 m width and 0.3 m length is suspended vertically in the superheated water vapor with $t_{\infty} = 227°\mathrm{C}$. The wall temperature is $t_{\mathrm{w}} = 90°\mathrm{C}$. Suppose the film condensation is laminar, please calculate the condensate heat and mass transfer on the plate.

Solution. The wall subcooled temperatures of the plate is $\Delta t_{\mathrm{w}} = t_{\mathrm{s}} - t_{\mathrm{w}} = 100-90 = 10°\mathrm{C}$, and then, the wall subcooled grade is $\Delta t_{\mathrm{w}}/t_{\mathrm{s}} = 10/100 = 0.1$. We have $\rho_{\mathrm{l,w}} = 965.32\,\mathrm{kg\,m^{-3}}$, and $\lambda_{\mathrm{l,w}} = 0.68\mathrm{W}\ (\mathrm{m°C^{-1}})$ for water at $t_{\mathrm{w}} = 90°\mathrm{C}$.

The vapor superheated temperature is $\Delta t_{\infty} = t_{\infty} - t_{\mathrm{s}} = 227-100 = 127°\mathrm{C}$, then, the vapor superheated grade is $\Delta t_{\infty}/t_{\mathrm{s}} = t_{\infty} - t_{\mathrm{s}}/t_{\mathrm{s}} = 127/100 = 1.27$. We have $\rho_{\mathrm{v}} = 0.4405\,\mathrm{kg\,m^{-3}}$ for water vapor at $t_{\infty} = 227°\mathrm{C}$.

Additionally, we have $\rho_{\mathrm{l,s}} = 958.1\,\mathrm{kg\,m^{-3}}$, $\nu_{\mathrm{l,s}} = 0.294\times10^{-6}\mathrm{m^2\,s^{-1}}$, $\mu_{\mathrm{l,s}} = 281.7\times10^{-6}\,\mathrm{kg\,(m\,s^{-1})}$ for saturated water at $100°\mathrm{C}$.

1. For heat transfer . From (14.57) and (14.59), the average heat transfer coefficient is evaluated as

$$\overline{\alpha_x} = -\frac{4}{3}\lambda_{\mathrm{l,w}}\left(\frac{1}{4}Gr_{x\mathrm{l,s}}\right)^{1/4} x^{-1}\left(\frac{\mathrm{d}\theta_{\mathrm{l}}}{\mathrm{d}\eta_{\mathrm{l}}}\right)_{\eta_{\mathrm{l}}=0}.$$

From (14.14) the local Grashof number $Gr_{x\mathrm{l,s}}$ of the film condensation should be evaluated as

$$\begin{aligned} Gr_{x\mathrm{l,s}} &= \frac{g(\rho_{\mathrm{l,w}} - \rho_{\mathrm{v},\infty})x^3}{\nu_{\mathrm{l,s}}^2\rho_{\mathrm{l,s}}} \\ &= \frac{9.8\times(965.32-0.4405)\times0.3^3}{(0.294\times10^{-6})^2\times958.1} \\ &= 3.08289\times10^{12}. \end{aligned}$$

From (14.61) the temperature gradient of the laminar film condensation of the superheated water vapor is calculated as

$$-\left(\frac{\mathrm{d}\theta_{\mathrm{l}}}{\mathrm{d}\eta_{\mathrm{l}}}\right)_{\eta_{\mathrm{l}}=0} = \left(-\left(\frac{\mathrm{d}\theta_{\mathrm{l}}}{\mathrm{d}\eta_{\mathrm{l}}}\right)_{\eta_{\mathrm{l}}=0}\right)_{\Delta t_{\infty}=0} + a\cdot\frac{\Delta t_{\infty}}{t_{\mathrm{s}}}.$$

From (14.62), the dimensionless temperature gradient $\left(-\left(\frac{\mathrm{d}\theta_{\mathrm{l}}}{\mathrm{d}\eta_{\mathrm{l}}}\right)_{\eta_{\mathrm{l}}=0}\right)_{\Delta t_{\infty}=0}$ for the film condensation of saturated water vapor is evaluated as

$$\begin{aligned} \left(-\left(\frac{\mathrm{d}\theta_{\mathrm{l}}}{\mathrm{d}\eta_{\mathrm{l}}}\right)_{\eta_{\mathrm{l}}=0}\right)_{\Delta t_{\infty}=0} &= \frac{1.74-0.19\frac{\Delta t_{\mathrm{w}}}{t_{\mathrm{s}}}}{\left(\frac{\Delta t_{\mathrm{w}}}{t_{\mathrm{s}}}\right)^{1/4}} \\ &= \frac{1.74-0.19\times0.1}{0.1^{1/4}} \\ &= 3.060419. \end{aligned}$$

From (14.63) the coefficient a is calculated as

$$\begin{aligned} a &= 10^{-2} \times \left(6.92 - 9.45\frac{\Delta t_{\mathrm{w}}}{t_{\mathrm{s}}} + 5.35\left(\frac{\Delta t_{\mathrm{w}}}{t_{\mathrm{s}}}\right)^2\right) \\ &= 10^{-2} \times (6.92 - 9.45 \times 0.1 + 5.35 \times 0.1^2) \\ &= 0.060285. \end{aligned}$$

Then, the temperature gradient of the laminar film condensation of the superheated water vapor is calculated as

$$\begin{aligned} -\left(\frac{\mathrm{d}\theta_{\mathrm{l}}}{\mathrm{d}\eta_{\mathrm{l}}}\right)_{\eta_{\mathrm{l}}=0} &= \left(-\left(\frac{\mathrm{d}\theta_{\mathrm{l}}}{\mathrm{d}\eta_{\mathrm{l}}}\right)_{\eta_{\mathrm{l}}=0}\right)_{\Delta t_\infty=0} + a\cdot\frac{\Delta t_\infty}{t_{\mathrm{s}}} \\ &= 3.060419 + 0.060285 \times 1.27 \\ &= 3.136981. \end{aligned}$$

Then, the average heat transfer coefficient is evaluated as

$$\begin{aligned} \overline{\alpha_x} &= -\frac{4}{3}\lambda_{\mathrm{l,w}}\left(\frac{1}{4}Gr_{x\mathrm{l,s}}\right)^{1/4} x^{-1}\left(\frac{\mathrm{d}\theta_{\mathrm{l}}}{\mathrm{d}\eta_{\mathrm{l}}}\right)_{\eta_{\mathrm{l}}=0} \\ &= \frac{4}{3} \times 0.68 \times \left(\frac{1}{4} \times 3.08289 \times 10^{12}\right)^{1/4} \times (0.3)^{-1} \times 3.136981 \\ &= 8883.064\,\mathrm{W}\ (\mathrm{m^2K})^{-1}. \end{aligned}$$

The total heat transfer rate of laminar film condensation of the superheated water vapor on the vertical plate is calculated

$$\begin{aligned} Q_x &= \overline{\alpha_x}(t_{\mathrm{w}} - t_{\mathrm{s}})A \\ &= 8883.064 \times (90 - 100) \times 0.3 \times 0.3 \\ &= -7994.76\,\mathrm{W}. \end{aligned}$$

The negative means that the heat transfer direction is to the plate from the condensate film.

2. For mass flow rate of the condensation . From (14.64), the total mass flow rate entering the liquid film for position $x = 0$ to x with width of b of the plate is evaluated as

$$G_x = \frac{4}{3}b\cdot\mu_{\mathrm{l,s}}\left(\frac{1}{4}Gr_{x\mathrm{l,s}}\right)^{1/4}(\eta_{\mathrm{l}\delta}W_{x\mathrm{l,s}} - 4W_{y\mathrm{l,s}}).$$

Here, from (14.65) the mass flow rate parameter film condensation of superheated vapor is evaluated as

$$\eta_{\mathrm{l}\delta}\cdot W_{x\mathrm{l},\delta} - 4W_{y\mathrm{l},\delta} = (\eta_{\mathrm{l}\delta}\cdot W_{x\mathrm{l},\delta} - 4W_{y\mathrm{l},\delta})_{\Delta t_\infty=0} - b\frac{\Delta t_\infty}{t_{\mathrm{s}}},$$

where the mass flow rate parameter of the laminar film condensation of saturated water vapor is calculated as

$$
\begin{aligned}
(\eta_{l\delta} \cdot W_{xl,\delta} - 4W_{yl,\delta})_{\Delta t_\infty = 0} &= \left(0.186 - 0.057\frac{\Delta t_{\mathrm{w}}}{t_{\mathrm{s}}}\right)\left(\frac{\Delta t_{\mathrm{w}}}{t_{\mathrm{s}}}\right)^{3/4} \\
&= (0.186 - 0.057 \times 0.1)\,(0.1)^{3/4} \\
&= 0.032062378.
\end{aligned}
$$

Additionally, the coefficient b is evaluated as

$$
\begin{aligned}
b &= 10^{-4} \times \left[2.756 + 121.4\frac{\Delta t_{\mathrm{w}}}{t_{\mathrm{s}}} - 60\left(\frac{\Delta t_{\mathrm{w}}}{t_{\mathrm{s}}}\right)^2\right] \\
&= 10^{-4} \times [2.756 + 121.4 \times 0.1 - 60 \times (0.1)^2] \\
&= 0.00143.
\end{aligned}
$$

The mass flow rate parameter of the laminar film condensation of superheated water steam is calculated as

$$
\begin{aligned}
(\eta_{l\delta} \cdot W_{xl,\delta} - 4W_{yl,\delta}) &= (\eta_{l\delta} \cdot W_{xl,\delta} - 4W_{yl,\delta})_{\Delta t_\infty = 0} - b\frac{\Delta t_\infty}{t_{\mathrm{s}}} \\
&= 0.032062378 - 0.00143 \times 1.27 \\
&= 0.030246.
\end{aligned}
$$

The mass flow rate G_x of the laminar film condensation of superheated water vapor is calculated as

$$
\begin{aligned}
G_x &= \frac{4}{3} b \cdot \mu_{l,s}\left(\frac{1}{4} Gr_{xl,s}\right)^{1/4} (\eta_{l\delta} W_{xl,s} - 4W_{yl,s}) \\
&= \frac{4}{3} \times 0.3 \times 281.7 \times 10^{-6} \times (\frac{1}{4} \times 3.08289 \times 10^{12})^{1/4} \times 0.030246 \\
&= 0.0031933\,\mathrm{kg\,s^{-1}} \\
&= 11.496\,\mathrm{kg\,h^{-1}}
\end{aligned}
$$

References

1. W. Nusselt, Die Oberflichenkondensation des Wasserdampfes, Z Ver. D Ing. 60, pp. 541–569, 1916
2. L.A. Bromley, Effect of heat capacity of condensate in condensing, Ind. Eng. Chem. 44, pp. 2966–2969, 1952
3. W.M. Rohsenow, Heat transfer and temperature distribution in laminar film condensation, Trans. Am. Soc. Mech. Engrs. 78, pp. 1645–1648, 1956
4. E.M. Sparrow and J.L. Gregg, A boundary-layer treatment of laminar film condensation, J. Heat Transfer 81, pp. 13–18, 1959

5. J.C. Koh, E.M. Sparow, and J.P. Hartnett, The two phase boundary layer in laminar film condensation, Int. J. Heat Mass Transfer 2, pp. 69–88, 1961
6. M.M. Chen, An analytical study of laminar film condensation–1, Flat plate, J. Heat Transfer 81, pp. 48–54, 1961
7. W.H. McAdams, Heat Transmission (3rd Edn), pp. 331–332, McGraw-Hill, New York, 1954
8. K.D. Voskresenskiy, Calculation of heat transfer in film condensation allowing for the temperature dependence of the physical properties of condensate (in Russian), (Inergy Inst.), U.S.S.R. Acad. Sci., 1948
9. D.A. Labuntsov, Effect of temperature dependence of physical parameters of condensate on heat transfer in film condensation of steam, Teploenergetika 4(2), pp. 49–51, 1957
10. G. Poots and R.G. Miles, Effects of variable physical properties, of laminar film condensation of saturated steam on a vertical flat plate, Int. J. Heat Mass Transfer 10, pp. 1677–1692, 1967
11. J. Stinnesbeck and H. Heawig, An asymptotic analisis of laminar film condensation on a vertical flat plate including veriable property effects, Proceedings of the Ninth International Heat Transfer Conference Jelusalem, Vol. 6, 1990
12. D.Y. Shang and T. Adamek, Study of laminar film condensation of saturated steam for consideration of variable thermophysical properties, Transport phenomena, Science and Technology, 1992 (Ed. by B.X. Wang), Higher Education Press, Beijing, 470–475, 1992
13. D.Y. Shang and T. Adamek, Study on laminar film condensation of saturated steam on a vertical flat plate for consideration of various physical factors including ariable thermophysical properties, Warme- und Stoffubertragung 30, Springer, Berlin Heidel berg, New York pp. 89–100, 1994
14. W.J. Minkowcs and E.M. Sparrow, Condensation heat transfer in the presence of noncondensable, interface resistance, superheating, variable properties, and diffusion, Int. J. Heat Mass Transfer 9, pp. 1125–1144, 1966
15. D.Y. Shang and B.X. Wang, An extended study on steady-state laminar film condensation of a superheated vapor on an isothermal vertical plate, Int. J. Heat Mass Transfer 40(4), pp. 931–941, 1997
16. J.Q. Zhang, Real Fluid Mechanics, Tsinghua University Press, Beijing, 1986. Giadwell and D.K. Sayers, Computational techniques for ordinary differential equations, Academic, London, 1980
17. VDI-Warmeatlas, Berechnungsbltter fur die Warmeubertragu, 5, ereiterte Auflage, VDI Verlage GmbH, Dusseldorf, 1988

Part III

Falling Film Flow of Non-Newtonian Fluids

15

Hydrodynamics of Falling Film Flow of Non-Newtonian Power-Law Fluids

Nomenclature

a	thermal diffusive coefficient, $\mathrm{m^2\,s^{-1}}$
b	width of plate, m
C_f	local skin-friction coefficient, $2Re_x^{-1/(n+1)}\left[\left(\frac{\mathrm{d}W_x}{\mathrm{d}\eta}\right)_{\eta=0}\right]^n$
c_p	specific heat at constant pressure, $\mathrm{J\,(kg\,K)^{-1}}$
g	gravitation acceleration, $\mathrm{m\,s^{-2}}$
g_x	local mass flow rate entering into the boundary layer of unit area at a certain position x, $\mathrm{kg\,(m^2s)^{-1}}$
G_x	total mass flow rate entering into the boundary layer from the inlet $x = 0$ to a stream downstream position x and with the width of b of the plate, $\mathrm{kg\,s^{-1}}$
K	coefficient of consistency, $\mathrm{kg\,s^{n-2}\,m^{-1}}$
n	power law index
V	volume flow rate of the falling film flow, $\frac{G_x(x_0)}{\rho}$, $\mathrm{m^3\,s^{-1}}$
Re_x	local Reynolds number, $\frac{x^n(w_{x,\infty})^{2-n}\rho}{K}$
w_x, w_y	velocity components in the x- and y- directions, respectively, $\mathrm{m\,s^{-1}}$
$W_x(\eta),\ W_y(\eta)$	dimensionless velocity components in the x- and y-directions, respectively
$w_{x,\infty}$	velocity of the fluid outside the boundary layer, $\sqrt{2gx\cos\alpha}$, $\mathrm{m\,s^{-1}}$
$-\mathrm{d}w_x/\mathrm{d}y$	shear rate, $1s^{-1}$
$x,\ y$	streamwise and cross-stream coordinates, m
x_0	length of the boundary layer region, m

Greek symbols

$\delta_l(x)$	boundary layer thickness at the x position, m
$\delta_l(x_0)$	critical film thickness related to x_0, m
η	dimensionless coordinate variable for boundary layer, $\frac{y}{x}Re_x^{1/(n+1)}$
Φ	mass flow rate parameter, $\frac{2(n+1)}{2n+1}\left(\frac{n\eta_{\delta_l}}{2(n+1)}W_{x,\delta_l}-W_{y,\delta_l}\right)$
ρ	density, $\mathrm{kg\,m^{-3}}$
μ	absolute viscosity, $\mathrm{kg\,(m\,s)^{-1}}$
μ_a	the apparent viscosity, $\mathrm{kg\,(m\,s)^{-1}}$
τ	shear stress, $\mathrm{N\,m^{-2}}$

Subscripts

i	inclined case
l	momentum boundary layer
v	vertical case or vapor
x	local value
w	at wall
α (or y)	angle of inclination
∞	far from the wall surface

15.1 Principal Types of Power-Law Fluids

15.1.1 Newtonian Fluids

In the previous chapters, we have presented the free convection, film boiling, and film condensation, where all the fluids dealt with are Newtonian fluids. Newtonian fluids are those which follow Newton's law, i.e.,

$$\tau = \mu \frac{\mathrm{d}w_x}{\mathrm{d}y} \tag{15.1}$$

for special coordinate. Here, μ is the absolute viscosity and is a constant independent of shear rate. If a fluid does not follow (15.1), it is a non-Newtonian fluid. Figure 15.1 shows shear stress τ is proportional to the shear rate $-\mathrm{d}w_x/\mathrm{d}y$. The line for a Newtonian fluid is straight, the slope being μ. However, a non-Newtonian fluid is a fluid in which the viscosity changes with the applied shear force. As a result, non-Newtonian fluids may not have a well-defined viscosity.

15.1.2 Power-Law Fluids

Power-law fluids can be subdivided into the following types according to the range of their power-law index:

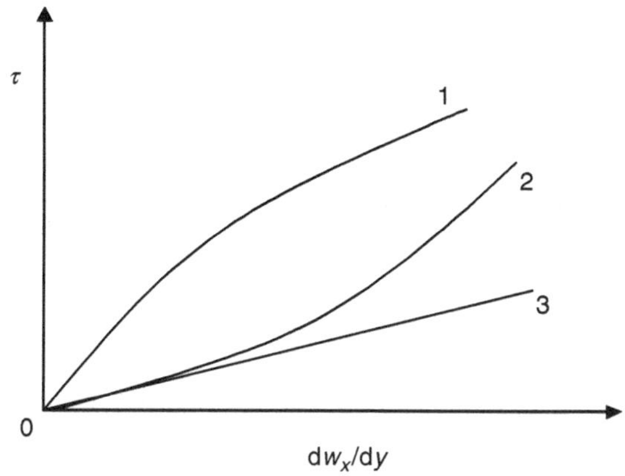

Fig. 15.1. Diagram for power-law fluids. (1) Pseudoplastic fluid; (2) Dilatant fluid; (3) Newtonian fluid

(i) Non-newtonian pseudoplastic fluids. For such fluids, the apparent viscosity will be reduced with rate of shear. The shape of the flow curve is shown in Fig. 15.1, and it generally can be represented by a power-law equation (sometimes called the *Ostwald-de Waele equation*).

$$\tau = K\left(\frac{\mathrm{d}w_x}{\mathrm{d}y}\right)^n \qquad (n < 1), \tag{15.2}$$

where K is coefficient of consistency, and n is the power-law index. The apparent viscosity μ_{a} in (15.3) is obtained from (15.1) and (15.2) and decreases with increasing shear rate:

$$\mu_{\mathrm{a}} = K\left(\frac{\mathrm{d}w_x}{\mathrm{d}y}\right)^{n-1}. \tag{15.3}$$

A common household example of a strongly shear-thinning fluid is styling gel, which primarily composed of water and a fixative such as a vinyl acetate/vinyl pyrrolidone copolymer (PVP/PA). The majority of non-Newtonian fluids is in this category and includes polymer solutions or melts, greases, starch suspensions, mayonnaise, biological fluids, detergent slurries, dispersion media in certain pharmaceuticals, and paints. Additionally, some colloids, clay, milk, gelatine, blood, and liquid cement also belong to pseudoplastic fluids.

(ii) Non-newtonian dilatant fluids. For dilatant fluids the power-law equation is often applicable, but with $n > 1$ as shown in (15.4), which means that their apparent viscosity will increase with rate of shear, i.e.,

$$\tau = K\left(\frac{\mathrm{d}w_x}{\mathrm{d}y}\right)^n \qquad (n > 1), \tag{15.4}$$

Table 15.1. Types of power-law fluids

name of the power-law fluid	range of power-law index n	example included
pseudoplastic fluid (majority of non-Newtonian fluids)	$n < 1$	polymer solutions or melts, greases, starch suspensions, mayonnaise, biological fluids, detergent slurries, dispersion media in certain pharmaceuticals, paints, styling gel, some colloids, clay, milk, gelatine, blood, and liquid cement
newtonian fluid (many of the most common fluids)	$n = 1$	such as water, most aqueous solutions, oils, corn syrup, glycerine, air, and other gases.
dilatant fluid (far less common than pseudoplastic fluid)	$n > 1$	some corn flour-sugar solutions, wet beach sand, starch in water, potassium silicate in water, some solutions containing high concentrations of powder in water, an uncooked paste of cornstarch and water, concentrated solution of sugar in water, and suspensions of rice starch or corn starch

These fluids, or shear-thickening fluids are far less common than pseudoplastic fluids, and their flow behavior (Fig. 15.1) shows an increase in apparent viscosity with increasing shear rate.

From (15.1) it is found that Newtonian fluid is the power-law fluid with a unit power-law index, where the shear stress is directly proportional to the shear rate. Therefore, Newtonian fluids can be regarded as special non-Newtonian power-law fluids. In addition, Newtonian fluids include many of the most common fluids, such as water, most aqueous solutions, oils, corn syrup, glycerine, air, and other gases. So far, the principal types of power-law fluids can be summarized as Table 15.1.

15.2 Introduction of Studies on Hydrodynamics of Gravity-Driven Film Flow of Non-Newtonian Power-Law Fluids (FFNF)

For heat-sensitive materials, short residence time, and close temperature controls during heat transfer process are essential, which can be achieved by allowing a liquid to flow in a thin falling film along a solid surface. Such cooling techniques have been widely used in many industrial applications, which are especially the chemical engineering operations, food and polymer processing industries, cooling systems, distillation, evaporators, ocean thermal energy conversion systems, molten plastics, pulps, coating equipment, etc. Such heat exchangers are characterized by high heat transfer coefficients at low mass flow rates and small temperatures differences, and invite a lot of work for the extensive studies.

Fully developed laminar film flow of non-Newtonian power-law fluids along a plane surface was conducted by Astarita et al. [1], who measured the film thickness for various inclinations and flow rates. Later, Therien et al. [2] conducted a similar study, in which experimental data for the film thickness were compared with an analytical expression for the thickness of fully developed films of power-law fluids. Sylvester et al. [3] also measured the film thickness as a function of the volumetric flow rate, but they primarily focused on the onset of rippling on the film surface and the characteristics of wavy film.

Yang and Yarbrough [4,5], Murthy and Sarma [6,7], Tekic et al. [8], Andersson and Irgens [9] among others have used the integral method approach to study the hydrodynamics of gravity-driven power law films. Theoretical analyses of hydrodynamics of gravity-driven power-law fluid films have been studied by means of similarity analysis by Andersson and Irgens [10,11], Yang and Yarbrough [4,5]. Murthy and Sarma [6] extended the conventional integral analysis for Newtonian films to cover power-law fluids. Later, Murthy and Sarma [7] included the effect of interfacial drag at the liquid–vapor interface in a similar analysis, while Tekic et al. [8] presented results accounted for the streamwise pressure gradient and surface tension. Andersson and Irgens [9] explored the influence of the rheology of the film on the hydrodynamic entrance length.

A different approach was adopted by Andersson and Irgens [10,11], namely to divide the accelerating film flow into three regions shown schematically in Fig. 15.1, the boundary layer region, the fully viscous region and the developed flow region. While the boundary layer region is divided into a developing viscous boundary layer and an external inviscid freestream. They further demonstrated that a similarity transformation exists, such that the boundary layer momentum equation for power-law fluids is exactly transformed into a Falkner–Skan type ordinary differential equation. The resulting two-point boundary-value problem was solved numerically with a standard shooting technique based on classical 4th-order Runge–Kutta integration in combination with a Newton iteration procedure. Numerical results were obtained for values of the power-law index n in the range $0.5 \leq n \leq 2.0$. Andersson and Irgens [11] provided a relative extensive review on the study of hydrodynamics of a falling film flow of power law fluids.

The dissolution of a soluble wall and the subsequent penetration of the solute into the non-Newtonian liquid film were considered by Astarita [12]. He provided the mass transfer rate between the wall and the hydrodynamically fully developed film, with an assumption of velocity near the wall to vary linearly with the distance from the wall. Mashelkher and Chavan [13] provided a more general solution of this problem.

More recently, Andersson and Shang [14] provided a development on formulation of a new similarity transformation for extensive studies of accelerating non-Newtonian film flow. The partial differential equations governing the hydrodynamics of the flow of a power-law fluid on an inclined plane surface are transformed into a set of two ordinary differential equations by means of the velocity component approach. Although the analysis is applicable for any angle of inclination $\alpha(0 \le \alpha \le \pi/2)$, the resulting one-parameter problem involves only the power-law index n. Nevertheless, physically essential quantities, like the velocity components and the skin-friction coefficient, do depend on α and the relevant relationships are deduced between the vertical and inclined cases. Accurate numerical similarity solutions are provided for n in the range from 0.1 to 2.0. The present method enables solutions to be obtained also for highly pseudoplastic films, i.e., for n below 0.5. The mass flow rate entrained into the momentum boundary layer from the inviscid freestream is expressed in term of a dimensionless mass flux parameter ϕ, which depends on the dimensionless boundary layer thickness and the velocity components at the edge of the viscous boundary layer, which is thus an integral part of the similarity solution, turns out to decrease monotonically with n. Using this new model, they were able to determine some difficult issues, such as the mass flow rate entrained into the boundary layer from the free stream and the length of boundary layer region, etc. In this chapter we focus on our recent developments on hydrodynamics analysis for the boundary layer region.

15.3 Physical Model and Governing Partial Differential Equations

Consider the accelerating laminar flow in the boundary layer region of a non-Newtonian power law liquid film down along an inclined plane surface, as shown schematically in Fig. 15.2. The incompressible and inelastic fluid is assumed to obey the Ostwald-de-Waele power-law model and the action of viscous stresses is confined to the developing momentum boundary layer adjacent to the solid surface. The basic conservation equations for mass and momentum within the viscous boundary layer are:

$$\frac{\partial w_x}{\partial x} + \frac{\partial w_y}{\partial y} = 0, \tag{15.5}$$

$$w_x \frac{\partial w_x}{\partial x} + w_y \frac{\partial w_x}{\partial y} = g\cos\alpha + n\frac{K}{\rho}\left(\frac{\partial w_x}{\partial y}\right)^{n-1} \frac{\partial^2 w_x}{\partial y^2} \tag{15.6}$$

with boundary conditions

$$y = 0: \qquad w_x = 0, \qquad w_y = 0, \tag{15.7}$$

$$y = \delta_l: \qquad w_x = w_{x,\infty}, \tag{15.8}$$

where w_x and w_y are velocity exponents in x and y directions, respectively, while g and α denote the gravitational acceleration and the angle of inclination of the plane wall. Here it has been anticipated that $\partial w_x/\partial y \ge 0$ throughout the entire the film. The fluid physical properties ρ, c_p, K, and n assumed to be constant in the present analysis are density, specific heat, coefficient of

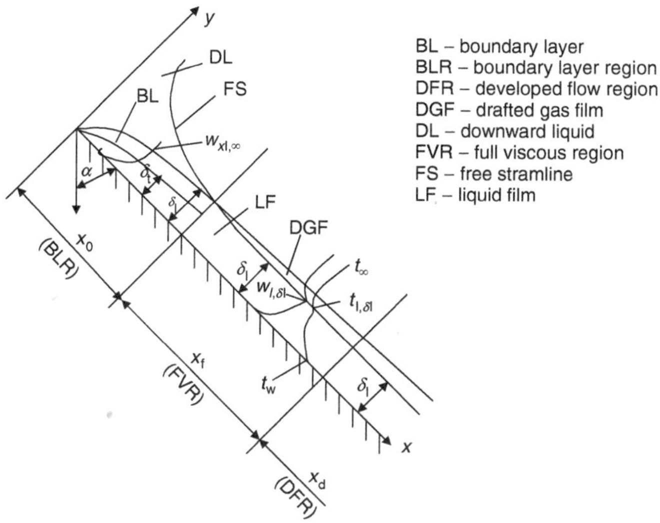

Fig. 15.2. Schematic representation of accelerating film flow, cited from Shang and Gu [15]

consistency, and power-law index, respectively. The two-dimensional rheological model represents pseudoplastic or shear-thing fluids if the power-law index n smaller than unity and dilatant or shear-thickening fluids for $n > 1$. The deviation of n from unity indicates the degree of deviation from Newtonian rheology and the particular case $n = 1$ represents a Newtonian fluid with dynamic coefficient of viscosity K.

No-slip and impermeability at the inclined wall $y = 0$ are expressed by the boundary conditions given by (15.7), while the outer condition, (15.8), assures that the velocity component w_x within the boundary layer approaches the external velocity

$$w_{x,\infty} = \sqrt{2gx\cos\alpha} \tag{15.9}$$

at the edge $y = \delta_l$ of the boundary layer. Since the friction flow between the viscous boundary layer and the free streamline bordering the constant-pressure atmosphere is quasi-one-dimensional, the simple solution given by (15.9) is readily derived from (15.6) by assuming $w_{x,\infty} = 0$ (and infinite film thickness) at the inlet $x = 0$.

It may be worthwhile to recall that the boundary layer theory conventionally adopted in the analysis of thin-film flow may be inadequate if the Reynolds number is too low. Wu and Thompson [16] compared boundary layer theory predictions with solutions of the full Cauchy equation for flow of a shear-thinning power-law fluid past a flat plate of length L. They found that the Reynolds number Re_x (with $x = L$), below which the boundary layer approximations become inaccurate, decreased from 120 for a Newtonian fluid ($n = 1$) to 4.5 for a highly pseudoplastic fluid ($n = 0.1$).

15.4 A New Similarity Transformation

Incidentally, as pointed out by Andersson and Irgens [10], the external velocity, (15.9), belong to the Falkner–Skan class of freestreams $w_{x,\infty} \propto x^m$ which permits a similarity transformation of the momentum boundary layer equation even for power-law fluids. A generalized Falkner–Skan type of transformation was therefore introduced by Andersson and Irgens [10, 11]. However, in the recent study, a new transformation is proposed by Andersson and Shang [14], which unlike the Falkner–Skan type of approach does not involve the stream function. Let us first introduce the related dimensionless similarity variables defined for similarity transformation of the governing partial equations.

Dimensionless similarity variables. According to the study in [14], a dimensionless similarity variable is defined as

$$\eta = \frac{y}{x} Re_x^{1/(n+1)} \tag{15.10}$$

where

$$Re_x = \frac{x^n (w_{x,\infty})^{2-n} \rho}{K} \tag{15.11}$$

is a generalized local Reynolds number. The dimensionless velocity components are defined as

$$W_x(\eta) = \frac{w_x}{\sqrt{2gx \cos\alpha}}, \tag{15.12}$$

$$W_y(\eta) = \frac{w_y}{\sqrt{2gx \cos\alpha}} Re_x^{1/(n+1)} \tag{15.13}$$

for the x and y directions, respectively. These dimensionless variables analogous to the similarity transformation used in the parts 1 and 2 of this book for free convection and film flows for the particular parameter value $n = 1$. Then, the partial differential equations given by (15.5) and (15.6) and their boundary equations given by (15.7) and (15.8) are now transformed as follows:

Derivation of (15.5). From (14.10) and (14.11), we have

$$\begin{aligned}
\frac{\partial \eta}{\partial x} &= \frac{\partial}{\partial x}\left[\frac{y}{x} Re_x^{1/(n+1)}\right] \\
&= \frac{\partial}{\partial x}\left[y\left(\frac{(2g\cos\alpha)^{(2-n)/2}\rho}{K}\right)^{1/(n+1)} x^{-n/(2(n+1))}\right] \\
&= -\frac{n}{2(n+1)}\left[y\left(\frac{(2g\cos\alpha)^{(2-n)/2}\rho}{K}\right)^{1/(n+1)} x^{-n/(2(n+1))-1}\right] \\
&= -\frac{n}{2(n+1)}\frac{y}{x}\left(\frac{x^n(2gx\cos\alpha)^{(2-n)/2}\rho}{K}\right)^{1/(n+1)} x^{-(n+2)/(2(n+1))} x^{-n/(2(n+1))} \\
&= -\frac{n}{2(n+1)}\eta x^{-1}
\end{aligned}$$

$$\frac{\partial \eta}{\partial y} = \frac{1}{x}(Re_x)^{1/(n+1)}$$

Then,

$$
\begin{aligned}
\frac{\partial w_x}{\partial x} &= \sqrt{2gx\cos\alpha}\frac{\mathrm{d}W_x}{\mathrm{d}\eta}\frac{\partial\eta}{\partial x} + \sqrt{\frac{g\cos\alpha}{2x}}W_x \\
&= \sqrt{2gx\cos\alpha}\frac{\mathrm{d}W_x}{\mathrm{d}\eta}\left(-\frac{n}{2(n+1)}\eta x^{-1}\right) + \sqrt{\frac{g\cos\alpha}{2x}}W_x \\
&= -\sqrt{\frac{g\cos\alpha}{2x}}\frac{\mathrm{d}W_x}{\mathrm{d}\eta}\frac{n}{(n+1)}\eta + \sqrt{\frac{g\cos\alpha}{2x}}W_x. \\
\frac{\partial w_y}{\partial y} &= \sqrt{2gx\cos\alpha}(Re_x)^{-1/(n+1)}\frac{\mathrm{d}W_y}{\mathrm{d}\eta}\frac{\partial\eta}{\partial y} \\
&= \sqrt{2gx\cos\alpha}(Re_x)^{-1/(n+1)}\frac{\mathrm{d}W_y}{\mathrm{d}\eta}\frac{1}{x}(Re_x)^{1/(n+1)} \\
&= \sqrt{\frac{2g\cos\alpha}{x}}\frac{\mathrm{d}W_y}{\mathrm{d}\eta}.
\end{aligned}
$$

Therefore, (15.5) is changed into

$$
= \sqrt{\frac{g\cos\alpha}{2x}}\frac{\mathrm{d}W_x}{\mathrm{d}\eta}\frac{n}{(n+1)}\eta + \sqrt{\frac{g\cos\alpha}{2x}}W_x + \sqrt{\frac{2g\cos\alpha}{x}}\frac{\mathrm{d}W_y}{\mathrm{d}\eta} = 0.
$$

Simplify the earlier equation we have

$$
W_x - \frac{n}{n+1}\eta\frac{\mathrm{d}W_x}{\mathrm{d}\eta} + 2\frac{\mathrm{d}W_y}{\mathrm{d}\eta} = 0. \tag{15.14}
$$

Derivation of (15.6). With (15.12) and (15.13) we have

$$
\begin{aligned}
\frac{\partial w_x}{\partial y} &= \sqrt{2gx\cos\alpha}\frac{\mathrm{d}W_x}{\mathrm{d}\eta}\frac{1}{x}(Re_x)^{1/(n+1)} = \sqrt{\frac{2g\cos\alpha}{x}}\frac{\mathrm{d}W_x}{\mathrm{d}\eta}(Re_x)^{1/(n+1)} \\
\frac{\partial^2 w_x}{\partial y^2} &= \sqrt{\frac{2g\cos\alpha}{x}}\frac{\mathrm{d}^2W_x}{\mathrm{d}\eta^2}\frac{1}{x}(Re_x)^{2/(n+1)}.
\end{aligned}
$$

Then, (15.6) is changed into

$$
\begin{aligned}
&\sqrt{2gx\cos\alpha}W_x\left(-\sqrt{\frac{g\cos\alpha}{2x}}\frac{\mathrm{d}W_x}{\mathrm{d}\eta}\frac{n}{(n+1)}\eta + \sqrt{\frac{g\cos\alpha}{2x}}W_x\right) + \\
&(\sqrt{2gx\cos\alpha}W_y)(Re_x)^{-1/(n+1)}\sqrt{\frac{2g\cos\alpha}{x}}\frac{\mathrm{d}W_x}{\mathrm{d}\eta}(Re_x)^{1/(n+1)} \\
&g\cos\alpha + n\frac{K}{\rho_l}\left(\sqrt{\frac{2g\cos\alpha}{x}}\frac{\mathrm{d}W_x}{\mathrm{d}\eta}(Re_x)^{1/(n+1)}\right)^{(n-1)}\sqrt{\frac{2g\cos\alpha}{x}}\frac{\mathrm{d}^2W_x}{\mathrm{d}\eta^2}\frac{1}{x}(Re_x)^{2/(n+1)}
\end{aligned}
$$

or

$$
\begin{aligned}
&W_x\left(-g\cos\alpha\frac{\mathrm{d}W_x}{\mathrm{d}\eta}\frac{n}{(n+1)}\eta + g\cos\alpha W_x\right) + (W_y)2g\cos\alpha\frac{\mathrm{d}W_x}{\mathrm{d}\eta} \\
&= g\cos\alpha + n\frac{K}{\rho_l}\left(\frac{\mathrm{d}W_x}{\mathrm{d}\eta}\right)^{(n-1)}\left(\frac{2g\cos\alpha}{x}\right)^{n/2}\frac{\mathrm{d}^2W_x}{\mathrm{d}\eta^2}\frac{1}{x}Re_x.
\end{aligned}
$$

The earlier equation is simplified to

$$W_x\left(-\frac{\mathrm{d}W_x}{\mathrm{d}\eta}\frac{n}{(n+1)}\eta+W_x\right)+(W_y)2\frac{\mathrm{d}W_x}{\mathrm{d}\eta}$$
$$=1+n\frac{K}{\rho_1}\left(\frac{\mathrm{d}W_x}{\mathrm{d}\eta}\right)^{(n-1)}\frac{2}{2gx\cos\alpha}\left(\frac{2gx\cos\alpha}{x^2}\right)^{n/2}\frac{\mathrm{d}^2W_x}{\mathrm{d}\eta^2}Re_x.$$

With (15.9) the earlier equation is changed into

$$W_x\left(-\frac{\mathrm{d}W_x}{\mathrm{d}\eta}\frac{n}{(n+1)}\eta+W_x\right)+(W_y)2\frac{\mathrm{d}W_x}{\mathrm{d}\eta}$$
$$=1+n\left(\frac{\mathrm{d}W_x}{\mathrm{d}\eta}\right)^{(n-1)}\frac{K}{\rho}\frac{2}{x^n w_{x,\infty}^{2-n}}\frac{\mathrm{d}^2W_x}{\mathrm{d}\eta^2}Re_x.$$

With the definition of local Reynolds number shown in (15.11), the earlier equation can be expressed as

$$W_x\left(-\frac{\mathrm{d}W_x}{\mathrm{d}\eta}\frac{n}{(n+1)}\eta+W_x\right)+(W_y)2\frac{\mathrm{d}W_x}{\mathrm{d}\eta}=1+n\left(\frac{\mathrm{d}W_x}{\mathrm{d}\eta}\right)^{(n-1)}2Re_x^{-1}\frac{\mathrm{d}^2W_x}{\mathrm{d}\eta^2}Re_x.$$

Finally, (15.6) is transformed as

$$W_x\left(-\frac{n}{(n+1)}\eta\frac{\mathrm{d}W_x}{\mathrm{d}\eta}+W_x\right)+2W_y\frac{\mathrm{d}W_x}{\mathrm{d}\eta}=1+2n\left(\frac{\mathrm{d}W_x}{\mathrm{d}\eta}\right)^{(n-1)}\frac{\mathrm{d}^2W_x}{\mathrm{d}\eta^2}. \tag{15.15}$$

Thus, the governing partial differential equations (15.5) and (15.6) are transformed to the dimensionless equations (15.14) and (15.15), respectively, with the related dimensionless boundary conditions

$$\eta=0:\quad W_x(\eta)=0,\quad W_y(\eta)=0, \tag{15.16}$$
$$\eta=\eta_{\delta_1}:\quad W_x(\eta)=1. \tag{15.17}$$

Evidently, the power-law index n is the only explicit parameter in the transformed problem.

15.5 Numerical Solutions

The nonlinear two-point boundary value problem defined by (15.14)–(15.17) was solved numerically for several values of the power-law index in the range $0.1 \le n \le 2.0$. Here, the shooting method was adopted. First, (15.14) and (15.15) were written as a system of three first-order differential equations, which was solved by means of fifth-order Runge–Kutta integration.

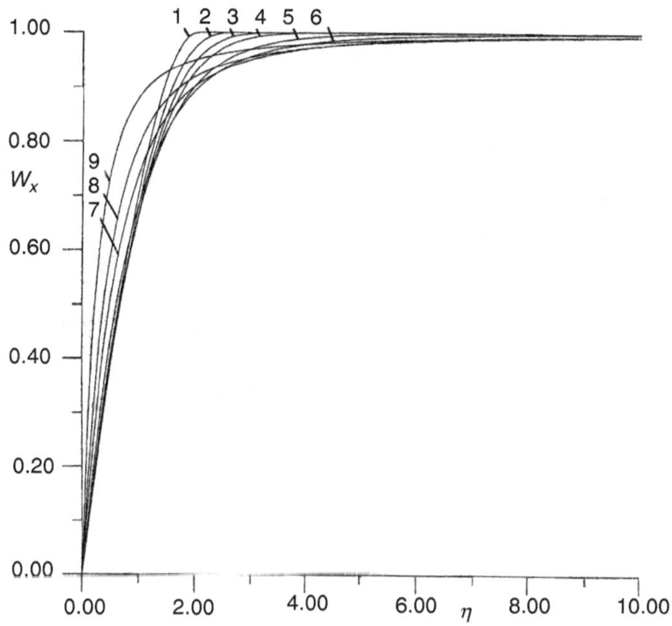

Fig. 15.3. Numerical solutions for the streamwise velocity component $W_x(\eta)$, cited from Andersson and Shang [14] (curves 1–9: $n = 2.0, 1.5, 1.2, 1.0, 0.7, 0.5, 0.3, 0.2,$ and 0.1)

Then, a Newton iteration procedure was employed to satisfy the outer boundary condition, (15.17). Concerning the numerical procedure, the present fifth-order scheme utilizes variable grid spacing.

Some of the velocity profiles $W_x(\eta)$ computed by Andersson and Shang [14] are shown in Fig. 15.3. The power-law index appears to have a substantial effect on the velocity distribution within the boundary layer and, as observed already by Andersson and Irgens [10], the most striking feature being the monotonic thinning of the boundary layer with increasing n-values. This is fully consistent with the findings for other two-dimensional plane flows, for instance the non-Newtonian analogue of the classical Blasius problem, i.e., flow past a semi-infinite that plate, which was first solved by Acrivos et al. [17] and more recently by Andersson and Toften [18].

The non-linearity of the highest-order derivative in (15.15) increases with increasing deviation of the power-law index n from unity. The thickening of the boundary layer with increasing pseudoplasticity $1 - n$, in combination with the steeper slope of the dimensionless velocity profile $W_x(\eta)$, makes the mathematical problem defined by (15.14)–(15.17) increasingly stiff. In fact, the shooting method becomes gradually less attractive as the distance (in boundary layer coordinates η) from the wall to the outer edge of the calculation domain, at which the condition $W_x(\eta) = 1$ should be satisfied. This is most likely the reason why Andersson and Irgens [10] failed to obtain converged solutions for highly pseudoplastic fluid with $n < 0.5$. In the present study, however, this difficulty was remedied by using variable grid spacing.

15.6 Local Skin-Friction Coefficient

The gradient of the dimensionless velocity $W_x(\eta)$ at the wall $\eta = 0$ is single most important characteristic of the solution. This is because the local skin-friction coefficient C_f is a dimensionless measure of the shear stress $\tau = K(\partial w_x/\partial y)^n$ at the wall, i.e.,

$$C_f \equiv \frac{\tau_{\mathrm{w}}}{\frac{1}{2}\rho w_{x,\infty}^2} = \frac{K\left[\left(\frac{\partial w_x}{\partial y}\right)_{\eta=0}\right]^n}{\frac{1}{2}\rho w_{x,\infty}^2}$$

With the dimensionless variables in (15.10)–(15.13), the earlier equation is changed into

$$\begin{aligned}C_f &= \frac{K\left[\sqrt{\frac{2g\cos\alpha}{x}}\left(\frac{\mathrm{d}W_x}{\mathrm{d}\eta}\right)_{\eta=0}(Re_x)^{1/(n+1)}\right]^n}{\frac{1}{2}\rho w_{x,\infty}^2}\\ &= 2\frac{K\left[\sqrt{\frac{2g\cos\alpha}{x}}(Re_x)^{1/(n+1)}\right]^n}{\rho w_{x,\infty}^2}\left[\left(\frac{\mathrm{d}W_x}{\mathrm{d}\eta}\right)_{\eta=0}\right]^n,\end{aligned}$$

where

$$\begin{aligned}&\frac{K\left[\sqrt{\frac{2g\cos\alpha}{x}}(Re_x)^{1/(n+1)}\right]^n}{\rho w_{x,\infty}^2}\\ &= \frac{K\left[\sqrt{\frac{2gx\cos\alpha}{x}}\right]^n}{\rho w_{x,\infty}^2}(Re_x)^{n/(n+1)}\\ &= \frac{K}{x^n(w_{x,\infty})^{2-n}\rho}(Re_x)^{n/(n+1)}\\ &= (Re_x)^{-1}(Re_x)^{n/(n+1)}\\ &= (Re_x)^{-1/(n+1)}.\end{aligned}$$

Then, with the present new dimensionless variables, the local skin-friction coefficient C_f is expressed as

$$C_f \equiv \frac{\tau_{\mathrm{w}}}{\frac{1}{2}\rho w_{x,\infty}^2} = 2Re_x^{-1/(n+1)}\left[\left(\frac{\mathrm{d}W_x}{\mathrm{d}\eta}\right)_{\eta=0}\right]^n. \quad (15.18)$$

The numerical results of the dimensionless velocity gradient at the wall $(\mathrm{d}W_x/\mathrm{d}\eta)_{\eta=0}$ are given in Table 15.2 [14] and shown in Fig. 15.4 from which it is observed that the wall gradient gradually decreases with increasing n. It is noteworthy, however, that since the local Reynolds number Re_x, as defined in (15.11), varies as $\propto x^{(n+2)/2}$, the streamwise variation of C_f becomes $C_f \propto x^{-(n+2)/(2(n+1))}$, i.e., the skin-friction coefficient decreases in the streamwise direction, irrespective of the value of the power-law index.

In order to assess the accuracy of the present numerical results, comparisons are made with the calculations by Andersson and Irgens [10] according to the relationship

$$\left(\frac{\mathrm{d}W_x}{\mathrm{d}\eta}\right)_{\eta=0} = \left(\frac{3}{4}\right)^{1/(n+1)} f''(0), \qquad (15.19)$$

Table 15.2. Computed variation of $(\mathrm{d}W_x/\mathrm{d}\eta)_{\eta=0}$ with power-law index n, cited from Andersson and Shang [14]

n	reference [14]	equation (14.19)	equation (14.20)
0.1	3.57308	–	3.6382
0.15	2.48411	–	–
0.2	1.96020	–	2.0010
0.25	1.65736	–	–
0.3	1.46275	–	1.4892
0.4	1.23218	–	–
0.5	1.10437	1.1047	1.1234
0.6	1.02613		–
0.7	0.97519	0.9753	–
1.0	0.89972	0.8997	0.9122
1.2	0.87902	0.8790	–
1.5	0.86592	0.8659	0.8749
2.0	0.86360	0.8636	0.8705

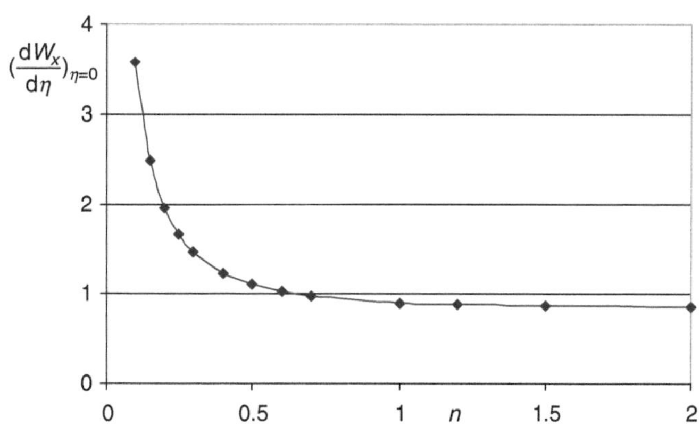

Fig. 15.4. Numerical solutions of the dimensionless velocity gradient $(\mathrm{d}W_x/\mathrm{d}\eta)_{\eta=0}$ for different values of the power-law index n

where f denotes the Falkner–Skan type function and the primes signify differentiation with respect to the similarity variable adopted in their analysis. Data obtained from the approximate interpolation formula

$$\left(\frac{\mathrm{d}W_x}{\mathrm{d}\eta}\right)_{\eta=0} = \frac{n+1/2}{n+1}\left[(f_0'')^{n+1} + \frac{(n+1)^2}{3n(n+1/2)}\right]^{1/(n+1)}, \tag{15.20}$$

derived by Acrivos et al. [18] are also included in Table 15.2. Here, f_0'' denotes the dimensionless wall shear stress for power-law boundary layer flow past a flat plate, for which data are tabulated by Acrivos et al. [18].

The comparison in Table 15.2 shows that the present numerical solutions are practically indistinguishable from the similarity solutions of Andersson and Irgens [10]. The data derived from the approximate formula given by (15.20) compares surprisingly well with the present similarity solutions and the velocity gradient at the wall is overpredicted by not more than 2% throughout the entire range of n-values considered. Here, it should be recalled that the similarity solutions could be considered as exact in the sense that they do not involve other approximations as those inherent in the boundary layer theory and the adoption of the power-law model.

15.7 Mass Flow Rate

Although the total mass flow rate within the film is constant, the partition of the mass flow rate between the viscous boundary layer and the external inviscid flow varies in the streamwise direction. As the boundary layer thickens, fluid is continuously being entrained from the freestream. Let set g_x denote the local mass flow rate entering into an element of the boundary layer of unit streamwise extent (and unit width) of a certain position x, it can be expressed as

$$g_x = \rho\left(w_{x,\delta_l}\frac{\mathrm{d}\delta_l}{\mathrm{d}x} - w_{y,\delta_l}\right), \tag{15.21}$$

where δ_l is the boundary layer thickness at the position x, and w_{x,δ_l} and w_{y,δ_l} are dimensionless velocity components in x and y directions at the edge of the boundary layer and at the position x.

Since the boundary layer thickness is given as $\delta_l = \eta_{\delta_l}(Re_x)^{-1/(n+1)}x$, then,

$$\frac{\mathrm{d}\delta_l}{\mathrm{d}x} = \eta_{\delta_l}\frac{n}{2(n+1)}(Re_x)^{\frac{-1}{(n+1)}}.$$

With (15.12) and (15.13) and the earlier equation, (15.21) can be expressed as in terms of dimensionless variables

$$\begin{aligned} g_x &= \rho\left[\sqrt{2gx\cos\alpha}W_{x,\delta_l}\eta_{\delta_l}\frac{n}{2(n+1)}(Re_x)^{-\frac{1}{(n+1)}} - \sqrt{2gx\ \cos\ \alpha}Re_x^{-1/(n+1)}W_{y,\delta_l}\right] \\ &= \rho\sqrt{2gx\cos\alpha}(Re_x)^{-1/(n+1)}\left(\frac{n\eta_{\delta_l}}{2(n+1)}W_{x,\delta_l} - W_{y,\delta_l}\right). \end{aligned} \tag{15.22}$$

Let G_x denote the total mass flow rate entering into the boundary layer for the area from the inlet $x = 0$ to a stream downstream position x and with the width of b of the plate, then, it should be the following integration:

$$\begin{aligned} G_x &= \iint\limits_A g_x \, \mathrm{d}A \\ &= b \int_0^x g_x \, \mathrm{d}x, \end{aligned} \tag{15.23}$$

where $A = b \cdot x$ is integrated area.

With (15.22), the earlier equation can be expressed in dimensionless form as

$$\begin{aligned} G_x &= b \int_0^x \left[\rho\sqrt{2gx\cos\alpha}(Re_x)^{-1/(n+1)} \left(\frac{n\eta_{\delta_1}}{2(n+1)} W_{x,\delta_1} - W_{y,\delta_1} \right) \right] \mathrm{d}x \\ &= \frac{2(n+1)}{2n+1} \left(\frac{n\eta_{\delta_1}}{2(n+1)} W_{x,\delta_1} - W_{y,\delta_1} \right) \rho w_{x,\infty} b \cdot x Re_x^{-1/(n+1)}. \end{aligned}$$

The earlier equation can be further expressed as

$$\frac{G_x Re_x^{1/(n+1)}}{\rho w_{x,\infty} b \cdot x} = \Phi \equiv \frac{2(n+1)}{2n+1} \left(\frac{n\eta_{\delta_1}}{2(n+1)} W_{x,\delta_1} - W_{y,\delta_1} \right). \tag{15.24}$$

Here, on the right-hand side, Φ defines the mass flow rate parameter, η_{δ_1} is the dimensionless boundary layer thickness, and W_{x,δ_1} and W_{y,δ_1} are dimensionless velocity components in x and y directions at the edge of the boundary layer. Since the dimensionless boundary layer thickness η_{δ_1} and the velocity components W_{x,δ_1} and W_{y,δ_1} may depend on the power-law index n, Φ turns out to be a function of n alone. The most frequently used definition of η_{δ_1} is the value of η_{δ_1} for which the dimensionless velocity component W_{x,δ_1} in Fig. 15.3 becomes equal to 0.99. Data for η_{δ_1}, W_{x,δ_1} and W_{y,δ_1} obtained from the numerical similarity solutions presented are shown in Fig. 15.5. The resulting variation of the mass flow rate parameter Φ is also included and it is observed that Φ is a monotonically decreasing function of the power-law index n.

To facilitate rapid estimate of the mass flow rate parameter Φ for any value of the power-law index in the interval $0.2 \le n \le 2$, accurate curve-fit formulas for η_{δ_1} and W_{y,δ_1}

$$\begin{aligned} \eta_{\delta_1} = {} & 4.9505 - 7.617(n-0.54) + 11.214(n-0.54)^2 + 8.703(n-0.54)^3 \\ & -0.37(n-0.54)^4 \ (0.2 \le n \le 1), \end{aligned} \tag{15.25}$$

$$\begin{aligned} \eta_{\delta_1} = {} & 2.3201 - 1.0623(n-1.425) + 0.9962(n-1.425)^2 - 0.7533(n-1.425)^3 \\ & (1 \le n \le 2), \end{aligned} \tag{15.26}$$

$$\begin{aligned} W_{y,\delta_1} = {} & -1.8675 + 3.9616(n-0.54) - 6.022(n-0.54)^2 - 3.22(n-0.54)^3 \\ & +16.946(n-0.54)^4 \ (0.2 \le n \le 1), \end{aligned} \tag{15.27}$$

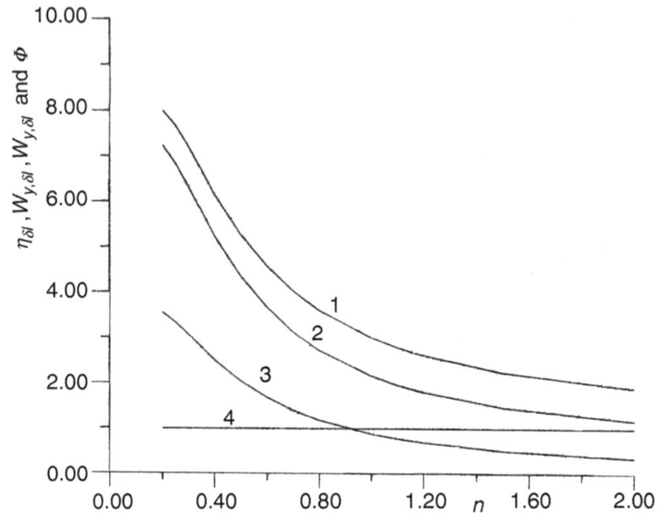

Fig. 15.5. Predicted results for mass flow rate parameter Φ and the related dimensionless variables η_{δ_l}, W_{x,δ_l}, and W_{y,δ_l} (1. η_{δ_l}, 2. Φ, 3. $-W_{y,\delta_l}$ and 4. W_{x,δ_l}), cited from Andersson and Shang [14]

$$W_{y,\delta_l} = -0.53954 + 0.5002(n-1.425) - 0.5078(n-1.425)^2 \\ +0.3946(n-1.425)^3 \ (1 \le n \le 2), \tag{15.28}$$

are shown in (15.25) - (15.28) [14] and can be used with $W_{x,\delta_l} = 1$ in (15.24). This curve-fit method turns out to be accurate to within 0.01%.

15.8 Length of Boundary Layer Region

Let us now denote the total flow rate within the film, ρV, where V is the volumetric flow rate of the falling film flow. Since the viscous boundary layer develops from $x = 0$, i.e., $\delta_l(0) = 0$, the entire mass flow is initially carried by the freestream. At a certain streamwise position $x = x_0$, on the other hand, the boundary layer extends all the way to the free surface of the film and the total mass flux is within the boundary layer, i.e.,

$$G_{x0} \equiv G_x(x_0) = \rho V. \tag{15.29}$$

This criterion, in combination with (15.24), can be rearranged to give the explicit relation as follows:

$$\frac{G_{x0} Re_{x0}^{1/(n+1)}}{\rho w_{x0,\infty} b \cdot x_0} = \Phi,$$

or

$$\frac{V Re_{x_0}^{1/(n+1)}}{w_{x_0,\infty} b \cdot x_0} = \Phi,$$

i.e.,

$$\frac{V \left[\frac{x^n (\sqrt{2 g x_0 \cos\alpha})^{2-n} \rho}{K}\right]^{1/(n+1)}}{b x_0 \sqrt{2 g x_0 \cos\alpha}} = \Phi,$$

or

$$\frac{V \left[\frac{(\sqrt{2 g \cos\alpha})^{2-n} \rho}{K}\right]^{1/(n+1)}}{b \sqrt{2 g \cos\alpha_0}} x_0^{-(2n+1)/(2(n+1))} = \Phi.$$

The earlier equation is transformed to

$$x_0 = \left\{ \frac{V}{\Phi b} \frac{\left[\frac{(\sqrt{2 g \cos\alpha})^{2-n}}{K} \rho\right]^{1/(n+1)}}{\sqrt{2 g \cos\alpha_0}} \right\}^{(2(n+1))/(2n+1)}$$

i.e.,

$$x_0 = \left\{ \left(\frac{V}{\Phi b}\right)^{n+1} \frac{\frac{(\sqrt{2 g \cos\alpha})^{2-n} \rho}{K}}{(\sqrt{2 g \cos\alpha})^{n+1}} \right\}^{2/(2n+1)},$$

or

$$x_0 = \left[\left(\frac{V}{\Phi b}\right)^{n+1} \frac{(2 g \cos\alpha)^{(1-2n)/2} \rho}{K} \right]^{2/(2n+1)}, \quad (15.30)$$

for the particular streamwise position x_0. Since the film inlet is at x = 0, cf. Fig. 15.1, the characteristic coordinate value x_0 defines the streamwise length of the boundary layer region.

15.9 Critical Film Thickness

When the boundary layer extends all the way to the free surface and the freestream disappears at $x = x_0$, the film thickness equals the boundary thickness $\delta_l(x_0)$. The latter can be obtained from the definition, (15.10), of the similarity variable at the outer edge of the viscous boundary layer, i.e., $y = \delta_l(x_0)$ for $\eta = \eta_{\delta_l}$ and $x = x_0$, and expressed as

$$\delta_l(x_0) = \eta_{\delta_l} x_0 Re_{x_0}^{-1/(n+1)}.$$

With the definition of local Reynolds number in (15.11), the earlier equation is changed into

$$\delta_l(x_0) = \eta_{\delta_l} x_0^{n/(2(n+1))} \left[\frac{(2 g \cos\alpha)^{(2-n)/2}}{K/\rho}\right]^{-1/(n+1)}. \quad (15.31)$$

It can therefore be concluded that both x_0 and $\delta_l(x_0)$ are completely determined as long as the problem characteristics n, K/ρ, Q, and $g\cos\alpha$ are known, along with the solution of the transformed problem, (15.14)–(15.17) or (15.24)–(15.28), which determines Φ.

The film thickness at the particular position $x = x_0$ is a critical quantity in film flow analysis since the boundary layer concept is applicable only for range $x \leq x_0$, and in this range the local film thickness $\delta_l(x_0)$ at $x = x_0$ is largest. Following (15.31) the boundary layer thickness $\delta_l(x)$ at any position of x in this range can be evaluated by the following equation:

$$\delta_l(x) = \eta_{\delta_l} x^{n/(2(n+1))} \left[\frac{(2g\cos\alpha)^{(2-n)/2}}{K/\rho}\right]^{-1/(n+1)} . \tag{15.32}$$

15.10 Effect of Wall Inclination

It is noteworthy that the angle of inclination α does not appear in the transformed problem defined by (15.14)–(15.17). Any solution W_x and W_y is accordingly independent of α but, nevertheless, valid for all inclinations $0 \leq \alpha \leq \pi/2$. Physically relevant quantities, on the other hand, do depend on α due to the similarity transformation, (15.10)–(15.13). For a given quantity, say p, the relationship

$$\frac{P_i}{P_v} = \left(\frac{\cos\alpha_i}{\cos\alpha_v}\right)^{\gamma} = \cos^{\gamma}\alpha_i, \tag{15.33}$$

between the inclined and vertical cases, identified by subscripts i and v, respectively, holds. Here, α_v, denotes the angle of inclination in the vertical case, i.e., $\alpha_v = 0$ and $\cos\alpha_v = 1$, and the exponent γ is derived as after:

For w_x

From (15.12) we can get the following equation

$$\frac{(w_x)_i}{(w_x)_v} = \left(\frac{\cos\alpha_i}{\cos\alpha_v}\right)^{1/2} = \cos^{1/2}\alpha_i.$$

Then, $\gamma = 1/2$

For w_y

From (15.13) we can obtain the following equation:

$$\frac{(w_y)_i}{(w_y)_v} = \left(\frac{\cos\alpha_i}{\cos\alpha_v}\right)^{1/2}\left(\frac{(Re_x)_i}{(Re_x)_v}\right)^{-1/(n+1)},$$

where with (15.9), we have

$$\left(\frac{(Re_x)_i}{(Re_x)_v}\right)^{-1/(n+1)} = \left(\frac{(w_{x,\infty})_i}{(w_{x,\infty})_v}\right)^{-(2-n)/(n+1)} = \left(\frac{\cos\alpha_i}{\cos\alpha_v}\right)^{-(2-n)/(2(n+1))}.$$

Table 15.3. Relationship between inclined and vertical film flow

P	w_x	w_y	C_f	x_0	$\delta_{\mathrm{l}}(x_0)$
γ	$\frac{1}{2}$	$\frac{2n-1}{2(n+1)}$	$\frac{n-2}{2(n+1)}$	$\frac{1-2n}{2n+1}$	$-\frac{1}{2n+1}$

Hence,

$$\frac{(w_y)_{\mathrm{i}}}{(w_y)_{\mathrm{v}}} = \left(\frac{\cos\alpha_{\mathrm{i}}}{\cos\alpha_{\mathrm{v}}}\right)^{1/2}\left(\frac{\cos\alpha_{\mathrm{i}}}{\cos\alpha_{\mathrm{v}}}\right)^{-(2-n)/(2(n+1))} = \left(\frac{\cos\alpha_{\mathrm{i}}}{\cos\alpha_{\mathrm{v}}}\right)^{(2n-1)/(2(n+1))}.$$

Then, $\gamma = \frac{2n-1}{2(n+1)}$.

For C_f

From (15.18) we can do the following derivation:

$$\frac{(C_f)_{\mathrm{i}}}{(C_f)_{\mathrm{v}}} = \left(\frac{(Re_x)_{\mathrm{i}}}{(Re_x)_{\mathrm{v}}}\right)^{-1/(n+1)} = \left(\frac{\cos\alpha_{\mathrm{i}}}{\cos\alpha_{\mathrm{v}}}\right)^{-(2-n)/(2(n+1))}.$$

Then, $\gamma = (n-2)/(2(n+1))$.

For x_0

From (15.30) we can do the following derivation:

$$\frac{(x_0)_{\mathrm{i}}}{(x_0)_{\mathrm{v}}} == \left(\frac{\cos\alpha_{\mathrm{i}}}{\cos\alpha_{\mathrm{v}}}\right)^{(1-2n)/(2n+1).}$$

Then, $\gamma = (1-2n)/(2n+1)$

For $\delta_{\mathrm{l}}(x_0)$

From (15.31) we can do the following derivation:

$$\begin{aligned}\frac{(\delta_{\mathrm{l}}(x_0))_{\mathrm{i}}}{(\delta_{\mathrm{l}}(x_0))_{\mathrm{v}}} &= \left(\frac{(x_0)_{\mathrm{i}}}{(x_0)_{\mathrm{v}}}\right)^{n/(2(n+1))}\left(\frac{\cos\alpha_{\mathrm{i}}}{\cos\alpha_{\mathrm{v}}}\right)^{-(2-n)/(2(n+1))} \\ &= \left(\frac{\cos\alpha_{\mathrm{i}}}{\cos\alpha_{\mathrm{v}}}\right)^{(n(1-2n))/(2(n+1)(2n+1))}\left(\frac{\cos\alpha_{\mathrm{i}}}{\cos\alpha_{\mathrm{v}}}\right)^{-(2-n)/(2(n+1))} \\ &= \left(\frac{\cos\alpha_{\mathrm{i}}}{\cos\alpha_{\mathrm{v}}}\right)^{-1/(2n+1)}.\end{aligned}$$

Then, $\gamma = -1/(2n+1)$.

For summary, exponent γ is provided in Table 15.3 for some quantities of particular interest.

The exponent γ in (15.32) depends on the physical quantity P under consideration

15.11 Summary

So far, we have presented our recent developments on hydrodynamics of falling film flow of non-Newtonian fluids. The related equations of hydrodynamics can be summarized in Table 15.4.

Table 15.4. Summary of the related equations of hydrodynamics of falling film flow of non-Newtonian fluids

term	equations
	governing partial differential equations
mass equation	$\frac{\partial w_x}{\partial x} + \frac{\partial w_y}{\partial y} = 0$
momentum equation	$w_x \frac{\partial w_x}{\partial x} + w_y \frac{\partial w_x}{\partial y} = g\cos\alpha + n\frac{K}{\rho}\left(\frac{\partial w_x}{\partial y}\right)^{n-1} \frac{\partial^2 w_x}{\partial y^2}$
boundary conditions	$y = 0: \quad w_x = 0, \quad w_y = 0$ $y = \delta_l: \quad w_x = w_{x,\infty}$
	defined transformation variables
η	$\frac{y}{x} Re_x^{1/(n+1)}$
Re_x	$\frac{x^n (w_{x,\infty})^{2-n} \rho}{K}$
$W_x(\eta)$	$\frac{w_x}{\sqrt{2gx\cos\alpha}}$
$W_y(\eta)$	$\frac{w_y}{\sqrt{2gx\cos\alpha}} Re_x^{1/(n+1)}$
	governing ordinary differential equations
dimensionless mass equation	$W_x - \frac{n}{n+1}\eta = \frac{dW_x}{d\eta} + 2\frac{dW_y}{d\eta} = 0$
dimensionless momentum equation	$W_x = \left(-\frac{n}{(n+1)}\eta\frac{dW_x}{d\eta} + W_x\right) + 2W_y\frac{dW_x}{d\eta} = 1 + 2n\left(\frac{dW_x}{d\eta}\right)^{(n-1)} \frac{d^2W_x}{d\eta^2}$
boundary conditions	$\eta = 0: \quad W_x(\eta) = 0, \quad W_y(\eta) = 0,$ $\eta = \eta_{\delta_l}: \quad W_x(\eta) = 1$
$w_{x,\infty}$	$\sqrt{2gx\cos\alpha}$

x_0	$\left[\left(\frac{V}{b\Phi}\right)^{n+1}\frac{(2g\cos\alpha)^{(1-2n)/2}}{K/\rho}\right]^{2/(2n+1)}$
η_{δ_1}	$4.9505-7.617(n-0.54)+11.214(n-0.54)^2+8.703(n-0.54)^3-0.37(n-0.54)^4$ $(0.2\le n\le 1)$
	$2.3201-1.0623(n-0.54)+0.9962(n-0.54)^2-0.7533(n-0.54)^3$ $(1\le n\le 2)$
W_{y,δ_1}	$-1.8675+3.9616(n-0.54)-6.022(n-0.54)^2-3.22(n-0.54)^3+16.946(n-0.54)^4$ $(0.2\le n\le 1)$
	$-0.53954+0.5002(n-0.54)-0.5078(n-0.54)^2+0.3946(n-0.54)^3$ $(0.2\le n\le 1)$
W_{x,δ_1}	1
C_f	$2Re_x^{-1/(n+1)}\left[\left(\frac{dW_x}{d\eta}\right)_{\eta=0}\right]^n$
Φ (defined as $\frac{G_x Re_x^{1/(n+1)}}{\rho w_{x,\infty} b\cdot x}$)	$\frac{2(n+1)}{2n+1}\left(\frac{n\eta_{\delta_1}}{2(n+1)}W_{x,\delta_1}-W_{y,\delta_1}\right)$
$\delta_1(x_0)$	$\eta_{\delta_1}x_0^{n/(2(n+1))}\left[\frac{(2g\cos\alpha)^{(2-n)/2}}{K/\rho}\right]^{-1/(n+1)}$

15.12 Remarks

In this chapter a new similarity transformation [14] has been used to study the gravity-driven flow of a non-Newtonian liquid film along inclined surface. The partial differential equations governing the hydrodynamics of the power-law fluid transform exactly into a set of two ordinary differential equations, which can be calculated numerically to an arbitrary degree of accuracy. The results are practically indistinguishable from those of Andersson and Irgens [10] in the parameter range $0.5 \leq n \leq 2$. With the present approach, however, calculations could be accomplished also for highly pseudoplastic liquids and the numerical results compared accurately with results deduced from an approximate interpolation formula due to Acrivos et al. [18] throughout the entire parameter range $0.1 \leq n \leq 2$. The nonlinearity of the momentum boundary layer problem for power-law fluid increases with increasing pseudoplasticity $1 - n$ and the variable grid spacing is therefore increasingly important for small n-values.

It is noteworthy that the resulting system of dimensionless ordinary differential equations provided in [14] depends only on the single parameter n. Furthermore, all other parameters, like the streamwise location x, the fluid properties K/ρ, and the component of the gravitational acceleration along the surface $g\cos\alpha$ have been combined into a generalized local Reynolds number Re_x and dimensionless velocity W_x and W_y. Various flow characteristics can thus be expressed only in term of n and Re_x, except the particular position x_0 at which the entire freestream has been entrained into the momentum boundary layer. In order to determine x_0 and the associated critical film thickness $\delta_l(x_0)$, knowledge about the total mass flow rate ρQ within the film is also required, together with the new dimensionless mass flux parameter Φ. The latter quantity, which depends on the dimensionless boundary layer thickness η_{δ_l} and the velocity components W_{x,δ_l} and W_{y,δ_l} at the edge of the boundary layer, is generally obtained as a part of the numerical solution of the transformed problem and turned out to be function only of the power-law index n. However, to facilitate rapid and accurate estimate of Φ, polynomial curve-fit formulas have been developed on the basis of the rigorous similarity solutions.

It should be indicated that, except a few works, such as of Andersson and Irgens [10, 11], Andersson and Shang [14], and Shang and Gu [15], in the most of current studies on hydrodynamics of FFNF system the hydraulic entrance region (i.e. the boundary layer region) was ignored in their analysis of modeling and simulation. With ignoring the existing boundary layer region of the FFNF system, it would be never possible to capture its adaptive remodeling process of hydrodynamics for conducting correct calculation on the hydrodynamic characteristics of FFNF system.

15.13 Calculation Example

Example A non-Newtonian power-law fluid having a density of 1,041 kg m^{-3} is flowing with volumetric flow rate of 0.02 m^3 s^{-1} along an inclined flat plate with angle of $\alpha = 30°$ and width of $b = 1$ m. The properties of the fluid are $K = 2.744\,\text{kg}\,(\text{s}^{\text{n}-2}\text{m}^{-1})$ and $n = 0.50$. Please calculate the followings:

(a) Length x_0 of the boundary layer region
(b) Critical film thickness $\delta_1(x_0)$
(c) Local skin-friction coefficient C_f at x_0
(d) w_{x0,δ_1} and w_{y_o,δ_1} corresponding to position x_o
(e) If the plate inclined angle is 0° (for vertical plate), calculate $x_0, \delta_1(x_0), C_f$, w_{x0,δ_1}, and w_{y0,δ_1}

Solution. The given data are as follows: volumetric flow rate $V = 0.02\,\text{m}^3\,\text{s}^{-1}$, density $\rho = 1041$ kg m^{-3}, plate angle $\alpha = 30°$ and width $b = 1$ m, coefficient of consistency $K = 2.744\,\text{kg}\,(\text{s}^{\text{n}-2}\text{m}^{-1})$ and $n = 0.50$.

(a) calculation of x_0 for $\alpha = 30°$. With (15.30), x_0 is evaluated as

$$x_0 = \left[\left(\frac{V}{b\Phi}\right)^{n+1} \frac{(2g\cos\alpha)^{(1-2n)/2}}{K/\rho}\right]^{2/(2n+1)}.$$

While, from (15.24) the mass flow rate parameter Φ can be evaluated as

$$\Phi = \frac{2(n+1)}{2n+1}\left(\frac{n\eta_{\delta_1}}{2(n+1)}W_{x,\delta_1} - W_{y,\delta_1}\right).$$

For $n = 0.5$, the boundary layer thickness η_{δ_1} and the velocity component W_{y,δ_1} at the edge of the boundary layer can be evaluated as:

$$\begin{aligned}
\eta_{\delta_1} &= 4.9505 - 7.617(n-0.54) + 11.214(n-0.54)^2 \\
&\quad +8.703(n-0.54)^3 - 0.37(n-0.54)^4 \\
&= 4.9505 - 7.617(0.5-0.54) + 11.214(0.5-0.54)^2 \\
&\quad +8.703(0.5-0.54)^3 - 0.37(0.5-0.54)^4 \\
&= 5.27256. \\
W_{y,\delta_1} &= -1.8675 + 3.9616(n-0.54) - 6.022(n-0.54)^2 \\
&\quad -3.22(n-0.54)^3 + 16.946(n-0.54)^4 \\
&= -1.8675 + 3.9616(0.5-0.54) - 6.022(0.5-0.54)^2 \\
&\quad -3.22(0.5-0.54)^3 + 16.946(0.5-0.54)^4 \\
&= -2.03535. \\
\Phi &= \frac{2(n+1)}{2n+1}\left(\frac{n\eta_{\delta_1}}{2(n+1)}W_{x,\delta_1} - W_{y,\delta_1}\right) \\
&= \frac{2\times(0.5+1)}{2\times 0.5+1}\left(\frac{0.5\times 5.27256}{2\times(0.5+1)}\times 0.99 + 2.03535\right) \\
&= 4.358.
\end{aligned}$$

(Note: W_{x,δ_1} is defined to be 0.99 in the earlier equation of ϕ)

Then,

$$x_0 = \left[\left(\frac{V}{b\Phi}\right)^{n+1} \frac{(2g\cos\alpha)^{(1-2n)/2}}{K/\rho}\right]^{2/(2n+1)}$$

$$= \left[\left(\frac{0.02}{4.358}\right)^{0.5+1} \frac{(2\times 9.8\times \cos\ 30^{\circ})^{(1-2\times 0.5)/2}}{(2.744/1041)}\right]^{2/(2\times 0.5+1)}$$

$$= 0.118\text{m}.$$

(b) For calculation of $\delta_1(x_0)$ *for* $\alpha = 30^{\circ}$. With (15.31), $\delta_1(x_0)$ can be calculated as

$$\delta_1(x_0) = \eta_{\delta_1} x_0^{n/(2(n+1))} \left[\frac{(2g\cos\alpha)^{(2-n)/2}}{K/\rho}\right]^{-1/(n+1)}$$

$$= 5.2726 \times 0.118^{0.5/(2\times(0.5+1))} \left[\frac{(2\times 9.8\times\cos 30^{\circ})^{(2-0.5)/2}}{2.744/1041}\right]^{-1/(0.5+1)}$$

$$= 0.0171\,\text{m}.$$

(c) For calculation of C_f *related to* x_0 *and for* $\alpha = 30^{\circ}$. From (15.18), C_f *related to* x_0 *and for* $\alpha = 30^{\circ}$ can be expressed as

$$C_f = 2Re_{x_0}^{-1/(n+1)} \left[\left(\frac{\mathrm{d}W_x}{\mathrm{d}\eta}\right)_{\eta=0}\right]^n.$$

From Table 15.4, the local Reynolds number at x_0 can be evaluated as

$$Re_{x_0} = \frac{x_0^n (w_{x0,\infty})^{2-n}\rho}{K}$$

$$= \frac{x_0^n (2gx_0\cos\alpha)^{(2-n)/2}\rho}{K}$$

$$= \frac{0.118^{0.5}\times(2\times 9.8\times 0.118\times\cos\ 30^{\circ})^{(2-n)/2}\times 1041}{2.744}$$

$$= 219.4.$$

From Table 15.2 we get

$$\left(\frac{\mathrm{d}W_x}{\mathrm{d}\eta}\right)_{\eta=0} = 1.10437 \text{ at } n = 0.5$$

Then,

$$C_f = 2Re_{x_0}^{-1/(n+1)} \left[\left(\frac{\mathrm{d}W_x}{\mathrm{d}\eta}\right)_{\eta=0}\right]^n$$

$$= 2\times 219.4^{-1/(0.5+1)}[1.10437]^{0.5}$$

$$= 0.0578.$$

(d) For calculation of w_{x0,δ_l} and w_{y_o,δ_l} for $\alpha = 30^o$. From (15.12) we have

$$W_{x0,\delta_l} = \frac{w_{x0,\delta_l}}{\sqrt{2gx\cos\alpha}}$$

i.e.,

$$\begin{aligned} w_{x0,\delta_l} &= \sqrt{2gx_0\cos\alpha}W_{x,\delta_l} \\ &= \sqrt{2\times 9.8\times 0.118\times \cos 30^o} \\ &= 1.415\,\mathrm{m\ s^{-1}}(W_{x,\delta_l}=1). \end{aligned}$$

From (15.13) we have

$$W_{y,\delta_l} = \frac{w_{y_o,\delta_l}}{\sqrt{2gx_o\cos\alpha}} Re_{x_o}^{1/(n+1)}$$

i.e.,

$$\begin{aligned} w_{y_o,\delta_l} &= \sqrt{2gx_0\cos\alpha}Re_{x0}^{-1/(n+1)}W_{y,\delta_l} \\ &= \sqrt{2\times 9.8\times 0.118\times \cos 30^o}\times 219.4^{-1/(0.5+1)}\times(-2.0354) \\ &= -0.079\,\mathrm{m\ s^{-1}}. \end{aligned}$$

(e) For calculation of $x_0, \delta_l(x_0), C_f, w_{x0,\delta_l}$ for vertical plate case. According to (15.33) and Table 15.3, we have the following expression

$$(x_0)_\mathrm{v} = (x_0)_\mathrm{i}/cos^\gamma\alpha,$$

where

For $x_0, \gamma = (1-2n)/(2n+1) = (1-2\times 0.5)/(2\times 0.5+1) = 0$, then,

$$\begin{aligned} (x_0)_\mathrm{v} &= 0.118/\cos^\circ 30^\circ \\ &= 0.118\,\mathrm{m}. \end{aligned}$$

For $\delta_l(x_0), \gamma = -1/(2n+1) = -1/(2\times 0.5+1) = -0.5$, then,

$$\begin{aligned} (\delta_l(x_0))_\mathrm{v} &= (\delta_l(x_0))_\mathrm{i}/\cos^\gamma\alpha \\ &= 0.0171/\cos^{-0.5}(30^o) \\ &= 0.0159\,\mathrm{m}. \end{aligned}$$

For $C_f, \gamma = (n-2)/(2(n+1)) = (0.5-2)/(2\times(0.5+1)) = -0.5$, then

$$\begin{aligned} (C_f)_\mathrm{v} &= (C_f)_\mathrm{i}/\cos^\gamma\alpha \\ &= 0.0578/\cos^{-0.5}(30^o) \\ &= 0.05379. \end{aligned}$$

For w_{x_0,δ_1}, $\gamma = 1/2$, then

$$\begin{aligned}(w_{x_0,\delta_1})_\mathrm{v} &= (w_{x_0,\delta_1})_\mathrm{i}/\cos^\gamma \alpha \\ &= 1.415/\cos^{0.5}(30^\circ) \\ &= 1.5205\,\mathrm{m\ s^{-1}}\end{aligned}$$

For w_{y_0,δ_1}, $\gamma = (2n-1)/(2(n+1)) = (2 \times 0.5 - 1)/(2 \times (0.5 + 1)) = 0$, then

$$\begin{aligned}(w_{y_0,\delta_1})_\mathrm{v} &= (w_{y_0,\delta_1})_\mathrm{i}/\cos^\gamma \alpha \\ &= -0.079/\cos^\circ \alpha \\ &= -0.079.\end{aligned}$$

References

1. G. Astarita, G. Marrucci, and G. Palumbo, Non-Newtonian gravity flow along inclined plane surface, Ind. Eng. Chem. Fundam. 3, pp. 333–339, 1964
2. N. Therien, B. Coupal, and J.L. Corneille, Verification experimentale de l'epaisseur du film pour des liquides non-Newtoniens s'ecoulant par gravite sur un plan incline, Can. J. Chem. Eng., 48, pp.17–20, 1970
3. N.D. Sylvester, J.S. Tyler, and A.H.P. Skelland, Non–Newtonian film fluids: theory and experiment, Can. J. Chem. Eng., 51, pp. 418–429, 1973
4. T.M.T. Yang and D.W. Yarbrough, A numerical study of the laminar flow of non-Newtonian fluids along a vertical wall, ASME J. App, Mech., 40, pp. 290–292, 1973
5. T.M.T. Yang and D.W. Yarbrough, Laminar flow of non-Newtonian liquid films inside a vertical pipe, Rheol. Acta, 19, pp. 432–436, 1980
6. Murthy V. Narayana and P.K. Sarma, A note on hydrodynamics entrance length of non-Newtonian laminar falling films, Chem. Eng. Ser. 32, pp. 566–567, 1977
7. Murthy V. Narayana, and P.K. Sarma, Dynamics of developing laminar non-Newtonian falling liquid films with free surface. ASME J. Appl. Mech., 45, pp. 19–24, 1978
8. M.N. Tekic, D. Posarac, and D. Petrovic, A note on the entrance region lengths of non-Newtonian laminar falling films, Chem. Sci. 41, pp. 3230–3232, 1986
9. H.I. Andersson and F. Irgens, Hydrodynamic entrance length of non-Newtonian liquid films, Chem. Eng. Sci., 45, pp. 537–541, 1990
10. H.I. Andersson and F. Irgens, Gravity-driven laminar film flow of power-law fluids along vertical walls, J. Non-Newtonian Fluid Mech. 27, pp. 153–172, 1988
11. H.I. Andersson and F. Irgens, Film flow of power law fluids, Encyclopaedia of Fluid Mechanics, Gulf Publishing Company, Houston, T, 9, pp. 617–648, 1990
12. G. Astarita, Mass transfer from a flat solid surface to a falling non-Newtonian liquid film, Ind. Eng. Chem. Fundamentals, 5, pp. 14–18, 1966
13. R.A. Mashelker and V.V. Chavan, Solid dissolution in falling films of non-Newtonian liquids, J. Chem. Jpn, 6, pp. 160–167, 1973

14. H.I. Andersson and D.Y. Shang, An extended study of hydrodynamics of gravity-driven film flow of power-law fluids, Fluid Dyn. Res., 22, pp. 345–357, 1998
15. D.Y. Shang and J. Gu, Analyses of pseudo-similarity and boundary layer thickness for non-Newtonian falling film flow, Heat Mass Transfer, Vol. 41, No.1, pp. 44–50, 2004
16. J. Wu and M.C. Thompson, 1996, Non-Newtonian shear-thinning flow past a flat plate, J. Non-Newtonian Fluid Mech., 66, pp. 127–144, 1998
17. A. Acrivos, M.J. Shah, and E.E. Peterson, Momentum and heat transfer in lamiar boundary-flows of non-Newtonian fluids past external surfaces, AICHE J. 6, pp. 312–317, 1960
18. H.I. Andersson and T.H. Toften, Numerical solution of the laminar boundary layer equations for power-law fluids, J. Non-Newtonian Fluid Mech. 32, pp. 175–195, 1989
19. A. Acrivos, M.J. Shah, and E.E. Peterson, On the solution of the two-dimensional boundary-flow equations for a non-Newtonian power-law fluid, Chem. Eng. Sci., 20, pp. 101–105, fluids past external surfaces, AICHE J. 6, pp. 312–317, 1965

16

Pseudosimilarity and Boundary Layer Thickness for Non-Newtonian Falling Film Flow

Nomenclature

a	thermal diffusive coefficient, $\frac{\lambda}{\rho c_p}$, $m^2\ s^{-1}$
c_p	specific heat at constant pressure, J $(kg\ K)^{-1}$
g	gravitation acceleration, m $(s^2)^{-1}$
$g(\eta,\zeta)$	defined temperature gradient, $\dfrac{\partial\theta(\eta,\zeta)}{\partial\zeta}$
K	coefficient of consistency, kg $s^{n-2}\ m^{-1}$
n	power law index
Pr_x	local Prandtl number, $\dfrac{xw_{x,\infty}}{a}Re_x^{-2/(n+1)}$
Pr_x^*	critical Prandtl number
Re_x	local Reynolds number, $\dfrac{x^n(w_{x,\infty})^{2-n}\rho}{K}$
t	temperature, °C
w_x, w_y	velocity components in the x- and y- directions, respectively, m s^{-1}
$W_x(\eta), W_y(\eta)$	dimensionless velocity components in the x- and y-directions, respectively
$w_{x,\infty}$	velocity of the fluid outside the boundary layer, m s^{-1}
x, y	streamwise and cross-stream coordinates, m
x_0	length of the boundary layer region, m

Greek symbols

η	dimensionless coordinate variable for boundary layer, $(y/x)Re_x^{1/(n+1)}$
η_{δ_l}	dimensionless momentum boundary layer thickness
η_{δ_t}	dimensionless thermal boundary layer thickness
$\eta_{\delta_l}*$	dimensionless critical boundary layer thickness

ζ	$\dfrac{x}{x_0}$
$\theta(\eta,\zeta)$	dimensionless temperature, $(t - t_\infty)/(t_\mathrm{w} - t_\infty)$
ρ	density, kg m^{-3}

Subscripts

i	inclined case
l	momentum boundary layer
v	vertical case or vapor
x	local value
w	at wall
α (or y)	angle of inclination
∞	far from the wall surface

16.1 Introduction

Efficient heating or cooling of liquids can be achieved by allowing the fluid to flow in a thin film along a solid surface kept at a constant temperature. While the hydrodynamics of thin film flow of Newtonian liquids has been extensively studied for several decades, only modest attention has been devoted to gravity-driven films of non-Newtonian liquids.

Heat transfer from a constant temperature wall to hydrodynamically fully developed power-law films was probably first considered by Yih and Lee [1], while the corresponding mass transfer problem (i.e., solid dissolution from the wall and diffusion into the film) has been studied by Astarita [2] and Mashelker and Chavan [3]. For the effect of injection/suction on the heat transfer, so far there has been study by Pop et al. [4] on the steady laminar gravity-driven film flow along a vertical wall for Newtonian fluids, which is based on Falkner–Skan type transformation. A mathematical model for heat transfer of non-Newtonian falling film flow was dealt with by Ouldhadda and Idrissi [5] on a horizontal circular cylinder. Meanwhile, Rao [6] measured experimentally the heat transfer in a fully developed non-Newtonian film flow falling down a vertical tube.

However, most of the previous theoretical models only solved the similarity momentum problem for the boundary layer region. The solution for the similarity thermal boundary layer encounters formidable difficulties when the boundary layer thickness of momentum and temperature differ significantly, which is a characteristic of non-Newtonian power-law fluids. The solution of similarity momentum boundary layer cannot be successfully applied for solution of the heat transfer in boundary layer region, since there is no similarity solution for the energy equation related to non-Newtonian power-law fluids. The determination of exact thermal boundary layer thickness is very important, otherwise, the hydrodynamics and heat transfer analyses for the boundary layer region, fully viscous region and the developed flow region will not produce reliable results.

Therefore, a reliable and convenient treatment approach of local non-similarity of thermal boundary layer is very important for solution for falling-film flow of non-Newtonian power-law fluids (FFNF). For this purpose, on the basis of [7], Shang and Andersson [8] and Shang and Gu [9] focused on an extensive study for a systematic solution on local nonsimilarity of thermal boundary layer for falling-film flow of non-Newtonian power-law fluids using a pseudosimilarity approach. They provided the similarity approach for the rigorous solution of heat transfer coefficient related to the nonsimilarity thermal boundary layer of the non-Newtonian power-law fluids for the boundary layer region.

In this chapter, the focus is on presentation of our recent developments on analyses of pseudosimilarity, and boundary layer thickness for non-Newtonian falling film flow. To this end, a mathematical model for thermal boundary layer in an accelerating liquid film of non-Newtonian power-law fluids is presented. A pseudosimilarity transformation method is induced for the thermal boundary layer equation. On this basis, the heat transfer problem can be solved by means of a local nonsimilarity approach with n and the induced local Prandtl number Pr_x being the only parameters. Based on the newly defined "local Prandtl number," proposed by Shang and Andersson [8], the dependence of the thickness both of the momentum boundary layer and thermal boundary layer is discussed. It is found that the momentum boundary layer thickness decreases monotonically with increasing power-law index; while the thermal boundary layer thickness keeps almost the same with variation of power-law index but decreases significantly with the increase of the "local Prandtl number." It shows that the adopted pseudosimilarity approach is capable of solving the problem of nonsimilarity thermal boundary layer in the falling film of a non-Newtonian power-law fluid. Meanwhile, a critical local Prandtl number Pr_x^* is introduced, which is a monotonically increasing function of n.

16.2 Physical Model and Governing Partial Differential Equations

Consider the accelerating laminar flow in the boundary layer region of a non-Newtonian liquid film down along an inclined plane surface, as shown schematically in Fig. 16.1. The incompressible and inelastic fluid is assumed to obey the Ostwald-de-Waele power-law model and the action of viscous stresses is confined to the solid surface. The basis boundary layer equation for mass, momentum, and thermal energy are:

$$\frac{\partial w_x}{\partial x} + \frac{\partial w_y}{\partial y} = 0, \tag{16.1}$$

$$w_x \frac{\partial w_x}{\partial x} + w_y \frac{\partial w_x}{\partial y} = g \cos \alpha + n \frac{K}{\rho} \left(\frac{\partial w_x}{\partial y} \right)^{n-1} \frac{\partial^2 w_x}{\partial y^2}, \tag{16.2}$$

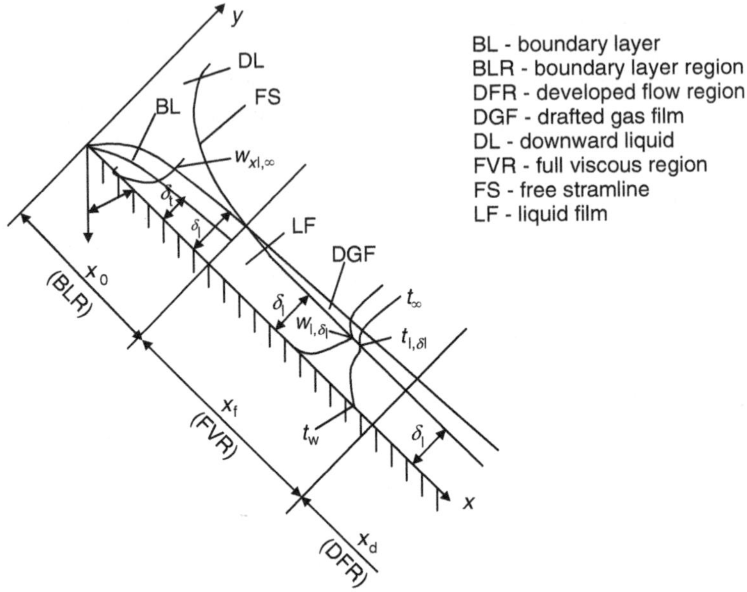

Fig. 16.1. Schematic representation of accelerating film flow, cited from Shang and Gu [9]

$$w_x \frac{\partial t}{\partial x} + w_y \frac{\partial t}{\partial y} = \frac{\lambda}{\rho c_p} \frac{\partial^2 t}{\partial y^2}, \tag{16.3}$$

and the boundary conditions are

$$y = 0 : \quad w_x = 0, \qquad w_y = 0, \qquad t = t_\mathrm{w}, \tag{16.4}$$

$$y = \delta_\mathrm{l}, \quad w_x = w_{x,\infty}, \tag{16.5}$$

$$y = \delta_\mathrm{t}, \quad t = t_\infty, \tag{16.6}$$

where w_x and w_y are velocity exponents in x and y directions, respectively, while g and α denote the gravitation acceleration and the angle of inclination of the plane wall. Here it has been anticipated that $\partial w_x/\partial y \geq 0$ throughout the entire of the film. δ_l and δ_t denote the thicknesses of the momentum and thermal boundary layers, respectively, while $w_{x,\infty}$ and t_∞ are velocity and temperature of the fluid outside the respective boundary layers. It is noteworthy that $w_{x,\infty}$ varies with x, and the wall temperature t_w and the external temperature t_∞ are constants as one kind of temperature conditions. The fluid physical properties λ, ρ, c_p, K, and n, which are assumed to be constant, are the thermal conductivity, density, specific heat, coefficient of consistency, and power-law index, respectively. The deviation of n from unity indicates the degree of deviation from Newtonian rheology and the particular case $n = 1$ represents a Newtonian fluid with dynamic coefficient of viscosity K.

No-slip and impermeability conditions at the inclined surface $y = 0$ are expressed by the boundary conditions (16.4), while the outer condition (16.5) assures that the velocity component w_x within the boundary layer approaches the external velocity

$$w_{x,\infty} = \sqrt{2gx\cos\alpha} \tag{16.7}$$

at the edge $y = \delta_l$ of the momentum boundary layer.

16.3 Similarity Transformation

Incidently, as pointed out by Andersson and Irgens [10], the external velocity (16.7) belongs to the Falkner–Skan class of freestreams $w_{x,\infty} \propto x^m$, which permits a similarity transformation of the momentum boundary layer equation even for power-law fluids. A generalized Falkner–Skan type of transformation was therefore introduced in [11, 12], while Andersson and Shang [7] devised an alternative similarity transformation. However, as we shall see, exact similarity solutions of the thermal energy equation exist only in the particular case when the power-law index n is equal to unity. In this case, Shang and Andersson derived a pseudosimilarity transformation for solution of thermal boundary layer problem in falling film flow with non-Newtonian power-law fluids [8].

According to [8], the new independent and dimensionless variables are introduced as follows:

$$\eta = \frac{y}{x} Re_x^{1/(n+1)}, \tag{16.8}$$

$$\zeta = \frac{x}{x_0}, \tag{16.9}$$

where x_0 is the length of the boundary layer region and

$$Re_x = \frac{x^n (w_{x,\infty})^{2-n} \rho}{K}, \tag{16.10}$$

where Re_x is a generalized local Reynolds number.

The dimensionless velocity components are defined as

$$W_x(\eta) = \frac{w_x}{\sqrt{2gx\cos\alpha}}, \tag{16.11}$$

$$W_y(\eta) = \frac{w_y}{\sqrt{2gx\cos\alpha}} Re_x^{1/(n+1)}, \tag{16.12}$$

which are independent of ζ.

The dimensionless temperature is defined as

$$\theta(\eta, \zeta) = \frac{t - t_\infty}{t_w - t_\infty}, \tag{16.13}$$

which will depend both on η and ζ.

According to the derivations presented in Chap. 15, the partial differential equations (16.1) and (16.2) are transformed into the following dimensionless equations, respectively:

$$W_x(\eta) - \frac{n}{(1+n)}\eta\frac{\mathrm{d}W_x(\eta)}{\mathrm{d}\eta} + 2\frac{\mathrm{d}W_y(\eta)}{\mathrm{d}\eta} = 0, \tag{16.14}$$

$$\begin{aligned} &W_x(\eta)\left[-\frac{n}{(1+n)}\eta\frac{\mathrm{d}W_x(\eta)}{\mathrm{d}\eta} + W_x(\eta)\right] \\ &+2W_y(\eta)\frac{\mathrm{d}W_x(\eta)}{\mathrm{d}\eta} = 1 + 2n\left(\frac{\mathrm{d}W_x(\eta)}{\mathrm{d}\eta}\right)^{n-1}\frac{\mathrm{d}^2W_x(\eta)}{\mathrm{d}\eta^2}. \end{aligned} \tag{16.15}$$

Additionally, the similarity transformation of (16.3) is done as follows:

At first, the derivative $\partial t/\partial x$ is expressed as

$$\frac{\partial t}{\partial x} = \frac{\partial t}{\partial \eta}\frac{\partial \eta}{\partial x} + \frac{\partial t}{\partial \xi}\frac{\partial \xi}{\partial x},$$

where

$$\frac{\partial t}{\partial \eta} = (t_\mathrm{w} - t_\infty)\frac{\partial \theta(\eta,\xi)}{\mathrm{d}\eta},$$

$$\begin{aligned} \frac{\partial \eta}{\partial x} &= \frac{\partial}{\partial x}\left[\frac{y}{x}Re_x^{1/(n+1)}\right] \\ &= \frac{\partial}{\partial x}\left[y\left(\frac{(2g\cos\alpha)^{(2-n)/2}\rho}{K}\right)^{1/(n+1)} x^{-n/(2(n+1))}\right] \\ &= -\frac{n}{2(n+1)}\eta x^{-1}, \end{aligned}$$

$$\frac{\partial t}{\partial \xi} = (t_\mathrm{w} - t_\infty)\frac{\partial \theta(\eta,\xi)}{\partial \xi},$$

$$\frac{\partial \xi}{\partial x} = \frac{1}{x_0}.$$

Therefore,

$$\frac{\partial t}{\partial x} = -(t_\mathrm{w} - t_\infty)\frac{\partial \theta(\eta,\xi)}{\mathrm{d}\eta}\frac{n}{2(n+1)}\eta x^{-1} + \frac{1}{x_0}(t_\mathrm{w} - t_\infty)\frac{\partial \theta(\eta,\xi)}{\partial \xi}.$$

The derivative $\partial t/\partial y$ is expressed as

$$\frac{\partial t}{\partial y} = \frac{\partial t}{\partial \eta}\frac{\partial \eta}{\partial y} + \frac{\partial t}{\partial \xi}\frac{\partial \xi}{\partial y},$$

where

$$\frac{\partial \eta}{\partial y} = \frac{1}{x}(Re_x)^{1/(n+1)}$$
$$\frac{\partial \xi}{\partial y} = 0.$$

Therefore

$$\frac{\partial t}{\partial y} = (t_{\mathrm{w}} - t_\infty)\frac{\partial \theta(\eta,\xi)}{\partial \eta}\frac{1}{x}(Re_x)^{1/(n+1)}.$$

Additionally,

$$\frac{\partial^2 t}{\partial y^2} = \frac{1}{x}(t_{\mathrm{w}} - t_\infty)(Re_x)^{1/(n+1)}\frac{\partial}{\partial \eta}\left(\frac{\partial \theta(\eta,\xi)}{\partial \eta}\right)\frac{1}{x}(Re_x)^{1/(n+1)}$$
$$= \frac{1}{x^2}(t_{\mathrm{w}} - t_\infty)(Re_x)^{2/(n+1)}\frac{\partial}{\partial \eta}\left(\frac{\partial \theta(\eta,\xi)}{\partial \eta}\right).$$

On these bases, (16.3) is changed into

$$\sqrt{2gx\cos\alpha}W_x(\eta)\left[-(t_{\mathrm{w}} - t_\infty)\frac{\partial \theta(\eta,\xi)}{\mathrm{d}\eta}\frac{n}{2(n+1)}\eta x^{-1} + \frac{1}{x_0}(t_{\mathrm{w}} - t_\infty)\frac{\partial \theta(\eta,\xi)}{\partial \xi}\right]$$
$$+\sqrt{2gx\cos\alpha}Re_x^{-1/(n+1)}W_y(\eta)(t_{\mathrm{w}} - t_\infty)\frac{\partial \theta(\eta,\xi)}{\partial \eta}\frac{1}{x}(Re_x)^{1/(n+1)}$$
$$= \frac{\lambda}{\rho c_p}\frac{1}{x^2}(t_{\mathrm{w}} - t_\infty)(Re_x)^{2/(n+1)}\frac{\partial}{\partial \eta}\left(\frac{\partial \theta(\eta,\xi)}{\partial \eta}\right).$$

The earlier equation is simplified to

$$\sqrt{2gx\cos\alpha}W_x(\eta)\left[-\frac{\partial \theta(\eta,\xi)}{\mathrm{d}\eta}\frac{n}{2(n+1)}\eta x^{-1} + \frac{1}{x_0}\frac{\partial \theta(\eta,\xi)}{\partial \xi}\right]$$
$$+\sqrt{2gx\cos\alpha}W_y(\eta)\frac{\partial \theta(\eta,\xi)}{\partial \eta}\frac{1}{x} = \frac{\lambda}{\rho c_p}\frac{1}{x^2}(Re_x)^{2/(n+1)}\frac{\partial}{\partial \eta}\left(\frac{\partial \theta(\eta,\xi)}{\partial \eta}\right),$$

or

$$\sqrt{2gx\cos\alpha}W_x(\eta)\left[-\frac{\partial \theta(\eta,\xi)}{\mathrm{d}\eta}\frac{n}{2(n+1)}\eta + \frac{x}{x_0}\frac{\partial \theta(\eta,\xi)}{\partial \xi}\right]$$
$$+\sqrt{2gx\cos\alpha}W_y(\eta)\frac{\partial \theta(\eta,\xi)}{\partial \eta} = \frac{\lambda}{\rho c_p}\frac{1}{x}(Re_x)^{2/(n+1)}\frac{\partial}{\partial \eta}\left(\frac{\partial \theta(\eta,\xi)}{\partial \eta}\right).$$

With (16.10), the earlier equation can be simplified into the following form

$$\left[-\frac{n}{2(n+1)}\eta W_x(\eta) + w_y(\eta)\right]\frac{\partial \theta(\eta,\xi)}{\partial \eta} + \xi W_x(\eta)\frac{\partial \theta(\eta,\xi)}{\partial \xi}$$
$$= \frac{1}{\frac{xw_{x,\infty}}{a}(Re_x)^{-2/(n+1)}}\frac{\partial^2 \theta(\eta,\xi)}{\partial \eta^2}, \tag{16.16}$$

subject to the boundary conditions

$$\eta = 0: \quad W_x(\eta) = 0, \quad W_y(\eta) = 0, \quad \theta(\eta, \zeta) = 1, \tag{16.17}$$

$$\eta = \eta_{\delta_l}: \quad W_x(\eta) = 1, \tag{16.18}$$

$$\eta = \eta_{\delta_t}, \quad \theta(\eta, \zeta) = 0. \tag{16.19}$$

16.4 Local Prandtl Number

The denominator $(xw_{x,\infty}/a)(Re_x)^{-2/(n+1)}$ in the diffusion coefficient in (16.16) can be defined as the local Prandtl number Pr_x, i.e.,

$$Pr_x = \frac{xw_{x,\infty}}{a} Re_x^{-2/(n+1)}, \tag{16.20}$$

where $a = \left(\frac{\lambda}{\rho c_p}\right)$ denotes the thermal diffusivity.

With (16.7) and (16.10), the earlier equation can be expressed as

$$\begin{aligned} Pr_x &= \frac{xw_{x,\infty}}{a} Re_x^{-2/(n+1)} \\ &= \frac{x\sqrt{2gx\cos\alpha}}{a}\left[\frac{x^n(\sqrt{2gx\cos\alpha})^{2-n}\rho}{K}\right]^{-2/(n+1)} \\ &= \frac{\sqrt{2g\cos\alpha}}{a}\left[\frac{(\sqrt{2g\cos\alpha})^{2-n}\rho}{K}\right]^{-2/(n+1)} x^{3/2}\left[x^{(n+2)2}\right]^{-2/(n+1)} \\ &= \frac{\sqrt{2g\cos\alpha}}{a}\left[\frac{(\sqrt{2g\cos\alpha})^{2-n}\rho}{K}\right]^{-2/(n+1)} x^{(n-1)/(2(n+1))}. \end{aligned}$$

Now it is readily seen that $Pr_x \to 0$ as $x \to 0$ if $n > 1$ and that $Pr_x \to \infty$ as $x \to 0$ if $n < 1$.

In the special case when the power-law index n is equal to unity, i.e., for a Newtonian liquid film, (16.20) can be simplified as

$$Pr_x = \frac{xw_{x,\infty}}{\frac{\lambda}{\rho c_p}}\left[\frac{x(w_{x,\infty})^{2-1}\rho}{K}\right]^{-2/(1+1)}.$$

For Newtonian fluids, the coefficient of consistency K is replaced by the absolute viscosity μ, and then the earlier equation is further simplified to

$$\begin{aligned} Pr_x &= \frac{xw_{x,\infty}}{\frac{\lambda}{\rho c_p}}\left[\frac{x(w_{x,\infty})\rho}{\mu}\right]^{-1} \\ &= \left(\frac{\rho c_p}{\lambda}\right)\left(\frac{\mu}{\rho}\right) \\ &= \frac{\mu c_p}{\lambda}, \end{aligned}$$

where μ, λ, and c_p are absolute viscosity, thermal conductivity, and specific heat of the Newtonian liquid, respectively. In this case, the local Prandtl number for non-Newtonian power law fluids is simplified to Prandtl number for Newtonian fluids, the diffusion coefficient in equation (16.16) becomes independent of x, i.e., $\partial\theta/\partial\xi = 0$ and similarity can be achieved also for the temperature field. This particular case has been explored by Andersson [13].

16.5 Pseudosimilarity for Energy Equation

Although the hydrodynamic problem admits similarity solutions, the accompanying thermal problem does not since the governing equation (16.16) for the temperature field exhibits explicit dependencies on both ζ and η. An accurate method for obtaining locally nonsimilar boundary layer solutions was suggested by Sparrow et al [14], and applied by Shang and Andersson [8] to solutions for thermal boundary layer of non-Newtonian power-law liquids. According to [8] local pseudosimilarity transformation for the thermal boundary layer is achieved by first introducing the new variable

$$g(\eta,\zeta) = \frac{\partial\theta(\eta,\zeta)}{\partial\zeta} \tag{16.21}$$

in the actual differential equation so that the energy equation (16.16) becomes

$$\left[-\frac{n}{2(1+n)}\eta W_x(\eta) + W_y(\eta)\right]\frac{\partial\theta(\eta,\zeta)}{\partial\eta} + \zeta W_x(\eta)g(\eta,\zeta) = \frac{1}{Pr_x}\frac{\partial^2\theta(\eta,\zeta)}{\partial\eta^2}. \tag{16.22}$$

Differentiating (16.22) with respect to ζ, we have

$$\left[-\frac{n}{2(n+1)}\eta W_x(\eta) + W_y(\eta)\right]\frac{\partial g(\eta,\zeta)}{\partial\eta} + W_x(\eta)g(\eta,\zeta) + \zeta W_x(\eta)\frac{\partial g(\eta,\zeta)}{\partial\eta}$$

$$= \frac{\partial\left(\frac{1}{Pr_x}\right)}{\partial\xi}\left(\frac{\partial^2\theta(\eta,\zeta)}{\partial\eta^2}\right) + \frac{1}{\mathrm{Pr}_x}\frac{\partial}{\partial\xi}\left(\frac{\partial^2\theta(\eta,\zeta)}{\partial\eta^2}\right), \tag{16.23}$$

where

$$\frac{\partial}{\partial\xi}\left(\frac{\partial^2\theta(\eta,\zeta)}{\partial\eta^2}\right) = \frac{\partial^2 g(\eta,\zeta)}{\partial\eta^2}.$$

With (16.7), (16.10), and (16.20), we have

$$\begin{aligned}
Pr_x &= \frac{x\sqrt{2gx\cos\alpha}}{a}\left[\frac{x^n(\sqrt{2gx\cos\alpha})^{2-n}\rho}{K}\right]^{-2/(n+1)} \\
&= \frac{x^{3/2}\sqrt{2g\cos\alpha}}{a}\left[\frac{x^n x^{(2-n)/2}(\sqrt{2g\cos\alpha})^{2-n}\rho}{K}\right]^{-2/(n+1)} \\
&= \frac{x^{3/2}\sqrt{2g\cos\alpha}}{a}\left[\frac{x^{(n+2)/2}(\sqrt{2g\cos\alpha})^{2-n}\rho}{K}\right]^{-2/(n+1)} \\
&= x^{(n-1)/(2(n+1))}\frac{\sqrt{2g\cos\alpha}}{a}\left[\frac{(\sqrt{2g\cos\alpha})^{2-n}\rho}{K}\right]^{-2/(n+1)}.
\end{aligned}$$

Then

$$\frac{\partial\left(\frac{1}{Pr_x}\right)}{\partial\xi}$$

$$= \frac{\partial}{\partial\xi}\left\{x^{(1-n)/(2(n+1))}\left(\frac{\sqrt{2g\cos\alpha}}{a}\right)^{-1}\left[\frac{(\sqrt{2g\cos\alpha})^{2-n}\rho}{K}\right]^{2/(n+1)}\right\}$$

$$= \frac{1-n}{2(n+1)}\left\{x_0 x^{(-1-3n)/(2(n+1))}\left(\frac{\sqrt{2g\cos\alpha}}{a}\right)^{-1}\left[\frac{(\sqrt{2g\cos\alpha})^{2-n}\rho}{K}\right]^{2/(n+1)}\right\}$$

$$= \frac{1-n}{2(n+1)}\left\{\xi^{-1}x\cdot x^{(-1-3n)/(2(n+1))}x^{3/2}x^{-(n+2)/(n+1)}\right.$$
$$\left.\left(\frac{x\sqrt{2gx\cos\alpha}}{a}\right)^{-1}\left[\frac{(x^n\sqrt{2gx\cos\alpha})^{2-n}\rho}{K}\right]^{2/(n+1)}\right\}$$

$$= \frac{1-n}{2(n+1)}\xi^{-1}Pr_x^{-1}.$$

Thus, (16.23) is changed into the following:

$$\left[-\frac{n}{2(n+1)}\eta W_x(\eta) + W_y(\eta)\right]\frac{\partial g(\eta,\zeta)}{\partial\eta} + W_x(\eta)g(\eta,\zeta)$$
$$= \frac{1-n}{2(n+1)}\xi^{-1}Pr_x^{-1}\left(\frac{\partial^2\theta(\eta,\zeta)}{\partial\eta^2}\right) + \frac{1}{Pr_x}\frac{\partial}{\partial\xi}\left(\frac{\partial^2\theta(\eta,\zeta)}{\partial\eta^2}\right),$$

or

$$\left[-\frac{n}{2(n+1)}\eta W_x(\eta) + W_y(\eta)\right]\frac{\partial g(\eta,\zeta)}{\partial\eta} + W_x(\eta)g(\eta,\zeta) + \xi W_x(\eta)\frac{\partial g(\eta,\zeta)}{\partial\xi}$$
$$= \frac{1}{Pr_x}\left[\frac{\partial^2 g(\eta,\zeta)}{\partial\eta^2} - \frac{n-1}{2(n+1)}\xi^{-1}\frac{\partial^2\theta(\eta,\zeta)}{\partial\eta^2}\right], \quad (16.24)$$

where the primes have been introduced to denote differentiation with respect to η. The final step is to neglect terms involving $(\partial/\partial\xi)$ in the subsidiary equation (16.24), whereas the primary equation (16.22) remains intact.

We introduce the new variable

$$h(\eta,\xi) = \xi\cdot g(\eta,\xi) = \xi\frac{\partial\theta(\eta,\xi)}{\partial\xi}. \quad (16.25)$$

Multiply (16.24) by ξ, and then (16.24) is simplified to

$$\left[-\frac{n}{2(n+1)}\eta W_x(\eta) + W_y(\eta)\right]\frac{\partial h(\eta,\zeta)}{\partial\eta} + W_x(\eta)h(\eta,\zeta)$$
$$= \frac{1}{Pr_x}\left[\frac{\partial^2 h(\eta,\zeta)}{\partial\eta^2} - \frac{n-1}{2(n+1)}\left(\frac{\partial^2\theta(\eta,\zeta)}{\partial\eta^2}\right)\right]. \quad (16.26)$$

Likewise, $\xi g(\eta,\xi)$ in (15.25) is replaced by h. Equation (16.26) is different from the general similarity equation, and named pseudosimilarity equation.

Thus, the two-equation local pseudosimilarity model consists of the coupled second-order differential equations (16.22) and (16.26) for the two unknowns $\theta(\eta,\xi)$ and $h(\eta,\xi)$. These equations can be treated as ordinary differential equations solved as a two-point boundary value problem in the single variables η with n and Pr_x being the only parameters. Boundary conditions for the subsidiary unknown h become

$$\eta = 0: \qquad h(\eta,\zeta) = 0, \tag{16.27}$$

$$\eta = \eta_{\delta_t}, \qquad h(\eta,\zeta) = 0, \tag{16.28}$$

after differentiation of the boundary conditions (16.17) and (16.19) for θ and respect ζ.

16.6 Critical Local Prandtl Number

The momentum boundary layer thickness η_{δ_l} and the thermal boundary layer thickness η_{δ_t} are different in most of the cases for falling-film flow of power-law fluids. From (16.14) and (16.15) it is found that the momentum boundary layer thickness η_{δ_l} only depends on the power-law index n, and it was observed in Chap. 15 that η_{δ_l} is a monotonically decreasing function of n throughout the parameter range $0.1 \le n \le 2$. In that study the momentum boundary layer thickness η_{δ_l} was defined in accordance with common practice in aerodynamic boundary layer theory, namely as the value of η for which the dimensionless velocity component $W_x(\eta)$ becomes equal to 0.99. For convenience, however, in the present investigation, the momentum boundary layer thickness η_{δ_l} is defined as the value of η for which $W_x(\eta)$ is practically equal to one (i.e., to within 10^{-4}%).

From (16.16) it is found that the thermal boundary thickness η_{δ_t} is observed as a part of the solution of a two-parameter problem, and it does not only depend on the power-law index n but varies also with the local Prandtl number Pr_x. In the study in [8,9], Shang, Andersson, and Gu found that thermal boundary thickness η_{δ_t} will increase with decreasing the local Prandtl number Pr_x, but almost keeps the same with variation of the power-law index n. For a given value of n, there should be a critical value of the local Prandtl number Pr_x, with which the thermal boundary layer thickness η_{δ_t} equals the momentum boundary layer thickness η_{δ_l}. This critical value of the local Prandtl number Pr_x is defined as the critical local Prandtl number, which is denoted by Pr_x^*. In this case, the completely indentical momentum boundary layer thickness η_{δ_l} and thermal boundary layer thickness η_{δ_t} are denoted by critical boundary layer thickness $\eta_{\delta_l}^*$.

The variation of the critical momentum boundary layer thickness $\eta_{\delta_l}^*$ and the critical value of the local Prandtl number Pr_x^* with n are displayed in Fig. 16.2. and 16.3 [8], respectively.

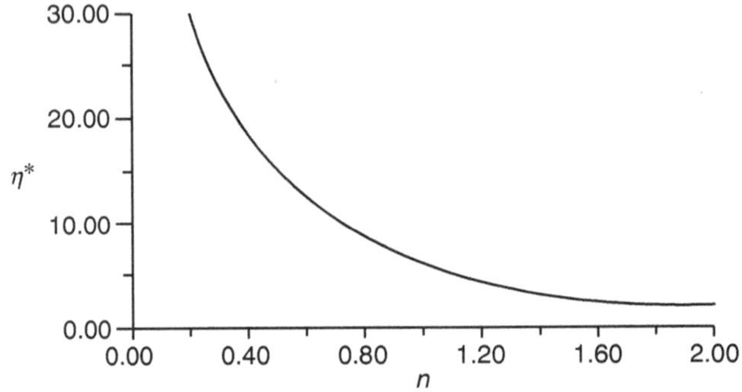

Fig. 16.2. Variation of critical momentum boundary layer thickness $\eta_{\delta_1}^*$ with power-law index n, cited from Shang and Andersson [8]

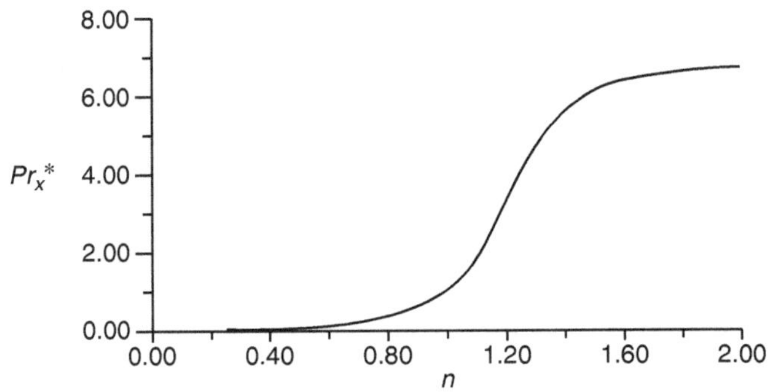

Fig. 16.3. Variation of critical Prandtl number Pr_x^* with power-law index n, cited from Shang and Andersson [8]

16.7 Analysis of Boundary Layer Thickness

16.7.1 Precautions for $Pr_x > Pr_x^*$

According to the study of Shang and Gu [9], the thickness of the thermal boundary layer η_{δ_t}, which is defined by $\partial\theta(\eta,\zeta)/\partial\eta = 0.00001$, has been determined for various local Prandtl numbers and power-law index. Figure 16.4 [9] shows a series of the related results, together with the variation of the thickness $\eta_{\delta_1}*$ with n. From Fig. 16.4, it is seen that, for a special power-law index n, the thermal boundary layer thickness η_{δ_t} is thinner than the critical momentum boundary layer thickness $\eta_{\delta_t}^*$ for $Pr_x > Pr_x^*$. The difference between the thickness increases significantly with the increase of Pr_x, so that the temperature gradients are only confined to the innermost part of the

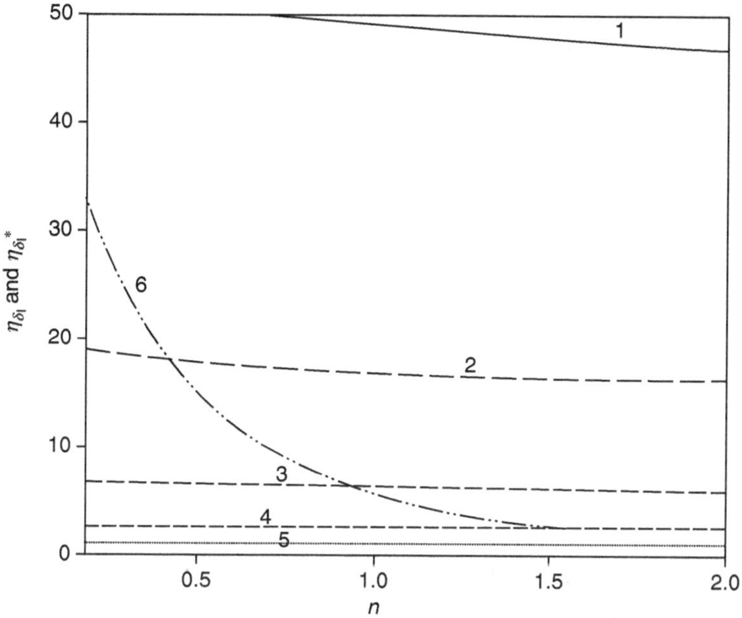

Fig. 16.4. Comparison of the thickness η_{δ_t} and $\eta_{\delta_l}*$ together with n and Pr_x (1–5: η_{δ_t} with $Pr_x = 0.01, 0.1, 1, 10, 100$, respectively; 6: $\eta_{\delta_l}^*$), cited from Shang and Gu [9]

velocity boundary layer for $Pr_x \gg Pr_x^*$. The numerical accuracy will accordingly deteriorate if the two boundary layer problems are solved simultaneously all the way from the wall ($\eta = 0$) to the edge of the momentum boundary layer ($\eta = \eta_{\delta_l}$), cf. Table 16.1. The remedy is to carry out the numerical calculation only sufficiently far so that the temperature gradient vanishes. To accomplish this, the external boundary condition for the velocity field in (16.18) is replaced with the accurately computed value of W_x at the particular position, which corresponds to the edge of the calculation domain for the temperature field. As for the specific example $n = 0.5$ and $Pr_x = 10$ in the Table 16.2, the numerical solution was obtained with $W_x(2.5) = 0.931076$ taken from Table 16.1 as outer condition for W_x, in spite of the fact that the momentum boundary layer extends all the way $\eta = 15$.

16.7.2 Precautions for $Pr_x < Pr_x^*$

For a special power-law index n, the thermal boundary layer becomes thicker than the viscous boundary layer if $Pr_x < Pr_x^*$ and in this case the ratio of the thermal boundary layer thichness η_{δ_t} to the momentum boundary layer thichness η_{δ_l}, i.e., $\eta_{\delta_t}/\eta_{\delta_l}$, increases with decreasing Pr_x. Temperature gradients thus extend far into the frictionless flow. To facilitate the numerical

Table 16.1. Similarity solution for the velocity field for power-law index $n = 0.5$, cited from Shang and Andersson [8]

η	W_y	W_x	$\frac{dW_x}{d\eta}$
0	0	0	1.104406
0.1	−0.001811	0.105273	1.000948
0.2	−0.007129	0.200581	0.905206
0.3	−0.015776	0.286522	0.8914695
0.4	−0.027575	0.363738	0.730722
0.5	−0.042347	0.432889	0.653416
0.6	−0.059914	0.494642	0.582748
0.7	−0.080097	0.549655	0.518561
0.8	−0.102721	0.598562	0.460595
0.9	−0.127614	0.64197	0.408512
1	−0.154608	0.680448	0.361926
1.2	−0.214262	0.744686	0.28355
1.4	−0.280483	0.794992	0.222044
1.6	−0.352205	0.83441	0.17416
1.8	−0.428503	0.865373	0.137048
2	−0.508594	0.889788	0.108331
2.2	−0.591817	0.909136	0.086094
2.5	−0.721362	0.931076	0.061694
3	−0.946498	0.955015	0.036572
3.5	−1.179419	0.969477	0.022584
4	−1.417354	0.978569	0.014495
4.5	−1.658602	0.984503	0.009636
5	−1.9021	0.988507	0.006611
6	−2.393348	0.99328	0.003375
8	−3.384437	0.99729	0.001132
10	−4.380367	0.998791	0.000484
12	−5.37828	0.999488	0.000248
15	−6.876808	1	0.000118

calculation of the thermal boundary layer problem and assure the numerical accuracy, the momentum boundary layer (16.14) and (16.15) are calculated only up to η_{δ_l}. Thereafter, the velocity field is taken as

$$W_x(\eta) = 1, \tag{16.29}$$

throughout the remaining η-range from η_{δ_l} to η_{δ_t}. Meanwhile, with (16.29) and by using the relationship $dW_x(\eta)/d\eta = 0$, the continuity (16.14) is changed into the following one:

$$\frac{dW_y(\eta)}{d\eta} = -\frac{1}{2}.$$

Integrating the earlier equation, we obtain the following relationship about $W_y(\eta)$:

$$W_y(\eta) = -\frac{1}{2}\eta + \text{const.} \tag{16.30}$$

Table 16.2. Local no-similarity (pseudosimilarity) solution of the heat transfer problem for $n = 0.5$ and $Pr_x = 10$, cited from Shang and Andersson [8]

η	W_y	W_x	$\frac{dW_x}{d\eta}$	$\theta(\eta,\zeta)$	$\frac{d\theta(\eta,\zeta)}{d\eta}$
0	0	0	1.104406	1.000000	−1.139345
0.1	−0.001811	0.105273	1.000948	0.886103	−1.137857
0.2	−0.007129	0.200581	0.905206	0.772716	−1.127787
0.3	−0.015776	0.286522	0.8914695	0.661083	−1.101675
0.4	−0.027575	0.363738	0.730722	0.553101	−1.054052
0.5	−0.042347	0.432889	0.653416	0.451082	−0.982272
0.6	−0.059914	0.494642	0.582748	0.357429	−0.887155
0.7	−0.080097	0.549655	0.518561	0.274286	−0.773095
0.8	−0.102721	0.598562	0.460595	0.203198	−0.647450
0.9	−0.127614	0.64197	0.408512	0.144881	−0.519255
1	−0.154608	0.680448	0.361926	0.099129	−0.397539
1.2	−0.214262	0.744686	0.28355	0.040563	−0.200474
1.4	−0.280483	0.794992	0.222044	0.013642	−0.081438
1.6	−0.352205	0.83441	0.17416	0.003704	−0.026229
1.8	−0.428503	0.865373	0.137048	0.000799	−0.006607
2	−0.508594	0.889788	0.108331	0.000135	−0.001286
2.2	−0.591817	0.909136	0.086094	0.000017	−0.000192
2.4	−0.677629	0.924559	0.068834	0.000001	−0.000022
2.5	−0.721362	0.931076	0.061694	0.000000	−0.000007

By using (16.29) and (16.30) the governing equations (16.14)–(16.16), and (16.26) are further calculated. A specific example $n = 1.5$ and $Pr_x = 1$ is given in Table 16.3 and shown graphically in Fig. 16.5. Here, the analytical continuation in the range $2.7 \leq \eta \leq 5.6$ is represented.

16.8 Remarks

The pseudosimilarity solutions of the thermal boundary layer of a falling film flow of power-law fluids are extensively developed and presented in this chapter. Based on a proposed "local Prandtl number," the dependence of the thickness of the momentum boundary layer and thermal boundary layer on the power-law index and local Prandtl number are discussed. Their changes with power-law index and local Prandtl number are also presented. The momentum layer thickness η_{δ_l} depends only on the power-law index n, while the thermal boundary layer thickness η_{δ_t} depends on the local Prandtl number Pr_x. The momentum boundary layer thickness η_{δ_l} decreases significantly with the increase of the parameter n. The thermal boundary layer thickness η_{δ_t} almost keeps the same with variation of the parameter n but decreases rapidly with increasing the parameter Pr_x, especially when $Pr_x < 1$. This analysis provides a clear identification for both $\eta_{\delta_t} > \eta^*_{\delta_l}$ and $\eta_{\delta_t} < \eta^*_{\delta_l}$.

Table 16.3. Local pseudosimilarity solution of the heat transfer problem for $n = 1.5$ and $Pr_x = 1.0$, cited from Shang and Andersson [8]

η	W_y	W_x	$\frac{\mathrm{d}W_x}{\mathrm{d}\eta}$	$\theta(\eta,\zeta)$	$\frac{\mathrm{d}\theta(\eta,\zeta)}{\mathrm{d}\eta}$
0	0	0	0.865908	1.000000	−0.485194
0.1	−0.000872	0.084788	0.829724	0.951482	−0.485141
0.2	−0.003513	0.165922	0.792848	0.902982	−0.484778
0.3	−0.007964	0.243336	0.755334	0.854547	−0.483804
0.4	−0.014265	0.316969	0.717243	0.806251	−0.481941
0.5	−0.022462	0.386768	0.678643	0.758197	−0.478929
0.6	−0.032598	0.452683	0.649610	0.710511	−0.474537
0.7	−0.044719	0.514678	0.600232	0.663342	−0.468567
0.8	−0.058871	0.572721	0.560604	0.616856	−0.460861
0.9	−0.075097	0.626794	0.520837	0.571231	−0.451305
1	−0.093438	0.676888	0.481052	0.526658	−0.439837
1.2	−0.136618	0.765174	0.401986	0.441431	−0.411192
1.4	−0.188649	0.827799	0.324677	0.362651	−0.375526
1.6	−0.249641	0.895264	0.250661	0.291595	−0.334268
1.8	−0.319506	0.938405	0.181772	0.229180	−0.289483
2	−0.397884	0.968456	0.120153	0.175877	−0.243570
2.2	−0.484046	0.987115	0.068278	0.131667	−0.198916
2.4	−0.566790	0.996604	0.028956	0.096088	−0.157574
2.6	−0.674316	0.999743	0.005337	0.068316	−0.121037
2.7	−0.724114	1	0.000493	0.057033	−0.104856
2.8	−0.774108	1	0	0.047296	−0.090137
3	−0.874108	1	0	0.031870	−0.065079
3.4	−1.074108	1	0	0.013323	−0.030914
3.8	−1.274108	1	0	0.004974	−0.012972
4	−1.374108	1	0	0.002909	−0.008021
4.4	−1.574108	1	0	0.000906	−0.002794
4.8	−1.774108	1	0	0.000243	−0.000858
5	−1.874108	1	0	0.000116	−0.000454
5.4	−2.074108	1	0	0.000016	−0.000114
5.6	−2.174108	1	0	0	−0.000054

Note: the velocity field beyond $\eta = 2.7$ is obtained from the analytical continuation in (16.29) and (16.30)

With the introduction of the "local Prandtl number," it is found that the heat transfer problem turned out to involve only two independent parameters, the power-law index and the local Prandtl number. In addition, the dependence of the power-law index and the local Prandtl number on the thermal boundary layer has been clarified.

The pseudosimilarity solution and the assumed true-similarity solution are presented for the investigation of nonsimilarity thermal boundary layer. The degree of nonsimilarity of thermal boundary layer has been determined for various values of power-law indices and local Prandtl numbers.

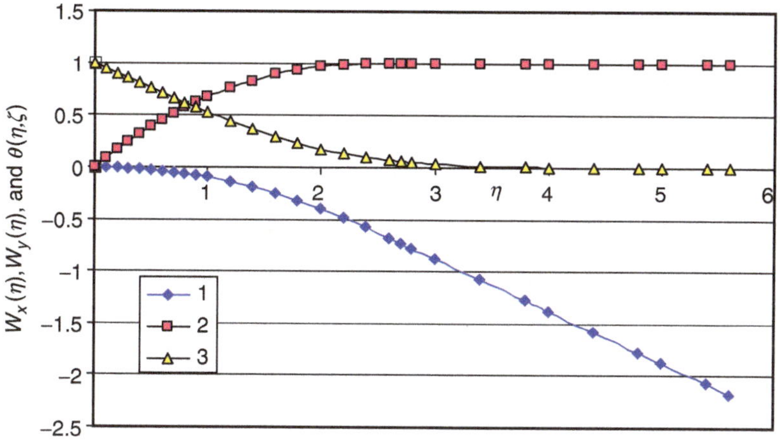

Fig. 16.5. Profiles of $W_x(\eta)$, $W_y(\eta)$, and $\theta(\eta,\zeta)$ for $Pr_x = 1$ and $n = 1.5$ Lines 1–3: for $W_y(\eta)$, $W_x(\eta)$, and $\theta(\eta,\zeta)$

References

1. S.M. Yih and M.W. Lee, Heating or evaporation in the thermal entrance region of a non-newtonian laminar falling liquid film, Int. J. Heat Mass Transfer, **29**, pp. 1999–2002, 1986
2. G. Astarita, Mass transfer from a flat solid surface to a falling non-Newtonian liquid film, Ind. Eng. Chem. Fundam. **5**, pp. 14–18, 1966
3. R.A. Mashelker and V.V. Chavan, Solid dissolution in falling films of non-Newtonian liquids, J. Chem. Jpn, **6**, pp. 160–167, 1973
4. I. Pop, T. Watanabe, and H. Konishi, Gravity-driven laminar film flow along a vertical wall with surface mass transfer, Int. Comm. Heat Mass Transfer, **23**, pp. 685–695, 1996
5. D. Ouldhadda and A. Idrissi, Laminar flow and heat transfer of non-Newtonian falling liquid film on a horizontal tube with variable surface heat flux, Int. Commun. Heat Mass Transfer, **28**, pp. 1125–1135, 2001
6. B.K. Rao, Heat transfer to falling power-law fluid film, Int. J. Heat Fluid Flow, 20, pp. 429–436, 1999
7. H. Andersson and D.Y. Shang, An extended study of hydrodynamics of gravity-driven film flow of power-law fluids, Fluid Dyn. Res., **22**, pp. 345–357, 1998
8. D.Y. Shang and H. Andersson, Heat transfer in gravity-driven film flow of power-law fluids, Int. J. Heat Mass Transfer, **42**, No. 11, pp. 2085–2099, 1999
9. D.Y. Shang and J. Gu, Analyses of pseudo-similarity and boundary layer thickness for non-Newtonian falling film flow, Heat Mass Transfer, **41**, No. 1, pp. 44–50, 2004
10. H.I. Andersson and F. Irgens, Hydrodynamic entrance length of non-Newtonian liquid films, Chem. Eng. Sci., **45**, pp. 537–541, 1990
11. H.I. Andersson and F. Irgens Gravity-driven laminar film flow of power-law fluids along vertical walls, J. Non-Newtonian Fluid Mech. **27**, pp. 153–172, 1988

12. H.I Andersson and F. Irgens, Film flow of power law fluids, Encyclopaedia of Fluid Mechanics, Gulf Publishing Company, Houston, Tx, **9**, pp. 617–648, 1990
13. H.I. Andersson, Diffusion from a vertical wall into an accelerating falling liquid film, Int. J. Heat Mass Transfer, **30**, pp. 683–689, 1987
14. E.M. Sparrow, H. Quack, and C.J. Boener, Local nonsimilarity boundary – layer solutions, AIAA J. **8**, pp. 1936–1942, 1970

17

Heat Transfer of the Falling Film Flow

Nomenclature

a	thermal diffusive coefficient, $m^2\ s^{-1}$
c_p	specific heat at constant pressure, $J\ (kg\ K)^{-1}$
g	gravitation acceleration, $m\ s^{-2}$
$g(\eta, \zeta)$	defined temperature gradient, $\frac{\partial\theta(\eta,\zeta)}{\partial\zeta}$
K	coefficient of consistency, $kg\ s^{n-2}\ m^{-1}$
n	power law index
Pr_x	local Prandtl number, $\frac{x w_{x,\infty}}{a} Re_x^{-2/(n+1)}$
Pr_x^*	critical local Prandtl number
q_x	local heat transfer rate at position x per unit area on the plate, $W\ m^{-2}$
Q_x	total heat transfer rate for position $x = 0$ to x with width of b on the plate
Re_x	local Reynolds number, $\frac{x^n (w_{x,\infty})^{2-n} \rho}{K}$
t	temperature, °C
V	volume flow rate of the falling film flow, $m^3\ s^{-1}$
w_x, w_y	velocity components in the x- and y-directions, respectively, $m\ s^{-1}$
$W_x(\eta), W_y(\eta)$	dimensionless velocity components in the x- and y-directions, respectively
$w_{x,\infty}$	velocity of the fluid outside the boundary layer, $m\ s^{-1}$
x, y	streamwise and cross-stream coordinates, m
x_0	length of the boundary layer region, m

Greek symbols

η	dimensionless coordinate variable for boundary layer, $\frac{y}{x} Re_x^{1/(n+1)}$
η_{δ_l}	dimensionless momentum boundary layer thickness
η_{δ_t}	dimensionless thermal boundary layer thickness
$\eta_{\delta_l}^*$	dimensionless critical boundary layer thickness
ζ	$\frac{x}{x_0}$

$\theta(\eta,\zeta)$	pseudo similarity solution of dimensionless temperature, $\frac{T-T_\infty}{T_\mathrm{w}-T_\infty}$
$-\left[\frac{\mathrm{d}\theta(\eta,\zeta)}{\mathrm{d}\eta}\right]_{\eta=0}$	pseudo similarity solution of dimensionless temperature gradient
$-\left[\frac{\mathrm{d}\theta(\eta,\zeta)}{\mathrm{d}\eta}\right]_{\eta=0}(1)$	local pseudo similarity solution
$-\left[\frac{\mathrm{d}\theta(\eta,\zeta)}{\mathrm{d}\eta}\right]_{\eta=0}(2)$	results evaluated by curve-fit formula equation (17.25)
$-\left[\frac{\mathrm{d}\theta(\eta,\zeta)}{\mathrm{d}\eta}\right]_{\eta=0}(3)$	local similarity solution
α_x	local heat transfer coefficient, W $(\mathrm{m}^2\mathrm{K})^{-1}$
ε_1	relative deviation evaluated by (17.26), $\frac{\left(-\frac{\mathrm{d}\theta(\eta,\zeta)}{\mathrm{d}\eta}\right)_{\eta=0}(1)-\left(-\frac{\mathrm{d}\theta(\eta,\zeta)}{\mathrm{d}\eta}\right)_{\eta=0}(2)}{\left(-\frac{\mathrm{d}\theta(\eta,\zeta)}{\mathrm{d}\eta}\right)_{\eta=0}(1)}$
ε_2	relative deviation evaluated by (17.28), $\frac{\left(-\frac{\mathrm{d}\theta(\eta,\zeta)}{\mathrm{d}\eta}\right)_{\eta=0}(1)-\left(-\frac{\mathrm{d}\theta(\eta,\zeta)}{\mathrm{d}\eta}\right)_{\eta=0}(3)}{\left(-\frac{\mathrm{d}\theta(\eta,\zeta)}{\mathrm{d}\eta}\right)_{\eta=0}(1)}$
ρ	density, kg m^{-3}

Subscripts

i	inclined case
v	vertical case or vapor
x	local value
w	at wall
α (or y)	angle of inclination
∞	far from the wall surface

17.1 Introduction

There have been a number of studies on heat transfer from a constant temperature wall to hydrodynamically fully developed power-law films, such as those by Yih and Lee [1], Astarita [2], Mashelker and Chavan [3], Pop et al. [4], Ouldhadda and Idrissi [5], Rao [6], etc. However, except a few works, such as of Shang and Andersson [7], and Shang and Gu [8], in the most of current studies of FFNF system, the hydraulic entrance region (i.e. the boundary layer region) was ignored in their heat transfer analysis of modeling and simulation. With ignoring the existing boundary layer region for the FFNF system, it is never possible to capture its adaptive remodeling process for conducting correct calculation for velocity and temperature fields and heat transfer coefficient. Meanwhile, only a few of recent studies of Shang, Andersson, and Gu, [7,8] focused on a system of solutions of thermal boundary layer by using a pseudosimilarity approach, including the rigorous solutions about the heat transfer coefficient of the falling-film flow of non-Newtonian power-law fluids.

Based on the presentation in Chap. 16 for pseudosimilarity analyses of the boundary layer thickness for the non-Newtonian falling film flow, in this present chapter, a mathematical model for the flow and heat transfer in accelerating liquid film of a non-Newtonian falling film flow is further presented. For the case that the local Prandtl number Pr_x is larger than the critical Prandtl number Pr_x^*, the predicted temperature field in the boundary layer region is controlled by the velocity field of the momentum boundary layer calculated by using the governing mass and momentum equations. For the case that the local Prandtl number Pr_x is smaller than the critical Prandtl number Pr_x^*, it is difficult to directly obtain the simultaneous solutions from the equations of the momentum and thermal boundary layers, and it is necessary to apply the perfect approach presented in Chap. 16 for overcoming such difficulty.

Since the thermal boundary layer permits exact similarity solution only in the particular case when the power-law index is equal to unity, i.e., for Newtonian films, the heat transfer problem is solved by means of a pseudosimilarity approach with power-law index n and local Prandtl number Pr_x being the only parameters. The pseudosimilarity heat transfer problem is calculated numerically in the ranges $0.2 \leq n \leq 2$ and $0.001 \leq Pr_x \leq 1000$, in which the calculations for $n = 1$ are compared favorably with earlier results for Newtonian fluid films. From the numerical solutions of the pseudosimilarity energy equation of the thermal boundary layer, it is found that the effect of the power-law index n on the wall gradient of the temperature field is slight except for the range both with smaller power-law index n and larger local Prandtl number Pr_x. However, the wall gradient of the temperature field will increase with increasing the local Prandtl number Pr_x. Curve-fit formulas for the temperature gradient at the wall are provided in order to facilitate rapid and yet accurate estimates for the heat transfer coefficient and the Nusselt number.

17.2 Governing Equations

Consider the accelerating laminar flow in the boundary layer region of a non-Newtonian power-law liquid film down along an inclined plane surface, as shown schematically in Fig. 15.1. According to Chaps. 15 and 16, the governing dimensionless differential equations for mass, momentum, and energy conservations are summarized as follows for the pseudosimilarity solutions:

$$W_x(\eta) - \frac{n}{(1+n)}\eta\frac{\mathrm{d}W_x(\eta)}{\mathrm{d}\eta} + 2\frac{\mathrm{d}W_y(\eta)}{\mathrm{d}\eta} = 0, \tag{17.1}$$

$$W_x(\eta)\left[-\frac{n}{(1+n)}\eta\frac{\mathrm{d}W_x(\eta)}{\mathrm{d}\eta} + W_x(\eta)\right] + 2W_y(\eta)\frac{\mathrm{d}W_x(\eta)}{\mathrm{d}\eta}$$
$$= 1 + 2n(\frac{\mathrm{d}W_x(\eta)}{d\eta})^{n-1}\frac{\mathrm{d}^2W_x(\eta)}{\mathrm{d}\eta^2}, \tag{17.2}$$

$$\left[-\frac{n}{2(1+n)}\eta W_x(\eta)+W_y(\eta)\right]\frac{\partial\theta(\eta,\zeta)}{\partial\eta}+\zeta W_x(\eta)g(\eta,\zeta)$$
$$=\frac{1}{Pr_x}\frac{\partial^2\theta(\eta,\zeta)}{\partial\eta^2}, \tag{17.3}$$

$$\left[-\frac{n}{2(n+1)}\eta W_x(\eta)+W_y(\eta)\right]\frac{\partial h(\eta,\zeta)}{\partial\eta}+W_x(\eta)h(\eta,\zeta)$$
$$=\frac{1}{Pr_x}\left[\frac{\partial^2 h(\eta,\zeta)}{\partial\eta^2}-\frac{n-1}{2(n+1)}(\frac{\partial^2\theta(\eta,\zeta)}{\partial\eta^2})\right], \tag{17.4}$$

subject to the boundary conditions

$$\eta=0: \quad W_x(\eta)=0, \quad W_y(\eta)=0, \quad \theta(\eta,\zeta)=1, \quad h(\eta,\zeta)=0, \tag{17.5}$$

$$\eta=\eta_{\delta_1}: \quad W_x(\eta)=1, \tag{17.6}$$

$$\eta=\eta_{\delta_t}, \quad \theta(\eta,\zeta)=0, \quad h(\eta,\zeta)=0. \tag{17.7}$$

Here, the dimensionless coordinate variables are

$$\eta=\frac{y}{x}Re_x^{1/(n+1)}, \tag{17.8}$$

$$\zeta=\frac{x}{x_0}, \tag{17.9}$$

where x_0 is a characteristic length scale in the streamwise direction (i.e. the length of the boundary layer region) and the generalized local Reynolds number is

$$Re_x=\frac{x^n(w_{x,\infty})^{2-n}\rho}{K}. \tag{17.10}$$

While the velocity component w_x within the boundary layer approaches the external velocity

$$w_{x,\infty}=\sqrt{2gx\cos\alpha}. \tag{17.11}$$

The dimensionless velocity components are defined as

$$W_x(\eta)=\frac{w_x}{\sqrt{2gx\cos\alpha}}, \tag{17.12}$$

$$W_y(\eta)=\frac{w_y}{\sqrt{2gx\cos\alpha}}Re_x^{1/n+1}, \tag{17.13}$$

which are independent of ζ. The dimensionless temperature and the related new variables are defined as

$$\theta(\eta,\zeta)=\frac{t-t_\infty}{t_\mathrm{w}-t_\infty}, \tag{17.14}$$

$$g(\eta,\zeta)=\frac{\partial\theta(\eta,\zeta)}{\partial\zeta}, \tag{17.15}$$

$$h(\eta,\xi)=\xi\cdot g(\eta,\xi)=\xi\frac{\partial\theta(\eta,\xi)}{\partial\xi}, \tag{17.16}$$

which depend both on η and ζ. The local Prandtl number Pr_x is defined by a dimensionless diffusion coefficient, i.e.,

$$Pr_x=\frac{xw_{x,\infty}}{a}(Re_x)^{-2/(n+1)}. \tag{17.17}$$

Since the thickness η_{δ_t} of the thermal boundary layer is observed as a part of the solution of a two-parameter problem, η_{δ_t} does not only depend on n but varies also with Pr_x. However, from the solutions it is found that the thermal boundary layer thickness almost keep the same with the variation of the power law index, but decrease obviously with increasing the local Prandtl number. For a given value of n, a critical local Prandtl number Pr_x^* is defined as the particular parameter value for which the thermal boundary layer thickness η_{δ_t} equals the momentum boundary layer thickness η_{δ_l}. This critical value is denoted by Pr_x^* and shown in Fig. 16.3, from which Pr_x^* can be seen to increase monotonically with n.

17.3 Heat Transfer Analysis

The heat transfer rate between the solid wall, which is maintained at temperature $\mathrm{t_w}$, and the liquid film is of particular significance in industrial applications. The local heat transfer rate q_x, which is defined as follows by Fourie's law:

$$q_x=-\lambda(\frac{\partial t}{\partial y})_{y=0}, \tag{17.18}$$

where λ is the thermal conductivity of the non-Newtonian liquid. With the earlier defined dimensionless coordinate variable η, (17.18) can be transformed to the following

$$q_x=-\lambda\left(\frac{\partial t}{\partial\eta}\right)_{\eta=0}\left(\frac{\partial\eta}{\partial y}\right)_{y=0}.$$

With (16.8) and (16.14), the local heat transfer rate is described as

$$q_x=-\lambda x^{-1}(t_\mathrm{w}-t_\infty)\left[\frac{\partial\theta(\eta,\zeta)}{\partial\eta}\right]_{\eta=0}(Re_x)^{1/(n+1)}, \tag{17.19}$$

where

$$\left[\frac{\partial\theta(\eta,\zeta)}{\partial\eta}\right]_{\eta=0}$$

is the dimensionless local temperature gradient on the wall. Then, with Newtonian-Cooling law $q_x=\alpha_x(t_\mathrm{w}-t_\infty)$, the local heat transfer coefficient α_x is expressed as follows:

$$\alpha_x=-\lambda x^{-1}\left[\frac{\partial\theta(\eta,\zeta)}{\partial\eta}\right]_{\eta=0}(Re_x)^{1/(n+1)}, \tag{17.20}$$

or, alternately, as local Nusselt number

$$Nu_x = \frac{\alpha_x x}{\lambda} = -(Re_x)^{1/(n+1)} \left[\frac{\partial\theta(\eta,\zeta)}{\partial\eta}\right]_{\eta=0}. \tag{17.21}$$

If Q_x is total heat transfer rate from the position 0 to x with the width of b on the plate, Q_x is the following integration:

$$Q_x = \iint\limits_A q_x \, \mathrm{d}A,$$

where $A = b \cdot x$. Then,

$$Q_x = b\int_0^x q_x \, \mathrm{d}x$$

$$= -b\int_0^x \lambda x^{-1}(t_\mathrm{w} - t_\infty) \left[\frac{\partial\theta(\eta,\zeta)}{\partial\eta}\right]_{\eta=0} (Re_x)^{1/(n+1)}\mathrm{d}x$$

$$= -b\int_0^x \lambda x^{-1}(t_\mathrm{w} - t_\infty) \left[\frac{\partial\theta(\eta,\zeta)}{\partial\eta}\right]_{\eta=0} \left(\frac{x^n (w_{x,\infty})^{2-n}\rho}{K}\right)^{1/(n+1)} \mathrm{d}x$$

$$= -b\int_0^x \lambda(t_\mathrm{w} - t_\infty) \left[\frac{\partial\theta(\eta,\zeta)}{\partial\eta}\right]_{\eta=0} \left(\frac{(2g\cos\alpha)^{\frac{2-n}{2}}\rho}{K}\right)^{1/(n+1)} x^{-n/(2(n+1))}\mathrm{d}x$$

$$= -b\frac{2(n+1)}{n+2}\lambda(t_\mathrm{w} - t_\infty) \left[\frac{\partial\theta(\eta,\zeta)}{\partial\eta}\right]_{\eta=0} \left(\frac{(2g\cos\alpha)^{2-n/2}\rho}{K}\right)^{1/(n+1)} x^{(n+2)/(2(n+1))}$$

$$= -b\frac{2(n+1)}{n+2}\lambda(t_\mathrm{w} - t_\infty) \left[\frac{\partial\theta(\eta,\zeta)}{\partial\eta}\right]_{\eta=0} \left(\frac{x^n(\sqrt{2gx\cos\alpha})^{2-n}\rho}{K}\right)^{1/(n+1)}.$$

Then,

$$Q_x = -b\frac{2(n+1)}{n+2}\lambda(t_\mathrm{w} - t_\infty) \left[\frac{\partial\theta(\eta,\zeta)}{\partial\eta}\right]_{\eta=0} (Re_x)^{1/(n+1)}. \tag{17.22}$$

The average heat transfer coefficient $\overline{\alpha}_x$, defined as $Q_x = \overline{\alpha}_x(t_\mathrm{w} - t_\infty)A$, is expressed as

$$\overline{\alpha}_x = -\frac{2(n+1)}{n+2}\lambda x^{-1} \left[\frac{\partial\theta(\eta,\zeta)}{\partial\eta}\right]_{\eta=0} (Re_x)^{1/(n+1)}. \tag{17.23}$$

The average Nusselt number, defined as $\overline{Nu_x} = \overline{\alpha}_x x/\lambda$, is expressed as

$$\overline{Nu_x} = -\frac{2(n+1)}{n+2} \left[\frac{\partial\theta(\eta,\zeta)}{\partial\eta}\right]_{\eta=0} (Re_x)^{1/(n+1)}. \tag{17.24}$$

From (17.19) to (17.24) it is found that the local temperature gradient $[\partial\theta(\eta,\zeta)/\partial\eta]_{\eta=0}$ is very important for evaluation of the heat transfer. From governing (17.3) and (17.4) it is follows that $[\partial\theta(\eta,\zeta)/\partial\eta]_{\eta=0}$ only depends on power-law index n and local Prandtl number Pr_x, i.e.,

$$\left[\frac{\partial\theta(\eta,\zeta)}{\partial\eta}\right]_{\eta=0} = f(n, Pr_x). \qquad (17.25)$$

17.4 Numerical Solution for Heat Transfer

The solutions for the pseudosimilarity equations of thermal boundary layer are obtained from (17.1)–(17.4) with the boundary condition equations (17.5)–(17.7). Fig. 17.1 shows a number of computed temperature profiles $\theta(\eta)$with the variations of power-law index n from 0.2 to 2 and local Prandtl number Pr_x from 0.001 to 1,000. While the wall temperature gradient $[(\partial\theta(\eta,\zeta))/\partial\eta]_{\eta=0}$, which is the most important heat transfer characteristic, is listed in Table 17.1 [7] and plotted in Fig. 17.2. It is shown that the wall temperature gradient increases significantly with increasing local Prandtl number

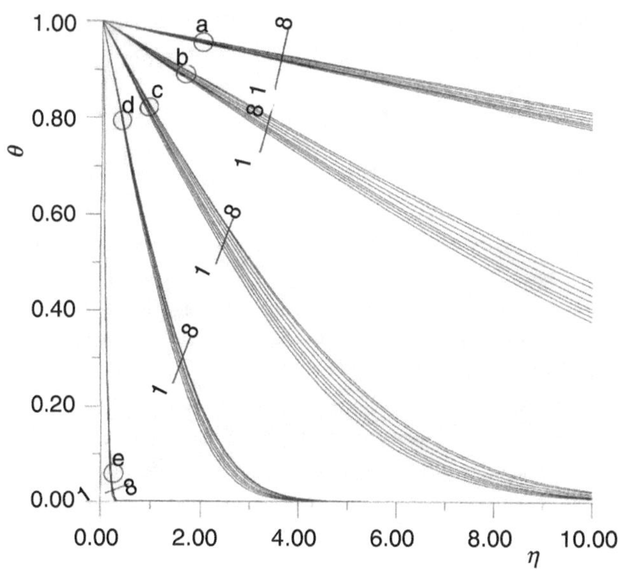

Fig. 17.1. Dimensionless temperature profile $\theta(\eta,\zeta)$for different values of Pr_x and n, cited from Shang and Andersson [7]. (a) $Pr_x = 0.001$; curves 1–8: n=0.2, 0.3, 0.5, 0.7, 1.0, 1.2, 1.5, and 2.0, (b) $Pr_x = 0.01$; curves 1–8: n=0.2, 0.3, 0.5, 0.7, 1.0, 1.2, 1.5, and 2.0, (c) $Pr_x = 0.1$; curves 1–8: n=0.2, 0.3, 0.5, 0.7, 1.0, 1.2, 1.5, and 2.0, (d) $Pr_x = 1.0$; curves 1–8: n=2.0, 1.5, 1.2, 1.0, 0.2, 0.7, 0.3, and 0.5, and (e) $Pr_x = 1,000$; curves 1–8: n=0.2,0.3, 2.0, 0.5, 1.5, 0.7,1.2,1.2, and 1.0

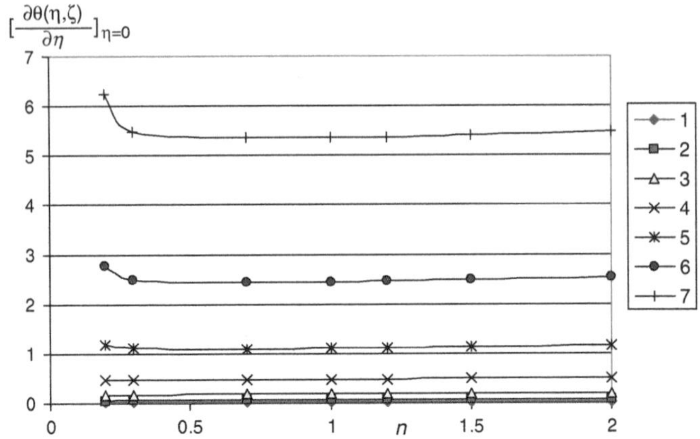

Fig. 17.2. Variation of dimensionless temperature gradient $-[(\partial\theta(\eta,\zeta))/\partial\eta]_{\eta=0}$ at the wall $\eta = 0$ with power-law index n for different values of the local Prandtl number Pr_x. Lines 1–8 note: $Pr_x = 0.001$, 0.01, 0.1, 1, 10, 100, and 1,000

Pr_x from 0.001 to 1,000. However, effect of the power-law index n on the wall gradient of the temperature field is slight except for smaller power-law index n together with larger local Prandtl number Pr_x. For the particular parameter value $n = 1$ the wall temperature gradient data in Table 17.1 agreed with the calculation for a Newtonian film by Andersson [9] throughout the entire Prandtl number range.

The most striking feature of Fig. 17.1 is that the local Prandtl number effect is more prominent than the influence of the rheological parameter n, except for larger local Prandtl number with smaller power-law index n. If the value of the power-law index n equals unity, the thickness of the thermal boundary layer is roughly the same as the thickness of the momentum boundary layer for $Pr_x = 1$. Moreover, for high local Prandtl number the thermal boundary layer is significantly thinner than the viscous boundary layer, while for $Pr_x \ll 1$ the thermal boundary layer extends far into the external free stream. Consequently, the thinning of the thermal boundary layer with increasing values of Pr_x makes the magnitude of the temperature gradient at the wall. The thick thermal boundary layer in the low local Prandtl number cases suggests that the temperature adjusts from T_{w} to T_∞ mainly in fluid with free stream velocity $w_{x,\infty}$. Thus, as a first approximation, the viscous boundary layer does not contribute to the heat flux and the temperature gradient at the wall should therefore be independent of n when $n < 1$. However, the data for $Pr_x = 0.001$ in Table 17.1 show that the wall temperature gradient increases slowly with n as n is varied from 0.2–2, the total increase being less than 8%. For $Pr_x \ll 1$ the principal effect of the viscous boundary layer on the temperature gradient at the wall stems from the displacement of the external inviscid flow away from the wall.

A qualitatively different situation occurs for high local Prandtl numbers. Due to the substantial thinning of the thermal boundary layer with

Table 17.1. Computed values of $-\left[\frac{\partial\theta(\eta,\zeta)}{\partial\eta}\right]_{\eta=0}$ for different values of Pr_x and n, cited from Shang and Andersson [7]

n		Pr_x						
		0.001	0.01	0.1	1	10	100	1,000
0.2	(1)	0.02053	0.06315	0.1863	0.5111	1.27986	2.98885	6.7598
	(2)	0.02078	0.06427	0.1838	0.5122	1.28135	2.98621	6.75929
	ε_1	−1.218%	−1.77%	1.34%	−0.215%	1.27%	0.088%	0.0075%
	(3)	0.01897	0.05837	0.1724	0.4742	1.19263	2.78664	6.23798
	ε_2	7.60%	7.57%	7.46%	7.226%	6.82%	6.77%	7.72%
0.5	(1)	0.020969	0.06391	0.1836	0.4793	1.13934	2.56412	5.62871
	(2)	0.020792	0.06494	0.1797	0.4782	1.13641	2.45958	5.63969
	ε_1	0.844%	−1.61%	2.12%	0.23%	0.257%	4.077%	−0.195%
	(3)	0.020222	0.06162	0.1773	0.4638	1.10503	2.49064	5.47372
	ε_2	3.56%	3.58%	3.43%	3.23%	3.01%	2.87%	2.75%
0.7	(1)	0.021202	0.06458	0.1847	0.4758	1.11569	2.49172	5.4554
	(2)	0.021161	0.06595	0.1812	0.4777	1.11644	2.49387	5.46885
	ε_1	0.193%	−2.12%	1.89%	−0.399%	−0.067%	−0.086%	−0.247%
	(3)	0.02081	0.06336	0.1814	0.468	1.09865	2.4545	5.37162
	ε_2	1.85%	1.89%	1.79%	1.64%	1.53%	1.49%	1.54%
1	(1)	0.021457	0.06536	0.186817	0.4776	1.10671	2.45403	5.35184
	(2)	0.021458	0.06673	0.182300	0.4773	1.10248	2.44926	5.351
	ε_1	−0.0047%	−2.096%	2.418%	0.063%	0.382%	0.194%	0.0157%
	(3)	0.021455	0.06536	0.186817	0.4776	1.10671	2.45403	5.35184
	ε_2	0%	0%	0%	0%	0%	0%	0%
1.2	(1)	0.021595	0.06583	0.1883	0.4804	1.10769	2.44844	5.33014
	(2)	0.021601	0.06715	0.1834	0.4803	1.10527	2.44816	5.33643
	ε_1	−0.0278%	−2%	2.6%	0.02%	0.2187%	0.0114%	−0.118%
	(3)	0.021786	0.06641	0.1899	0.4842	1.11582	2.46617	5.37031
	(5)	−0.88%	−0.88%	−0.85%	−0.79%	−0.73%	−0.72%	−0.75%
1.5	(1)	0.021763	0.06641	0.1903	0.4853	1.11297	2.45196	5.32975
	(2)	0.021788	0.06776	0.1851	0.4847	1.10965	2.45074	5.33106
	ε_1	−0.115%	−2.03%	2.73%	0.124%	0.298%	0.05%	−0.0246%
	(3)	0.022179	0.06767	0.1939	0.4934	1.13046	2.4897	5.41286
	ε_2	−1.91%	−1.90%	−1.89%	−1.67%	−1.57%	−1.54%	−1.56%
2	(1)	0.022111	0.06717	0.193	0.4928	1.1246	2.46821	5.35668
	(2)	0.022099	0.06877	0.1879	0.492	1.12048	2.46717	5.35603
	ε_1	0.0543%	−2.38%	2.64%	0.162%	0.367%	0.042%	0.012%
	(3)	0.022654	0.06922	0.1989	0.5064	1.15317	2.5297	5.49016
	ε_2	−2.46%	−3.05%	−3.06%	−2.76%	−2.54%	−2.49%	−2.49%

Note: (i) $-\left[\frac{\partial\theta(\eta,\zeta)}{\partial\eta}\right]_{\eta=0}$ (1)(for short as (1)) local pseudosimilarity solution

(ii) $-\left[\frac{\partial\theta(\eta,\zeta)}{\partial\eta}\right]_{\eta=0}$(2) (for short as (2)), result evaluated by curve-fit formula (17.26)

(iii) Deviation ε_1 defined as (17.27)

(iv) $-\left[\frac{\partial\theta(\eta)}{\partial\eta}\right]_{\eta=0}$ (3) (for short (3)), local similarity solution

(v) Deviation ε_2 defined as (17.29)

increasing Pr_x, the temperature gradients are contained within the innermost part of the momentum boundary layer. Thus, the wall gradient of the temperature field is controlled by the velocity gradient $\mathrm{d}W_x/\mathrm{d}\eta$ at the wall. The accurate numerical solution of the hydrodynamic problem in [10] showed that $\mathrm{d}W_x/\mathrm{d}\eta$ is practically independent of n for dilatant film but increases significantly with increasing pseudoplasticity 1–n for $n < 1$. It is therefore interesting to observe that exactly the same n-dependency is carried over to the wall gradients of the temperature field in Table 17.1.

To facilitate rapid estimates of the local heat transfer coefficient α_x or the local Nusselt number Nu_x, accurate curve-fit formulas for the wall gradient of the temperature field are provided. The optimized expressions for the coefficients a, b, and c, as obtained by Shang and Andersson [7] for matching the formula (17.26) to the data in Table 17.1, are given in Table 17.2. Predictions by means of this short-cut method are also included in Table 17.1.

$$-\left[\frac{\partial\theta(\eta,\zeta)}{\partial\eta}\right]_{\eta=0} = a + bPr_x^c. \tag{17.26}$$

If we take the corresponding pseudosimilarity solution of temperature gradient from (17.3) and (17.4) as $-\left[\frac{\mathrm{d}\theta(\eta,\zeta)}{\mathrm{d}\eta}\right]_{\eta=0}(1)$, the result evaluated by using (17.26) as $-\left[\frac{\mathrm{d}\theta(\eta,\zeta)}{\mathrm{d}\eta}\right]_{\eta=0}(2)$, the result of deviations ε_1 of $-\left[\frac{\mathrm{d}\theta(\eta,\zeta)}{\mathrm{d}\eta}\right]_{\eta=0}(2)$ from $-\left[\frac{\mathrm{d}\theta(\eta,\zeta)}{\mathrm{d}\eta}\right]_{\eta=0}(1)$ can be expressed as

$$\varepsilon_1 = \frac{-\left[\frac{\mathrm{d}\theta(\eta,\zeta)}{\mathrm{d}\eta}\right]_{\eta=0}(1) - \left\{-\left[\frac{\mathrm{d}\theta(\eta,\zeta)}{\mathrm{d}\eta}\right]_{\eta=0}(2)\right\}}{-\left[\frac{\mathrm{d}\theta(\eta,\zeta)}{\mathrm{d}\eta}\right]_{\eta=0}(1)}. \tag{17.27}$$

The evaluated results $-\left[\mathrm{d}\theta(\eta,\zeta)/\mathrm{d}\eta\right]_{\eta=0}(2)$ and their deviations ε_1 are listed in Table 17.1. It is found that the evaluated values of $-\left[\mathrm{d}\theta(\eta,\zeta)/\mathrm{d}\eta\right]_{\eta=0}(2)$ are very well agreement with the pseudosimilarity solutions $-\left[\mathrm{d}\theta(\eta,\zeta)/\mathrm{d}\eta\right]_{\eta=0}(1)$ over the entire parameter ranges $0.2 \le n \le 2$ and $0.001 \le Pr_x \le 1,000$.

In addition, the thickness of temperature boundary layer increases with the decrease of local Prandtl numberPr_x. The related numerical solutions $h(\eta)$ are plotted in Figs. 17.3 (a) and (b), which correspond to Pr_x= 1 and 100, respectively. In each figure the five curves correspond to n = 0.2, 0.5, 1, 1.5, and 2, respectively. The solutions $h(\eta)$ reflect the nonsimilarity of the thermal boundary layer. It is shown in Fig. 17.3 that with variation of local Prandtl number Pr_x, the nonsimilarity of the thermal boundary layer will be different slightly [8]. However with the increase of the value $|n-1|$, non similarity of the thermal boundary layer increases obviously. Only when the power-law index n equals unity, true similarity solution exists. The variable $h(\eta)$ is positive for $n < 1$ and negative for $n > 1$.

Table 17.2. Coefficients a, b, and c in the curve-fit formula (17.26), cited from Shang and Andersson [7]

Pr_x	n	a	b	c
$0.001 \le Pr_x \le 1$	$0.2 \le n \le 1$	$-0.0086 + \frac{0.0009}{n}$	$0.485 + 0.00001 \times \left(\frac{1}{n}\right)^{\frac{1}{n}}$	$0.399 + \frac{0.008}{n}$
	$1 \le n \le 2$	$-0.0074-0.00028n$	$0.47+0.015n$	0.407
$1 \le Pr_x \le 1,000$	$0.2 \le n \le 1$	$0.0205 - 0.0845 \times \left(\frac{1}{n}\right)^{\frac{1}{3}}$	$0.518 + \frac{0.0234}{n}$	$0.3324 + 0.00096 \times \left(\frac{1}{n}\right)^{1.6}$
	$1 \le n \le 2$	$-0.12 + 0.079 \times \left(\frac{n}{n+1}\right)^{\frac{1}{2}}$	$0.541 + 0.0009n^3$	$0.3306 + 0.0027 \times \left(\frac{1}{n}\right)^{1.65}$

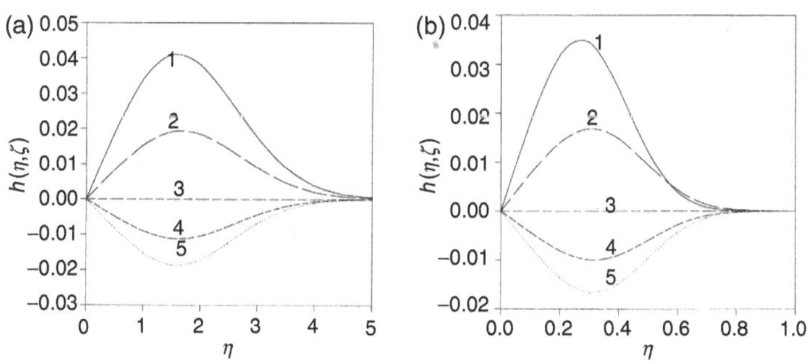

Fig. 17.3. Numerical solutions $h(\eta)$(curves 1–5: for n = 0.2, 0.5, 1, 1.5, and 2), cited from Shang and Gu [8]. (a) for Pr_x= 1 and (b) for Pr_x = 100

17.5 Local Similarity vs. Local Pseudosimilarity

If a true-similarity solution were assumed to exist for the thermal boundary layer, the dimensionless energy equations (17.3) and (17.4) would be simplified to

$$\left[-\frac{n}{2(1+n)}\eta W_x(\eta) + W_y(\eta)\right]\frac{\mathrm{d}\theta(\eta)}{\mathrm{d}\eta} = \frac{1}{Pr_x}\frac{\mathrm{d}^2\theta(\eta)}{\mathrm{d}\eta^2}. \tag{17.28}$$

A simple approach to the nonsimilar heat transfer problem associated with the gravity-driven power-law film would be the local similarity scheme. In that approach ξ is regarded as a known constant at any streamwise position and the dimensionless energy equations (17.3) and (17.4) would be turned to (16.28). The numerical solution of the simplified version of (17.28) can then be obtained locally with Pr_x and n as parameters by means of the same calculation technique as for the local nonsimilar problem defined in Chap. 16.

The two-parameter local similarity problem was solved numerically for various combinations of n and Pr_x in the intervals $0.2 \le n \le 2$ and $0.001 \le Pr_x \le 1{,}000$. Results for the magnitude of the wall gradient of the dimensionless temperature field i.e., $-\left[\partial\theta(\eta,\zeta)/\partial\eta\right]_{\eta=0}$ are reported in Table 17.1. To facilitate the comparison between the local pseudosimilarity and the local similarity approach the deviation between the two sets of data, normalized with the latter, has been defined as ε_2 and included in the Table 17.1.

If we take the assumed true-similarity solution of the temperature gradient from (17.28) as $-\left[\mathrm{d}\theta(\eta)\mathrm{d}\eta\right]_{\eta=0}(3)$, the deviations ε_2 of $-\left[\mathrm{d}\theta(\eta)\mathrm{d}\eta\right]_{\eta=0}(3)$ from pseudosimilarity solution, $-\left[\partial\theta(\eta,\zeta)/\partial\eta\right]_{\eta=0}(1)$ can be expressed as in (17.29). The value of ε_2 for each n and Pr_x is shown in Table 17.1 and Fig. 17.4.

$$\varepsilon_2 = \frac{-\left[\dfrac{\partial\theta(\eta,\zeta)}{\partial\eta}\right]_{\eta=0}(1) - \left[-\left[\dfrac{\mathrm{d}\theta(\eta)}{\mathrm{d}\eta}\right]_{\eta=0}(3)\right]}{-\left[\dfrac{\partial\theta(\eta,\zeta)}{\partial\eta}\right]_{\eta=0}(1)} \tag{17.29}$$

Let us emphasize that the two approaches become identical for $n = 1$ since the thermal boundary layer problem admits exact similarity solutions for Newtonian fluids. It is therefore not surprising that the relative deviation ε_2 in the local similarity and local pseudosimilarity solutions increases with deviation of the power-law index n from unity, i.e., with increasing non-Newtonian rheology. The deviation is moreover more significant for the most pseudoplastic liquids ($n = 0.2$) than for the highly dilatant film ($n = 2$), whereas the local Prandtl number Pr_x turned out to have only a minor effect on ε_2. It is concluded that the results obtained with the local pseudosimilarity approach are of good accuracy.

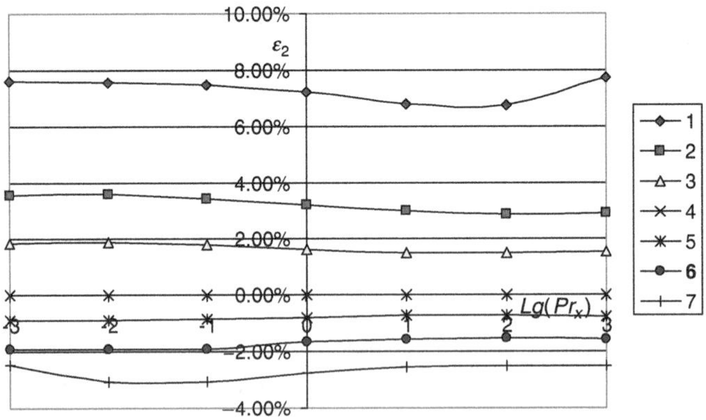

Fig. 17.4. The relative deviation ε_2 between local pseudosimilarity and local similarity solutions in the wall temperature gradient $-\left[\partial\theta(\eta,\zeta)/\partial\eta\right]_{\eta=0}$ variation with power-law index n and local Prandtl number Pr_x (curves 1–7: $n = 0.2$, 0.5, 0.7, 1, 1.2, 1.5, and 2)

It is noteworthy to emphasize that the absolute value of deviation rate ε_2 equals zero for $n = 1$, increases with the increase of the absolute value $|n-1|$. The value of ε_2 is a measure of the degree of nonsimilarity of the thermal boundary layer, and it is seen from Fig. 17.4 that such nonsimilarity for pseudoplastic fluid (with $n < 1$) is more obvious than that of *dilatant* fluid (with $n > 1$).

17.6 Summary

So far, we have presented our recent developments on heat transfer of falling film flow of non-Newtonian power-law fluids. The related governing equations and heat transfer expressions are summarized in Tables 17.3 and 17.4, respectively.

17.7 Remarks

This chapter has focused on the heat transfer from an inclined plane surface to an accelerating liquid film of a non-Newtonian power-law fluid. Although the thermal boundary layer equation of the non-Newtonian fluid flows generally fails to permit similarity solutions, a novel similarity transformation devised by Shang and Andersson [7] for the accompanying hydrodynamical problem was adopted in combination with a local pseudosimilarity method due to Sparrow et al. [11]. The resulting transformed problem turned out to involve only two independent parameters, namely the power-law index n and the local Prandtl number Pr_x. It is noteworthy that all other parameters, like the streamwise location x, the fluid properties ρ, K, n, λ and the component of the gravitational acceleration along the wall $g\cos\alpha$, have been combined into the induced Pr_x, the local Reynolds number, and the velocity components $W_x(\eta)$ and $W_y(\eta)$. Accurate numerical results were obtained for various combinations of Pr_x and n covering the range of the local Prandtl numbers Pr_x from 0.001–1,000 and for the power-law index n in the range $0.2 \le n \le 2$. Special treatment for the low and high Prandtl number cases was essential in order to maintain the numerical accuracy. The calculated results obtained both by using local similarity and local pseudosimilarity methods were practically indistinguishable for $n = 1$ over the entire Pr_x range. The main findings can be summarized as follows:

The thickness of the thermal boundary layer decreases monotonically with increasing Pr_x. The thermal boundary layer extends far out in the free stream for $Pr_x \ll 1$ and is on the other hand confined to the innermost part of the momentum boundary layer for $Pr_x \gg 1$. The local heat transfer coefficient and the local Nusselt number depend on the local Reynolds number Re_x and the wall gradient of the dimensionless temperature.

The critical local Prandtl number Pr_x^* is a very important concept for solution of thermal boundary layer, which is closely related to the solution

Table 17.3. Summary of governing equations of of falling film flow of non-Newtonian power-law fluids

term	equations
	governing partial differential equations
mass equation	$\frac{\partial w_x}{\partial x} + \frac{\partial w_y}{\partial y} = 0$
momentum equation	$w_x \frac{\partial w_x}{\partial x} + w_y \frac{\partial w_x}{\partial y} = g \cos\alpha + n \frac{K}{\rho} \left(\frac{\partial w_x}{\partial y}\right)^{n-1} \frac{\partial^2 w_x}{\partial y^2}$
energy equation	$w_x \frac{\partial t}{\partial x} + w_y \frac{\partial t}{\partial y} = \frac{\lambda}{\rho c_p} \frac{\partial^2 t}{\partial y^2}$
boundary conditions	$y = 0: \quad w_x = 0, \quad w_y = 0, \quad t = t_{\mathrm{w}}$
	$y = \delta_{\mathrm{l}}, \quad w_x = w_{x,\infty}$
	$y = \delta_{\mathrm{t}}, \quad t = t_\infty$
	similarity variables
η	$\frac{y}{x} Re_x^{1/(n+1)}$
Re_x	$\frac{x^n (w_{x,\infty})^{2-n} \rho}{K}$
Pr_x	$\frac{x w_{x,\infty}}{a} (Re_x)^{-2/(n+1)}$
$W_x(\eta)$	$\frac{w_x}{\sqrt{2gx\cos\alpha}}$
$W_y(\eta)$	$\frac{w_y}{\sqrt{2gx\cos\alpha}} Re_x^{1/(n+1)}$
	governing ordinary differential equations
mass equation	$W_x - \frac{n}{n+1} \eta \frac{\mathrm{d}W_x}{\mathrm{d}\eta} + 2 \frac{\mathrm{d}W_y}{\mathrm{d}\eta} = 0$
momentum equation	$W_x \left(-\frac{n}{(n+1)} \eta \frac{\mathrm{d}W_x}{\mathrm{d}\eta} + W_x\right) + 2W_y \frac{\mathrm{d}W_x}{\mathrm{d}\eta} = 1 + 2n \left(\frac{\mathrm{d}W_x}{\mathrm{d}\eta}\right)^{(n-1)} \frac{\mathrm{d}^2 W_x}{\mathrm{d}\eta^2}$
pseudosimilarity energy equation	$\left[-\frac{n}{2(1+n)} \eta W_x(\eta) + W_y(\eta)\right] \frac{\partial \theta(\eta,\zeta)}{\partial \eta} + \zeta W_x(\eta) g(\eta,\zeta) = \frac{1}{Pr_x} \frac{\partial^2 \theta(\eta,\zeta)}{\partial \eta^2}$
pseudosimilarity energy equation	$\left[-\frac{n}{2(n+1)} \eta W_x(\eta) + W_y(\eta)\right] \frac{\partial h(\eta,\zeta)}{\partial \eta} + W_x(\eta) h(\eta,\zeta) = \frac{1}{Pr_x} \left[\frac{\partial^2 h(\eta,\zeta)}{\partial \eta^2} - \frac{n-1}{2(n+1)} \left(\frac{\partial^2 \theta(\eta,\zeta)}{\partial \eta^2}\right)\right]$
boundary conditions	$\eta = 0: \quad W_x(\eta) = 0, \quad W_y(\eta) = 0, \quad \theta(\eta,\zeta) = 1, \quad h(\eta,\zeta) = 1$
	$\eta = \eta_{\delta_{\mathrm{l}}}: \quad W_x(\eta) = 1, \quad \theta(\eta,\zeta) = 0, \quad h(\eta,\zeta) = 0$

Table 17.4. Summary of the related expressions on heat transfer of the falling film flow of non-Newtonian power-law fluids

α_x, defined as $\alpha_x = \frac{q_x}{(t_w - t_\infty)}$	$-\lambda x^{-1}(Re_x)^{1/(n+1)} \left[\frac{\partial\theta(\eta,\zeta)}{\partial\eta}\right]_{\eta=0}$	
$\overline{\alpha}_x$ defined as $\frac{Q_x}{(t_w - t_\infty)A}$	$\frac{2(n+1)}{n+2}\alpha_x$	
Nu_x, defined as $\frac{\alpha_x x}{\lambda}$	$-(Re_x)^{1/(n+1)} \left[\frac{\partial\theta(\eta,\zeta)}{\partial\eta}\right]_{\eta=0}$	
$\overline{Nu}_x$, defined as $\frac{\overline{\alpha}_x x}{\lambda}$	$\frac{2(n+1)}{n+2} Nu_x$	
$-\left[\frac{\partial\theta(\eta,\zeta)}{\partial\eta}\right]_{\eta=0}$	$a + bPr_x^c$	
a	$-0.0086 + \frac{0.0009}{n}$	$(0.2 \le n \le 1 \quad 0.001 \le Pr_x \le 1)$
	$-0.0074 - 0.00028n$	$(1 \le n \le 2 \quad 0.001 \le Pr_x \le 1)$
	$0.0205 - 0.0845 \times \left(\frac{1}{n}\right)^{1/3}$	$(0.2 \le n \le 1 \quad 1 \le Pr_x \le 1,000)$
	$-0.12 + 0.079 \times \left(\frac{n}{n+1}\right)^{1/2}$	$(1 \le n \le 2 \quad 1 \le Pr_x \le 1,000)$
b	$0.485 + 0.00001 \times \left(\frac{1}{n}\right)^{1/n}$	$(0.2 \le n \le 1 \quad 0.001 \le Pr_x \le 1)$
	$0.47 + 0.015n$	$(1 \le n \le 2 \quad 0.001 \le Pr_x \le 1)$
	$0.518 + \frac{0.0234}{n}$	$(0.2 \le n \le 1 \quad 1 \le Pr_x \le 1,000)$
	$0.541 + 0.0009n^3$	$(1 \le n \le 2 \quad 1 \le Pr_x \le 1,000)$
c	$0.399 + \frac{0.008}{n}$	$(0.2 \le n \le 1 \quad 0.001 \le Pr_x \le 1)$
	0.407	$(1 \le n \le 2 \quad 0.001 \le Pr_x \le 1)$
	$0.3324 + 0.00096 \times \left(\frac{1}{n}\right)^{1.6}$	$(0.2 \le n \le 1 \quad 1 \le Pr_x \le 1,000)$
	$0.3306 + 0.0027 \times \left(\frac{1}{n}\right)^{1.65}$	$(1 \le n \le 2 \quad 1 \le \mathrm{Pr}_x \le 1,000)$

convergence of the thermal boundary layer equation. With increasing the power-law index n, the momentum boundary layer thichness η_{δ_l} will decrease. With increasing the local Prandtl number Pr_x, the thermal boundary layer thickness η_{δ_T} will decrease too. But, with variation of the power-law index n, the thermal boundary layer thickness η_{δ_t} will almost keep the same. With increasing the power-law index n, the critical boundary layer thickness $\eta_{\delta_l}*$ with decrease, meanwhile, the critical local Prandtl number Pr_x* will increase. It is also seen that if the power-law index n tends to unity, the critical Prandtl number Pr_x* tends to unity too with the critical boundary layer thickness $\eta_{\delta_l}* \rightarrow 5$. It means, when $n = 1$ and $Pr_x = 1$, the thermal boundary layer thickness η_{δ_t} is identical to the momentum boundary layer thickness η_{δ_l}, and $\eta_{\delta_t} = \eta_{\delta_l} \rightarrow 5$.

For $Pr_x \gg 1$ the wall temperature gradient is controlled by the velocity gradient at the wall, which was practically independent of n for dilatant fluids ($n > 1$) but increased with increasing pseudoplasticity for pseudoplastic fluids ($n < 1$). For $Pr_x \ll 1$ the temperature gradient increased slightly with increasing n and this modest variation was ascribed to the displacing influence of the momentum boundary layer on the external frictionless flow.

A set of accurate curve-fit formulas for the wall temperature gradient is provided in order to enable rapid estimates of the heat transfer rate for any combination of n and Pr_x within the parameter ranges considered.

A special case of the transformation method, velocity component method, used herein has been applied in the previous analysis in Parts 1 and 2 of this book for the heat transfer in Newtonian liquid films with variable fluid properties. The successful generalization on non- Newtonian fluids makes us believe that the present approach should be applicable also to the analysis of heat transfer in power-law films with variable thermophysical properties.

17.8 Calculation Example

Example. Continue the example of Chap. 15, a non-Newtonian power-law fluid having a density of 1,041 kg m^{-3} is flowing with volumetric flow rate of 0.02 m^3 s^{-1} along an inclined flat plate with angle of $\alpha = 30°$ and width of $b =$ 1 m. The properties of the fluid are thermal diffusivity $a = 13.6 \times 10^{-8}$ m^2 s^{-1}, thermal conductivity $\lambda = 0.41$ W (m °C^{-1}), coefficient of consistency $K =$ 2.744 kg (s^{n-2} m^{-1}) and power-law index $n = 0.50$. The plate temperature is kept for $t_w = 60$°C, and the fluid bulk temperature t_∞ is 20°C. Calculate the total heat transfer rate Q_x in the boundary layer region. Consider the cases for $\alpha = 30°$ and 0°.

Solution. Given conditions: volumetric flow rate $V = 0.02$ m^3 s^{-1}, fluid thermal diffusivity $a = 13.6 \times 10^{-8}$ m^2 s^{-1}, fluid thermal conductivity $\lambda =$ 0.41 W (m °C)$^{-1}$, fluid density $\rho = 1{,}041$ kg m^{-3}, plate inclined angle $\alpha = 30°$, fluid coefficient of consistency $K = 2.744$ kg (s^{n-2} m^{-1}), power-law index

$n = 0.50$, wall temperature $t_{\mathrm{w}} = 60°\mathrm{C}$, fluid bulk temperature $t_\infty = 20°\mathrm{C}$, and plate with $b = 0.3\,\mathrm{m}$. The necessary values calculated from the example in Chap. 15: the boundary layer region length $x_0 = 0.118\,\mathrm{m}$ both for $\alpha = 30°$ and $0°$.

1. In the case for $\alpha = 30°$. The total heat transfer rate Q_x from wall to fluid in the boundary layer region is calculated as

$$Q = \overline{\alpha}_{x0}(t_{\mathrm{w}} - t_\infty) \times b \times x_0,$$

where $\overline{\alpha}_{x0}$ is average heat transfer coefficient for the boundary layer region, b is width of the plate, and x_0 is the length of the boundary layer region.

With (17.23) the average heat transfer coefficient in the boundary layer region is expressed as

$$\overline{\alpha}_{x0} = -\frac{2(n+1)}{n+2}\lambda x_0^{-1}\left[\frac{\partial\theta(\eta,\zeta)}{\partial\eta}\right]_{\eta=0}(Re_{x0})^{1/(n+1)}.$$

According to the previous calculated result in Chap. 15, the local Reynolds number is $Re_{x0} - 219.4$.

With (17.17) the local Prandtl number at x_0 is calculated as

$$\begin{aligned} Pr_{x0} &= \frac{x_0 w_{x0,\infty}}{a}(Re_{x0})^{-2/(n+1)} \\ &= \frac{x_0\sqrt{2gx_0\cos\alpha}}{a}(Re_{x0})^{-2/(n+1)} \\ &= \frac{0.118 \times \sqrt{2 \times 9.8 \times 0.118 \times \cos 30°}}{13.6 \times 10^{-8}} \times (219.4)^{-2/(0.5+1)} \\ &= 927.96. \end{aligned}$$

According to (17.26), the temperature gradient on the plate is expressed as follows

$$-\left[\frac{\partial\theta(\eta,\zeta)}{\partial\eta}\right]_{\eta=0} = a + bPr_{x0}^c.$$

According to Table 17.2, the coefficients a and b and exponent c can be evaluated as later with $Pr_{x0} = 927.96$ and $n = 0.5$

$$\begin{aligned} a &= 0.0205 - 0.0845 \times \left(\frac{1}{n}\right)^{1/3} \\ &= 0.0205 - 0.0845 \times \left(\frac{1}{0.5}\right)^{1/3} \\ &= -0.08596 \\ b &= 0.518 + \frac{0.0234}{n} \\ &= 0.518 + \frac{0.0234}{0.5} \\ &= 0.5648. \end{aligned}$$

$$
\begin{aligned}
c &= 0.3324 + 0.00096 \times \left(\frac{1}{n}\right)^{1.6} \\
&= 0.3324 + 0.00096 \times \left(\frac{1}{0.5}\right)^{1.6} \\
&= 0.33531.
\end{aligned}
$$

Then,

$$
\begin{aligned}
-\left[\frac{\partial\theta(\eta,\zeta)}{\partial\eta}\right]_{\eta=0} &= a + bPr_{x0}^{c} \\
&= -0.08596 + 0.5648 \times 927.96^{0.33531} \\
&= 5.4979.
\end{aligned}
$$

Therefore, average heat transfer coefficient from $x = 0$ to x_0 is evaluated as

$$
\begin{aligned}
\overline{\alpha}_{x0} &= -\frac{2(n+1)}{n+2}\lambda x_0^{-1}\left[\frac{\partial\theta(\eta,\zeta)}{\partial\eta}\right]_{\eta=0}(Re_{x_0})^{1/(n+1)} \\
&= \frac{3}{2.5} \times 0.41 \times 0.118^{-1} \times 5.4979 \times (219.4)^{1/(0.5+1)} \\
&= 833.88\ \mathrm{W(m^2\ K)^{-1}}.
\end{aligned}
$$

The total heat transfer rate Q_x at $x = 0$ to x_0 for the boundary layer region is calculated as

$$
\begin{aligned}
Q_x &= \overline{\alpha}_{x_0}(t_\mathrm{w} - t_\infty) \times b \times x_0 \\
&= 833.88 \times (60 - 20) \times 1 \times 0.118 \\
&= 3935.9\ \mathrm{W}.
\end{aligned}
$$

2. In the case for $\alpha = 0°$. Local Reynolds number Re_{x_0} related to x_0 is evaluated as

$$
\begin{aligned}
Re_{x_0} &= \frac{x_0^n(w_{x0,\infty})^{2-n}\rho}{K} \\
&= \frac{x_0^n(2gx_0\cos\alpha)^{2-n/2}\rho}{K}.
\end{aligned}
$$

According to the previous calculated results in Chapt. 15, $x_0 = 0.118$, then,

$$
\begin{aligned}
Re_{x_0} &= \frac{0.118^{0.5}(2 \times 9.8 \times 0.118\cos 0°)^{2-0.5/2} \times 1041}{2.744} \\
&= 245.11.
\end{aligned}
$$

Also

$$
-\left[\frac{\partial\theta(\eta,\zeta)}{\partial\eta}\right]_{\eta=0} = a + bPr_{x0}^{c}
$$

While,

$$
\begin{aligned}
Pr_{x0} &= \frac{x_0 w_{x0,\infty}}{a}(Re_{x0})^{-2/n+1} \\
&= \frac{x_0\sqrt{2gx_0\cos\alpha}}{a}(Re_{x0})^{-2/n+1} \\
&= \frac{0.118\times\sqrt{2\times 9.8\times 0.118\times\cos 0^\circ}}{13.6\times 10^{-8}}\times(245.11)^{-2/0.5+1} \\
&= 860.2.
\end{aligned}
$$

According to the earlier calculations, coefficients a and b and exponent c are $a = -0.08596, b = 0.5648$, and $c = 0.33531$

Then,

$$
\begin{aligned}
-\left[\frac{\partial\theta(\eta,\zeta)}{\partial\eta}\right]_{\eta=0} &= a + bPr_{x0}^c \\
&= -0.08596 + 0.5648\times 860.2^{0.33531} \\
&= 5.3578.
\end{aligned}
$$

The average local heat transfer coefficient $\overline{\alpha}_{x0}$ from $x = 0$ to x_0 is evaluated as

$$
\begin{aligned}
\overline{\alpha}_{x0} &= -\frac{2(n+1)}{n+2}\lambda x_0^{-1}\left[\frac{\partial\theta(\eta,\zeta)}{\partial\eta}\right]_{\eta=0}(Re_{x0})^{1/(n+1)} \\
&= \frac{3}{2.5}\times 0.41\times 0.118^{-1}\times 5.3578\times 245.11^{1/(0.5+1)} \\
&= 874.94\ \mathrm{W}\ (\mathrm{m}^2\,{}^\circ\mathrm{C})^{-1}.
\end{aligned}
$$

The total heat transfer rate Q_x is calculated as

$$
\begin{aligned}
Q_x &= \overline{\alpha}_{x0}(t_\mathrm{w} - t_\infty)\times b\times x_0 \\
&= 874.94\times(60-20)\times 1\times 0.118 \\
&= 4129.72\ \mathrm{W}.
\end{aligned}
$$

References

1. S.M. Yih and M.W. Lee, Heating or evaporation in the thermal entrance region of a non-newtonian laminar falling liquid film, Int. J. Heat Mass Transfer, **29**, pp. 1999–2002, 1986
2. G. Astarita, Mass transfer from a flat solid surface to a falling non-Newtonian liquid film, Ind. Eng. Chem. Fundam. **5**, pp. 14–18, 1966
3. R.A. Mashelker and V.V. Chavan, Solid dissolution in falling films of non-Newtonian liquids, J. Chem. Jpn, **6**, pp. 160–167, 1973
4. I. Pop, T. Watanabe and H. Konishi, Gravity-driven laminar film flow along a vertical wall with surface mass transfer, Int. Commun. Heat Mass Transfer, **23**, pp. 685–695, 1996

5. D. Ouldhadda and A. Idrissi, Laminar flow and heat transfer of non-Newtonian falling liquid film on a horizontal tube with variable surface heat flux, Int. Comm. Heat Mass Transfer, **28**, pp. 1125–1135, 2001
6. B. K. Rao, Heat transfer to falling power-law fluid film, Int. J. Heat Fluid Flow, **20**, pp. 429–436, 1999
7. D. Y. Shang and H. Andersson, Heat transfer in gravity-driven film flow of power-law fluids," Int. J. Heat Mass Transfer, **42**, No. 11, pp. 2085–2099, 1999
8. D. Y. Shang and J. Gu, Analyses of pseudo-similarity and boundary layer thickness for non-Newtonian falling film flow, Heat Mass Transfer, **41**, No. 1, pp. 44–50, 2004
9. H.I. Andersson, Diffusion from a vertical wall into an accelerating falling liquid film, Int. J. Heat Mass Transfer, **30**, pp. 683–689, 1987
10. H. Andersson and D.Y. Shang, An extended study of hydrodynamics of gravity-driven film flow of power-law fluids, Fluid Dyn. Res., **22**, pp. 345–357, 1998
11. E.M. Sparrow, H. Quack, and C.J. Boener, Local nonsimilarity boundary – layer solutions, AIAA J., **8**, pp. 1936–1942, 1970

A

Tables with Thermophysical Properties

Appendix A.1 Thermophysical Properties of Gases at Atmospheric Pressure

air [1]						
T [K]	ρ [Kg m^{-3}]	c_p [kJ (kg°C)$^{-1}$]	$\mu \times 10^7$ [m^2 s^{-1}]	λ [W (m°C)$^{-1}$]	$a \times 10^6$ [m^2 s^{-1}]	Pr
100	3.5562	1.032	71.1	0.00934	2.54	0.786
150	2.3364	1.012	103.4	0.0138	5.84	0.758
200	1.7458	1.007	132.5	0.0181	10.3	0.737
250	1.3947	1.006	159.6	0.0223	15.9	0.720
300	1.1614	1.007	184.6	0.0263	22.5	0.707
350	0.9950	1.009	208.2	0.0300	29.9	0.7
400	0.8711	1.014	230.1	0.0338	38.3	0.69
450	0.7740	1.021	250.7	0.0373	47.2	0.686
500	0.6964	1.030	270.1	0.0407	56.7	0.684
550	0.6329	1.040	288.4	0.0439	66.7	0.683
600	0.5804	1.051	305.8	0.0469	76.9	0.685
650	0.5356	1.063	322.5	0.0497	87.3	0.69
700	0.4975	1.075	338.8	0.05240	98	0.695
750	0.4643	1.087	354.6	0.0549	109	0.702
800	0.4354	1.099	369.8	0.0573	120	0.709
850	0.4097	1.11	384.3	0.0596	131	0.716
900	0.3868	1.121	398.1	0.0620	143	0.720
950	0.3666	1.131	411.3	0.0643	155	0.723
1,000	0.3482	1.141	424.4	0.0667	168	0.726
1,100	0.3166	1.159	449.0	0.0715	195	0.728
1,200	0.2920	1.175	473.0	0.0763	224	0.728

monoxide, CO [1]

T [K]	ρ [Kg m^{-3}]	c_p [kJ (kg°C)$^{-1}$]	$\mu \times 10^7$ [m^2 s^{-1}]	λ x [W (m °C)$^{-1}$]	$a \times 10^6$ [m^2 s^{-1}]	Pr
200	1.6888	1.045	127	0.017	9.63	0.781
220	1.5341	1.044	137	0.0190	11.9	0.753
240	1.4055	1.043	147	0.0206	14.1	0.744
260	1.2967	1.043	157	0.0221	16.3	0.741
280	1.2038	1.042	166	0.0236	18.8	0.733
300	1.1233	1.043	175	0.025	21.3	0.730
320	1.0529	1.043	184	0.0263	23.9	0.730
340	0.9909	1.044	193	0.0278	26.9	0.725
360	0.9357	1.045	202	0.0291	29.8	0.729
380	0.8864	1.047	210	0.0305	32.9	0.719
400	0.8421	1.049	218	0.0318	36.0	0.719
450	0.7483	1.055	237	0.0350	44.3	0.714
500	0.67352	1.065	254	0.0381	53.1	0.710
550	0.61226	1.076	271	0.0411	62.4	0.710
600	0.56126	1.088	286	0.0440	72.1	0.707
650	0.51806	1.101	301	0.0470	82.4	0.705
700	0.48102	1.114	315	0.0500	93.3	0.702
750	0.44899	1.127	329	0.0528	104	0.702
800	0.42095	1.140	343	0.0555	116	0.705

helium, He [1]

T [K]	ρ [Kg m^{-3}]	c_p [kJ (kg °C)$^{-1}$]	$\mu \times 10^7$ [m^2 s^{-1}]	λ [W (m°C)$^{-1}$]	$a \times 10^6$ [m^2 s^{-1}]	Pr
100	0.4871	5.193	96.3	0.073	28.9	0.686
120	0.4060	5.193	107	0.0819	38.8	0.679
140	0.3481	5.193	1‘18	0.0907	50.2	0.676
160		5.193	129	0.0992		
180	0.2708	5.193	139	0.1072	76.2	0.673
200		5.193	150	0.1151		
220	0.2216	5.193	160	0.1231	107	0.675
240		5.193	170	0.130		
260	0.1875	5.193	180	0.137	141	0.682
280		5.193	190	0.145		
300	0.1625	5.193	199	0.152	180	0.680
350		5.193	221	0.170		
400	0.1219	5.193	243	0.187	295	0.675
450		5.193	263	0.204		
500	0.09754	5.193	283	0.220	434	0.668
600		5.193	320	0.252		
700	0.06969	5.193	350	0.278	768	0.654
800		5.193	382	0.304		0
900		5.193	414	0.330		
1,000	0.04879	5.193	446	0.354	1, 400	0.654

hydrogen, H_2 [1]						
T [K]	ρ [Kg m^{-3}]	c_p [kJ (kg °C)$^{-1}$]	$\mu \times 10^7$ [m^2 s^{-1}]	λ [W (m °C)]	$a \times 10^6$ [m^2 s^{-1}]	Pr
100	0.24255	11.23	42.1	0.067	24.6	0.707
200	0.12115	13.54	68.1	0.131	79.9	0.704
300	0.08078	14.31	89.6	0.183	158	0.701
400	0.06059	14.48	108.2	0.226	258	0.695
500	0.04848	14.52	126.4	0.266	378	0.691
600	0.04040	14.55	142.4	0.305	519	0.678
700	0.03463	14.61	157.8	0.342	676	0.675
800	0.03030	14.70	172.4	0.378	849	0.670
900	0.02694	14.83	186.5	0.412	1, 030	0.671
1,000	0.02424	14.99	201.3	0.448	1, 230	0.673
1,100	0.02204	15.17	213.0	0.488	1, 460	0.662
1,200	0.02020	15.37	226.2	0.528	1, 700	0.659
1,300	0.01865	15.59	238.5	0.569	1, 955	0.655
1,400	0.01732	15.81	250.7	0.610	2, 230	0.650
1,500	0.01616	16.02	262.7	0.655	2, 530	0.643
1,600	0.01520	16.28	273.7	0.697	2, 815	0.639

nitrogen, N_2 [1]						
T [K]	ρ [Kg m^{-3}]	c_p [kJ (kg °C)$^{-1}$]	$\mu \times 10^7$ [m^2 s^{-1}]	λ [W (m°C)$^{-1}$]	$a \times 10^6$ [m^2 s^{-1}]	Pr
100	3.4388	1.070	68.8	0.00958	26	0.768
150	2.2594	1.050	100.6	0.0139	586	0.759
200	1.6883	1.043	120.2	0.0183	104	0.736
250	1.3488	1.042	154.9	0.0222	158	0.727
300	1.1233	1.041	178.2	0.0259	221	0.716
350	0.9625	1.042	200.0	0.0293	292	0.711
400	0.8425	1.045	220.4	0.0327	371	0.704
450	0.7485	1.050	239.6	0.0358	456	0.703
500	0.6739	1.056	257.7	0.0389	547	0.700
550	0.6124	1.065	274.7	0.0417	639	0.702
600	0.5615	1.075	290.8	0.0416	739	0.701
700	0.4812	1.098	321.0	0.0499	944	0.706
800	0.4211	1.122	349.1	0.0548	116	0.715
900	0.3743	1.146	375.3	0.0597	139	0.721
1,000	0.3368	1.167	399.9	0.0647	165	0.721
1,100	0.3062	1.187	423.2	0.0700	193	0.718
1,200	0.2807	1.204	445.3	0.0758	224	0.707
1,300	0.2591	1.219	466.2	0.0810	256	0.701

Continued

oxygen, O_2 [1]						
T [K]	ρ [Kg m^{-3}]	c_p [kJ (kg °C)$^{-1}$]	$\mu \times 10^7$ [m^2 s^{-1}]	λ [W (m °C)$^{-1}$]	$a \times 10^6$ [m^2 s^{-1}]	Pr
100	3.945	0.962	76.4	0.00925	2.44	0.796
150	2.585	0.921	114.8	0.0138	5.80	0.766
200	1.930	0.915	147.5	0.0183	10.4	0.737
250	1.542	0.915	178.6	0.0226	16.0	0.723
300	1.284	0.920	207.2	0.0268	22.7	0.711
350	1.100	0.929	233.5	0.0296	29.0	0.733
400	0.962	0.942	258.2	0.0330	36.4	0.737
450	0.8554	0.956	281.4	0.0363	44.4	0.741
500	0.7698	0.972	303.3	0.0412	55.1	0.716
550	0.6998	0.988	324.0	0.0441	63.8	0.726
600	0.6414	1.003	343.7	0.0473	73.5	0.729
700	0.5498	1.031	380.8	0.0523	93.1	0.744
800	0.4810	1.054	415.2	0.0589	116	0.743
900	0.4275	1.074	447.2	0.0649	141	0.740
1,000	0.3848	1.090	477.0	0.0710	169	0.733
1,100	0.3498	1.103	505.5	0.0758	196	0.736
1,200	0.3206	1.115	532.5	0.0819	229	0.725
1,300	0.2060	1.125	588.4	0.0871	262	0.721

carbon dioxide, CO_2 [1]						
T [K]	ρ [Kg m^{-3}]	c_p [kJ (kg °C)$^{-1}$]	$\mu \times 10^7$ [m^2 s^{-1}]	λ [W (m °C)$^{-1}$]	$a \times 10^6$ [m^2 s^{-1}]	Pr
220	2.4733	0.783	111.05	0.010805	5.92	0.818
250	2.1675	0.804	125.9	0.012884	7.401	0.793
300	1.7973	0.871	149.58	0.016572	10.588	0.77
350	1.5362	0.900	172.05	0.02047	14.808	0.755
400	1.3424	0.942	193.2	0.02461	19.463	0.738
450	1.1918	0.980	213.4	0.02897	24.813	0.721
500	1.0732	1.013	232.6	0.03352	30.84	0.702
550	0.9739	1.047	250.8	0.03821	37.50	0.695
600	0.8938	1.076	268.3	0.04311	44.83	0.668

ammonia, NH_3 [2]						
T [K]	ρ [Kg m^{-3}]	c_p [kJ (kg °C)$^{-1}$]	$\mu \times 10^7$ [m^2 s^{-1}]	λ [W (m °C)$^{-1}$]	$a \times 10^6$ [m^2 s^{-1}]	Pr
220	0.9828	2.198	72.55	0.0171	20.54	0.93
273	0.7929	2.177	93.53	0.0220	13.08	0.9
323	0.6487	2.177	110.35	0.0270	19.20	0.88
373	0.5590	2.236	128.86	0.0327	26.19	0.87
423	0.4934	2.315	146.72	0.0391	34.32	0.87
473	0.4405	2.395	164.9	0.0476	44.21	0.84

water vapor [2]						
T [K]	ρ [Kg m^{-3}]	c_p [kJ (kg°C)$^{-1}$]	$\mu \times 10^7$ [m^2 s^{-1}]	λ [W (m °C)$^{-1}$]	$a \times 10^6$ [m^2 s^{-1}]	Pr
380	0.5863	2.060	127.1	0.0246	20.36	1.060
400	0.5542	2.014	134.4	0.0261	23.38	1.040
450	0.4902	1.980	152.5	0.0299	30.7	1.010
500	0.4405	1.985	170.4	0.0339	38.7	0.996
550	0.4005	1.997	188.4	0.0379	47.5	0.991
600	0.3652	2.026	206.7	0.0422	57.3	0.986
650	0.3380	2.056	224.7	0.0464	66.6	0.995
700	0.3140	2.085	242.6	0.0505	77.2	1.000
750	0.2931	2.119	260.4	0.0549	88.3	1.05
800	0.2739	2.152	278.6	0.0592	100.1	1.010
850	0.2579	2.186	296.9	0.0637	113.0	1.019

gas mixture(CO_2 = 0.13, H_2O = 0.11, N_2 = 0.76) [3]						
t [°C]	ρ [Kg m^{-3}]	c_p [kJ (kg °C)$^{-1}$]	$\mu \times 10^7$ [m^2 s^{-1}]	λ [W (m °C)$^{-1}$]	$a \times 10^6$ [m^2 s^{-1}]	Pr
0	1.295	1.042	158	0.0228	12.2	0.72
100	0.950	1.068	204	0.0313	21.54	0.69
200	0.748	1.097	245	0.0401	32.8	0.67
300	0.617	1.122	282	0.0484	45.81	0.65
400	0.525	1.151	317	0.057	60.38	0.64
500	0.457	1.185	348	0.0656	76.3	0.63
600	0.405	1.214	379	0.0742	93.61	0.62
700	0.363	1.239	407	0.0827	112.1	0.61
800	0.330	1.264	434	0.0915	131.8	0.6
900	0.301	1.290	459	0.1	152.5	0.59
1,000	0.275	1.306	484	0.109	174.3	0.58
1,100	0.257	1.323	507	0.1175	197.1	0.57
1,200	0.240	1.340	530	0.1262	221.0	0.56

Appendix A.2 Physical Properties of Some Saturated Liquids

ammonia, NH_3 [4]						
t [°C]	ρ [Kg m^{-3}]	c_p [kJ (kg °C)$^{-1}$]	$\nu \times 10^6$ [m^2 s^{-1}]	λ [W (m °C)$^{-1}$]	$a \times 10^7$ [m^2 s^{-1}]	Pr
−50	703.69	4.463	0.435	0.547	1.742	2.60
−40	691.68	4.467	0.406	0.547	1.775	2.28
−30	679.34	4.467	0.387	0.549	1.801	2.15
−20	666.69	4.509	0.381	0.547	1.819	2.09
−10	653.55	4.564	0.378	0.543	1.825	2.07
0	640.10	4.635	0.373	0.540	1.819	2.05
10	626.16	4.714	0.368	0.531	1.801	2.04
20	611.75	4.798	0.359	0.521	1.775	2.02
30	596.37	4.890	0.349	0.507	1.742	2.01
40	580.99	4.999	0.340	0.493	1.701	2.00
50	564.33	5.116	0.330	0.476	1.654	1.99

Continued

carbon dioxide, CO_2 [4]

t [°C]	ρ [Kg m^{-3}]	c_p [kJ (kg °C)$^{-1}$]	$\nu \times 10^6$ [m^2 s^{-1}]	λ [W (m °C)$^{-1}$]	$a \times 10^7$ [m^2 s^{-1}]	Pr
−50	1,156.34	1.84	0.119	0.085	0.4021	2.96
−40	1,117.77	1.88	0.118	0.1011	0.4810	2.45
−30	1,076.76	1.97	0.117	0.1116	0.5272	2.22
−20	1,032.39	2.05	0.115	0.1151	0.5445	2.12
−10	983.38	2.18	0.113	0.1099	0.5133	2.20
0	926.99	2.47	0.108	0.1045	0.4578	2.38
10	860.03	3.14	0.101	0.0971	0.3608	2.80
20	772.57	5.0	0.091	0.0872	0.2219	4.10
30	597.81	36.4	0.080	0.0703	0.0279	28.7

sulphur dioxide, SO_2 [4]

t [°C]	ρ [Kg m^{-3}]	c_p [kJ (kg °C)$^{-1}$]	$\nu \times 10^6$ [m^2 s^{-1}]	λ [W (m °C)$^{-1}$]	$a \times 10^7$ [m^2 s^{-1}]	Pr
−50	1,560.84	1.3595	0.484	0.242	1.141	4.24
−40	1,536.81	1.3607	0.424	0.235	1.130	3.74
−30	1,520.64	1.3616	0.371	0.230	1.117	3.31
−20	1,488.60	1.3624	0.324	0.225	1.107	2.93
−10	1,463.61	1.3628	0.288	0.218	1.097	2.62
0	1,438.46	1.3636	0.257	0.211	1.081	2.38
10	1,412.51	1.3645	0.232	0.204	1.066	2.18
20	1,386.40	1.3653	0.210	0.199	1.050	2.00
30	1,359.33	1.3662	0.190	0.192	1.035	1.83
40	1,329.22	1.3674	0.173	0.185	1.019	1.70
50	1,299.10	1.3683	0.162	0.177	0.999	1.61

freon 12, CCl_2F_2 [4]

t [°C]	ρ [Kg m^{-3}]	c_p [kJ (kg °C)$^{-1}$]	$\nu \times 10^6$ [m^2 s^{-1}]	λ [W (m °C)$^{-1}$]	$a \times 10^7$ [m^2 s^{-1}]	Pr
−50	1,546.75	0.8750	0.310	0.067	0.501	6.2
−40	1,518.71	0.8847	0.279	0.069	0.514	5.4
−30	1,489.56	0.8956	0.253	0.069	0.526	4.8
−20	1,460.57	0.9073	0.235	0.071	0.539	4.4
−10	1,429.49	0.9203	0.221	0.073	0.550	4.0
0	1,397.45	0.9345	0.214	0.073	0.557	3.8
10	1,364.30	0.9496	0.203	0.073	0.560	3.6
20	1,330.18	0.9659	0.198	0.073	0.560	3.5
30	1,295.10	0.9835	0.194	0.071	0.560	3.5
40	1,257.13	1.019	0.191	0.069	0.555	3.5
50	1,215.96	1.0216	0.190	0.067	0.545	3.5

$C_2H_4(OH)_2$ [4]						
t [°C]	ρ [Kg m^{-3}]	c_p [kJ (kg °C)$^{-1}$]	$\nu \times 10^6$ [m^2 s^{-1}]	λ [W (m °C)$^{-1}$]	$a \times 10^7$ [m^2 s^{-1}]	Pr
0	1,130.75	2.294	57.53	0.242	0.934	615
20	1,116.65	2.382	19.18	0.249	0.939	204
40	1,101.43	2.474	8.69	0.256	0.939	93
60	1,087.66	2.562	4.75	0.260	0.932	51
80	1,077.56	2.650	2.98	0.261	0.921	32.4
100	1,058.50	2.742	2.03	0.263	0.908	22.4

mercury, Hg [4]						
t [°C]	ρ [Kg m^{-3}]	c_p [kJ (kg °C)$^{-1}$]	$\nu \times 10^6$ [m^2 s^{-1}]	λ [W (m °C)$^{-1}$]	$a \times 10^7$ [m^2 s^{-1}]	Pr
0	1,3628.22	0.1403	0.124	8.2	42.99	0.0288
20	1,3579.04	0.1394	0.114	8.69	46.04	0.0249
50	1,3505.84	0.1386	0.104	9.40	50.22	0.0207
100	1,3384.58	0.1373	0.0928	10.51	57.16	0.0162
150	1,3264.28	0.1365	0.0853	11.49	63.54	0.0134
200	1,3144.94	0.1570	0.0802	12.34	69.08	0.0116
250	1,3025.60	0.1357	0.0765	13.07	74.06	0.0103
315.5	1,2847.00	0.134	0.0673	14.02	81.50	0.0083

water, H_2O [2]						
t [°C]	ρ [Kg m^{-3}]	c_p [kJ (kg °C)$^{-1}$]	$\nu \times 10^6$ [m^2 s^{-1}]	λ [W (m °C)$^{-1}$]	$a \times 10^6$ [m^2 s^{-1}]	Pr
0	1,002.28	4.2178	1.788	0.552	0.1308	13.6
20	1,000.52	4.1818	1.006	0.597	0.143	7.02
40	994.59	4.1784	0.658	0.628	0.1512	4.34
60	985.46	4.1843	0.478	0.651	0.1554	3.02
80	974.08	4.1964	0.364	0.668	0.1636	2.22
100	960.63	4.2161	0.294	0.680	0.168	1.74
120	945.25	4.25	0.247	0.685	0.1708	1.446
140	928.27	4.283	0.214	0.684	0.1724	1.241
160	909.69	4.342	0.190	0.680	0.1729	1.099
180	889.03	4.417	0.173	0.675	0.1724	1.004
200	866.76	4.505	0.160	0.665	0.1706	0.937
220	842.41	4.610	0.150	0.652	0.168	0.891
240	815.66	4.756	0.143	0.635	0.1639	0.871
260	785.87	4.949	0.137	0.611	0.1577	0.874
280	752.55	5.208	0.135	0.580	0.1481	0.910
300	714.26	5.728	0.135	0.540	0.1324	1.019

References

1. F.P. Incropera and D.P. Dewitt, Fundamentals of Heat Transfer, Wiley, New York, 1981.
2. T. Cebeci and P. Brabshaw, Physical and Computational Aspects of Convective Heat Transfer, Springer, Berlin Heidelberg New York, 1984.
3. S. Yang, Heat Transfer, 2nd edition, Higher Education, Beijing, 1987.
4. E.R.G. Eckert and R.M. Drake, Heat and Mass Transfer, 2nd edition, Mcgraw-Hill, New York, 1959.

Index

MIX

Papier | Fördert gute Waldnutzung

FSC® C083411

Kolibri 360 GmbH
Ferdinand-Jühlke-Straße 7
99095 Erfurt, Deutschland
produktsicherheit@kolibri360.de